INTRODUCTORY MINING ENGINEERING

INTRODUCTORY MINING ENGINEERING

SECOND EDITION

Howard L. Hartman

Jan M. Mutmansky

JOHN WILEY & SONS, INC.

This book is printed on acid-free paper. ♾

Published by John Wiley & Sons, Inc., Hoboken, New Jersey
Published simultaneously in Canada.

Library of Congress Cataloging-in-Publication Data:

Hartman, Howard L.
Introductory mining engineering / Howard L. Hartman, Jan M. Mutmansky.--2nd ed.
p. cm.
Includes index.
ISBN 0-471-34851-1 (cloth : alk. paper)
1. Mining engineering. I. Mutmansky, Jan M. II. Title.

TN275.H35 2002
622--dc21

2002071315

Printed in the United States of America.

20 19 18 17 16 15 14 13 12 11

Much of the credit for this book goes to our former teachers, who inspired us when we were young, and to our former students, particularly those whose efforts made teaching and publishing books a worthwhile profession. Their influence on us is much appreciated. We dedicate the book to our wives, Bonnie and Diane, for their help and their willingness to sacrifice some of their freedom to aid our academic careers.

H.L.H.
J.M.M.

Note to readers: Dr. Howard L. Hartman passed away January 12, 2002, in Carmichael, California. Although he suffered from the effects of Parkinson's disease for a number of years, he maintained his commitment to the writing of this book until his projected contribution was nearly completed. The dedication and foreword of this book were written in the summer of 2001. Accordingly, they are printed here as he and I wrote them.

J.M.M.

CONTENTS

PREFACE

It is not a frequent experience for a writer to receive the opportunity to revise and update a previous work. Such is the case, however, with this revised second edition of *Introductory Mining Engineering*, originally authored by Howard L. Hartman. Further, the planned revision took an unexpected turn when Jan M. Mutmansky joined the undertaking as coauthor of the second edition. In fact, Dr. Mutmansky contributed to the original work, serving as a reviewer of the outline, as reader, and as critic. He was acknowledged in the preface of the first edition and thus was the logical choice as coauthor of the second edition.

In the best sense of the word, the second edition is a collaborative one. Coauthors of an earlier mining textbook,[1] the two writers of this work have evolved a common style that reinforces ideas and clarifies subject matter. The objective of the course that this book is intended to serve as a text, as stated in the first edition, "is to educate beginning students in mining engineering, and to bring them to an understanding of the minerals industry and the challenges that it faces" (Preface, p. vii).

As in the first edition, the teaching strategy emphasizes students' observing and generating ideas: They are encouraged to conduct case studies, visit mines, see films and videos, solve problems, compile mining methods notebooks, and design specific mining systems (Preface, p. viii). A new chapter on environmental considerations has been added, and chapters on the choice of mining methods have been condensed. Some of the case studies and problems are new. However, the basic objectives of the book remain the same: (1) familiarization with the minerals industry and its challenges and (2) the selection of the optimal mining method for the specified conditions.

The authors extend their sincere thanks to those whose help was received in the production of this book. In particular, we thank the technical societies, equipment manufacturers, mining companies, and individuals who provided us with permission to use diagrams, photos, and computer files to illustrate the book. A special note of thanks to Otto L. Schumacher, Western Mine Engineering, Inc., Spokane, Washington, for his generosity in sharing valuable cost information and providing insights into mining costs.

[1] H. L. Hartman, J. M. Mutmansky, R. V. Ramani, and Y. J. Wang, *Mine Ventilation and Air Conditioning*, 3d ed. (New York: John Wiley & Sons, Inc., 1997), 730 pp.

We especially thank the Society for Mining, Metallurgy, and Exploration, Inc. (SME), Littleton, Colorado, which allowed us unlimited use of illustrations and information from its many publications related to mining engineering. We encourage all of our student readers to join as student members of this technical society. Information on SME can be found at its website (www.smenet.org).

Problems are included at the end of most chapters. Answers to some of the problems can be found after the Appendix.

A solutions manual for university instructors who use this text in their classes is available. Persons requesting this manual should be willing to establish that they have ordered this text for their courses. For more information, contact Jan Mutmansky at The Pennsylvania State University via mail or using the e-mail address j93@psu.edu.

HOWARD L. HARTMAN
Sacramento, California

JAN M. MUTMANSKY
University Park, Pennsylvania

1

INTRODUCTION TO MINING

1.1 MINING'S CONTRIBUTION TO CIVILIZATION

Mining may well have been the second of humankind's earliest endeavors granted that agriculture was the first. The two industries ranked together as the primary or basic industries of early civilization. Little has changed in the importance of these industries since the beginning of civilization. If we consider fishing and lumbering as part of agriculture and oil and gas production as part of mining, then agriculture and mining continue to supply all the basic resources used by modern civilization.

From prehistoric times to the present, mining has played an important part in human existence (Madigan, 1981). Here the term *mining* is used in its broadest context as encompassing the extraction of any naturally occurring mineral substances — solid, liquid, and gas — from the earth or other heavenly bodies for utilitarian purposes. The most prominent of these uses for minerals are identified in Table 1.1.

The history of mining is fascinating. It parallels the history of civilization, with many important cultural eras associated with and identified by various minerals or their derivatives: the Stone Age (prior to 4000 B.C.E.), the Bronze Age (4000 to 5000 B.C.E.), the Iron Age (1500 B.C.E. to 1780 C.E.), the Steel Age (1780 to 1945), and the Nuclear Age (1945 to the present). Many milestones in human history — Marco Polo's journey to China, Vasco da Gama's voyages to Africa and India, Columbus's discovery of the New World, and the modern gold rushes that led to the settlement of California, Alaska, South Africa, Australia, and the Canadian Klondike — were achieved with minerals providing a major incentive (Rickard, 1932). Other interesting aspects of mining and metallurgical history can be found by referring to the historical record provided by Gregory (1980), Raymond (1984), and Lacy and Lacy (1992).

Table 1.1 Humans' Uses of Minerals

Need or Use	Purpose	Age
Tools and utensils	Food, shelter	Prehistoric
Weapons	Hunting, defense, warfare	Prehistoric
Ornaments and decoration	Jewelry, cosmetics, dye	Ancient
Currency	Monetary exchange	Early
Structures and devices	Shelter, transport	Early
Energy	Heat, power	Medieval
Machinery	Industry	Modern
Electronics	Computers, communications	Modern
Nuclear fission	Power, warfare	Modern

The abundance of minerals also provides a method of creating wealth. Minerals can be marketed on the open market, enabling the countries that possess them to obtain valuable currency from countries that do not. This generally results in the minerals-rich countries being the great civilizations of the world while the 'have-not' countries generally suffer from a lower standard of living. For more on this topic, see Section 1.6. The ability to use mineral resources as a means of creating wealth opens the possibility that a given country or countries will attempt to control the entire market in a particular mineral, that is, to create an economic cartel in that mineral. In 1973, the Organization of Petroleum Exporting Countries (OPEC) attempted to control oil prices in a bold maneuver to obtain windfall profits from the oil they produced. Although successful in the short run, the cartel eventually lost effectiveness because of increased oil production elsewhere and difficulty in controlling their own member countries. For a few years, OPEC was successful at regulating petroleum prices in an awesome display of the value of possessing and producing some of the world's most valuable minerals. Other cartels have likewise been attempted. However, the greater freedom in international trade now makes such an attempt less likely to succeed.

1.2 MINING TERMINOLOGY

There are many terms and expressions unique to mining that characterize the field and identify the user of such terms as a "mining person." The student of mining is thus advised to become familiar with all the terms used in mining, particularly those that are peculiar to either mines or minerals. Most of the mining terminology is introduced in the sections of this book where they are most applicable. Some general terms are best defined at the outset; these are outlined here. For a complete list of mining terminology, see a standard reference (Gregory, 1980; American Geological Institute, 1997). The following three terms are closely related:

Mine: an excavation made in the earth to extract minerals

Mining: the activity, occupation, and industry concerned with the extraction of minerals

Mining engineering: the practice of applying engineering principles to the development, planning, operation, closure, and reclamation of mines

Some terms distinguish various types of mined minerals. Geologically, one can distinguish the following mineral categories:

Mineral: a naturally occurring inorganic element or compound having an orderly internal structure and a characteristic chemical composition, crystal form, and physical properties

Rock: any naturally formed aggregate of one or more types of mineral particles

Economic differences in the nature of mineral deposits is evident in the following terms:

Ore: a mineral deposit that has sufficient utility and value to be mined at a profit.

Gangue: the valueless mineral particles within an ore deposit that must be discarded.

Waste: the material associated with an ore deposit that must be mined to get at the ore and must then be discarded. Gangue is a particular type of waste.

A further subdivision of the types of minerals mined by humankind is also common. These terms are often used in the industry to differentiate between the fuels, metals, and nonmetallic minerals. The following are the most common terms used in this differentiation:

Metallic ores: those ores of the ferrous metals (iron, manganese, molybdenum, and tungsten), the base metals (copper, lead, zinc, and tin), the precious metals (gold, silver, the platinum group metals), and the radioactive minerals (uranium, thorium, and radium).

Nonmetallic minerals (also known as industrial minerals): the nonfuel mineral ores that are not associated with the production of metals. These include phosphate, potash, halite, trona, sand, gravel, limestone, sulfur, and many others.

Fossil fuels (also known as *mineral fuels*): the organic mineral substances that can be utilized as fuels, such as coal, petroleum, natural gas, coalbed methane, gilsonite, and tar sands.

It should be noted that the mining engineer is associated with the extraction of nearly all these mineral resources. However, the production of petroleum and natural gas has evolved into a separate industry with a specialized technology of its own. These mineral products will not be discussed in any detail here.

The essence of mining in extracting mineral wealth from the earth is to drive an excavation or excavations from the surface to the mineral deposit. Normally, these openings into the earth are meant to allow personnel to enter into the underground deposit. However, boreholes are at times used to extract the mineral values from the earth. These fields of boreholes are also called mines, as they are the means to mine a mineral deposit, even if no one enters into the geologic realm of the deposit. Note that when the economic profitability of a mineral deposit has been established with some confidence, *ore* or *ore deposit* is preferred as the descriptive term for the mineral occurrence. However, coal and industrial mineral deposits are often not so designated, even if their profitability has been firmly established. If the excavation used for mining is entirely open or operated from the surface, it is termed a *surface mine*. If the excavation consists of openings for human entry below the earth's surface, it is called an *underground mine*. The details of the procedure, layout, and equipment used in the mine distinguish the *mining method*. This is determined by the geologic, physical, environmental, economic, and legal circumstances that pertain to the ore deposit being mined.

Mining is never properly done in isolation, nor is it an entity in itself. It is preceded by *geologic investigations* that locate the deposit and *economic analyses* that prove it financially feasible. Following extraction of the fuel, industrial mineral, or metallic ore, the run-of-mine material is generally cleaned or concentrated. This preparation or beneficiation of the mineral into a higher-quality product is termed *mineral processing*. The mineral products so produced may then undergo further concentration, refinement, or fabrication during *conversion*, *smelting*, or *refining* to provide consumer products. The end step in converting a mineral material into a useful product is *marketing*.

Quite frequently, excavation in the earth is employed for purposes other than mining. These include *civil* and *military works* in which the object is to produce a stable opening of a desired size, orientation, and permanence. Examples are vehicular, water, and sewer tunnels, plus underground storage facilities, waste disposal areas, and military installations. Many of these excavations are produced by means of standard mining technology.

Professionally, the fields of endeavor associated with the mineral industries are linked to the phase or stage in which an activity occurs. Locating and exploring a mineral deposit fall in the general province of *geology* and the earth sciences. *Mining engineering*, already defined, encompasses the proving (with the geologist), planning, developing, and exploiting of a mineral deposit. The mining engineer may also be involved with the closure and reclamation of the mine property, although he or she may share those duties with those in the environmental fields. The fields of processing, refining, and fabricating are assigned to *metallurgy*, although there is often some overlap in the mineral processing area with mining engineering.

1.3 ADVANCEMENTS IN MINING TECHNOLOGY

As one of humanity's earliest endeavors—and certainly one of its first organized industries—mining has an ancient and venerable history (Gregory, 1980). To understand modern mining practices, it is useful to trace the evolution of mining technology, which (as pointed out earlier in this chapter) has paralleled human evolution and the advance of civilization.

Mining in its simplest form began with Paleolithic humans some 450,000 years ago, evidenced by the flint implements that have been found with the bones of early humans from the Old Stone Age (Lewis and Clark, 1964). Our ancestors extracted pieces from loose masses of flint or from easily accessed outcrops and, using crude methods of chipping the flint, shaped them into tools and weapons. By the New Stone Age, humans had progressed to underground mining in systematic openings 2 to 3 ft (0.6 to 0.9 m) in height and more than 30 ft (9 m) in depth (Stoces, 1954). However, the oldest known underground mine, a hematite mine at Bomvu Ridge, Swaziland (Gregory, 1980), is from the Old Stone Age and believed to be about 40,000 years old. Early miners employed crude methods of ground control, ventilation, haulage, hoisting, lighting, and rock breakage. Nonetheless, mines attained depths of 800 ft (250 m) by early Egyptian times.

Metallic minerals also attracted the attention of prehistoric humans. Initially, metals were used in their native form, probably obtained by washing river gravel in placer deposits. With the advent of the Bronze and Iron Ages, however, humans discovered smelting and learned to reduce ores into pure metals or alloys, which greatly improved their ability to use these metals.

The first challenge for early miners was to break the ore and loosen it from the surrounding rock mass. Often, their crude tools made of bone, wood, and stone were no match for the harder rock, unless the rock contained crevices or cracks that could be opened by wedging or prying. As a result, they soon devised a revolutionary technique called *fire setting*, whereby they first heated the rock to expand it and then doused it with cold water to contract and break it. This was one of the first great advances in the science of rock breakage and had a greater impact than any other discovery until dynamite was invented by Alfred Nobel in 1867.

Mining technology, like that of all industry, languished during the Dark Ages. Notably, a political development in 1185 improved the standing of mining and the status of miners, when the bishop of Trent granted a charter to miners in his domain. It gave miners legal as well as social rights, including the right to stake mineral claims. A milestone in the history of mining, the edict has had long-term consequences that persist to this day.

The greatest impact on the need for and use of minerals, however, was provided by the Industrial Revolution at the close of the eighteenth century. Along with the soaring demand for minerals came spectacular improvements in mining technology, especially in scientific concepts and mechanization, that have continued to this day.

During the last two centuries, there has been great progress in mining technology in many different areas. Such progress is often made in an evolutionary rather than a revolutionary manner. Yet every once in a while, a revolutionary discovery comes along and changes the process of mining profoundly. During the nineteenth century, the invention of dynamite was the most important advance. In the twentieth century, the invention of continuous mining equipment, which extracts the softer minerals like coal without the use of explosives, was perhaps the most notable of these acccomplishments. The first continuous miner was tested in about 1940, with its usefulness greatly enhanced by the development of tungsten carbide inserts in 1945 by McKenna Metals Company (now Kennametal). By 1950 the continuous miner had started to replace other coal mining methods. The era of mechanized mining had begun.

It is not possible to chronicle all of the developments that made mining what it is today. A more complete chronology of the important events is outlined in Table 1.2. It has been prepared using the references cited in Section 1.1, as well as those by Stack (1982) and Molloy (1986). These sources can be used to obtain a more comprehensive list of the crucial events in the development of mining technology.

1.4 STAGES IN THE LIFE OF A MINE

The overall sequence of activities in modern mining is often compared with the five stages in the life of a mine: *prospecting*, *exploration*, *development*, *exploitation*, and *reclamation*. Prospecting and exploration, precursors to actual mining, are linked and sometimes combined. Geologists and mining engineers often share responsibility for these two stages—geologists more involved with the former, mining engineers more with the latter. Likewise, development and exploitation are closely related stages; they are usually considered to constitute mining proper and are the main province of the mining engineer. Reclamation has been added to these stages since the first edition, to reflect the times. Closure and reclamation of the mine site has become a necessary part of the mine life cycle because of the demands of society for a cleaner environment and stricter laws regulating the abandonment of a mine. The overall process of developing a mine with the future uses of the land in mind is termed *sustainable development*. This concept was defined in a book entitled *Our Common Future* (World Commission on Environment and Development, 1987) as "development that meets the needs of the present without compromising the ability of future generations to meet their own needs." The ideas presented therein have been widely endorsed as a practical means of providing for future generations. The five stages in the life of a mine are summarized in Table 1.3 and are discussed in the following sections. There will be more extensive discussion of these stages in Chapters 2, 3, and 4.

TABLE 1.2 Chronological Development of Mining Technology

Date	Event
450,000 B.C.E.	First mining (at surface), by Paleolithic humans for stone implements.
40,000	Surface mining progresses underground, in Swaziland, Africa.
30,000	Fired clay pots used in Czechoslovakia.
18,000	Possible use of gold and copper in native form.
5000	Fire setting, used by Egyptians to break rock.
4000	Early use of fabricated metals; start of Bronze Age.
3400	First recorded mining, of turquoise by Egyptians in Sinai.
3000	Probable first smelting, of copper with coal by Chinese; first use of iron implements by Egyptians.
2000	Earliest known gold artifacts in New World, in Peru.
1000	Steel used by Greeks.
100 C.E.	Thriving Roman mining industry.
122	Coal used by Romans in present-day United Kingdom.
1185	Edict by bishop of Trent gives rights to miners.
1524	First recorded mining in New World, by Spaniards in Cuba.
1550	First use of lift pump, at Joachimstal, Czechoslovakia.
1556	First mining technical work, *De Re Metallica*, published in Germany by Georgius Agricola.
1585	Discovery of iron ore in North America, in North Carolina.
1600s	Mining commences in eastern United States (iron, coal, lead, gold).
1627	Explosives first used in European mines, in Hungary (possible prior use in China).
1646	First blast furnace installed in North America, in Massachusetts.
1716	First school of mines established, at Joachimstal, Czechoslovakia.
1780	Beginning of Industrial Revolution; pumps are first modern machines used in mines.
1800s	Mining progresses in United States; gold rushes help open the West.
1815	Sir Humphrey Davy invents miner's safety lamp in England.
1855	Bessemer steel process first used, in England.
1867	Dynamite invented by Nobel, applied to mining.
1903	Era of mechanization and mass production opens in U.S. mining with development of first low-grade copper porphyry, in Utah; although the first modern mine was an open pit, subsequent operations were underground as well.
1940	First continuous miner initiates the era of mining without explosives.
1945	Tungsten carbide bits developed by McKenna Metals Company (now Kennametal).

1.4.1 Prospecting

Prospecting, the first stage in the utilization of a mineral deposit, is the search for ores or other valuable minerals (coal or nonmetallics). Because mineral deposits may be located either at or below the surface of the earth, both direct and indirect prospecting techniques are employed.

TABLE 1.3 Stages in the Life of a Mine

Stage/ (Project Name)	Procedure	Time	Cost/Unit Cost
	Precursors to Mining		
1. Prospecting (Mineral deposit)	Search for ore a. Prospecting methods Direct: physical geologic Indirect: geophysical, geochemical b. Locate favorable loci (maps, literature, old mines) c. Air: aerial photography, airborne geophysics, satellite d. Surface: ground geophysics, geology e. Spot anomaly, analyze, evaluate	1–3 yr	\$0.2–10 million or \$0.05–1.00/ton (\$0.05–1.10/tonne)
2. Exploration (Ore body)	Defining extent and value of ore (examination/evaluation a. Sample (drilling or excavation), assay, test b. Estimate tonnage and grade c. Valuate deposit (Hoskold formula or discount method): present value = income − cost Feasibility study: make decision to abandon or develop.	2–5 yr	\$1–15 million or \$0.20–1.50/ton (\$0.22–1.65/tonne)
	Mining Proper		
3. Development (Prospect)	Opening up ore deposit for production a. Acquire mining rights (purchase or lease), if not done in stage 2 b. File environmental impact statement, technology assessment, permit c. Construct access roads, transport system d. Locate surface plant, construct facilities e. Excavate deposit (strip or sink shaft)	2–5 yr	\$10–500 million or \$0.25–10.00/ton (\$0.275–11.00/tonne)

Stage/ (Project Name)	Procedure	Time	Cost/Unit Cost
4. Exploitation (Mine)	Large-scale production of ore a. Factors in choice of method: geologic, geographic, economic, environmental, societal safety b. Types of mining methods Surface: open pit, open cast, etc. Underground: room and pillar, block caving, etc. c. Monitor costs and economic payback (3–10 yr)	10–30 yr	\$5–75 million/yr or \$2.00–150/ton (\$2.20–165/tonne)
	Post-mining		
5. Reclamation (Real estate)	Restoration of site a. Removal of plant and buildings b. Reclamation of waste and tailings dumps c. Monitoring of discharges	1–10 yr	\$1–20 million \$0.20–4.00/ton (\$0.22–4.40/tonne)

The *direct method* of discovery, normally limited to surface deposits, consists of visual examination of either the exposure (outcrop) of the deposit or the loose fragments (float) that have weathered away from the outcrop. Geologic studies of the entire area augment this simple, direct technique. By means of aerial photography, geologic maps, and structural assessment of an area, the geologist gathers evidence by direct methods to locate mineral deposits. Precise mapping and structural analysis plus microscopic studies of samples also enable the geologist to locate the hidden as well as surface mineralization.

The most valuable scientific tool employed in the *indirect search* for hidden mineral deposits is geophysics, the science of detecting anomalies using physical measurements of gravitational, seismic, magnetic, electrical, electromagnetic, and radiometric variables of the earth. The methods are applied from the air, using aircraft and satellites; on the surface of the earth; and beneath the earth, using methods that probe below the topography. Geochemistry, the quantitative analysis of soil, rock, and water samples, and geobotany, the analysis of plant growth patterns, can also be employed as prospecting tools.

1.4.2 Exploration

The second stage in the life of a mine, *exploration*, determines as accurately as possible the size and value of a mineral deposit, utilizing techniques similar to but more refined than those used in prospecting. The line of demarcation between prospecting and exploration is not sharp; in fact, a distinction may not be possible in some cases. Exploration generally shifts to surface and subsurface locations, using a variety of measurements to obtain a more positive picture of the extent and grade of the ore body. Representative samples may be subjected to chemical, metallurgical, X ray, spectrographic, or radiometric evaluation techniques that are meant to enhance the investigator's knowledge of the mineral deposit. Samples are obtained by chipping outcrops, trenching, tunneling, and drilling; in addition, borehole logs may be provided to study the geologic and structural makeup of the deposit. Rotary, percussion, or diamond drills can be used for exploration purposes. However, diamond drills are favored because the cores they yield provide knowledge of the geologic structure. The core is normally split along its axis; one half is analyzed, and the other half is retained intact for further geologic study.

An evaluation of the samples enables the geologist or mining engineer to calculate the tonnage and grade, or richness, of the mineral deposit. He or she estimates the mining costs, evaluates the recovery of the valuable minerals, determines the environmental costs, and assesses other foreseeable factors in an effort to reach a conclusion about the profitability of the mineral deposit. The crux of the analysis is the question of whether the property is just another mineral deposit or an ore body. For an ore deposit, the overall process is called *reserve estimation,* that is, the examination and valuation of the ore body. At the conclusion of this stage, the project is developed, traded to another party, or abandoned.

1.4.3 Development

In the third stage, *development*, the work of opening a mineral deposit for exploitation is performed. With it begins the actual mining of the deposit, now called the ore. Access to the deposit must be gained either (1) by stripping the overburden, which is the soil and/or rock covering the deposit, to expose the near-surface ore for mining or (2) by excavating openings from the surface to access more deeply buried deposits to prepare for underground mining.

In either case, certain preliminary development work, such as acquiring water and mineral rights, buying surface lands, arranging for financing, and preparing permit applications and an environmental impact statement (EIS), will generally be required before any development takes place. When these steps have been achieved, the provision of a number of requirements—access roads, power sources, mineral transportation systems, mineral processing facilities, waste disposal areas, offices, and other support facilities—must precede actual mining in most cases. Stripping of the overburden will then proceed if the minerals are to be mined at the surface. Economic considerations determine the *stripping ratio*, the ratio of waste removed to ore recovered; it may range from as high as 45 yd^3/ton (38 m^3/tonne) for coal mines to as low as 1.0 yd^3/ton (0.8 m^3/tonne) in metal mines. Some nonmetallic mines have no overburden to remove; the mineral is simply excavated at the surface.

Development for underground mining is generally more complex and expensive. It requires careful planning and layout of access openings for efficient mining, safety, and permanence. The principal openings may be shafts, slopes, or adits; each must be planned to allow passage of workers, machines, ore, waste, air, water, and utilities. Many metal mines are located along steeply dipping deposits and thus are opened from shafts, while drifts, winzes, and raises serve the production areas. Many coal and nonmetallic mines are found in nearly horizontal deposits. Their primary openings may be drifts or entries, which may be distinctly different from those of metal mines. These differences are outlined in Chapters 4, 7, 8, and 10 to 14.

1.4.4 Exploitation

Exploitation, the fourth stage of mining, is associated with the actual recovery of minerals from the earth in quantity. Although development may continue, the emphasis in the production stage is on production. Usually only enough development is done prior to exploitation to ensure that production, once started, can continue uninterrupted throughout the life of the mine.

The mining method selected for exploitation is determined mainly by the characteristics of the mineral deposit and the limits imposed by safety, technology, environmental concerns, and economics. Geologic conditions, such as the dip, shape, and strength of the ore and the surrounding rock, play a key role in selecting the method. *Traditional exploitation methods* fall into two broad categories based on locale: surface or underground. *Surface mining* includes mechanical excavation methods such as open pit and open cast (strip mining), and aqueous methods such as placer and solution mining. *Underground mining* is usually classified in three categories of methods: unsupported, supported, and caving.

1.4.4.1 Surface Mining. Surface mining is the predominant exploitation procedure worldwide, producing in the United States about 85% of all minerals, excluding petroleum and natural gas. Almost all metallic ores (98%), about 97% of the nonmetallic ores, and 61% of the coal in the United States are mined using surface methods (U.S. Geological Survey, 1995; Energy Information Administration, 2000), and most of these are mined by open pit or open cast methods. In *open pit mining*, a *mechanical extraction method*, a thick deposit is generally mined in benches or steps, although thin deposits may require only a single bench or face. Open pit or open cast mining is usually employed to exploit a near-surface deposit or one that has a low stripping ratio. It often necessitates a large capital investment but generally results in high productivity, low operating cost, and good safety conditions.

The *aqueous extraction methods* depend on water or another liquid (e.g., dilute sulfuric acid, weak cyanide solution, or ammonium carbonate) to extract the mineral. *Placer mining* is used to exploit loosely consolidated deposits like common sand and gravel or gravels containing gold, tin, diamonds, platinum, titanium, or coal. *Hydraulicking* utilizes a high-pressure stream of water that is directed against the mineral deposit (normally but not always a placer), undercutting it, and causing its removal by the erosive actions of the water. *Dredging* performed from floating vessels, accomplishes the extraction of the minerals mechanically or hydraulically. *Solution mining* includes both *borehole mining*, such as the methods used to extract sodium chloride or sulfur, and *leaching*, either through drillholes or in dumps or heaps on the surface. Placer and solution mining are among the most economical of all mining methods but can only be applied to limited categories of mineral deposits.

1.4.4.2 Underground Mining. Underground methods—unsupported, supported, and caving—are differentiated by the type of wall and roof supports used, the configuration and size of production openings, and the direction in which mining operations progress. The *unsupported methods* of mining are used to extract mineral deposits that are roughly tabular (plus flat or steeply dipping) and are generally associated with strong ore and surrounding rock. These methods are termed *unsupported* because they do not use any artificial pillars to assist in the support of the openings. However, generous amounts of roof bolting and localized support measures are often used. *Room-and-pillar mining* is the most common unsupported method, used primarily for flat-lying seams or bedded deposits like coal, trona, limestone, and salt. Support of the roof is provided by natural pillars of the mineral that are left standing in a systematic pattern. *Stope-and-pillar mining* (a stope is a production opening in a metal mine) is a similar method used in noncoal mines where thicker, more irregular ore bodies occur; the pillars are spaced randomly and located in low-grade ore so that the high-grade ore can be extracted. These two methods account for almost all of the underground mining in horizontal deposits in the United States and a very high proportion of the underground tonnage as well. Two other methods applied to steeply dipping deposits are also included in the unsupported category. In *shrinkage stoping*, mining progresses upward, with horizontal slices of ore being blasted along the length of the stope. A portion of the broken ore is allowed to accumulate in the stope to provide a working platform for the miners and is thereafter withdrawn from the stope through chutes. *Sublevel stoping* differs from shrinkage stoping by providing sublevels from which vertical slices are blasted. In this manner, the stope is mined horizontally from one end to the other. Shrinkage stoping is more suitable than sublevel stoping for stronger ore and weaker wall rock.

Supported mining methods are often used in mines with weak rock structure. *Cut-and-fill stoping* is the most common of these methods and is used primarily in steeply dipping metal deposits. The cut-and-fill method is practiced both in the overhand (upward) and in the underhand (downward) directions. As each horizontal slice is taken, the voids are filled with a variety of fill types to support the walls. The fill can be rock waste, tailings, cemented tailings, or other suitable materials. Cut-and-fill mining is one of the more popular methods used for vein deposits and has recently grown in use. *Square-set stoping* also involves backfilling mine voids; however, it relies mainly on timber sets to support the walls during mining. This mining method is rapidly disappearing in North America because of the high cost of labor. However, it still finds occasional use in mining high-grade ores or in countries where labor costs are low. *Stull stoping* is a supported mining method using timber or rock bolts in tabular, pitching ore bodies. It is one of the methods that can be applied to ore bodies that have dips between 10° and 45°. It often utilizes artificial pillars of waste to support the roof.

Caving methods are varied and versatile and involve caving the ore and/or the overlying rock. Subsidence of the surface normally occurs afterward. *Longwall mining* is a caving method particularly well adapted to horizontal seams, usually coal, at some depth. In this method, a face of considerable length (a long face or wall) is maintained, and as the mining progresses, the overlying strata are caved, thus promoting the breakage of the coal itself. A different method, *sublevel caving*, is employed for a dipping tabular or massive deposit. As mining progresses downward, each new level is caved into the mine openings, with the ore materials being recovered while the rock remains behind. *Block caving* is a large-scale or bulk mining method that is highly productive, low in cost, and used primarily on massive deposits that must be mined underground. It is most applicable to weak or moderately strong ore bodies that readily break up when caved. Both block caving and longwall mining are widely used because of their high productivity.

In addition to these conventional methods, *innovative methods* of mining are also evolving. These are applicable to unusual deposits or may employ unusual techniques or equipment. Examples include automation, rapid excavation, underground gasification or liquifaction, and deep-sea mining.

1.4.5 Reclamation

The final stage in the operation of most mines is *reclamation*, the process of closing a mine and recontouring, revegetating, and restoring the water and land values. The best time to begin the reclamation process of a mine is before the first excavations are initiated. In other words, mine planning engineers should plan the mine so that the reclamation process is considered and the overall cost of mining plus reclamation is minimized, not just the cost of mining itself. The new philosophy in the mining industry is *sustainability*, that is, the meeting of economic and environmental needs of the present while enhancing the ability of future generations to meet their own needs (National Mining Association, 1998).

In planning for the reclamation of any given mine, there are many concerns that must be addressed. The first of these is the safety of the mine site, particularly if the area is open to the general public. The removal of office buildings, processing facilities, transportation equipment, utilities, and other surface structures must generally be accomplished. The mining company is then required to seal all mine shafts, adits, and other openings that may present physical hazards. Any existing highwalls or other geologic structures may require mitigation to prevent injuries or death due to geologic failures.

The second major issue to be addressed during reclamation of a mine site is restoration of the land surface, the water quality, and the waste disposal areas so that long-term water pollution, soil erosion, dust generation, or vegetation problems do not occur. The restoration of native plants is often a very important part of this process, as the plants help build a stable soil structure and naturalize the area. It may be necessary to carefully place any rock or tailings with acid-producing properties in locations where rainfall has little effect on the material and acid production is minimized. The same may be true of certain of the heavy metals that pollute streams. Planning of the waste dumps, tailings ponds, and other disturbed areas will help prevent pollution problems, but remediation work may also be necessary to complete the reclamation stage of mining and satisfy the regulatory agencies.

The final concern of the mine planning engineer may be the subsequent use of the land after mining is completed. Old mine sites have been converted to wildlife refuges, shopping malls, golf courses, airports, lakes, underground storage facilities, real estate developments, solid waste disposal areas, and other uses that can benefit society. By planning the mine for a subsequent development, mine planners can enhance the value of the mined land and help convert it to a use that the public will consider favorable. The successful completion of the reclamation of a mine will enhance public opinion of the mining industry and keep the mining company in the good graces of the regulatory agencies. The fifth stage of the mine is thus of paramount importance and should be planned at the earliest possible time in the life of the mine.

1.5 UNIT OPERATIONS OF MINING

During the development and exploitation stages of mining when natural materials are extracted from the earth, remarkably similar unit operations are normally employed. The *unit operations* of mining are the basic steps used to produce mineral from the deposit, and the auxiliary operations that are used to support them. The steps contributing directly to mineral extraction are *production operations*, which constitute the production cycle of operations. The ancillary steps that support the production cycle are termed *auxiliary operations.*

The production cycle employs unit operations that are normally grouped into *rock breakage* and *materials handling*. Breakage generally consists of drilling and blasting, and materials handling encompasses loading or excavation and haulage (horizontal transport) and sometimes hoisting (vertical or inclined transport). Thus, the basic production cycle consists of these unit operations:

$$\text{Production cycle} = \text{drill} + \text{blast} + \text{load} + \text{haul}$$

Although production operations tend to be separate and cyclic in nature, the trend in modern mining and tunneling is to eliminate or combine functions and to increase continuity of extraction. For example, in coal and other soft rock mines, continuous miners break and load the mineral to eliminate drilling and blasting; boring machines perform the same tasks in medium-hard rock. The cycle of operations in surface and underground mining differs primarily by the scale of the equipment. Specialized machines have evolved to meet the unique needs of the two regimes.

In modern surface mining, blastholes of 3 to 15 in. (75 to 380 mm) in diameter are produced by rotary or percussion drills for the placement of explosives when consolidated rock must be removed. The explosive charge is then inserted and detonated to reduce the overburden or ore to a size range suitable for excavation. The broken material is loaded by shovel, dragline, or wheel loader into haulage units—generally trucks—for transport. Railroad cars are also used for haulage, and belt conveyors are often used after the material is crushed. Soil and coal are often moved in the same manner, though blasting is sometimes unnecessary. In the quarrying of dimension stone, the blocks are often freed without blasting, using wire saws or other mechanical devices.

In underground mining, the production cycle is similar, although the equipment used may be scaled down in size. Smaller drillholes are used, trucks are sometimes replaced by shuttle cars, and belt conveyors are more prevalent. Coal, salt, potash, and trona are often mined without the use of explosives or mined after undercutting the face to reduce the consumption of explosives.

In addition to the operations of the production cycle, certain auxiliary operations must be performed in many cases. Underground, these usually include roof support, ventilation and air-conditioning, power supply, pumping, maintenance, lighting, communications, and delivery of compressed air, water, and supplies to the working sections. In surface mining, the primary auxiliary operations include those providing slope stability, pumping, power supply, maintenance, waste disposal, and supply of material to the production centers.

1.6 ECONOMICS OF THE MINERAL INDUSTRIES

1.6.1 Mineral Production

It has been estimated that only a fraction of 1% of the earth's surface is underlain with mineral deposits of commercial value. From this resource, the United States extracted nearly 60 billion in mineral values in 1997: $27 billion

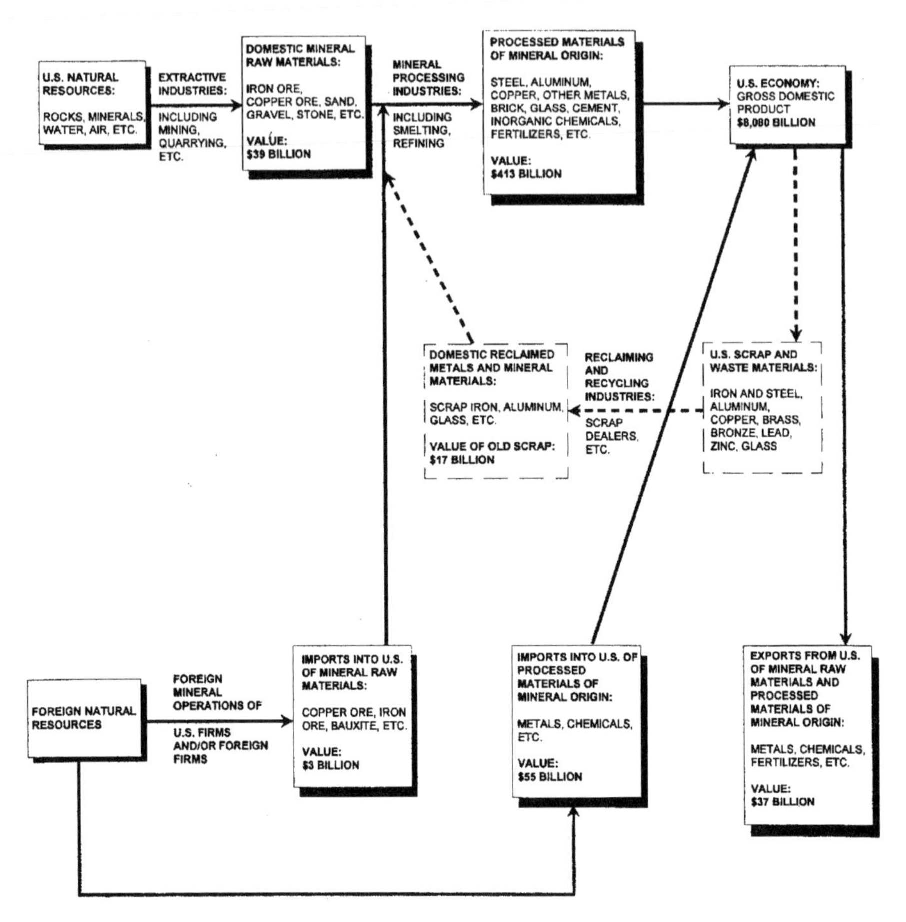

FIGURE 1.1. The role of nonfuel minerals in the U.S. economy (estimated values in 1998). *Source:* U.S. Geological Survey (1999).

worth of industrial minerals, $20 billion worth of coal, and $12 billion worth of metals (National Mining Association, 1998). Additional information on the value of minerals production in the United States can be found in Figure 1.1 (U.S. Geological Survey, 1999). This figure shows that the net value of exports of nonfuel mineral raw materials and materials processed from minerals has a value of about $35 billion per annum. In addition, about 355,000 people are employed by the mining industry in the United States, and every man, woman, and child in this country requires 23 tons (21 tonnes) of minerals, including nearly 4 tons (3.6 tonnes) of coal, each year to maintain his or her modern lifestyle (National Mining Association, 1998).

World consumption of minerals has increased to such an extent in modern times that more minerals were used in the twentieth century than were used since the beginning of history. This has occurred because we are now a society that depends on automobiles, trains, and airplanes for transportation; telephones, television, and computers for communications; fertilizers and heavy machinery for our agricultural output; industrial minerals for home building products; and coal-fired and nuclear plants for our electrical power. These human and industrial services in turn depend on the production of minerals and mineral products in great amounts.

The United States produces a very large tonnage of mineral products, but it has nevertheless become a major and growing importer of minerals, as shown in Figure 1.2. This country now imports more than 70% of its potash, chromium, tungsten, tin, and cobalt, plus almost all (better than 90%) of its fluorspar, manganese, and aluminum ore. Although it has increased imports of

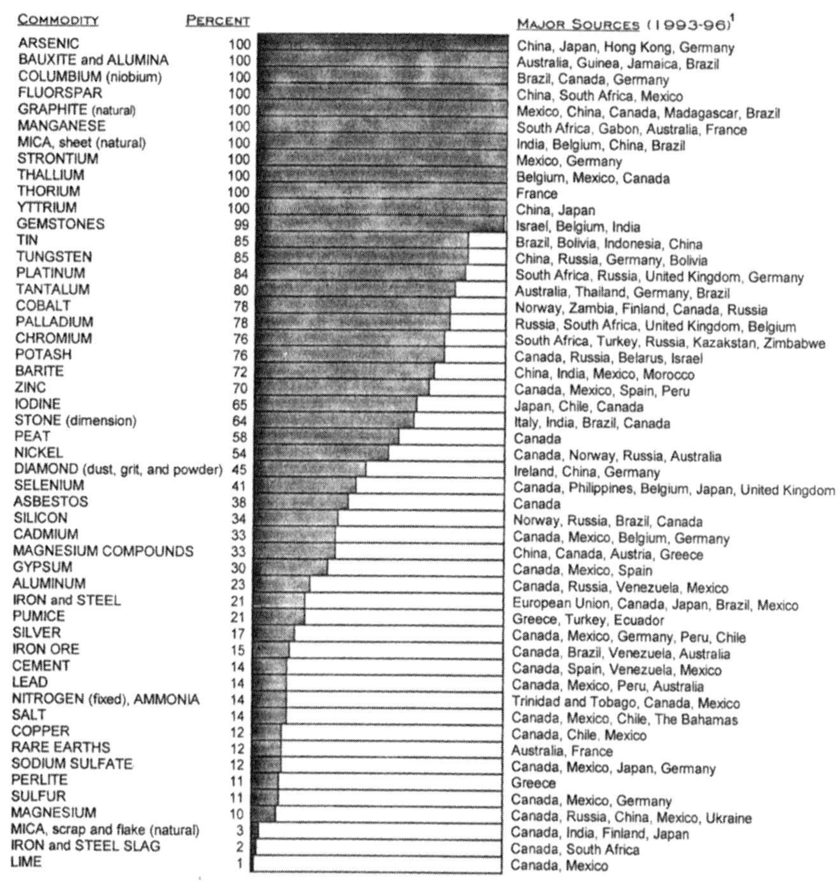

FIGURE 1.2. U.S. net import reliance for selected nonfuel mineral materials in 1998. *Source:* U.S. Geological Survey (1999).

many minerals, the United States is also producing more of the platinum group metals, gold, and diamonds.

Conservation of mineral resources is currently an important issue to the general public. Society is becoming much more cognizant of the need to conserve energy, minerals, and the environment. Accordingly, the mining industry has endorsed a policy that favors extraction of minerals in a more sustainable manner (National Mining Association, 1998). This policy favors mining as long as the effects on the environment or the economic welfare are not a burden on future generations. Extreme conservationist groups often attempt to push for punitive regulatory controls of mining activities and even a ban on mining. However, societal needs dictate that a compromise between the most strident conservationist viewpoint and the most open mineral development concept be adopted for the present and the future. This text attempts to present a balanced approach to mining in the environmental world in which we live.

1.6.2 Mineral Economics

The uniqueness of minerals as economic products accounts for the complexity of mineral economics and the business of mining (Vogely, 1985; Strauss, 1986). Minerals are unevenly distributed and, unlike agricultural or forest products, cannot reproduce or be replaced. A mineral deposit may therefore be considered a depleting asset whose production is restricted to the area in which it occurs. These factors impose limitations on a mining company in the areas of business practices, financing, and production practices. Because its mineral assets are constantly being depleted, a mining company must discover additional reserves or acquire them by purchase to stay in the mining business.

Other peculiar features of the mineral industries are associated with operations. Production costs tend to increase with depth and declining grade. Thus, low-cost operations are mined first, followed by the harder-to-mine deposits. In addition, commodity prices are subject to market price swings in response to supply and demand, which can make the financial risk of a long-term minerals project quite risky. A change in mining or processing technology can also drastically alter the economic landscape. The pattern of usage, in terms of intensity of use (lb/capita or kg/capita) and total consumption of metals on the world market for the nonferrous metals, shows that the intensity of usage of many of these metals continues to go down while overall consumption goes up (Crowson, 1998). Any swing in intensity of use due to substitution or recycling can greatly affect the market price of a metal. Mining companies must therefore keep their prices low by further improvements in productivity, or market price drops can easily create great economic hardships.

Some minerals, such as precious metals, iron, and most of the base metals, can be recycled economically, thereby affecting the markets for freshly mined metal. This is good practice and favorable for the future of humankind, but it can create economic problems if the market price is adversely affected.

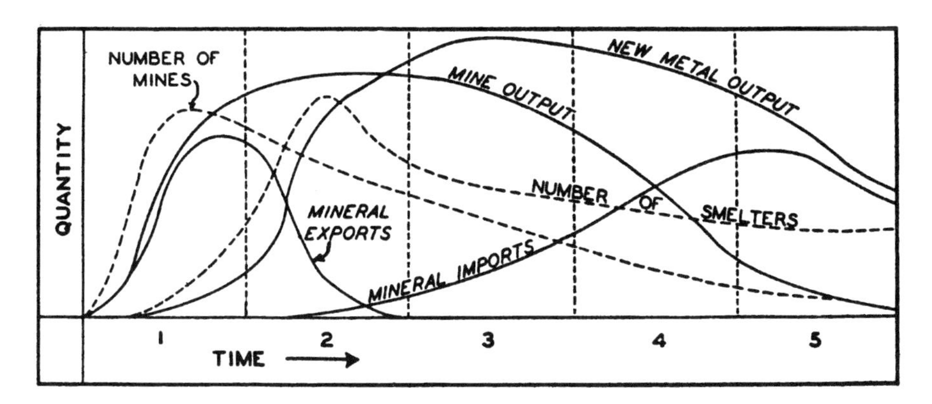

FIGURE 1.3. Periods in the growth of a country's mineral industries. (After Hewett, 1929; Lovering, 1943. By permission from the Society for Mining, Metallurgy, and Exploration, Inc., Littleton, CO.)

Substitutes for a particular mineral may be developed, particularly if the price of the mineral remains high. For example, aluminum and plastics have often been substituted for copper, and plastics have been substituted for a variety of other metals as well as for glass.

At times in recent history certain minerals have been exceptions to economic laws because their prices have been fixed by government decree or cartels. Official prices of gold, silver, and uranium have been regulated by government action, although they now fluctuate in free world markets. Cartels controlling industrial diamonds, oil, mercury, and tin have strongly influenced their market prices during certain time periods. However, many of these cartels have weakened or collapsed because of competition from new suppliers.

It has long been recognized that a given mineral-rich country can follow a fairly predictable pattern in its economic development, based on the state of its minerals industry (Hewett, 1929; Lovering, 1943). Five periods have been identified that reflect the periods of discovery, exploitation, and exhaustion of its mineral industry. The characteristic periods of mineral industry development are shown in Figure 1.3; each is described in the following list.

1. *Period of mine development:* Exploration, discovery of new districts, many small mines working; first recognition of large deposits and development of larger mines; rapidly increasing production of metals.
2. *Period of smelter development:* Fewer new discoveries; small mines becoming exhausted, but increasing output from large mines; many smelters competing for ore.
3. *Period of industrial development:* Decreasing costs, increasing standard of living; rapid accumulation of wealth; expanding internal and external markets; approaching the zenith of commercial power.

4. *Period of rapid depletion of inexpensive raw materials at home:* Ever-increasing costs of mining and materials produced; more and more energy is required to get the same amount of raw material. Mines and smelters continue to decline. Some foreign markets are lost; foreign imports, both raw materials and manufactured products, invade the home market.
5. *Period of decreasing internal and external markets:* Increasing dependence on foreign sources of raw materials brings increasing costs to manufacturers. This period can be characterized by a decreasing standard of living with its accompanying social and political problems; quotas, tariffs, subsidies, cartels, and other artificial expedients are used in the effort to maintain a competitive price in the domestic and world markets. This is a period of decreasing commercial power.

As current-day examples of the cycle in Figure 1.3, many less-developed countries fall into period 1, Australia into period 2, Russia in period 3, the United States in period 4, and the United Kingdom and Japan in period 5. Some economists maintain that this cycle is oversimplified and that it can be modified by technology or dedication to efficient production. For example, Japan after World War II arose from the ashes to become a world power without the benefit of extensive mineral resources. The Japanese had previously extracted nearly all of their inexpensive minerals. However, they purchased raw materials on the open market, created efficient automobile and electronics industries, and prospered by means of the "value added" to the mineral products.

At times during its history, the U.S. government has maintained a stockpile of strategic minerals to guard against shortages in case of war or economic blackmail. The practice became common after 1939 and sharply increased after 1946 because of Cold War tensions. More recently, the government has been reducing its stockpiles. This shift is the result of better access to minerals around the world and a less threatening world political scene since the breakup of the Soviet Union. Mineral companies have often been critical of the stockpiling policy, inasmuch as sudden large purchases from the stockpile can have drastic effects on the price of a given commodity. An informative history of the U.S. strategic minerals stockpile is given by Perkins (1997).

A further aspect of mineral economics concerns the financing and marketing of mines and mineral properties. Mining enterprises are financed in much the same manner as other businesses (Wanless, 1984; Tinsley et al., 1985). Because of greater financial risks, however, the expected return on investment is higher and the payback period shortened in a mining enterprise. Mineral properties as well as mines are marketable. The selling price is determined generally by a valuation based on the report of an engineer or geologist; the value of future earnings is usually discounted to the date of purchase in computing the present value of the property. Mineral properties may be determined to have worth because they are very rich, very large, easily accessible, in great demand, favorably located, cheaply mineable, or militarily strategic.

1.7 COMPUTER APPLICATIONS AND USAGE

In today's world, computers are becoming more useful and finding more application to composition, engineering calculations, library research, and general communications than ever before. It is perhaps worthwhile to compare the progress in the area of computations over the last few decades. Both authors of this text were trained as engineers to perform their computations on a slide rule in their respective university educations. You, the student, were probably introduced to computers in elementary school and most likely became quite proficient in their use before entering your first college course. It is important, therefore, that you continue to develop your skills on computers for every aspect of your future as professionals.

The first aspect of computers considered here is their use in retrieval of information of value in your studies. To provide useful information, the following World Wide Web (WWW) sites are provided as an introduction to useful sources of minerals data and other information of interest to mineral engineers. The web addresses are current at the time of publication; however, web sites are constantly changing. Your web browser may be necessary to provide access to the sites listed here at some future time. It should be noted that the prefix "http://" is omitted from all the WWW sites listed in this book. You must prefix that to each address.

U.S. Geological Survey www.usgs.gov

Energy Information Administration, U.S. Department of Energy www.eia.doe.gov

National Institute of Occupational Safety and Health www.cdc.gov/niosh/homepage.html

Mine Safety and Health Administration www.msha.gov

Society for Mining, Metallurgy, and Exploration, Inc. www.smenet.org

Office of Surface Mining, Department of the Interior www.osmre.gov

These sites will be of use to you in your upcoming assignments and in your professional life as well.

1.8 SPECIAL TOPIC: ECONOMIC ANALYSIS OF A MINERAL COMMODITY

Prepare a summary analysis of pertinent production and economic data for several important mineral commodities, using Table 1.4 as a format. Obtain statistics for mine production (world and domestic) and domestic imports, exports, scrap, stockpile changes (private and government), and net consumption (total and by use). Determine data for the most recent year available. State the commodity price, preferably on an as-mined basis ($/ton, or $/tonne, of mineral, F.O.B., mine).

TABLE 1.4 Economic Analysis of Mineral Commodity

Mineral ____________ Year ____ Average price ________

Mine Production

World	Country	Tonnage	Percentage	Value
	____	____	____	____
	____	____	____	____
	____	____	____	____
	____	____	____	____
	Total	____	____	____
Domestic (P)	State			
	____	____	____	____
	____	____	____	____
	____	____	____	____
	____	____	____	____
	Total	____	____	____
	District			
	____	____	____	____
	____	____	____	____
	____	____	____	____
	____	____	____	____
	Total	____	____	____
	Company		Mine	
	____	____	____	____
	____	____	____	____
	____	____	____	____
	____	____	____	____
	Total	____	____	____

Domestic Use

Imports (I)		Exports (E)		Scrap (S)
Country	Tonnage	Country	Tonnage	Tonnage ________
____	____	____	____	
____	____	____	____	Stockpile (SP)
____	____	____	____	Private
____	____	____	____	Tonnage ________
Total		Total		Government
	____		____	tonnage ________

Net Consumption (C)		Consumers	
		Industry or Use	Tonnage or Percentage
Actual tonnage	____	____	____
Apparent tonnage	____	____	____
		____	____
$C = P + S + I - E \pm SP$		____	____

= ______________________________

Sources: ______________________________

Select the metallic and nonmetallic commodities you wish to study by referring to and utilizing the following references:

1. U.S. Geological Survey web page (www.usgs.gov). This reference provides a two-page summary of more than 30 of the primary metallic and nonmetallic minerals produced in the United States. It may not supply you, however, with all the information you need to complete the summary in Table 1.4.
2. U.S. Geological Survey, *Mineral Commodities Survey* (1999 or latest edition). This volume is produced annually. It covers more than 80 metallic and nonmetallic mineral products. Each product is summarized in approximately two pages using statistics that are about a year or two old.
3. U.S. Geological Survey, *Minerals Yearbook* (1995 or latest edition). The most complete mineral production analysis is provided in this yearbook, which is printed every several years. It is produced in three volumes, with the largest (Vol. 3) covering the U.S. production statistics for all the nonfuel minerals. More information is available here than in any other reference, though the data may be two to five years old.
4. *Engineering and Mining Journal* (monthly), "Prices, American Metal Market." This monthly journal provides the prices being paid for metals on a specific date each month; it is useful for more current price information.

Similar information on coal production can be found in the following references. The statistics contained therein are generally a year or two old.

1. Energy Information Administration, *Coal Industry Annual* (2000 or most recent edition). This resource provides several hundred pages each year on the primary statistics of coal production in the United States, including state-by-state information, production, employment, prices, distribution, exports, and imports.
2. Energy Information Administration web page (www.eia.doe.gov). This resource provides exactly the same information as the previously cited publication, as well as many other sources of information available for downloading.

PROBLEMS

1.1 Select one of the web sites listed in Section 1.7 (or use one selected for you by your instructor) and write a one- or two-page summary of the important elements of the web site. Evaluate the usefulness of the information found therein. Your instructor may ask you to share these with your classmates for future use.

1.2 Go to the web site of the U.S. Geological Survey (www.usgs.gov) and find the list of the metallic and nonmetallic minerals produced in the United States. Pick the one you know the least about and write a one-page summary of the mineral and its production picture in the United States.

2

MINING AND ITS CONSEQUENCES

2.1 INTRODUCTION

Although mining is one of only two resource industries and occupies a strategic position in our modern technological society, it does not enjoy a lofty status in the eyes of the general public (Bingham, 1994; Prager, 1997). Reasons for this dichotomy include a lack of understanding by the public of the source of all of our material goods, the tendency of the news media to report more readily the negative stories related to mining, and the visible consequences associated with mining operations. In many cases, the public image of the mining industry is compromised because of factors beyond the industry's control; in others, the industry can only blame the shortsighted mining practices of the past for its negative image. Currently, about 80% of Americans are reported (Prager, 1997) to consider themselves "environmentalists." Clearly, we live in an environmental age. It is therefore important that mining engineers also consider themselves environmentalists. After all, our own personal needs are similar to those of other human beings and we are in a position to positively (or negatively) affect the world in which we live. It is important that we, the mining engineers of the world, be inclined to join others in protection of the environment.

Students should take note of the fact that the most highly regarded mining companies today make a positive effort to present the public with an image of leadership in safeguarding the environment and in other activities. They strive to produce minerals with the least amount of environmental damage, to provide for the health and safety of their workers, to affect in a positive manner the geographic regions in which they conduct their operations, and to help the people who are their neighbors. They support the concept of sustainability in the minerals sector and do their best to educate the public on their role in society, how they contribute to the general welfare of the public, and how they are helping to build a better future.

Analysis of the public perception of the mining industry is worthwhile, so that the problems of the past are not perpetuated in the future. It is important to realize that there is a link between the consequences of mining and the laws that have been passed to regulate the industry. Several of the significant mining laws discussed in the next section fall in this category. This chapter summarizes some of the regulations that apply to mining, the health and safety consequences of the mining industry, and the environmental issues that are of importance to the industry and to society in general.

2.2 GOVERNMENT REGULATIONS APPLIED TO MINING

The laws and regulations that apply to the mining industry are many and varied. However, the most important of these are the General Mining Law of 1872, which specifies the use of public lands for mining; the Mineral Leasing Act of 1920, which outlines the methods for leasing certain minerals on federal lands; the Coal Mine Health and Safety Act of 1969, which authorizes the health and safety rules for mining coal; the Federal Mine Safety and Health Admendments Act of 1977, which dictates safety standards for metal and nonmetallic mines; and the Surface Mining Control and Reclamation Act of 1977, which regulates the surface mining of coal and the reclamation of surface coal mines. A long list of other federal statutes that apply to mining is provided by Schanz (1992) and Marcus (1997a, pp. 31, 45). Many of these are oriented toward air and water pollution control, antiquities, wilderness lands, and endangered species. Because of the many statutes that a mining organization must observe in its daily operations, mining companies often have lawyers or other specialists on their staffs to interpret and manage compliance with these laws and regulations. Summaries of the aforementioned five Acts are provided in the following subsections. A more complete review of the five sets of laws and regulations can be found in Hartman (1992, pp. 123–201).

2.2.1 Mining Law of 1872

The General Mining Law of 1872 was enacted to provide a means of regulating the discovery and claiming of mineral deposits on federal lands. The law has been amended somewhat from its original form, but it exists today in essentially the same form as when it was enacted. It applies to claiming lands for mining purposes in 19 states where federal lands exist and have not been withdrawn from mining use. The types of minerals that are locatable under the law have been reduced over time to metallic deposits and certain industrial minerals that have distinct and special value (Parr, 1992). Common sedimentary minerals such as coal and hydrocarbons such as oil and gas must be leased under the Mineral Leasing Act of 1920. A more complete description of the various ramifications of the law can be found in Parr (1992); significant additional information on the various court rulings that apply to the law are contained in Maley (1996). A summary of important provisions of the law follows.

A *lode claim* is locatable on vein (lode) deposits and is subject to the definition that the deposit must be in its original geologic location with rock on both sides (to differentiate it from a placer deposit). The claim must have a "discovery" with the long dimension of the claim (1500 ft or 457 m) along the length of the vein and the short dimension of the claim (600 ft or 183 m) perpendicular to the vein. The specific rules for posting and recording notice of the claim and marking its boundary are set by the state in which the claim is located. Filing a lode claim is a procedure for obtaining the legal right to explore a given mineral deposit to prove its value without interference from other parties.

Placer claims are generally 20 ac (8 ha) in area and are located on any deposit that is not in its original point of deposition or on a deposit that has had its surface outcrop badly distorted by erosion. Unlike a lode claim, a placer claim need not possess a particular shape. However, conformity to public land surveys is recommended (Maley, 1996). Each person is limited to a single placer claim, but an individual can form a group and consolidate up to 160 ac (64 ha) into a single claim as long as no one person has more than a 20 ac (8 ha) share and each individual in the group is a true partner. Placer claims do not establish claim to veins below the placer deposit and cannot be located on a patented lode claim.

Establishing a *tunnel site* is a method of establishing rights to any vein deposit found within 3000 ft (914 m) of the mouth of the tunnel (actually, an *adit* in mining terminology). The person who establishes the tunnel site must pursue the development of the tunnel, and if that person pursues the tunnel faithfully, no surface claim may be made that takes away his or her rights to discovery. This also means that the person can drive the tunnel even though he or she does not have a valid discovery at the time of initiation of the tunnel. The person must not cease advancing the tunnel for more than six months, however, as that will constitute an abandonment of the claim. Any surface claim established before the tunnel is initiated is still valid, but a surface claim established after the tunnel is begun will be legally junior to the tunnel claim if the tunnel claim discovers the same vein as that of the surface claim holder. For this reason, a person making a surface claim must obtain knowledge of any tunnel claims in the area.

The final type of claim under the mining law is termed a *mill site*. The site is claimed like a lode claim, but has no specific shape and must be less than 5 ac (2 ha) in area and located on land that is not mineralized. A mill site cannot be utilized for other purposes, in general, and patent (permanent ownership) can be obtained only if clear justification for the mill site is evident.

The legal rights to a mining claim can be maintained indefinitely by performing annual assessment work. The assessment work year begins on September 1 each year. Valid assessment work normally includes labor to excavate or prove mineral values, drilling exploration holes or excavating trenches, conducting geophysical or geochemical surveys, constructing buildings or facilities directly connected to the mining endeavor, construction of roads that are necessary for developing the mine, and other exploration or development activities that can readily be seen to be necessary for the development of a mining operation.

After a valid discovery has been made and the owner of the claim or claims has performed more than $500 worth of assessment and improvement work on each claim, the owner is eligible to file a patent application to obtain a patent on the claims. Historically, this was routinely performed without great emphasis on requiring the claim owner to provide unassailable proof of the value of the mineral discovery. This resulted in patents being awarded on property that had little in the way of mineral value. In more recent times, the government agencies have required that the claim owners provide proof that the minerals contained on the claim are minable by a prudent person, that is, that they can be mined at a profit. More information on the process of patenting and problems that may occur can be found in Parr (1992) and Maley (1996).

The consequences of the General Mining Law of 1872 have been favorable to the U.S. economy. The law encourages the discovery, development, and exploitation of minerals on public lands. This, in turn, generates jobs for mine workers, profits for investors, and tax revenue for the government. In other words, it creates wealth and general economic prosperity. However, such consequences have not been without conflict. Modern society values forests, wild areas, streams, recreational activities, and other aspects of public lands. Thus, many public areas are withdrawn from mining use for other purposes. In addition, the public has reacted negatively to the practice of allowing mining companies to mine public lands without royalties being paid on the minerals and has expressed concern that federal lands can be converted to private ownership by the patenting process. In recent years, the U.S. Congress has been considering a major overhaul of the mining law, which may significantly change the way minerals are mined on public lands in the future.

2.2.2 Mineral Leasing Act of 1920

The Mineral Leasing Act of 1920 was passed to set aside oil, gas, coal, potassium, phosphate, sodium, oil shale, gilsonite, native asphalt, bitumens, bituminous rock, and sulfur from discovery under the General Mining Law. These minerals thereafter became minable on public lands only on a lease basis. Where the mineral values are known to exist, the government may periodically offer leases on blocks of land based on a competitive bidding process. Tracts of many mineral lease sales are arranged in 2560 ac (1035 ha) blocks, but coal lands are often offered for lease in larger blocks to enhance their value to the mining company and to the public. The competitive bid arrangement is only one of the ways of obtaining a lease on mineral land under the Act. Where mineral occurrence is possible but not proven to exist, a person or company wishing to explore for leasable minerals may obtain a two-year prospecting permit. If such minerals of sufficient value are discovered, the permit holder is then entitled to a preference-right lease; that is, he or she has the right to lease the land for mining. Most leases are valid for a period of 20 years.

Coal, oil, and gas leases result in significant royalties being paid to the federal government, with a share going to the state in which the lease is located. Coal lands are leased only after a comprehensive land use plan for the area has been formulated by pertinent government agencies and reviewed by other interested parties. The coal lessor pays a minimum of $3.00/ac ($7.50/ha) per year rental and royalties of no less than 12.5% of the coal's value from a surface mine and no less than 8% from an underground mine. The lessor is also required to develop the mine within a ten-year period with continuing production after development. The requirements for continuing production are outlined in Parr (1992). The Mineral Law of 1920 provides for opportunities for mineral production from public lands and offers benefits to the public as well. Because most minerals administered under this law do not require discovery, the royalty system is appropriate.

2.2.3 Federal Coal Mine Health and Safety Act of 1969

The regulation of coal mining activity in the United States was, for much of the twentieth century, left primarily up to the states. Most states had reasonably safe mining practices built into their laws, but enforcement was not strict enough to overcome the many hazards inherent in the mining of coal. In 1968 an explosion at the Consol No. 9 Mine in West Virginia killed 78 miners and provided impetus for the U.S. Congress to enact a stronger federal law to overcome the problem. The Federal Coal Mine Health and Safety Act of 1969 authorized the Mine Safety and Health Administration (MSHA) to set up regulations to control both the safety and health effects of coal mining in the United States. MSHA was also given the power necessary to enforce the regulations that were established.

The mining industry found it difficult at first to accept the law because it resulted in significant increases in costs. However, members of the industry adjusted their thinking over time and soon were better integrating the new regulations into their operations. As time went by, they overcame the productivity and cost problems associated with the law, and the regulations became an accepted part of the task of mining coal. With improved cooperation between the operators and the government, the law and its regulations have been a tremendous success. More will be said about the success of the law in Section 2.3.

It is difficult to outline the many aspects of the law. A quick summary of some of the provisions of the regulations is given in the following paragraphs.

The regulations and their revisions are published each year in Title 30 of the Code of Federal Regulations (Code of Federal Regulations, 2000). Parts 70-12, 74-75, 77, and 90 of Title 30 cover health and safety standards that apply to both surface and underground mines. The easiest way to access the regulations is to use the Internet web site http://www.msha.gov/REGDATA/MSHA/0.0.HTM (Mine Safety and Health Administration, 2000). Because the regulations have frequent minor changes and occasional major changes, it is important to access the web site whenever an up-to-date set of regulations is desired.

A major change that resulted from the Federal Coal Mine Health and Safety Act of 1969 was the institution of strict dust control regulations. Part 70 of the regulations specified the maximum average coal dust exposure to which an underground miner can be subjected during a workday and, in addition, specified a maximum exposure to silica in the coal mine atmosphere. This part of the regulations is helping to reduce the incidence of pneumoconiosis (black lung) and silicosis among coal miners. A second major change was the emphasis on extensive education of new miners and retraining of all miners on an annual basis. This strategy was aimed at keeping safety constantly on the minds of miners so that they do not forget the hazardous environment in which they work.

Safety regulations for underground coal mining are quite voluminous and must be studied in detail by anyone engaged in the underground mining of coal. They cover the following general topics (among others): certified persons, roof support, ventilation, rock dusting, electrical systems, fire protection, maps, explosives and blasting, hoisting, mantrips, escapeways, emergency shelters, and communications. It is beyond the scope of this book to outline the regulations more extensively. Interested persons should consult the regulations for more details.

2.2.4 Federal Mine Safety and Health Amendments Act of 1977

Although the work in metal and nonmetal mines is usually not as hazardous as that in coal mines because of the general absence of explosive strata gases, there are still many dangers inherent in the mining of any mineral, particularly in underground mines. In 1972 a fire in the Sunshine silver mine in Idaho killed 91 miners. This incident showed the need for better escape plans and better training of miners in the use of respirators for escape attempts. There was great impetus for an improved set of regulations for metal and nonmetal mines. The Federal Mine Safety and Health Amendments Act of 1977 came about as a result of concern for the safety of metal and nonmetal miners. This act, combined with previous legislation, provides the regulations for all metal and nonmetallic mines.

Parts 56 through 58 of the Code of Federal Regulations cover the regulations that pertain to the mining of metal and nonmetal deposits in the United States, both on the surface and underground, that were authorized by the

Federal Mine Safety and Health Amendments Act of 1977. Part 56 covers the operation of surface mines and specifies both safety and health regulations. In a similar manner, Part 57 covers the operation of underground metal and nonmetal mines, and Part 58 pertains to the health standards for these mines.

The regulations for metal and nonmetal mines are not as voluminous as those pertaining to coal mines because of the somewhat less hazardous environment. However, they do require annual retraining and cover many of the same topics found in the coal mine regulations. Important sections of the regulations for underground mines include ground control, fire prevention, air quality, ventilation, explosives usage, drilling, haulage, electrical power, compressed air usage, equipment, materials storage, hoisting, and personal protection. Regulations for surface mining of metal and nonmetal deposits include sections on air quality, fire prevention, explosives, drilling, haulage, compressed air, machinery, and personal protections. Education of new miners and annual retraining are also included under the law. Details of the regulations applicable to metal and nonmetal mines can also be found at http://www.msha.gov/REGDATA/MSHA/0.0.HTM (Mine Safety and Health Administration, 2000). These regulations are likewise enforced by MSHA.

In September 1999, MSHA promulgated training rules for all persons engaged in the general areas of dredging, sand and gravel mining, stone mining, and clay, phosphate and surface limestone mines. These regulations became effective in October 2000 and are oriented toward providing 24 hours of training to new miners and 8 hours of annual refresher training to keep all engaged in this industry aware of the dangers they face on the job. In addition, the regulations require that a miner assigned to a new task must be trained in safe work procedures specific to that new task. Each company must have a training plan that is implemented by a competent person. MSHA is responsible for enforcing these regulations. They can be found in Part 46 of the Code of Federal Regulations (Code of Federal Regulations, 2000). The detailed regulations can also be found at http://www.msha.gov/REGDATA/MSHA/0.0.HTM (Mine Safety and Health Administration, 2000).

2.2.5 Surface Mining Control and Reclamation Act

The Surface Mining Control and Reclamation Act (SMCRA) was adopted in 1977 to establish the Office of Surface Mining, Reclamation, and Enforcement (often called the Office of Surface Mining and abbreviated OSM) and to provide a uniform and comprehensive set of surface mine regulations throughout the country. The Act was aimed at previous problems of coal mine reclamation in which the mine sites were not adequately reclaimed or were abandoned when companies went bankrupt. A summary of the major aspects of the law may be found in Kaas (1992); a complete text of the law can be ordered at http://www.osmre.gov/ (Office of Surface Mining, 2000).

A number of important measures were instituted in this law to protect the environment and to reclaim previous sites that had not been properly reclaimed. First, a system of permitting all surface coal mines was set up to ensure that mining companies had a mining and reclamation plan that met the standards of the law. Second, the states were restricted from enforcing the law unless they had a suitable program in place that met the federal reclamation requirements. Third, a performance bond was required as insurance that reclamation could be completed in the event that a company itself could not reclaim the mine site to the required standards. Fourth, a system of taxes on coal production was instituted to establish a reclamation fund to reclaim mining sites that had not been adequately reclaimed in the past. This provision established taxes of $0.35/ton ($0.386/tonne) for surface-mined coal and $0.15/ton ($0.165/tonne) for underground coal to be paid into a reclamation fund for abandoned mine lands.

The requirements outlined within SMCRA for conducting a surface coal mining operation include many performance standards (Kaas, 1992). Some of the noteworthy goals include restoring the original contour of the land, reclaiming to a higher or better use than before mining, segregating and replacing the soil and subsoil to provide adequate growing conditions, controlling runoff water from the site, treating or burying acid-bearing strata, and revegetating the land with native species. Many other objectives, too numerous to mention here, make the law a rather comprehensive system of restoring mined lands to an acceptable form for the future.

SMCRA has generated tremendous impacts, both economically and socially. The regulations require significant planning, extensive preventive measures, and tightly controlled reclamation of all lands that are mined by means of surface methods for coal in the United States. The expense of such a program is great in many mining situations, which drives up the costs associated with surface-mined coal. The social benefits are also great, with many of the mined lands being returned to a more aesthetic and useful form. The reclamation fund set up under the law has helped to restore many of the lands mined in the past, with emphasis on those that present recognizable hazards to people or to the environment.

2.3 HEALTH AND SAFETY ISSUES

Historically, mining has been an extremely hazardous industry in which to work. Mines can be dangerous because of falls of earth, strata gases that are emitted into the mine atmosphere, the explosive nature of mineral fuels when in the form of dust, and the many types of heavy equipment used in the mining process. The hazardous nature of mining work continued into the twentieth century, with coal mines being more of a threat to the safety of miners than metal and nonmetal mines. However, there was greater progress in the control of the hazards of mining in the twentieth century than in all previous centuries of mining combined. Several reasons for this progress are evident.

First, the U.S. Bureau of Mines (USBM) was formed in 1910 when the U.S.

Congress realized that fatalities in the mining industry were not being controlled. At the time, the number of deaths in the coal mining industry was nearly 3000 per year, and the fatalities in the metal/nonmetal industry were nearly 1000 per year. The number of fires and explosions with multiple fatalities was quite high, prompting the new bureau to seek an end to the hazards associated with methane gas and coal dust. In addition, deaths and injuries due to roof falls were targeted with better methods for roof control. During the 85-year history of the USBM, safety equipment and industry practices greatly improved the safety conditions in mines. The USBM was disbanded in 1995, with much of its activity taken over by the National Institute for Occupational Safety and Health (NIOSH). NIOSH is currently carrying forth the more important health and safety research and educational activities of the previous agency, and future improvements in this important area should continue.

Second, the laws controlling the practice of mining became more stringent as scientific and engineering methods improved and better monitoring instruments and safety equipment became available. The Federal Coal Mine Health and Safety Act of 1969 was a watershed event in improving the mining industry's safety record as it authorized stronger regulation, provided for continued enforcement, and instituted educational efforts to train all miners in safe work habits. When the companies mining coal made adherence to the new regulations part of their everyday production objectives, the effects were felt and mine safety improved significantly. Similarly, in metal and nonmetal mines, the Federal Mine Safety and Health Amendments Act of 1977 better controlled hazards in the mining industry to reduce the incidence of accidents. The number of mining-related fatalities and injuries has been reduced significantly since those Acts were passed.

The last major influence on mine safety was the sociological and political support for providing each worker with a safe working environment. Modern society demands much more of an employer than it has in the past. In part, this is the result of news events of the past century to which the public and the union organizations could react. Mine disasters spurred much of the legislation of the past century, as the public and the media pushed the U.S. Congress to act in the legislative area. Industry has also changed its thinking about health and safety. It has become more cognizant of the high costs associated with accidents and considers safety to be economically justifiable. The sociological landscape has thus changed to be much more beneficial to the health and safety of the worker.

2.3.1 Safety Performance

Records of the twentieth century in the area of mine safety show a striking reduction in the number of deaths and injuries in mines. The marked improvement in the number of fatalities in mines, in both the coal and the metal/nonmetal mining industries, is shown in Figure 2.1. This graph indicates the success of the mining industry in overcoming fatalities but is not a very logical

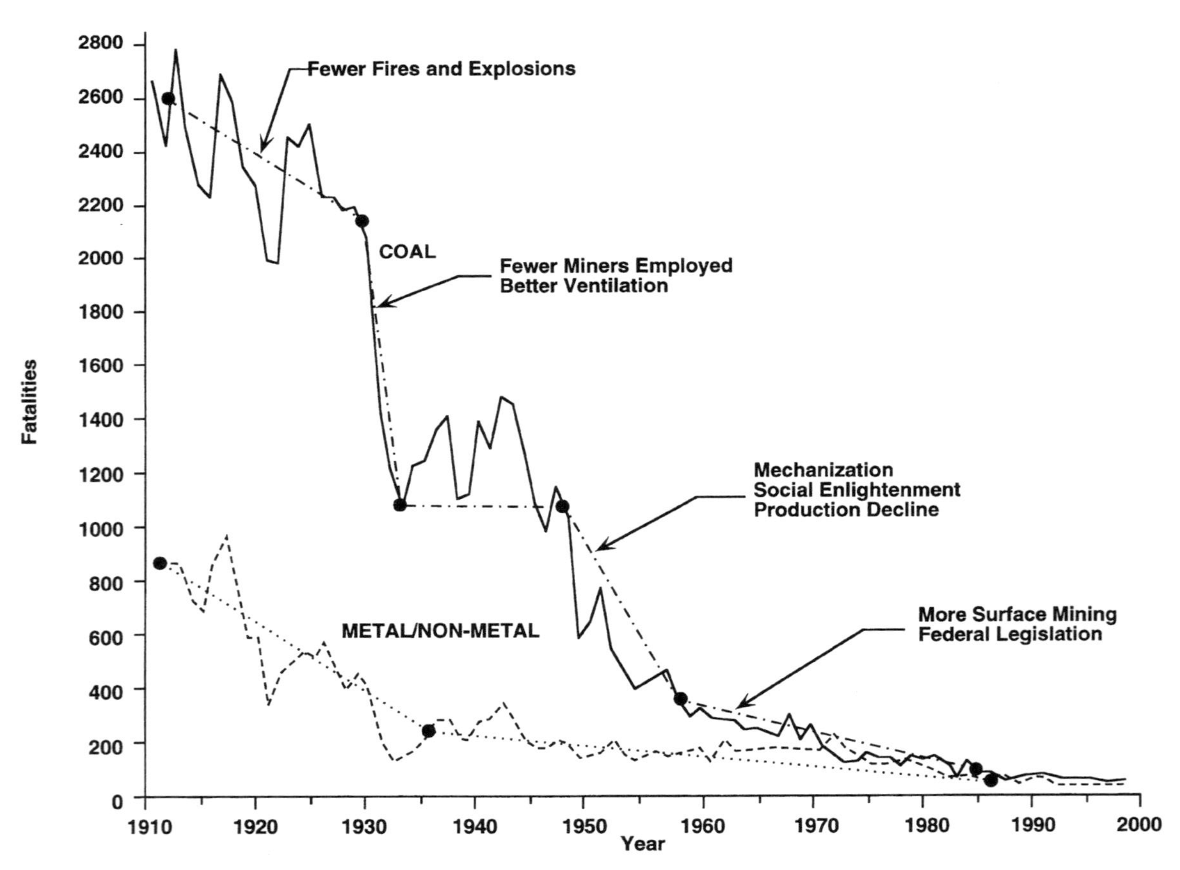

FIGURE 2.1. U.S. mine fatalities, 1910 – 1999. *Sources:* Katen (1992); Mine Safety and Health Administration (2000).

means of judging the relative improvement. For example, the drop in fatalities in the early 1930s resulted primarily from the smaller employment in the industry during that period. Thus, it is important to study the incidence rate (number of fatalities per million hours worked) in addition to the number of fatalities. This rate is provided in Figure 2.2 for the coal and the metal/nonmetal mining groups. The figure also shows a steady improvement in the safety statistics for both coal and metal/nonmetal mines.

The improvement has been especially striking since the adoption of the Federal Coal Mine Health and Safety Act of 1969 and the Federal Mine Safety and Health Amendments Act of 1977. Clearly, Figures 2.1 and 2.2 indicate a tremendous improvement in the accident statistics associated with mining during the twentieth century. At the beginning of the century, mining occupations were at the top of the list of most dangerous occupations. Mining has greatly improved its standing; a number of other common occupations are now higher on the scale. However, there is still much to be done to reduce accidents further. Any fatality or injury must be considered unacceptable. Only by using this assumption will the mining industry be able to continue to improve in this important area of endeavor.

2.3.2 Health Performance

Mankind has known health effects related to the mining industry for centuries (Hartman et al., 1997, pp. 88–89). The most important of these, respiratory disease, was recognized to some degree in the first century C.E. However, it was not until the invention of the X ray in 1896 that health scientists began to investigate such diseases in detail. During the twentieth century, the studies of lung disease allowed medical personnel to identify silicosis, coal workers' pneumoconiosis (black lung), and asbestosis as diseases of the mining profession caused by silica, coal dust, and asbestos, respectively. Regulation of exposure to these minerals came about when the incidences of the diseases began to alarm the public. The regulation of silica in the mine atmosphere began in the 1930s after a tunnel driven near Gauley Bridge, West Virginia, resulted in hundreds of workers acquiring silicosis (Seaton, 1975, p. 82). The mechanization of the coal mining industry after 1950 resulted in significant exposures of miners to coal dust and an increasing incidence of black lung, prompting the regulation of coal dust exposure by the Coal Mine Health and Safety Act of 1969. Asbestos minerals were of even greater concern because they were determined to be the cause of lung cancer in those who were exposed to asbestos during mining, processing, and utilization. Thus, very stringent regulation of the asbestos industry was instituted.

The result of this legislation to control mining-related health effects has been a notable decrease in the incidence of lung diseases in the mining industry. However, the problem persists for a number of reasons. One reason is the reluctance of mine workers to protect themselves by using proper respirators and other protective measures. A second is the difficulty of completely

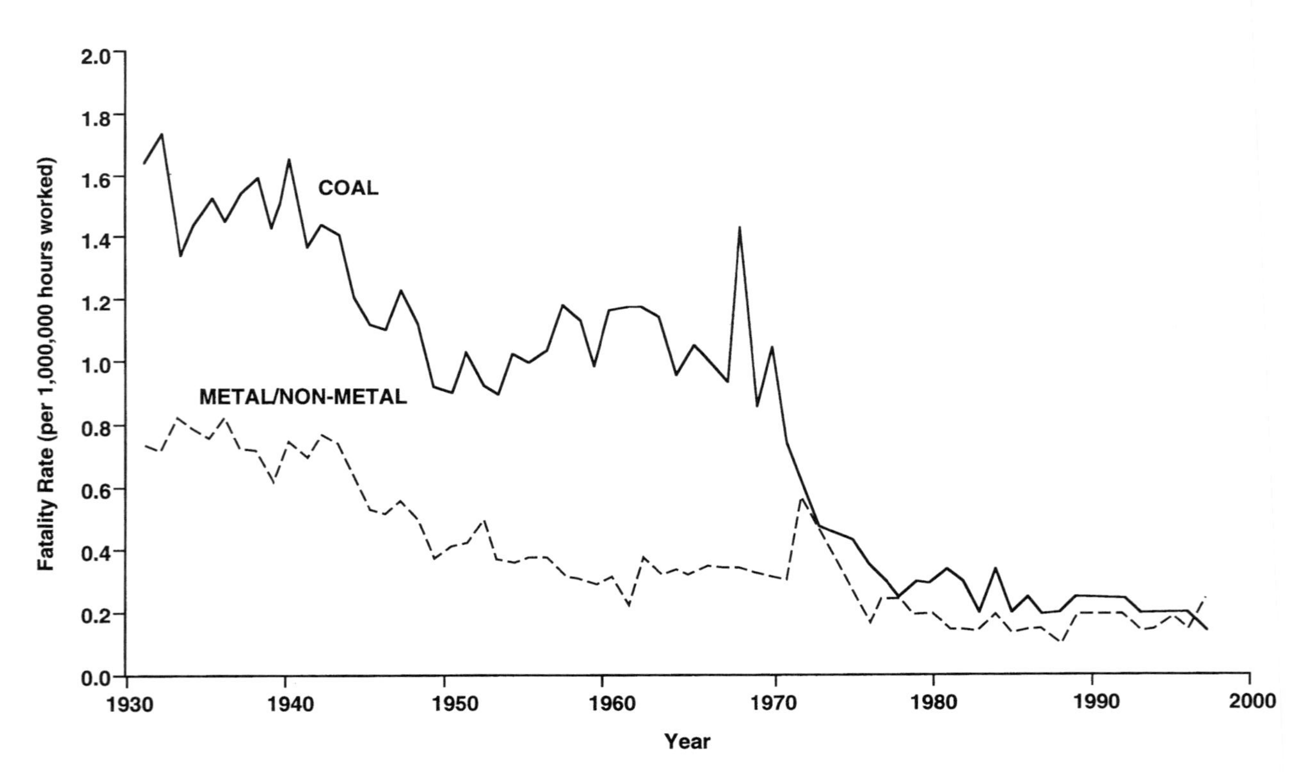

FIGURE 2.2. U.S. mine fatality rates per 1,000,000 hours worked, 1931 – 1997. *Sources:* Katen (1992); Mine Safety and Health Administration (2000).

eliminating dust from the mine atmosphere. In addition, there are new fears arising from recent studies of the possibility that silica dust and diesel particulate matter may cause cancer after long periods of exposure. These studies require the mining industry to continue its efforts to reduce exposure of mine workers to these substances wherever possible. Although much has been accomplished during the past century, there is still more to be done to eradicate the health hazards related to the mining industry.

2.4 ENVIRONMENTAL RESPONSIBILITIES

Among the duties and responsibilities required of the mining engineer today, the responsibility toward the environment was virtually unknown at the beginning of the twentieth century. During the early development of our country, one of the major concerns of the established industrial complex was to utilize our natural resources to provide raw materials to advance the industrial revolution. It was not a major concern of industry to consider the environmental consequences of mining or of the effects of other heavy industry on the earth's environment. Yet, although concern for the environment was of low priority, it was not absent. Naturalists and others expressed their concern from time to time. John Muir, an early naturalist and conservationist, spoke eloquently of nature and wilderness and was instrumental in forming the Sierra Club (Wolfe, 1946). Muir's expressions of his love of the natural world were important in getting others to appreciate our world and its preservation. However, many say that Rachel Carson provided the clarion call to modern environmentalism with her book *Silent Spring* (Carson, 1964). The book, an exposé of the biological effects of the pesticide DDT, became a best-seller. It also opened the eyes of the public to the possible consequences of indifference to the state of the environment. Since her book was published, we have lived in a world increasingly aware of environmental issues. The mining industry has made significant progress in environmental protection since that time, but much remains to be done. In the next two subsections, we will look at the progress made in the environmental area within the mining industry, in both land reclamation and control of the effects of chemicals used in the mining process. We then examine the process of developing an environmental orientation in mining and the advantages of such a policy.

2.4.1 Surface Mine Reclamation

The history of surface mine reclamation began in earnest in the 1960s. At that time various states where coal mining was common enacted the first mine reclamation laws with any impact. However, the state laws were often ineffective because companies in violation, but intent on beating the system, would often move to a new state to conduct their mining. The Surface Mining Control and Reclamation Act, enacted in 1977, ended many of these reclamation abuses by having a federal law applicable in all states, with strong control over the issuing of surface mine permits, control over the minimum requirements that the states could enforce, and a tracking system for violators. A summary of the various aspects of the law is provided by Kaas (1992). Among the important emphases of the law are the positive control of the water and sediments generated on a surface mine, the removal of highwalls and the restoration of the original contour unless a higher use for the land can be found, proper disposal of waste and acid-bearing materials, revegetation of the land to a proper standard, and the posting of bonds to ensure reclamation of the mined lands. These aims are discussed in summary form in the following paragraphs.

One of the continuing problems of coal mines, particularly in Appalachia, has been water pollution resulting from old mining operations. SMCRA attempts to overcome this problem in modern coal mining operations (1) through control of the water generated on a mine site by precipitation and from the groundwater system during mining and (2) by restoring the land in such a manner that water pollution will not be a factor after mining is terminated. This requires a suitable plan before mining begins so that all aspects of the water and sediment flows are controlled during and after the mining operation. General requirements of the law are provided by Kaas (1992) and Filas (1997); some of the engineering aspects of this problem have been outlined by Warner (1992).

Although the pollution of streams with acid drainage and sediment is an important environmental consideration, the removal of highwalls under the regulations was aimed at both safety and the aesthetic aspects of these topographic features. At many sites in Appalachia fatalities have resulted from falls off old mine highwalls, which was a primary reason for the requirement that mine sites be restored to original contour (Kaas, 1992). Such restoration eliminates both the danger and the unnatural state of the topography. However, SMCRA will permit a departure from the original contour if the land is dedicated to a higher use after mining. Mountaintop removal is a good example. In hilly or mountainous terrain, old mountaintop removal operations have been utilized after mining for airports, shopping malls, public schools and athletic fields, hospitals, and other facilities that are beneficial to the public. In some parts of the country a sizable area of level land can be provided inexpensively only by this method. Some of the issues of mountaintop removal are outlined by Sweigard (1992).

The disposal of waste materials from coal mines is another important issue in the design of the mining operation. It becomes of primary importance in mountaintop operations, where extra overburden material must be placed on the topography with minimum impact on the streams and without the possibility of future slope failures. A variety of waste placement strategies and contaminant control procedures can be found in the literature. Those provided by Zahl et al. (1992) and Dwyer (1997) are of particular note.

FIGURE 2.3. Mine land reclamation. (a) Land during the mining process. (b) Same land after reclamation and revegetation. *Source:* Office of Surface Mining (2000).

Among the primary concerns of any company operating a surface mine is the revegetation of the landscape after mining. The degree of difficulty in providing vegetation on mined lands varies with the type of land and the postmining use. Sweigard (1992) and Filas (1997) discuss the general strategies for revegetating mined lands, as well as the special needs inherent in the restoration of prime farmland. Additional insights into reclamation and agricultural activities on prime farmland are offered by Dunker et al. (1992). The strategies applied to this process are too numerous to be outlined in detail here. However, it is generally appropriate to use both native and nonnative species, to use effective methods of soil amendment, to implement adequate erosion controls, and to use fertilizer only in the initial stages of revegetation. The mining company involved in any surface mine reclamation has an economic incentive to restore the land so that its reclamation bonds can be released. If the company's program of revegetation is not effective, it only increases the costs of restoring the vegetation and the costs associated with the posting of reclamation bonds.

In assessing the progress of surface mine reclamation over the previous few decades, it is clear that much has been accomplished since the first reclamation laws with any impact were enacted in the 1960s. The federalization of reclamation under SMCRA has resulted in the reclamation process being more uniformly enforced. In addition, it has provided better regulations aimed at restoration of the landscape, prevention of erosion, use of mined land for higher purposes, and proper revegetation of mined lands. It can easily be seen from Figure 2.3 that success can be achieved. SMCRA sets tough standards for reclamation of mined lands, and restoration of the type shown in Figure 2.3 is now the norm rather than the exception.

2.4.2 Solution Mining Operations

The changes in the mining industry that are taking place at the present time include a strong movement toward solution methods of mining. Bartlett (1998) defines *solution mining* as "all forms of extraction of materials from the Earth by leaching and fluid recovery, both by *in situ* methods and by heap leaching of excavated ore." This is in general agreement with the definitions found in Hartman (1987) and established by the American Geological Institute (1997). In contrast, the definitions in Marcus (1997a, p. 4) differentiate between extraction using water alone (solution mining) and extraction using water plus other chemicals (leaching). Marcus's definitions have a certain appeal but are not as commonly used in the literature. We will therefore adhere to the most utilized definitions in this book.

In assessing the environmental hazards associated with solution mining methods, surface mining methods and underground mining methods are considered separately. On the surface, the environmental problems are associated primarily with the contamination of soil and water and possible damage to the flora and fauna arising from the ore or chemical agents used in the solution mining process. Underground, several variations of solution mining are used to dissolve valuable mineral constituents from the ore. These methods all use solutions in the extraction process that may be difficult to contain within the deposit and therefore have the possible ramification of polluting surrounding formations. Potential problems are outlined in the sections that follow; all of these methods are discussed in more detail in Chapter 8.

2.4.2.1 Surface Solution Mining Problems. The most significant potential chemical problems in solution mining on the surface are associated with heap or dump leaching of metallic minerals. *Heap/dump leaching* is the process of recovering mineral elements from a mass of broken ore material that has been piled on the surface. The metal ions from the ore are normally leached out of the heaps or dumps by water and a chemical lixiviant. The heaps or dumps ordinarily consist of low-grade metal ores with copper, gold, or uranium content. When the chemicals are introduced into the ore materials, they represent a threat to the soil, water, flora, and fauna if the system is not well designed and properly controlled. Consider, for example, the general diagram shown in Figure 2.4. In this process, a certain quantity of water is circulated continuously through a gold-bearing ore heap, returned to the carbon columns for extraction of the gold content, and then returned to the system to be used again.

Once the sodium cyanide (NaCN) lixiviant is added to the water, the solution poses a hazard to the environment, even though the solution is rather weak in terms of its cyanide content. As the solution permeates through the heap, it must be reliably collected at the bottom of the heap so that it does not contaminate the soil or the groundwater. This is normally accomplished by the use of geomembranes and a drainage layer that are carefully designed and

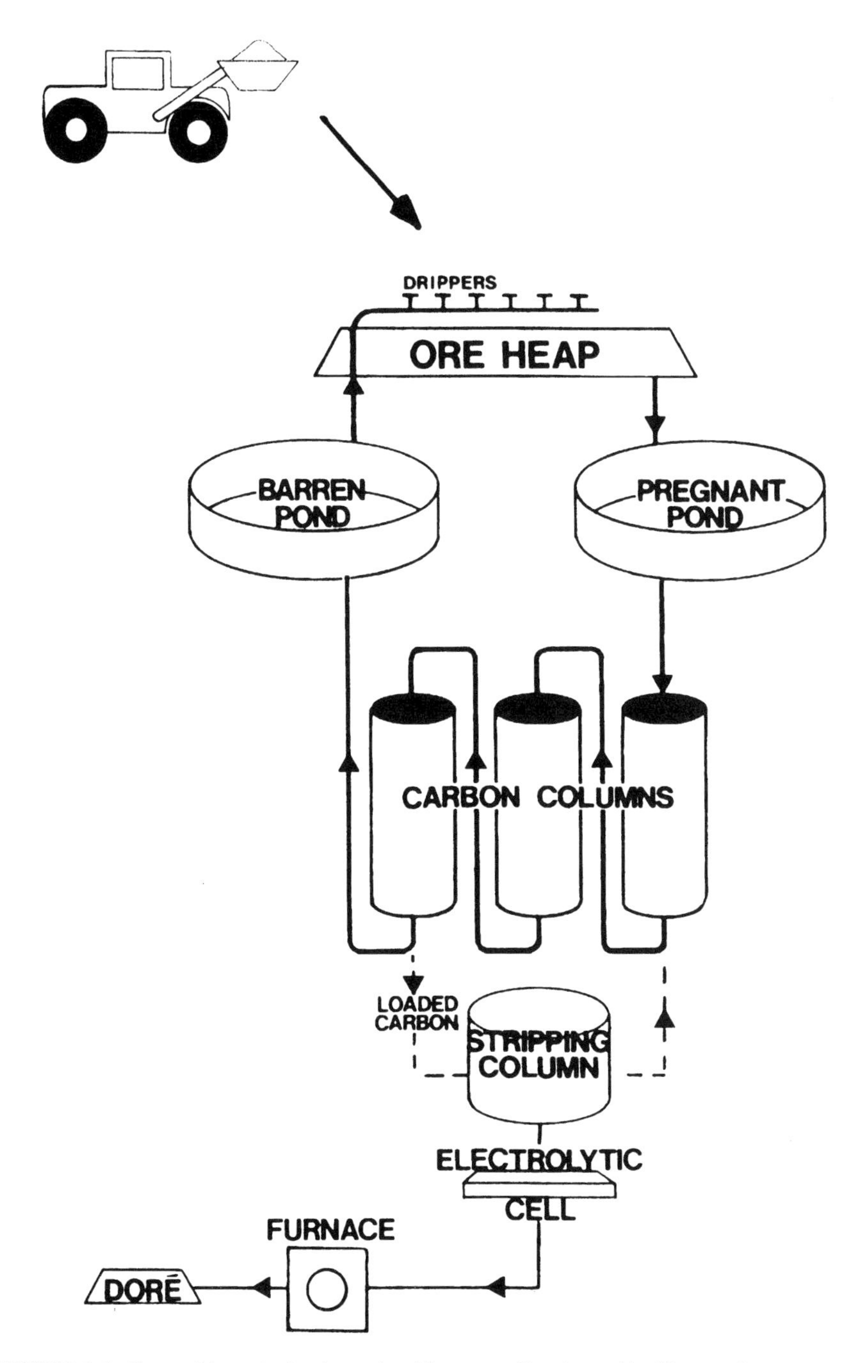

FIGURE 2.4. General layout of a heap leaching operation for gold with a carbon recovery system. *Source:* Bartlett (1998). This figure is under copyright ownership by OPA (Overseas Publishers Association) N.V. By permission of Taylor & Francis Ltd., Reading, United Kingdom.

placed at the bottom of the heap to ensure flow of the leach solution out of the heap and into the pond for the pregnant solution. More complete descriptions of these methods are found in Hutchinson and Ellison (1992), Henderson (1997), and Bartlett (1998). In addition, the existence of weak solutions of NaCN in both the barren and the pregnant solution ponds is a threat to the migratory birds that may land in them. This hazard is normally overcome by placing nets over the ponds or floating a layer of plastic balls over their entire surfaces. This precaution has significantly reduced the threat to bird life.

Another dilemma occurs in some areas as a result of water control. The leaching system is normally designed to recirculate the water over and over again. Evaporation tends to reduce the water volume, and rainfall can increase the volume. However, the design must be such that the water volume is under control of the system operators. One of the major failures contributing to the now-famous Summitville Mine environmental disaster (Jones, 1993; Filas and Gormley, 1997) was that the amount of water within the system was increasing because of precipitation, a variable that was not adequately addressed in the initial design of the system. Heap leaching operations are now used throughout the West, sometimes in strategic locations in regard to the watershed, which creates a problem in many mining operations.

2.4.2.2 Underground Solution Mining Problems. Underground solution mining operations are normally of two types: mines that use boreholes to access the deposit and those that solution mine a broken mass of ore left or created in an underground mine. Copper, uranium, gold, sulfur, and evaporites can be mined in this fashion. An example of the process is shown in Figure 2.5 (Roman, 1973). In this example, an existing mine with low-grade ore in underground stopes is mined by the solution method. This method of mining is favorable as it reduces the cost of extraction and eliminates the waste material on the surface after mining. As may be expected in this type of operation, the major worry is control of the solutions that are used in the mining process. Whether the deposit is mined using water alone or water combined with a lixiviant, the spread of the solution into surrounding formations represents a threat to the groundwater and streams in the general area. As a result, a careful evaluation of the fluid paths through the deposit must be made, and monitor wells are often drilled to ensure that the design is carried out as planned. Further discussions of the measures taken to protect the environment using these methods are presented in Chapter 8.

2.4.2.3 Other Environmental Problems. A number of other environmental ramifications are sometimes of concern in the mining industry. These include hazards from old mine workings, heavy metals leached from old mine dumps, dusts blown from old mine dumps and tailings impoundments, the failure of tailings dams, the radioactivity of some mine wastes, and the production of acid mine drainage from old mines. These occurrences can have a great impact in specific locales. The good news is that they will not, in general, constitute significant hazards under the current mining and environmental regulations.

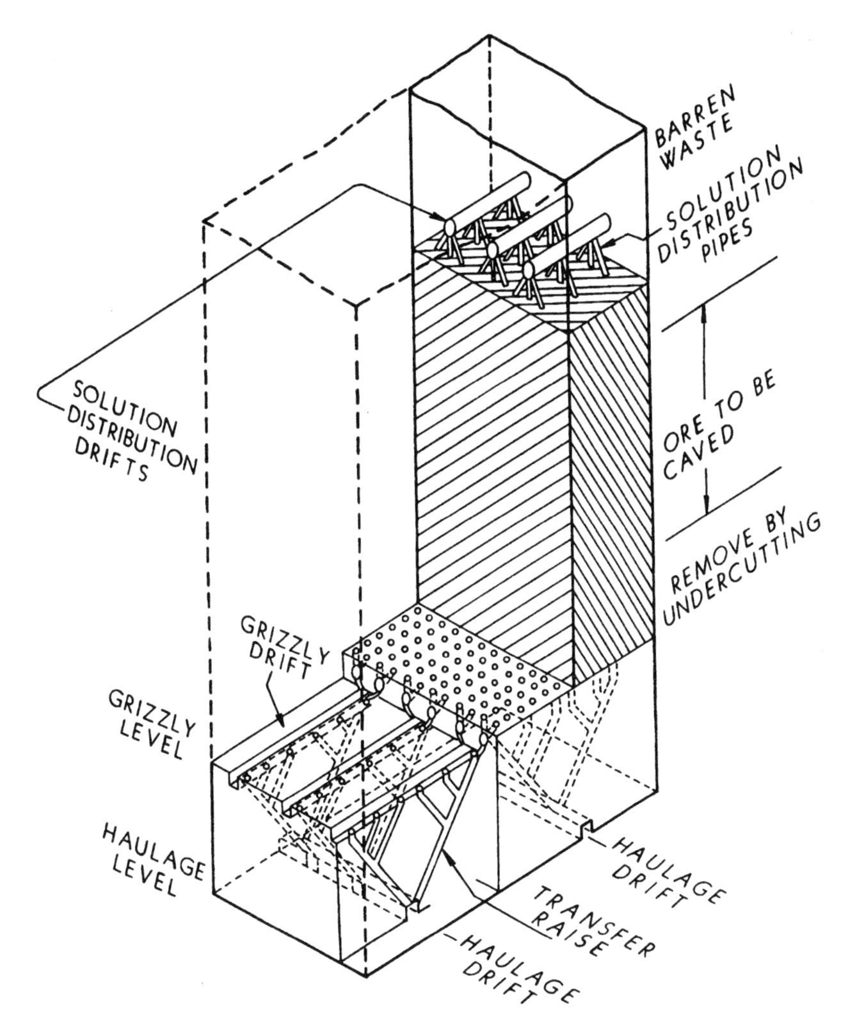

FIGURE 2.5. Underground leaching of low-grade ore after block caving. *Source:* Roman (1973). By permission of the New Mexico Bureau of Mines and Geology.

Such problems are all solvable with careful planning and the use of established mine design procedures. In addition, many of these problems of the past have been mitigated by reclamation efforts.

2.5 SOCIOENGINEERING

Engineers characteristically, by personal nature and by the nature of their work, may not think outwardly to the world around them as they perform their jobs. This is more likely to be true in certain categories of engineering positions than in others. However, the tendency to work without considering the effect of our actions on society should be avoided. *Socioengineering* is defined as the study of the interactions between engineering and society and how each affects the other (Hartman, 1970, 1974). All engineering programs have a selection of elective courses. These are often the only opportunity engineering students have to consider the effects of their profession. It is important that these courses be used to establish a life-long commitment to understanding and utilizing the interaction of engineering and society for the joint benefit of both. Lacking this commitment, the potential benefits of engineering and technology cannot be optimized.

There have been a number of important developments in this area. Many institutions of higher learning have established measures to better promote the optimal use of science, engineering, and technology in society. A number of large universities have courses of study that go by the name of science, technology, and society (STS), socioengineering, or some similar combination of words. *Science, technology, and society* has been defined as the study of the ways in which technical and social phenomena interact and influence each other (McGinn, 1991). These studies can help engineers understand the effects of their work on the general public and how the general public can endorse a technology or doom it to failure. It is important that all engineers understand the nature of this interaction between their work and society so that they can reap the rewards of acceptance by the public and optimal effects of their engineering work.

Modern engineers and managers can use this knowledge in many ways to enhance their image with the public and pave the way for new technology and new projects. Mining operations are particularly important in this regard, as they can affect the public even though no direct individual negative impact can be detected. This can occur because even the thought of a mine impacting the public is an effect. When people expect a mine to change their everyday existence, that in itself is an impact. Consequently, every informed mining manager will work with the public to minimize undesirable effects on society. The manager will make certain that members of the public are well informed about every new project so that they are not surprised by what happens. Meetings with local politicians and concerned citizens will be held prior to every new development so that everyone involved will have an opportunity to comment. The manager will provide honest answers about the advantages and disadvantages of the project. In addition, he or she will not be reticent to trumpet the organization's contributions to the community and to its people. This is how the engineer and the manager can use socioengineering to advantage so that each project achieves its optimal effect on the mining organization and on society.

2.6 SUMMARY OF CONSEQUENCES

It is important to review and analyze the past consequences of mining. Such an analysis can point out the difficulty the mining industry has had in convincing the general public of its contribution to society in providing low-cost material to supply consumer goods. In part, this situation is the result of the separation of mining and those consumer products in the mind of the consumer and the past performance of the mining industry in regard to health, safety, and environmental matters. The performance of the mining industry has improved considerably in the last century in these important areas. However, improvement in the public image of mining has not automatically followed. Enhancement of the image of the mining industry is a goal that should be sought by every person associated with the minerals industry.

Health and safety is an area in which the minerals industry has advanced dramatically in recent decades. This development is a result of the combined efforts of the federal government, the state governments, the mining companies, the miners' unions, and the many individuals who work in the general area of mine health and safety. It takes the cooperation of all personnel to make significant improvements in safety in an industry where hazards are common. The mining industry's outstanding record of improvements in this area over the last century can be continued into the future as well. Although we have accomplished much through better regulations, better adherence to safe practices, and effective education, there are many other procedures that will help in the future. For example, the high-technology processes of automation, remote control of equipment, and computer monitoring of hazardous conditions are still in their infancy. The next century will clearly see the advantages that these technologies will produce in the health and safety of workers. We should be able to further reduce the dangers of mining and improve the public image of the mining industry. This is a goal that must be pursued by every generation of mining personnel.

Protection of the environment is another area of great importance to the public image of mining. The environmental movement in mining began in earnest in the 1960s. Since that period, the minerals industry has made great strides to better protect our land, our air, and our water resources. Many new techniques to protect the environment and mitigate previous problems have been developed over the last four decades. In addition, many useful books outlining environmental protection procedures have been published in recent years (Hutchinson and Ellison, 1992; Ripley et al., 1996; Marcus, 1997a; Bartlett, 1998). However, the public perception of the mining industry has clearly not been corrected (Bingham, 1994; Prager, 1997). The difficulties of elevating that image are due to the strong commitment of the general citizenry toward the environment, the tendency of the media to report any environmental problem, and the general lack of exposure to the positive environmental acts of forward-thinking mining companies. As a result, the mining industry must not only perform its environmental duties well, but must also tell the world of its contributions to the environment. The National Mining Association (1998) has adopted a frank, straightforward approach to environmental problems, admitting that there were problems in the past and celebrating the progress evident in current mining activity. This is a philosophy that all personnel and corporate entities in the minerals industry must adopt. The building of a positive environmental ethic in mining has a solid foundation; we must all play a part in completing its construction.

PROBLEMS

2.1 Search the Mine Safety and Health Administration web site, http://www.msha.gov/ (Mine Safety and Health Administration, 2000), to determine the number of fatalities that have occurred in the last three recorded years in the U.S. mining industry in the categories of coal mining and the mining of metal and nonmetallic minerals. Is the downward trend in fatalities in each of these two mine categories continuing? If the answer is positive, provide a few reasons that you think the trend is continuing. If the answer is negative, provide some insights into why you think this has occurred. Write a one- or two-page response.

2.2 Go to the web page of the National Society of Professional Engineers, http://www.nspe.org/ (National Society of Professional Engineers, 2000), and peruse the site. Then proceed to the information on ethics and review the fundamental canons, the rules of practice, and the professional obligations of engineers. Which sections provide guidance to the engineer on the ethics of dealing with health and safety problems? Is the guidance clear, in your opinion? Which sections provide guidance to the engineer on the ethics of dealing with environmental problems? Is the guidance clear in this case? Write a one- or two-page response to these questions.

3

STAGES OF MINING: PROSPECTING AND EXPLORATION

3.1 PRECURSORS TO MINING

The stages in the life of a mine were outlined in Section 1.4. This chapter discusses in detail the first two stages, prospecting and exploration. It should be noted that we define *prospecting* as the search for ore or for geologic occurrences that have the potential of being an ore body. Building on this definition, we must therefore use the narrow meaning for the second stage of mining. *Exploration* is defined as the activities of evaluating a geologic prospect to determine its size, shape, grade, and profit potential. Every mine must normally go through these first two stages so that its economic viability can be established. Although it is helpful to separate them in a basic course on mining, the two stages are often conducted together, making it difficult to distinguish between prospecting and exploration activities. When the two activities are joined together within a corporate structure, the combined activity is normally referred to as exploration.

The basic background of prospecting and exploration professionals is often geology or geophysics. The reason is clearly apparent—a detailed understanding of geology and the physical properties of the earth are normally required to analyze the often complicated nature of mineral occurrences. The mining engineer need not be an expert in these areas, but a good working knowledge of the geological sciences is imperative if the process is to be understood. It is therefore assumed that mining students will have a good fundamental knowledge of physical geology, rocks and minerals, structural geology, and mineral deposits.

The discovery of a new mineral deposit has been likened to the search for the proverbial needle in a haystack. Most mineral deposits mined for their economic values are geologic anomalies; that is, they are clearly something out of the ordinary, freaks of nature. The scarcity of these mineral occurrences is a worthwhile study in itself. MacKenzie and Bilodeau (1984) defined an "economic discovery" as one that realizes revenue of $20 million and a rate of return of at least 10%. Cook (1986) defines a "significant discovery" as one that will yield more than $500 million in gross revenue, based on the average metal prices for the previous five years. Cook also outlines the dimensions of a "world-class discovery" as one (1) that is capable of being operated continuously and consistently returns a profit even during periods of low metal prices, (2) that is large enough to significantly affect a medium-sized corporation's profits, and (3) that ranks in the lower one-third of similar mines in terms of the cost per metal unit produced. In assessing the chances of discovering an economic deposit, previous statistics are of use. One publication (Placer Development Ltd., 1980) indicates that only one deposit out of 1000 in Canada ever moves from the status of a prospect to an operating mine, at a cost of $30 million for each economic deposit. Cook (1983) estimates that only one deposit out of 1500 ever becomes a highly profitable mine. Cook also cites the statistic that only 2.3 world-class deposits are found each year, at an average cost of $290 million per deposit.

A more recent review of the number and types of deposits that are being discovered in worldwide exploration can be found in the annual project reviews in the *Engineering and Mining Journal* (1995a, 1996a, 1997, 1998, 1999a). The journal reviews mining projects under development each year. Coal, basic industrial minerals, crushed stone, and other industrial minerals projects are excluded. Thus, the review essentially outlines the projects developed in the metals industry. For the five-year period from 1995 to 1999, about 150 projects worldwide are listed each year as being under development. Although it is difficult to pinpoint how many of these projects come on line each year, somewhat more than 50 mining projects seem to be started each year, with initial project investments ranging from $2.4 million to $2.8 billion. The average investment for a project in the 1999 survey was more than $400 million. Ore deposits are not only hard to find, they are costly to develop. Thus, the overall cost of finding and developing a viable metal mine is extremely high, making it primarily the province of large mining corporations.

A primary reason for the difficulty in finding a new deposit is that most deposits are truly rare occurrences. Further, the geologic forces that have affected the mantle of our planet control them. Using the average crustal abundance of any given metallic element, one can compute the *enrichment factor*, the multiplier by which the elemental concentration must be increased on average for the element to be minable at a profit. Based on the prevailing *cutoff grade*, or minimum percentage by mass in an economic deposit, enrichment factors for several common metals are calculated in Table 3.1 (Peters, 1980). These are similar to the grades in Thomas (1973), except for some metals for which the technology has changed considerably. The enrichment factors point out the highly unusual nature of some of the metal deposits. Couple this with the fact that most of the earth's surface has been prospected in the past,

TABLE 3.1 Enrichment Factors for Ore Bodies

Metal	Crustal Abundance (%)	Minimum (Cutoff) Grade (%)	Enrichment Factor
Mercury	0.0000089	0.2	22,500
Lead	0.0013	4	3,100
Tin	0.00017	0.5	2,900
Tungsten	0.00011	0.2	1,800
Gold	0.00000035	0.0003	900
Molybdenum	0.00013	0.1	800
Uranium	0.00017	0.1	600
Zinc	0.0094	3	300
Copper	0.0063	0.3	50
Nickel	0.0089	0.3	35
Iron	5.80	30	5
Aluminum	8.30	30	4

Source: Peters, 1980. (By permission from Rocky Mountain Mineral Law Foundation, Denver, CO.)

and one can conclude that the search for ore deposits is clearly a highly complicated business. The days of a solitary prospector with a burro (or a four-wheel-drive vehicle) are largely over; today's mineral deposits are typically discovered by geologists and geophysicists armed with an array of high-technology tools.

Nearly all ore bodies found today are discovered through the process of prospecting and exploration. The manner in which these activities are organized and relate to each other is shown in Figure 3.1. Notice that Stage 1 (prospecting) emphasizes reconnaissance or identification of favorable targets and Stage 2 (exploration) focuses on target evaluation. Each stage in turn consists of two phases, proceeding from regional appraisal to deposit evaluation. Following alternative paths on the flowchart leads to favorable, untimely, or unfavorable decisions as to whether a given target area is to be evaluated further.

It is evident, in progressing from left to right in the flowchart, that the first objective of prospecting and exploration is to narrow the search, that is, to reduce the size and number of the targets under consideration. Typically, search areas decrease from a range of 1000 to 100,000 mi^2 (2,500 to 250,000 km^2) in Phases 2 and 3, narrowing to 0.1 to 20 mi^2 (0.25 to 50 km^2) in Phase 4 (Bailly, 1966). The second objective is to increase the probability that the remaining target areas will yield ore deposits. If both of these objectives are met, the third objective, reducing risk, is automatically achieved. Systematic progression through the prospecting and exploration stages, with logical evaluation of alternatives at each phase of the process, will help the exploration team achieve all three of the objectives.

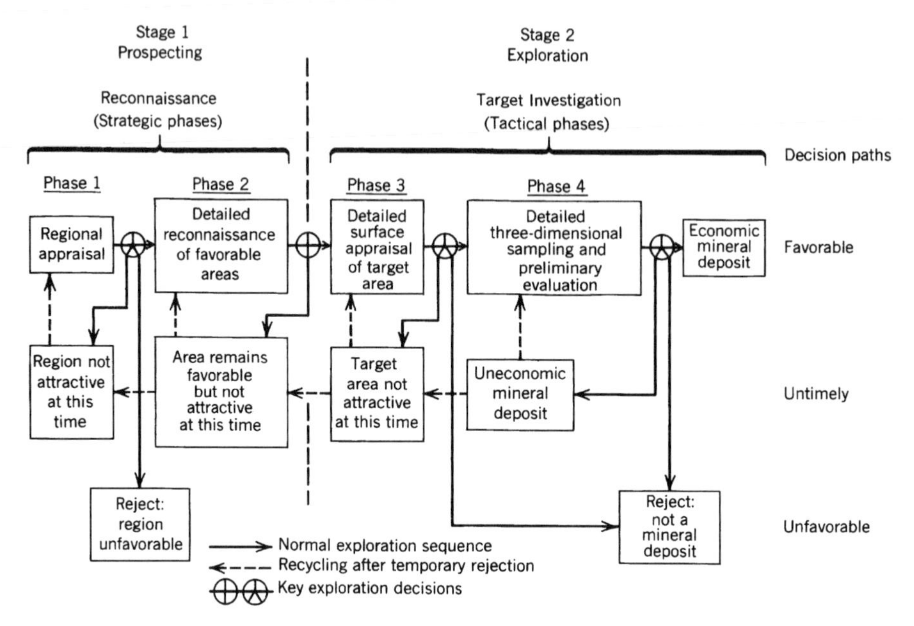

FIGURE 3.1. Phases in regional prospecting and exploration. Sequential paths and decision points are arranged in a flowchart. (After Bailly, 1968; Payne, 1973. By permission from the Society for Mining, Metallurgy, and Exploration, Inc., Littleton, CO.)

3.2 PROSPECTING: GENERAL

The conduct of a good prospecting program, as defined previously and outlined in Figure 3.1, is aimed at the discovery of a maximum number of mineral deposits with the potential to be ore deposits. This must also be accomplished at minimum cost. The type of occurrence that these procedures are seeking is an *anomaly*, a geologic incongruity that has the possibility of being an ore deposit. Obviously, an anomaly is not necessarily an ore deposit. However, every ore deposit is an anomaly, that is, something out of the ordinary.

Prospecting may be commodity- or site-specific; that is, the search may be limited to a particular mineral or metal or to a particular geographic area. In today's world the former is seldom the case. The large minerals corporation is multinational in scope and often energy-oriented as well. Thus, it is common for the targets in prospecting to encompass multiple sites around the world and a wide suite of metals, nonmetals, and fuels. A recent tally of the new projects around the world (*Engineering and Mining Journal*, 1999a) indicates the following distribution of exploration targets:

37% in North and Central America
24% in South America and the Caribbean
19% in Africa
14% in Asia
13% in Australia and Oceania

The choice of targets may seem to be a function of the mineral potential. However, the political climate plays a large part in the process of choosing areas in which prospecting and exploration will be performed. The political variables to be considered (*Engineering and Mining Journal*, 1999b) include taxation, environmental regulations, duplication and administration of regulations, native land claims, protected areas, infrastructure, labor variables, and socioeconomic agreements. As outlined in the previous reference, a better index of favorableness for exploration would be some combination of the mineral potential index and the political policy index, that is, the investment attractiveness index. This type of index is clearly in the thinking of many companies as they choose exploration sites.

A remarkable aspect of current prospecting and exploration is that the amount of previous exploration in an area may not be an impediment to prospecting success in that area. In highly explored areas, nearly all the surface expressions (outcrops, gossan, and float) have already been discovered and investigated. However, new deposits can be found in these areas if efforts are concentrated on the so-called hidden deposits, that is, deposits below the surface. Several good examples can be found in the United States. Nevada is clearly the most favorable state for exploration at the current time (*Engineering and Mining Journal*, 1999a), even though it has been explored over and over through the years. Many new gold deposits have been located there in recent decades. Another example can be found in Colorado, which has also been extensively explored. The finding of diamonds in recent years (Coopersmith, 1997) in a kimberlite deposit in northern Colorado indicates that better search techniques and analysis methods can locate new ore deposits, even in a heavily explored region.

Because many targets today are concealed, *direct search* techniques (physical examination, geologic study and mapping, and sampling) must be supplemented by *indirect methods* (geophysics, geochemistry, geobotany). Search techniques may be applied on the surface of the earth, in the subsurface formations, in the atmosphere above the earth, or from a satellite in space. Direct techniques are still successful in prospecting for coal and nonmetal deposits that often outcrop or occur under shallow overburden, but for metallic deposits, indirect techniques are almost always required. Generally, the prospecting procedure follows these steps:

1. Search geologic reports and other technical literature.
2. Study available geologic maps and surface maps.
3. Study aerial and satellite photographs.
4. Prepare photogeologic maps from available information and new aerial data.
5. Conduct airborne geophysical survey of the area under study.
6. Establish base of operations, set up mapping control, and organize ground prospecting parties.
7. Conduct preliminary ground geologic, geophysical, and/or geochemical surveys.
8. Assemble and analyze findings.

Phase 1 (in Figure 3.1) proceeds through steps 4 or 5; Phase 2 encompasses the rest. Certain steps may be omitted or performed out of sequence. Further, overlap between prospecting and exploration is likely in steps 7 and 8.

A useful summary of all the methods that may be employed in both prospecting and exploration is given in Table 3.2. The phase(s) in which they normally find use, their class (direct or indirect), and their detection capability for base metals are indicated. Many of these methods are discussed further in the sections that follow. Remember that these methods have application in both prospecting and exploration.

3.3 PROSPECTING: GEOLOGIC

Geologic prospecting applies knowledge of the genesis and occurrence of mineral deposits, structural mapping, and mineralogic and petrographic analyses to discover, define, and appraise mineral prospects. It remains the keystone of both prospecting and exploration; all other methods depend on it and utilize it for interpretation.

One of the most important tools in the geologic search for minerals is knowledge of the forms that ore deposits take and how these formations can affect the search for new deposits. A rather comprehensive treatment of mineral deposits can be found in Jensen and Bateman (1981) and Guilbert and Park (1986). These references outline a classification system for ore deposits that dictates the genesis of the ore deposit. Deposits are either *primary* (formed directly from magmas), *secondary* (altered through chemical or mechanical weathering), or *metamorphic* (formed from other minerals and rocks when subjected to heat and pressure). A further distinction is that ore deposits are syngenetic if they are formed contemporaneously with the host rock and epigenetic if formed subsequently.

The geologist uses his or her knowledge of ore deposits and their genesis to locate favorable geologic venues in a variety of ways. General geologic surface mapping of formations showing the composition, distribution, age, and relationship of the various formations is a starting point. The structural geology can be a clue to the existence of ore bodies; this becomes a second major mapping concern. Callahan (1982) notes the following as promising sites or environments for ore deposits: erosional unconformities, plutonic intrusions, enrichment and alteration zones, overthrust zones where tectonic plates collide, stratiform deposits in volcanics and sediments, and the interface between the ocean and its floor. These promising targets can be found through detailed geologic mapping of the land surface. This can be accomplished through conventional geologic mapping or through the use of remote sensing.

Remote sensing is defined as the science of acquiring, processing, and interpreting images and related data obtained from aircraft, satellites, and underwater instrumentation systems (Sabins, 1997). This term is used for methods that employ electromagnetic energy, such as light, heat, and radio waves, as the means of mapping and measuring the properties of the earth or other astronomical entity, generally over wide areas. In most references, this technique is differentiated from the geophysical methods and considered as a separate evaluation technique. Aerial photography is one form of remote sensing that can be used for mapping purposes. It utilizes images from the visible region of the electromagnetic spectrum. However, most of the data processed from remote sensing systems are in the infrared and microwave regions of the spectrum. Satellites can be utilized to obtain images in a variety of spectral bands in order to obtain geotechnical information that will be of use in many areas of geologic, botanical, and environmental science. Thus, prospecting is only one of the many uses of remote sensing in the natural sciences.

In geologic prospecting, remote sensing images can be used in a variety of ways to locate promising targets. First, the 2.08 to 2.35 μm spectral window can be used for the mapping of hydroxyl-bearing minerals such as sheet silicates (*Engineering and Mining Journal*, 1991; Gupta, 1991). This enables remote sensing to detect alteration minerals including kaolinite, montmorillonite, mica, talc, and chlorite. Because ore bodies are often associated with alteration halos, these can be favorable targets in a prospecting campaign.

A second major ability of remote sensing is identification of important lineaments and other structural features that may help locate ore deposits. Sabins (1997) has a chapter devoted to mineral exploration that outlines some of the world's mineral-producing regions and the relationship of the ore deposits to the lineaments, alteration zones, and other features that can be detected by remote sensing. This work makes a strong case for identifying lineament zones as favorable prospecting targets. Coupled with knowledge of alteration zones, the lineaments in an area can be utilized to indicate a number of promising targets for further geologic investigation.

Remote sensing has several limitations. One is the fact that no current remote sensing methods can see deep into the surface of the ground. In addition, many of the methods are not usable in areas with heavy vegetation. However, in the areas where vegetation covers the ground surface, remote sensing still has an application. It can be used to locate altered vegetation that may be indicative of ore deposits. This is an application of geobotanical prospecting, which will be discussed in Section 3.5.

Traditional procedures for geologic mapping, photogeologic interpretation, structural analysis, and field reconnaissance are provided by Ahrens (1983) and Peters (1987). The information gained from these methods is often combined with remote sensing data to guide the search process. Data compilation, analysis, and display are crucial steps in geologic prospecting. Essential to each of these is the computer, which has revolutionized the handling of information and greatly reshaped prospecting and exploration strategy.

3.4 PROSPECTING: GEOPHYSICAL

Geophysics may be defined simply as the science of measuring and interpreting the physical properties of the earth or other astronomical bodies. *Geophysical*

TABLE 3.2 Compilation of Prospecting and Exploration Methods

Methods and Techniques	Usable at Phase 1: Regional Appraisal	2: Detailed Reconn.	3: Detailed Surface Study	4: Detailed 3-D Study	Detection Capability for Nonferrous Metallic Deposits[a] — Direct Detection: Good	Direct Detection: Questionable	Indirect Detection: High Discrimination Capability[b]	Indirect Detection: Low Discrimination Capability
Geologic								
Office compilation	X							—
Photogeologic study	X	X					—	—
Aerial examination	X	X					—	—
Outcrop examination	X	X	X		—	—	—	—
Geologic mapping and investigations	X	X	X	X	—	—	—	—
Geologic logging				X	—	—		
Boulder tracking		X	X		—	—		
Geochemical								
Stream sediment sampling	X	X					—	—
Water sampling	X	X					—	—
Rock sampling	X	X	X				—	—
Specialized sampling	X	X	X				—	—
Assaying	X	X	X	X	—			

Geophysics — airborne					
Aeromagnetic surveys	X	X			
Electromagnetic	X	X			
Radiometric surveys	X				
Remote sensing surveys	X				
Geophysics — ground					
Gravity	X	X	X		
Magnetic	X	X	X		
Radiometric		X	X		
Seismic		X	X		
Resistivity		X	X		
Self-potential		X	X		
Induced polarization		X	X		
Downhole electrical				X	
Sampling and evaluation					
Trenching			X	X	
Rotary drilling				X	
Core drilling				X	
Tunnel-shaft work				X	
Mineral dressing tests				X	Not a detection method
Economic evaluation				X	Not a detection method

[a] *Detection* refers to the ability to detect a deposit if it is there. *Indirect detection* refers to a geological, chemical, or physical response, showing a deposit may be the cause of the response; this is in opposition to direct evidence of the presence of a deposit.
[b] *Discrimination* with regard to indirect methods refers to the ability to determine whether a certain response (anomaly) is due to a deposit or to another cause.

Sources: Bailly, 1966; Payne, 1973. (By permission from the Society for Mining, Metallurgy, and Exploration, Inc., Littleton, CO.)

prospecting is the process of using geophysical measurements for the purpose of identifying anomalies in the physical properties that may indicate the presence of ore deposits. Geophysics may be applied at the ground surface, in boreholes or mine openings, or using airborne or satellite instrumentation. In most prospecting efforts, surface and airborne methods predominate. Geophysical prospecting has been more successful in oil and gas prospecting than in searching for minerals, but continued improvements have made geophysical methods increasingly useful in mineral exploration as well.

For geophysics to be successful, there must be some detectable differences in the physical properties of the ore body and the host rock. The selection of the proper geophysical method(s) to use for a suspected deposit is based on the best geologic and physical data available. It is too costly to use all the methods of geophysics on a given target area; therefore, methods are chosen on the basis of their appropriateness (singly or jointly) in detecting the minerals that are being sought. In some cases, geophysics may be employed for indirect detection; that is, a gangue mineral associated with the ore may be more detectable than the ore mineral itself. Geophysics may also be used to define a deposit's spatial relations (depth, shape, etc.) or structural features (folds, faults, etc.).

The selection of geophysical methods is determined in part by their suitability for different locales (airborne, ground surface, or subsurface). For this reason, some are preferred for prospecting (air or ground) and others for exploration (ground or subsurface). Commonly, the most cost-effective scheme is to package several methods for airborne surveys and then to repeat the process with other methods suited to a surface or subsurface locale.

Geophysical methods are usually classified on the basis of the principle of operation or property measured. Seven categories of geophysical methods are commonly recognized: electrical, electromagnetic, magnetic, gravity, radiometric, seismic, and thermal. Some prospecting personnel may also consider remote sensing as a geophysical technique, but others consider it a separate geologic mapping and geobotanical prospecting tool. Van Blaricom (1992) provides an excellent review of the theoretical and practical aspects of the aforementioned methods. Table 3.3, derived from this author (1992), outlines the applications of each category of geophysical method and various aspects of its logistics and cost requirements. The costs for the methods are listed according to the predominant units used, with the cost per day and the cost per line-mile (line-kilometer) being the most common. Costs for Australian exploration activities, a decade older but rather complete, are given by Emerson (1982). The procedures more commonly used in mineral prospecting are discussed in more detail in the following sections.

It should be noted that the end product of a geophysical prospecting campaign is usually a series of maps or profiles portraying the variation in the measured physical properties of the area investigated. In an airborne survey, readings are taken continuously along flight lines; on the ground, they are made at stations located along profile lines or at geometric grid points. Examples of profiles resulting from geophysical surveys over a copper sulfide ore body are shown in Figure 3.2. In this case, the three methods used all indicate an anomaly at the location of the ore body, even though the deposit is buried below 20 ft (6 m) of overburden. This is one of the primary reasons that geophysics is so helpful in prospecting. Geophysical methods can detect ore bodies several hundred feet (a hundred meters) below the surface. Naturally, not all mineral deposits will produce distinct indications of ore; thus, data processing and analysis becomes extremely important. Detailed analysis of geophysical information is quite theoretical and beyond the scope of this book. However, the following sections discuss the basic concepts of each major method used in mineral exploration. For more complete descriptions, see Payne (1973), Sumner (1992), and Van Blaricom (1992).

3.4.1 Electrical: Self-Potential

Applied only on the ground, the self-potential method is aimed at the tendency of certain mineral deposits (massive sulfides, manganese ore, graphite, cobalt minerals, magnetite, and anthracite) to produce a weak electric current when interacting with groundwater. The deposits in this case act like a weak battery. Systematic measurements of voltages at the surface may show a significant change when one of these target minerals is located beneath the surface. Figure 3.2 illustrates the effect over a massive sulfide deposit.

3.4.2 Electrical: Induced Polarization

An electrical field can be created in the ground by passing a measured amount of current through it using two electrodes inserted into the ground and a generator to drive the system. The voltage caused by this field around the two electrodes is then determined by using a second pair of electrodes to measure the resistivity of the surrounding formations. Where metallic minerals are present, even in concentrations as low as 0.5%, the ground can become charged by the electrical field. This charging phenomenon, called induced polarization (IP), can be measured in several ways (Hallof, 1992). Used mainly with sulfide and oxide deposits, the IP method can also locate water and deposits of sand, gravel, and petroleum. The method is applied only on the ground.

3.4.3 Electrical: Electromagnetic

Electromagnetic methods use very low frequency (VLF) electromagnetic waves generated either on the ground or from an airborne platform to detect mineral deposits. The primary field is generated by a transmitter; a secondary wave that can be measured by a receiver coil is then used to detect anomalies that indicate a conductive body below the ground surface. This method is capable of locating sulfide and oxide ore bodies, graphite, and associated pyrite or pyrrhotite.

When using the electromagnetic method from the air, the procedure is normally applied as shown in Figure 3.3. The platform can be a fixed-wing plane or a helicopter with a transmitter mounted on the structure of the aircraft. The receiver (often called a bird) is towed below and behind the

TABLE 3.3. Classification, Characteristics, and Costs of Geophysical Exploration Methods

	Method (A) Active (P) Passive	Parameter Measured			
		Designation	Unit	Characteristic Physical Property	Main Causes of Anomalies
Electrical	Resistivity (A)	Apparent resistivity	Ω-m	Resistivity Conductivity	Conductive veins, sedimentary layers, volcanic intrusions, shear zones, faults, weatherings, hot waters
	Induced Polarization (A)	Time domain: chargeability polarizability Frequency domain: frequency effect Phase domain: phase shift	ms pfe pfe mr	Ionic-electronic Over voltage	Conductive mineralizations: disseminated or massive (graphite, sulfides), clay
	Self-potential (P)	Natural potential	mV	Conductivity Oxydability	Massive conductive ores Graphite Electro-filtration Faults
	Mise-a-la-masse (A)	Applied potential	mV	Conductivity	Extension of previously located conductive ore bodies
	Telluric (P)	Relative ellipse area Ratios — apparent resistivity	Ω-m	Conductance	Basin and range studies Conductance of sedimentary series Salt domes, geothermal
	Magnetotelluric (P) MT, AMT (A) CS-AMT	Apparent impedance (resistivity and phase)	Ω-m degrees	Resistivity Conductivity	Conductive veins, sedimentary layers, shear zones, faults, weatherings, resistive basements, bedded ores
	Electromagnetic (A)	Phase difference Tilt angle Amplitude ratio Sampling of decay curve induced in receiving coil by eddy currents In-phase, out-of-phase components	degrees Ω-m σt	Electrical conductivity	Conductive mineralizations Surficial conductors Shear zones

	Magnetic (P)	Earth magnetic field Vertical component Z Total intensity Horizontal gradient Vertical gradient	$1\gamma = 10^{-5}$ gauss	Magnetic susceptibility	Contrasts of magnetization Magnetite content of the materials
	Gravity (P)	Gravity field	milligal ($1 \text{ gal} = 1 \text{ cm/s}^2$)	Density	Deposits of heavy ores Salt domes (light) Basement rocks
	Radioactivity (P)	Intensity and spectral composition of gamma rays	cps	Radioactivity	Radioactive elements Uranium – Thorium – K_{40}
Seismic	Refraction (A)	Traveling time of seismic waves	milliseconds feet per second	Seismic wave velocity Dynamic modulus	Contrasts of velocity: markers at variable depth
	Reflection (A)				Fissured rocks
	Thermometry (P)	Temperature	°C	Geothermal gradient and temperature	Abnormal flux of heat Thermal inertia of rocks

TABLE 3.3 (Continued)

	Method (A) Active (P) Passive	Applications: Direct Detection	Applications: Indirect Detection	Geophysical Crew: Production	Geophysical Crew: Survey Cost ($)	Interpretation
Electrical	Resistivity (A)	Massive sulfides Oil shales Clays Geothermal reservoirs	Bulk materials Base metals Coal and natural steam	Depending upon depth of investigation El. soundings: 20 to 250 E.S./month Resistivity profiling: 20 to 100 line km/month Resistivity maps: 5 to 20 sq km/month	650/day	Curve matching Computer modeling
	Induced Polarization (A)	Conductive: sulfides oxides mn oxides	Associated minerals (zinc, gold, silver, tin, uranium, etc.)	800 to 2000 stations/month 20 to 70 line km/month	750/day	Curve matching Computer modeling
	Self potential (P)	Sulfides: pyrite pyrrhotite copper Mn ore	Associated minerals (lead, gold, silver, zinc, nickel)	2500 to 3500 stations/month 60 to 200 line km/month	200/day	Rules of thumb Curve matching Visual interpretation
	Mise-a-la-masse (A)	Conductive ores	Associated minerals (zinc, tin)	5 to 15 sq km/month	400/day	Computer reduced, computer modeling, visual interpretation
	Telluric (P)	Structural studies Steam	Regional exploration	20 to 500 sq km/month	450/day	Computer reduced Curve matching Visual
	Magnetotelluric (P) MT, AMT (A) CS-AMT	Massive sulfides Clays Natural steam	Shear zones General tectonics General structure	Up to 3 stations/day Up to 20 stations/day	1200/day to 1000/station	Computer reduced Computer inversion Visual interpretation
	Electromagnetic (A)	Conductive: sulfides oxides Mn oxides	Kimberlites Associated minerals Ground follow-up (lead, nickel), shear zones, weathered zones, conductivity maps	50 to 150 line km/month 6000 to 10,000 line km/month	350/day (ground) 35–65 km air	Curve matching Visual interpretation Computer modeling

	Magnetic (P)	Magnetite Pyrrhotite Titano – magnetite	Molybdenum Iron ore Chromite Copper ore Kimberlites Asbestos Geological mapping in terms of magnetic changes (basic rocks...) and/or discontinuities inventory of mineral resources	4000 to 8000 stations/month 80 to 300 line km/month (on shore) 20 to 80 line km/working day (offshore) 1000 to 15,000 line km/month	200/day (ground) 25 – 30 km air	Computer reduced Rule of thumb Computer modeling Computer inversion Visual interpretation
	Gravity (P)	Chromite Pyrite Chalcopyrite Lead	Placer configuration Karstic cavities Basement topography Structure	400 to 1500 stations/month 20 to 50 stations/working day Surface meter combined with seismic turnover	450/day	Computer reduced Computer modeling Computer inversion Curve matching
	Radioactivity (P)	Uranium Thorium Phosphates	Ground follow-up Geological structural mapping (differentiation in granites)	80 to 400 line km/month (ground) 1000 to 15,000 line km/month (air)	350/day 25 – 60 km	Computer reduced Rations of responses Geologic correlation
Seismic	Refraction (A)	Buried channels Faults	Tin, diamonds Heavy minerals	10 to 50 line km/month (on shore)	350/day	Computer interpreted
Seismic	Reflection (A)	Morphological traps Basement topography	Natural steam Uranium	10 to 40 line km/working day (offshore)	1200/day	Computer inverted
	Thermometry (P)	Thermal springs	Natural steam Boron, sulfur, Subterranean volcanism		1000/day	Computer reduced

Source: Van Blaricom (1992). Reprinted by permission from the Northwest Mining Association, Spokane, WA.

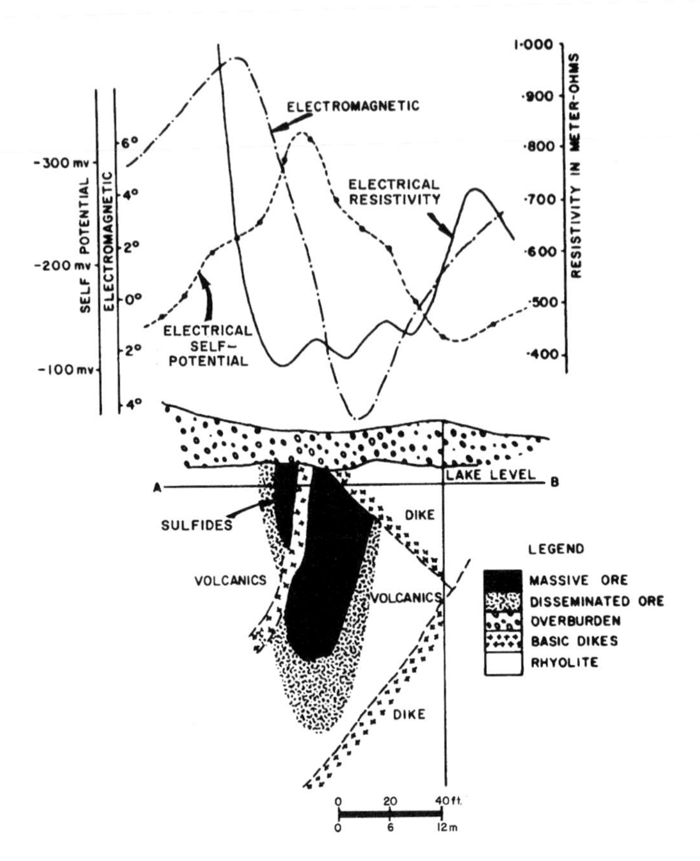

FIGURE 3.2. Geophysical measurements made by three different methods over a copper sulfide ore body. (After Bruce, 1982. By permission from the Society for Mining, Metallurgy, and Exploration, Inc., Littleton, CO.)

aircraft. It allows the system to measure the properties of the secondary field in the earth. The electromagnetic method has been one of the most successful methods of geophysical prospecting in recent decades.

3.4.4 Gravity

The gravitational field associated with the earth varies from place to place because of the localized effects of the variation in mineral densities. The gravitational force will therefore increase slightly when dense formations are located under the point of measurement. The instrument that detects the changes in gravity—a gravimeter—can register a variation in gravitational pull on the order of one part per hundred million, which means that measurable change is registered if the instrument is raised or lowered a few inches. In mining, the gravity method is used to detect heavy minerals like galena, chromite, pyrite, and chalcopyrite, as well as to locate geologic structures such as faults, anticlines, salt domes, intrusions, and buried channels. The gravity method can be applied either on the ground or from an aircraft.

3.5 GEOCHEMICAL AND RELATED PROSPECTING METHODS

Geochemical prospecting may be defined as the process of systematically analyzing the chemical content of rock, soil, stream sediments, plants, water, or air samples to detect anomalous occurrences that may indicate the existence of ore minerals. Although few ore bodies have been discovered by geochemistry alone, it is a valuable adjunct to direct search by geophysical means. Rose et al. (1979, p. 7) list a number of important mineral deposits in which geochemical methods played a major part in discovery. Geochemistry, an indirect method, is carried out on the ground and is often used where geologic and geophysical methods have been relatively ineffective. It does not serve as a tool for pinpointing drilling targets in exploration, but it does serve as a useful reconnaissance procedure, particularly for very large areas of remote terrain.

Through systematic collection and precise trace analysis of appropriate samples, geochemical anomalies can be detected. When interpreted properly and integrated with geologic and other data, geochemical analyses are useful in indicating or helping to confirm promising target areas. They have been used on sulfides of copper, lead, zinc, nickel, and molybdenum, and, to a lesser extent, uranium, diamonds, tungsten, tin, mercury, gold, and silver minerals.

The costs of geochemical prospecting activities are a function of the number of samples to be analyzed and the cost of maintaining a geochemical sample collection team in the field. Peters (1987) lists such costs, including general sample preparation costs plus single- and multiple-element analyses of a variety of types. Commercial laboratory analysis rates and the cost data (Peters 1987) indicate that geochemical analyses have the following costs.

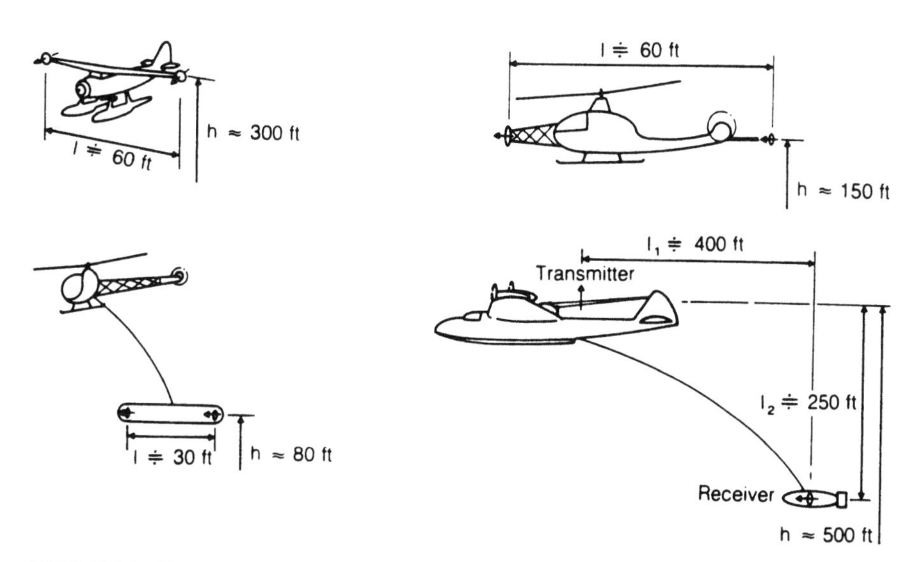

FIGURE 3.3. Examples of airborne electromagnetic installations and configurations. (Klein and Lajoie, 1992; by permission of the Northwest Mining Association, Spokane, WA.)

Sample preparation: $1 to $10 per sample, depending on the steps required
Gold or silver analyses: $3 to $30 per sample, depending on the method used
Single-element geochemistry: $3 to $25 per sample, depending on the element
Trace element geochemistry: $7 to $28 per sample, for up to 30 elements

Note that the different methods have different detection limits on the various elements, and that the proper method must be chosen to be most effective on a cost basis. Also note that the cost of putting a field crew on the job of collecting the samples is in addition to the analysis costs.

Geobotanical prospecting is the science of employing the changes in the patterns of vegetation growth in an area as a visual or analytical guide to mineralization (Brooks, 1995). Although inexact and only an indicator, geobotany can nonetheless be used to detect areas of the ground surface where minerals have affected the characteristics of plant species. There are recognized species that grow in conjunction with ore deposits, and there are plants that change their growing characteristics when located in the vicinity of certain ore minerals. These relationships can help pinpoint the presence of mineral deposits. One of the important applications of remote sensing is in the detection of patterns in botanical systems that may indicate differences in the geologic hosts for the plants (Gupta, 1991; Brooks, 1995).

In recent years other biological systems have shown promise in helping to detect mineral anomalies. Geozoological, geomicrobiological, and biogeochemical prospecting methods have been outlined by Brooks et al. (1995). These methods all use measurements of one or more biological systems to evaluate the state of the ground surface and to help in the discovery of valuable mineral deposits. The methods are not very likely to be used today, but scientific developments may make them more useful in the future.

3.6 EXPLORATION: GENERAL

If the goal of prospecting is to locate anomalies due to mineral deposits, then the goal of exploration is to define and evaluate them. As mentioned in Section 1.4, exploration determines the geometry, extent, and worth of an ore deposit using techniques similar to, but more precise than, those used in prospecting. As Stage 2 in the life of a mine, exploration continues the mining attempt through the tactical phases of detailed appraisal and evaluation, culminating in preparation of a feasibility report that either accepts or rejects the deposit(s) for further consideration. Exploration normally involves both geology and geophysics, but geochemistry and geobotany may have little utility beyond Phase 2. The distinct differences between prospecting and exploration are as follows:

1. *Locales.* As the search area decreases and the favorability of the targets improves, the locale shifts from air (or space) to the ground and subsurface locations. Airborne geophysics is normally replaced by

ground-based geophysics, geology is increasingly subsurface oriented, and additional subsurface exploration techniques are utilized.

2. *Physical samples.* As the site shifts from surface to underground, indirect methods are replaced by direct methods to provide data. Because most ore bodies today are hidden, subsurface excavation methods to obtain actual mineral samples must be employed. The most commonly used is drilling.
3. *Data.* To diminish risk during the exploration stage, much more substantial data about the target are required. The data are characterized by greater precision, specificity, and certainty.

Generally, the progression of steps in exploration is as follows. First, the favorable area identified by prospecting must be delineated by exploration techniques. Second, once located, the mineral deposit is sampled thoroughly and impartially and the samples analyzed. Third, the sampling data are utilized to prepare an estimate of the tonnage and grade, from which the present worth of the deposit can be calculated and recommendations made regarding the economic feasibility of mining.

In dealing with the economic evaluation of mineral deposits, standard terminology has acquired accepted usage and should be used in all reporting. For the discussion here, we will use the definitions adopted by the Society for Mining, Metallurgy, and Exploration, Inc. (SME), as our standard (Society for Mining, Metallurgy, and Exploration, Inc., 1999a, 1999b). These definitions were based on the recommended terms and definitions adopted by the U.S. Bureau of Mines and the U.S. Geological Survey (U.S. Bureau of Mines, 1980), with some refinements added to the subcategories of resources and reserves. Note that a *resource* is a concentration or occurrence of material in such a form and quantity that there are reasonable prospects for eventual economic extraction. A mineral *reserve* is the economically minable part of a measured or indicated mineral resource. The two terms have the same relationship as a similar pair of terms, *mineral deposit* and *ore deposit*; that is, all reserves are resources, just as all ore deposits are mineral deposits. Notice also that a reserve can become a resource overnight if the market price for the mineral in the deposit drops suddenly. More detailed definitions of these terms, paraphrased from the SME publications, are presented in the following paragraphs.

A *resource* is a concentration of naturally occurring solid, liquid, or gaseous material in or on the earth's crust in such a form and amount that economic extraction of a commodity from the concentration is currently or potentially feasible. Location, grade, quality, and quantity are known or estimated for specific geologic evidence. Resources are subdivided into the following categories, with a decreasing degree of certainty as one progresses down the list.

1. *Measured mineral resource:* That part of a mineral resource for which tonnage, densities, shape, physical characteristics, grade, and mineral content can be estimated with a high level of confidence. The data must be based on detailed and reliable information from outcrops, trenches,

pits, workings, and drillholes. The locations of the data points must be spaced closely enough to confirm geological and/or grade continuity.

2. *Indicated mineral resource:* That part of a mineral resource for which the tonnage, densities, shape, physical characteristics, grade, and mineral content can be measured with a reasonable level of confidence. Data are gathered from similar sample points as used for measured mineral resources, but the locations are too widely spaced to confirm geological and/or grade continuity. However, the spacing is close enough for continuity to be assumed.
3. *Inferred mineral resource:* That part of a mineral resource for which tonnage, grade, and mineral content can be estimated with a low level of confidence. It is inferred from geological evidence and assumed, but not verified.

A *mineral reserve* is the portion of a measured or indicated mineral resource that is economically minable at a given moment in time. Two subcategories of reserves have been defined as follows:

1. *Proved mineral reserve:* That part of a measured resource that can be economically mined.
2. *Probable mineral reserve:* That part of an indicated resource and, in some circumstances, measured mineral resource that can be economically mined. A probable mineral reserve entails a lower level of confidence than a proved mineral reserve.

The logical structure of the resource and reserve definitions is shown in Figure 3.4. Note that all that is needed to move vertically down the graph is more detailed knowledge of the mineral deposit. This can normally be achieved through the use of better exploration tools. To move from left to right on the graph requires better economic variables. This can occur as a result of a better commodity price, an improved mining technology, a more favorable mining plan, or some other enhancement to the bottom line.

The use of standard definitions in reporting exploration information provides more useful information to stockholders, potential property buyers, and the general public. It is important that such information be trustworthy and useful for evaluation purposes. Mining-related scams occasionally sully the reputation of the industry, and truthful reporting is therefore imperative. Although such activity is not common in the mining field, a recent swindle regarding a gold exploration project in Indonesia has reminded those in the industry of the potential for deception in exploration activities. Danielson and Whyte (1997) provide a summary of what is known of this event. Because of potential scams of this type, mining personnel must ensure against salting (fraudulent enhancing of grade) and other actions that may affect the grade determined from samples. It is equally important that slipshod collection or handling of samples does not decrease the grade of the samples.

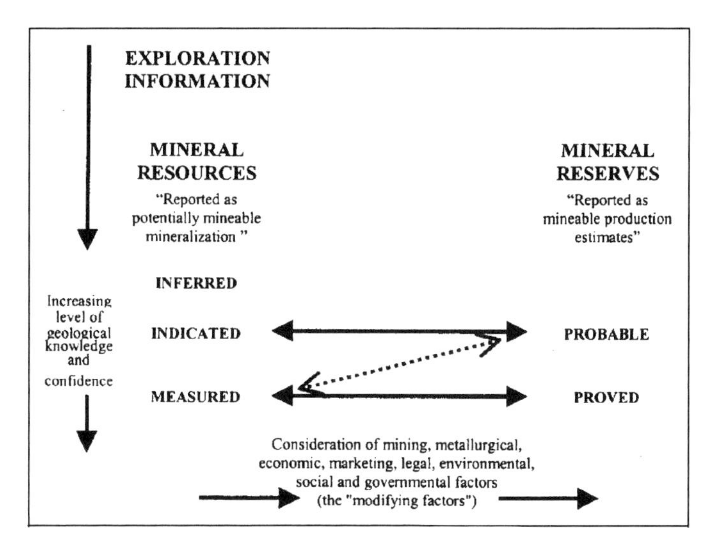

FIGURE 3.4. General relationship between exploration information, mineral resources, and mineral reserves. (Society for Mining, Metallurgy, and Exploration, Inc., 1999a; used with permission.)

3.7 EXPLORATION: METHODS

Although geologic and geophysical methods are used for exploration as well as prospecting, they are confined to Phase 3 of the process and are more tightly focused to obtain reliable data. They will not be discussed further here. Instead, we will describe the methods that are used solely in Phase 4 of the process. These include excavation, analysis, and logging of the deposits to provide more data on the deposit itself and metallurgical or process testing to determine the recoverability of the desired mineral components. These steps are outlined here to define the process of further proving the value of the mineral occurrence.

3.7.1 Excavation, Logging, and Analysis

As noted in Table 3.2, sampling and evaluation techniques provide the emphasis in this phase of exploration. For hidden deposits, excavation or drilling must be performed to access the mineralization zone. Even for deposits that outcrop, excavation or drilling must be used to establish the three-dimensional nature of the deposit. In addition, final geologic logging, analysis of samples, and mapping of the deposit must be completed.

3.7.1.1 Excavation and Drilling. Conclusive evidence of the existence of an ore body requires that samples representative of the whole deposit (or a portion that in itself is economic to mine) be collected and analyzed. Portions of the samples are generally held for later study; suffice it to say that the samples must be extensive, reliable, and statistically meaningful.

Samples are normally obtained by excavation or drilling. If the deposit is exposed at the surface or occurs at a shallow depth, *openings* can be excavated at the surface. In most cases these are in the form of pits or channels. Channels (or trenches) are often excavated parallel to each other and so that they extend through the entire deposit. For large-scale work, a bulldozer or similar machine is used, preceded by blasting, if necessary. If the deposit occurs at moderate or greater depths, then samples are recovered by drilling or by excavating small openings—adit, tunnel, or shaft—in the ore body.

Drilling small-diameter boreholes into the deposit is by far the most expeditious and economical means of sampling most prospects. Drilling today performs more than 95% of the mineral sampling during exploration. The three common drilling methods used for exploration purposes are diamond, auger or roller-bit rotary, and percussion or rotary-percussion. *Diamond drilling* is the standard procedure for consolidated rock, moderate to great depths, and conditions that require an intact core sample (as opposed to cuttings or sludge) for geologic logging, structural geology, rock mechanics, or other analyses. The cutting action is rotary in nature, with the cutting accomplished by many small diamonds mounted or impregnated into the tungsten carbide bit matrix. Drill cuttings are flushed from the hole with water, air, or mud. Diamond drills have reached depths of more than 10,000 ft (3000 m) (Bruce, 1982). Drill bits normally range from 1.5 to 6.5 in. (38 to 165 mm) in diameter, with most cores recovered from the smaller diameters. The cost of a diamond-drilled hole (Peters, 1987) is about $15 to $25/ft ($50 to $80/m) for depths to about 1000 ft (300 m) and $20 to $40/ft ($70 to $150/m) for depths to about 3000 ft (1000 m).

A typical diamond-drilling rig is shown in Figure 3.5. These drills are normally of the wire-line type that allows the operator to retrieve sections of the core without removing the drill rods from the hole. There are many factors involved in the proper use of a diamond-drilling tool in an exploration program. For this reason, many companies employ contractors to perform the drilling and have their own personnel recover the core and perform the analysis. More information on standard diamond drill sizes, rods and couplings, drilling fluids, steering of diamond drillholes, and other diamond drilling topics are provided by Metz (1992) and Heinz (1989).

Rotary drilling is also employed in exploration, generally for soft to medium-hard rock and soil. Most rotary exploration holes are relatively shallow, but holes up to 3000 ft (900 m) have been utilized and oil well boreholes have been drilled to depths of 25,000 ft (7500 m). Hole sizes for explorations are generally 3.5 to 6 in. (89 to 152 mm) (Bruce, 1982). One of the uses for rotary drilling is to penetrate overburden (soil or rock) prior to diamond drilling. Either an auger bit for soil or soft rock or a roller bit for soft to medium-hard rock is used. For shallow depths, rigs are truck-mounted; for deeper holes, larger, crawler-mounted rigs are used.

For noncoring drilling operations in shallow holes and medium to hard rock, *percussion* or *rotary-percussion* drills are used. These methods are also used for general-purpose drilling during an exploration program because of their favorable costs. Both traditional percussion drills, in which the percussion

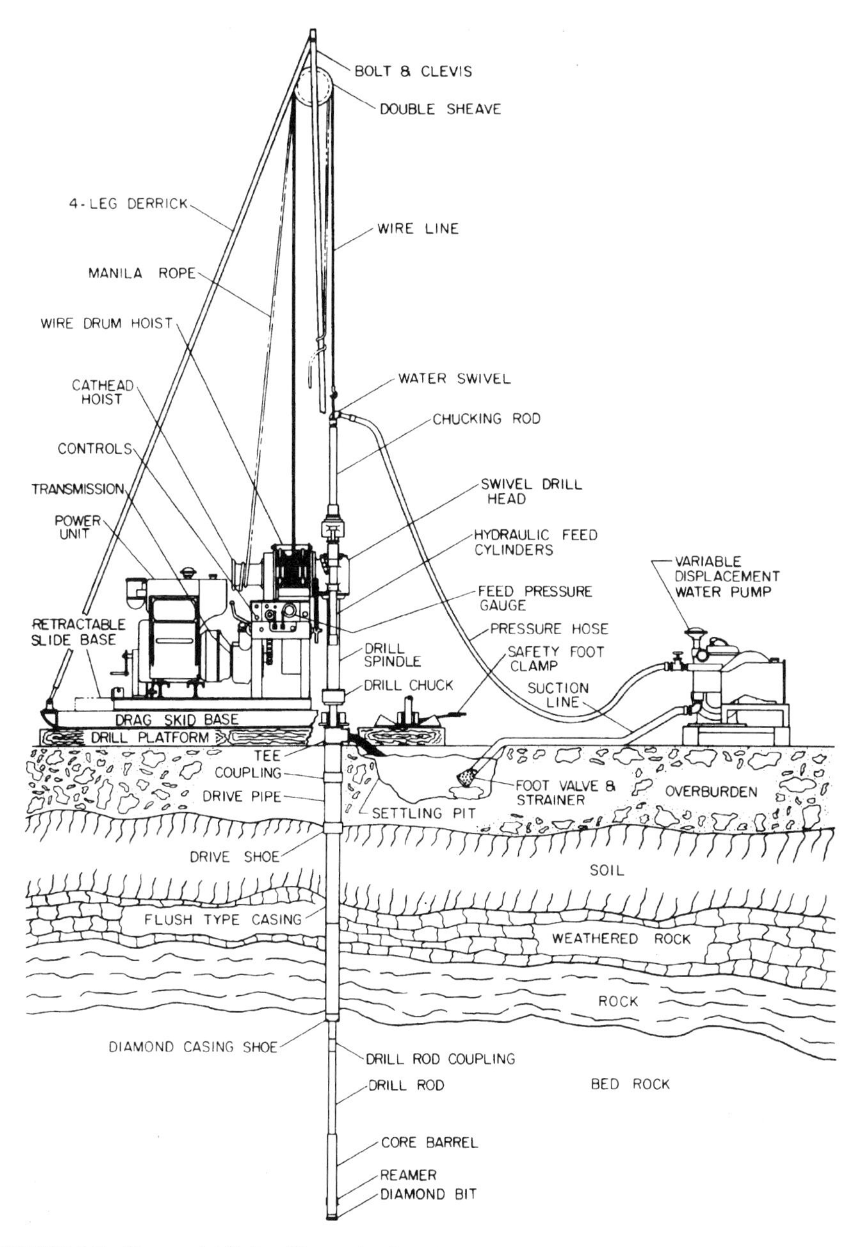

FIGURE 3.5. Diamond drill rig with mechanical drive, used for exploration from the surface. (By permission from Acker Drill Co., Inc., Clarks Summit, PA.)

energy is transmitted along a drill rod string, and downhole percussion drills, in which the drill follows the bit down the hole, are used. Depths are generally confined to 300 ft (90 m) for traditional drills and 650 ft (200 m) for downhole drills, although new methods can at times push downhole drills much deeper. Holes sizes are 1.5 to about 4.0 in. (38 to 102 mm). The method is popular because of its speed of up to 825 ft/day (250 m/day) and its low direct cost of $5 to $15/ft ($15 to $45/m) (Metz, 1992).

In both rotary and percussion drilling, careful recovery of the drill cuttings is essential for analyzing the geologic formations drilled. Recovery of the cuttings is better if reverse circulation (with the cuttings coming up the drill rod) is used; 90% to 95% recovery is achievable with careful handling of the cuttings. Problems of recovery and handling of the cuttings are discussed by Metz (1992).

For most exploration programs, drilling is favored over excavation for both practical and economic reasons. Selection of the drilling method depends on the type of deposit. Coring will almost always be chosen for deep metallic deposits; therefore, diamond drilling is the preferred drilling method. However, in coal mine exploration drilling, a large percentage of the holes are normally produced by rotary drilling (Martens, 1982).

3.7.1.2 Logging. Logging of the samples taken during exploration is an important part of the data collection process. Two types of logging are generally practiced: logging of the cores and cuttings, and logging of the boreholes. If drill core is collected, it is placed carefully in core boxes in 5 ft (1.5 m) lengths in the exact sequence in which they are removed from the hole. The core may then be split longitudinally, one half being preserved for geologic study and the other half used for compositional analysis. A magnifying glass may be all the geologist needs to make petrographic and mineralogic identifications and rough distribution estimates. More accurate results require laboratory analyses and microscopic studies of thin and polished sections. The analysis of cuttings proceeds in a somewhat similar fashion, but the field geologist can gather less information. Erickson (1992) provides detailed descriptions of logging procedures.

Borehole logging is the process of investigating the rocks that occur over the length of the borehole. Normally, this is conducted by sending geophysical probes down the hole to record certain of the geologic properties of the rocks. For a variety of reasons, this is ordinarily needed only in coal and uranium exploration. In these cases, it is used to identify the coal or uranium horizon and establish its thickness and certain of its properties. In addition, it can identify some of the surrounding rock formation in terms of lithologic type. Ash, sulfur, and washability analyses, however, still require core samples. See Elkington et al. (1983) and Erickson (1992) for a summary of the types of geophysical logs and their ramifications in exploration.

3.7.1.3 Analytical Determinations. Excavation and drillhole samples must be quantitatively analyzed (*assayed*, in mining terms) in the laboratory. Here precision, accuracy, sensitivity, and stability are the watchwords. When detection and quantification to a level of 1 ppb (parts per billion) may be necessary with trace elements—and 1 ppm is the standard for minor constituents of the sample—the selection of the proper analytical method is of critical importance (Cardwell, 1984). Traditional methods like chemical analysis (for inorganic compounds) and fire assaying (for gold and silver) are seldom used today

TABLE 3.4 Analytical Laboratory Methods

Method	Elements, Applications
Atomic absorption	Gold, silver, mercury, molybdenum, copper, lead, zinc, tin, nearly all others
Colorimetry	Arsenic, tungsten, molybdenum, titanium
Fluorimetry	Uranium
Emission spectrometry	70 elements
Inductively coupled plasma	50 elements, e.g., barium, manganese, boron
X ray fluorescence	Minor elements, major oxides in geochemical samples
Neutron activation analysis	Gold (nondestructive)
Electron microprobe	Many elements, useful for small quantities
Radiometric	Uranium, thorium
Fire assaying	Gold, silver, platinum

Source: Cardwell, 1984. (By permission from the Society for Mining, Metallurgy, and Exploration, Inc., Littleton, CO.)

because they are slow, expensive, limited, and insensitive. Many sophisticated, versatile instruments have lately found application for the kind of exact determination so essential in mineral exploration, of which atomic absorption is the most popular. In order of availability, Table 3.4 lists the analytical methods in current use and their application for the common elements in ores.

The analysis step for coal ordinarily consists of determining the proximate analysis (moisture, volatile matter, fixed carbon, and ash percentages by weight), the chemical analysis (moisture, carbon, hydrogen, sulfur, nitrogen, oxygen, and ash percentages by weight), the calorific value in Btu/lb (kJ/kg), and perhaps some of the properties that affect the metallurgical properties of the coal. Most of the tests are performed according to procedures outlined by the American Society for Testing and Materials (ASTM). One can find a list of those that apply to coal in Hower and Parekh (1991, p. 29).

3.7.2 Metallurgical/Process Testing

Because most mineral deposits (ore, coal, and stone) require processing—also known as preparation or beneficiation—before their products can be marketed, metallurgical tests (or washability studies, for coal) must be initiated during the exploration stage. The purpose of the tests is to determine an optimal beneficiation process or processes for the mineral, an approximate recovery of the valuable minerals, and a preliminary flow sheet for the treatment. Although the specifics of testing on limited sample sizes are not found readily in modern processing textbooks, the publications by McQuiston and Bechaud (1968) and Aplan (1973) are helpful in understanding this process. Tests may include additional geologic exams, principally mineralogic and petrographic, and analytical determinations, screen analyses, crushing and grinding studies, and gravity and heavy-media separation processes. More specialized investigations of suitable concentration processes, such as magnetic, electrostatic, and flotation methods, will follow if they are useful. Tests may be conducted at bench scale, first using 50 to 500 lb (25 to 225 kg) of material, and then in a pilot plant using samples of 25 to 1000 tons (20 to 900 tonnes). A consulting company with specialized equipment is often engaged to perform the investigation.

3.8 EXPLORATION: RESERVE ESTIMATES

Providing a reserve estimate is part of the overall process of valuation of a mineral deposit, leading (it is hoped) to the conclusion that it is indeed an ore body worthy of exploitation. In this process, the reserve establishes the tonnage of minerals that is considered to be minable, whereas the overall valuation process also considers the marketing parameters, the transportation issues, the taxes to be paid, and the expectation of profit margin. This section concentrates on the sampling patterns used to define the boundaries of mineral deposits and the methods used to produce a reserve estimate, that is, the tonnage and grade of the deposit.

3.8.1 Sampling Patterns

One of the many challenges facing exploration personnel in determining the value of a property is the optimization of the sampling pattern for a given mineral deposit. The deposit can be defined in terms of a *grid*, a three-dimensional array of points in the deposit. Ideally, it would be advantageous to obtain samples at equally spaced points throughout the grid system to define the boundary and grade of the ore body. However, practical constraints often get in the way of this objective. Thus, basic sampling grids are of two geometric forms: *regular* and *irregular*. Wherever possible, a regular pattern of points is used, as equally spaced samples will normally produce the maximum amount of information at minimum cost.

If surface excavations are employed — shallow channels or deeper pits or trenches — then the deposit geometry generally dictates the pattern and spacing (Readdy et al., 1982). When the excavations are arranged at equally spaced intervals, the data recovered are most useful; however, these excavations normally provide only a two-dimensional distribution of sample points. Where exploration openings are located underground, the sampling points are usually chosen based on the knowledge of the mineral distribution. As shown in Figure 3.6, this may be done by orienting diamond drillholes outward in a radial pattern that provides the most information. Choosing an optimal array of drillholes is a crucial and complex decision for this situation because most of the important factors — location, shape, and attitude of the mineral deposit — may be established only sketchily at this point in the exploration program. However, an irregular pattern of sample locations is the norm for this situation.

Often the sampling pattern progresses in steps, from a few holes widely spaced to many holes at closer spacing (Bailly, 1968). Figure 3.7 illustrates this concept applied to a massive sulfide deposit. Drilling is performed in three steps, each based on the knowledge generated by the preceding step. The

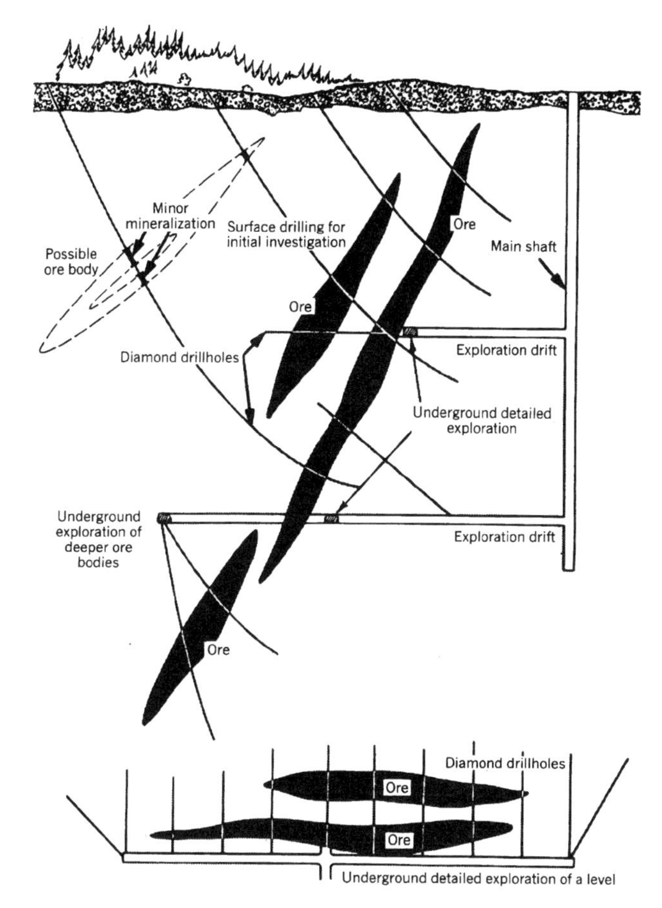

FIGURE 3.6. Exploration of hidden ore bodies by drilling and underground excavation (shafts and drifts). Diamond drillholes originate initially at the surface but later are continued from drifts as the ore bodies deepen. (After Hamrin, 1982. By permission from the Society for Mining, Metallurgy, and Exploration, Inc., Littleton, CO.)

diagram shows 5 holes drilled initially, 15 more in the second step, and 31 additional holes in the third. Each step is justified by the previous step; in this case the results continued to be favorable. Note how the pattern emerges: irregular holes in the first two steps, and a regular (square) pattern in step 3.

For steeply dipping veins or lenses, the drilling pattern may assume a different form, as shown in Figure 3.6. In this situation, the diamond drillholes for the surface are nonvertical so that they can intersect the ore bodies at right angles. The pattern of holes may be irregular, but an attempt is normally made to space the ore intercepts in a somewhat regular fashion if the topography or other variables do not interfere. The underground exploration drillholes shown at the bottom of Figure 3.6 are equidistant and parallel, in an attempt to obtain a regular pattern of sample values.

Where the mineral deposit is horizontal or massive in form, a regular grid of samples is employed whenever possible. This establishes a uniform level of

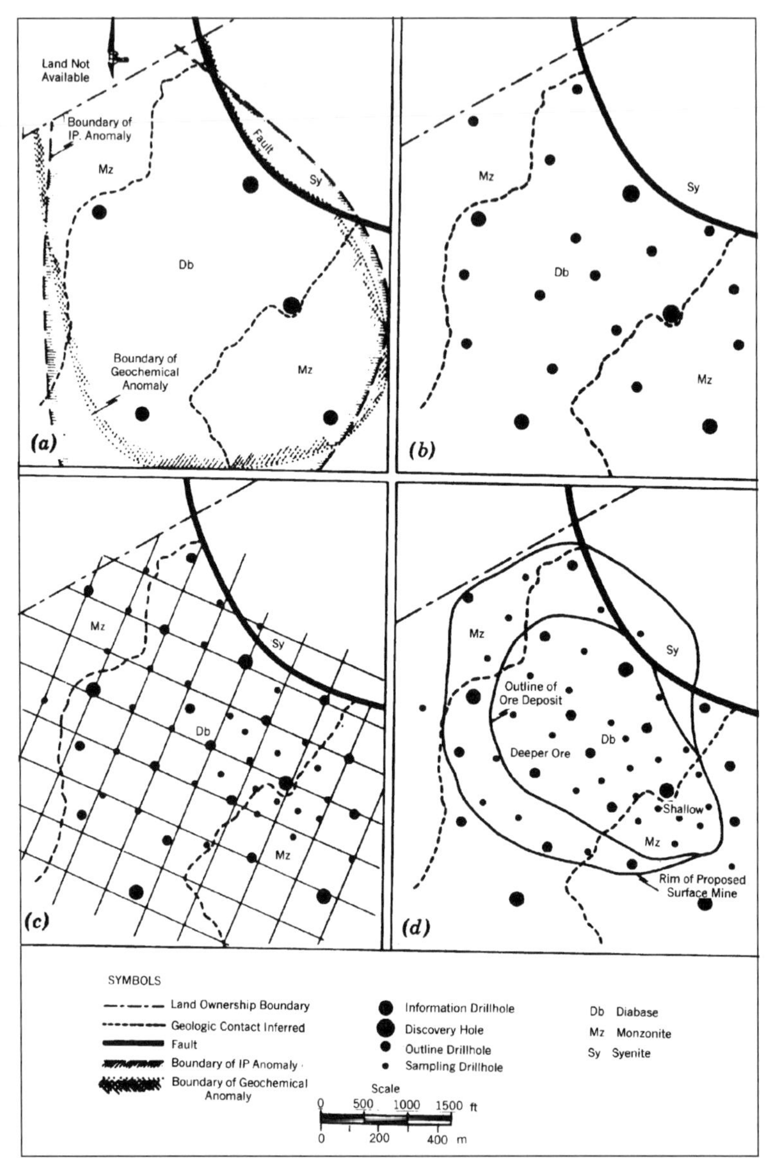

FIGURE 3.7. Successive steps in an exploration program to detect, outline, and sample a massive sulfide deposit. The result is an outline of the ore body and a tentative mine layout. (After Bailly, 1968. By permission from the Society for Mining, Metallurgy, and Exploration, Inc., Littleton, CO.)

knowledge of all parts of the mineral deposit and enhances visual interpretation of the results. Where a regular grid is viewed through a horizontal plane, two patterns are normally used: (1) *rectangular* and (2) *triangular*. Rectangular patterns are often *square* patterns, and the triangular patterns normally utilize equilateral triangles. Hamilton and Trasker (1984) discuss the advantages of each; they favor the triangular pattern for tabular deposits.

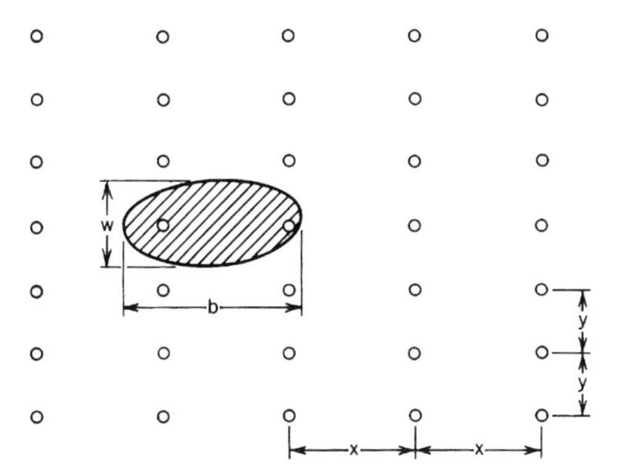

FIGURE 3.8. Principle of drillhole layout (plan view) to ensure detection of flat-bedded deposit. Rectangular grid spacing $x \times y$. Deposit horizontal dimension $b \times w$. (After Payne, 1973. By permission from the Society for Mining, Metallurgy, and Exploration, Inc., Littleton, CO.)

In addition to hole depth and pattern, the horizontal spacing of the holes is a major decision. As a general rule, the spacing should be at the maximum value that will still detect the smallest minable ore body. For example, in Figure 3.8, the rectangular pattern of holes is attempting to locate a minimum-sized ore body that is about b units in length and w units in width. The rectangular grid pattern is thus chosen such that the horizontal dimension $x < b$ and the vertical spacing $y < w$. This ensures the discovery of any ore body of length b and width w, provided that the long dimension of the ore body is oriented as shown. The following are typical grid spacings used for different types of mineral deposits:

Porphyry copper	150 × 250 ft	(45 × 75 m)
	200 × 400	(60 × 120)
Taconite	200 × 300	(60 × 90)
	100 × 200	(30 × 60)
Bauxite	50 × 100	(15 × 30)
Molybdenite	200 × 200	(60 × 60)
Bituminous coal	3000 × 3000	(900 × 900) maximum
	500 × 500	(150 × 150) minimum

The maximum and minimum values are chosen based on knowledge of the continuity and variation in grade in the deposit. For example, for coal seams that are known to vary little over large areas, the maximum drillhole spacing will be used. However, for coal seams that are known to have "*want areas*" (areas where the seam has been washed out by erosion) or extensive faults, the minimum spacing will be used to ensure greater confidence in the estimated value of the deposit. Similar variations in the spacing will be made for metallic and other deposit types.

3.8.2 Reserve Estimates

We have already defined (Section 3.6) two terms—*reserve* and *resource*–that now acquire special significance. The quantitative and qualitative classification of ore (mineral, coal) reserves into three or more geoeconomic categories is one of the objectives of Stage 2, exploration, and especially of Phase 4, deposit evaluation (Figure 3.1). Mineral reserve estimation, or *ore estimation*, involves the calculation of the *extent* (units: tons, tonnes, yd^3, etc.) and average *grade* or tenor (units: percent, oz/ton, g/tonne, $/ton, $/million Btu, etc.) of reserves in a deposit. Ore estimation provides the basic data that determine whether the mine is to be developed; it begins during exploration and continues throughout the life of the mine.

Two general categories of methods are normally employed in the ore estimation phase of exploration. The first are the *classical methods* that use the area of influence (volume of influence) principle or the included area (included volume) principle in determining the tonnage and grade. These methods utilize a limited number of samples in assigning a grade to an area or a volume. The second category consists of the *weighting methods* that assign mathematical weights to numerous surrounding samples to provide a statistically more accurate estimation of the grade in a given area or block. Two weighting methods are commonly employed: (1) the *inverse distance weighting* method and (2) the *geostatistics* method. Both of the weighting methods give a higher weight to samples in the immediate vicinity to determine the assigned grade of the area or block on a statistical basis. The procedures used in this category all assume that the influence of any given sample in determining the grade of an area or block is a function of how distant that sample is from the area or block to be estimated. The weights assigned are reduced as the samples used for estimation get farther from the block to be estimated. These methods have no defined limit on how many samples can be used for a single block estimate. They are statistically superior to the classical methods for most reserve estimation problems because they use more of the data to come up with estimates.

In each of the methods discussed here, the grade and tonnage associated with a specific block of material within the deposit must be determined. To this end, it is often necessary to use the following equations, where SG is the specific gravity, w is the specific weight in lb/ft^3 (kg/m^3), and TF represents the tonnage factor in ft^3/ton ($m^3/tonne$).

Specific weight $$w = 62.4\ \text{lb/ft}^3 \times \text{SG} \qquad \text{lb/ft}^3 \tag{3.1}$$

$$w = 1000\ \text{g/kg} \times \text{SG} \qquad \text{kg/m}^3 \tag{3.1a}$$

Tonnage factor $$\text{TF} = (2000\ \text{lb/ton})/w \qquad \text{ft}^3\text{/ton} \tag{3.2}$$

$$\text{TF} = (1000\ \text{kg/tonne})/w \qquad \text{m}^3\text{/tonne} \tag{3.2a}$$

Note that where two versions of the same equation appear throughout the book, the second (denoted by the letter *a* in the figure number) is for use with SI units.

3.8.2.1 Classical Methods. The two major categories of classical ore estimation methods are known by a variety of names. The first method, known primarily as the *area of influence method*, is also known as the *polygon method* or the *extended area* method. The principle of the area of influence method is that every area constructed in the analysis contains points that are closer to the sample collection point within the area than they are to any other sample point. Therefore, the entire area of influence of that sample is assumed to have a value based on the grade at that one sample point. This principle can be seen in the diagrams in Figure 3.9(a). A drillhole like the one shown in Figure 3.9(a)(1) represents one sample. The area of influence of that sample is derived on the diagram by first drawing lines from the sample to the surrounding drillholes, as shown by dotted lines in Figure 3.9(a)(2). A perpendicular bisector for each dotted line is then drawn and placed on the diagram, as shown in 3.9(a)(3). The perpendicular bisectors define the area of influence of the drillhole, as illustrated in 3.9(a)(4). For any pattern of drillholes through a simple bedded or planar deposit, areas of influence can be determined in the same fashion, with some freedom given to logically determining the individual areas of influence at the boundary. This is shown in Figure 3.9(b) for 13 randomly spaced holes. Note that one alternative name for the method comes from the fact that each area of influence is a polygon. Depending on the grade and weight or mass units used, the total amount of metal or other mineral content of each area of influence is calculated as:

$$\text{Total content} = \text{grade at sample location} \times \text{weight inside the area} \quad (3.3)$$

Note that any appropriate grade units (oz/ton, % by weight, g/tonne, etc.) can be used if compatible units are used for the weight or mass (lb, kg, tons, tonnes, etc.) within the area of influence.

The *triangle method*, also known as the *included area method* or the *valence method*, uses a similar procedure, but each triangular area is associated with three sample locations, as shown in Figure 3.9(c). The individual areas are valued based on the weighted average of the three corner-point samples. The total mineral content of each area is then calculated as:

$$\begin{aligned}\text{Total content} = {} & \text{weighted average grade at the three sample points} \\ & \times \text{weight or mass inside the area} \quad (3.4)\end{aligned}$$

The weighted average grade is obtained using the length of the boreholes in the deposit as the weighting factors. Again, the units for grade must be matched with compatible weight or mass units to achieve the total content.

In the examples shown so far, the samples were assumed to be collected using drillholes through a planar deposit, that is, a deposit that has a minimal third dimension. Many deposits are of the massive type in which the ore body is three-dimensional. To apply these two methods to such deposits, the *section method* is used. In this method, cross sections are utilized in a systematic manner. Each sample gathered from a cross section is used to determine the weighted average grade for that section. The section averages are then combined, using the basic logic of the extended area or the included area method, which is based on using volumes instead of areas. For example, in Figure 3.9(d), each section through the ore body defines an extended volume of influence, indicated by the dotted lines. In the extended volume method, each volume of influence is valued based on the weighted average grade of the

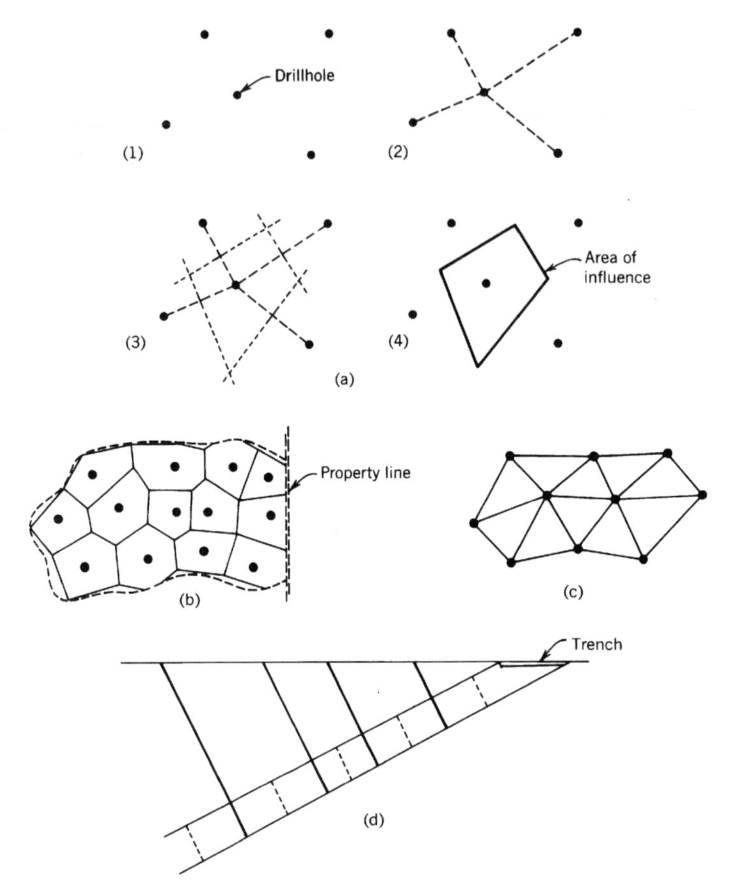

FIGURE 3.9. Traditional procedures for calculating reserve estimates. (a) Polygon method: (1) drillhole plan, (2) connecting lines between drillholes, (3) perpendicular bisectors of connecting lines, (4) final polygon. (b) Polygon method: plan view of completed diagram. (c) Triangle method: plan view of completed diagram. (d) Section method: cross-sectional view of completed diagram, with four drillholes and one trench sample. (After Readdy et al., 1982. By permission from the Society for Mining, Metallurgy, and Exploration, Inc., Littleton, CO.)

section. In the included volume method, the volume of the deposit contained between any two adjacent sections is valued by calculating the weighted average grade of the two sections that bound the volume. For example, the volume of the deposit between section 1 and section 2 in Figure 3.9(d) is determined by taking the weighted average of the cross-sectional weighted averages associated with section 1 and section 2. The remaining volume of the deposit is evaluated in an identical fashion.

Employing one of these methods, we can complete calculation of the tonnage and grade for an ore body. If the deposit contains more than one valuable component, such as lead and zinc, then estimates are prepared for each constituent. If estimates of different overall grades, calculated using different cutoff grades, are desired, then a curve rather than a single answer results. If mineral variables other than grade of the ore mineral are important (such as sulfur or ash in coal, phosphorus in iron ore, iron oxide in limestone, etc.), they can be determined in the same manner. Finally, if the deposit is being considered for surface mining, then separate estimates of ore tonnage, volume of overburden, and stripping ratio (units: yd^3/ton or m^3/tonne) are prepared, at times for several alternative pit limits. A numerical example of a reserve estimation problem is provided in Section 3.11.

3.8.2.2 Weighting Methods. Each weighting method provides a logical procedure for determination of a block of material based on samples gathered within a given region of the block. All known samples within that region are used to determine the estimate of grade for the block, with the weights assigned based on the distance from the block. The first of these methods is the *inverse distance weighting method* (Noble, 1992), which determines the weights according to the relationship:

$$w_i = \frac{d_i^{-\alpha}}{\sum_{i=1}^{n} d_i^{-\alpha}} \quad \text{for } i = 1, 2, \ldots, n \tag{3.5}$$

where w_i is the weight for the ith sample, d_i the distance to the ith sample, n is the number of samples, and α is the exponent used in the equation to reflect the strength of the relationship between holes in the deposit. In this method, the distance d_i must not be very close to zero, as that will create an overflow in the division. This is overcome by adding a small fixed constant to every distance. A trial-and-error procedure is normally used to set the value of α and the radius to be used in the sample search.

The second method used for estimation using weights is geostatistics. *Geostatistics* may be defined as the three-dimensional spatial statistics of the earth. Geostatistics derives its usefulness from a study of the variations in grade in the individual deposit. This is done by the plotting of a *semivariogram*, a plot that describes the statistical relationship of samples as a function of the distance between samples (Clark, 1979; Noble, 1992). The semivariogram (often called a variogram in the literature) is defined by the function:

$$\gamma(h) = \frac{1}{2N} \sum_{i=1}^{N} (g_i + g_{i+h})^2 \tag{3.6}$$

where $\gamma(h)$ represents the semivariogram value for the distance h, g_i is the grade at sample i, g_{i+h} is the grade at a sample located a distance h from sample i, and N represents the number of samples that are a distance h from another sample. A semivariogram is illustrated in Figure 3.10 with labels showing the range, sill, and nugget values. A small nugget value is desirable, as it indicates that the semivariogram will be more useful in predicting grades. The range gives an idea of the distance beyond which the values are no longer useful in prediction. Finally, the sill value should be close to 1.0 if the best results are to be obtained. This indicates that the distribution of values is relatively free of statistical complications.

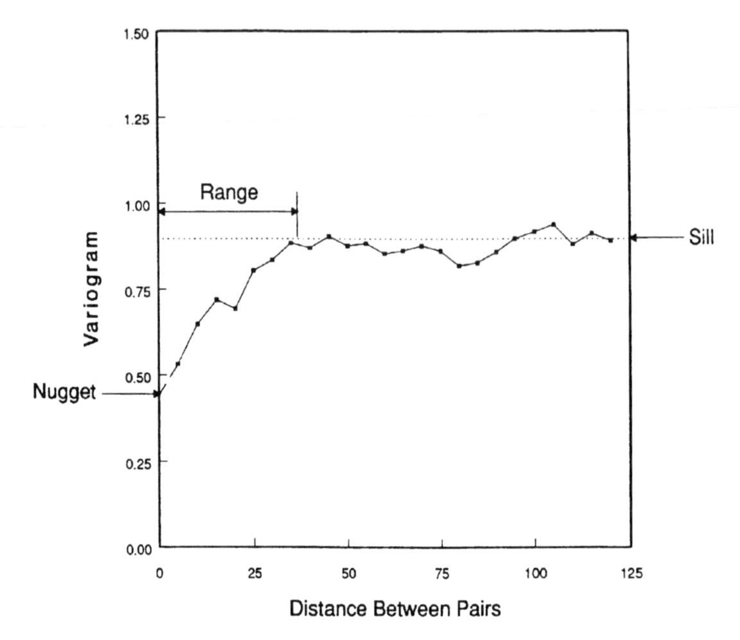

FIGURE 3.10. Typical experimental variogram plot. In this case, the vertical axis is γ (h) divided by the variance. (Noble, 1992. By permission from the Society for Mining, Metallurgy, and Exploration, Inc., Littleton, CO.)

The use of geostatistics is complicated by the possibility of different grade distributions within the ore body and with the semivariograms varying in different directions. In addition, the provision of weights to most accurately determine the block grade values can be quite complicated. A summary of methods for producing the weights for block estimation is provided by Noble (1992, p. 355). The geostatistics method is used primarily after a large mass of data has been collected on the deposit and the characteristics of the ore-grade distribution are well known. Geostatistics is now the primary procedure used for grade estimation in large surface-mine ore bodies. A geostatistical reserve estimate generally entails these steps (Readdy et al., 1982):

1. Study of geologic controls on mineralization and identification of any zoning within the deposit.
2. Computation of semivariograms for each of the zones within the deposit.
3. Division of the ore body into geometric blocks for *kriging* (calculation of the block estimates based on a weighted average of adjacent samples); blocks are then classified as proven or probable reserves.
4. Estimation of tonnage and grade of each block at a given cutoff grade.
5. Mapping of recoverable grade distribution plans by level or bench in the mine.

Although a detailed description of all the geostatistical methods and estimation techniques is beyond the scope of this text, students should acquire an understanding of this technique because they may need to apply it in future mine planning work. For additional treatment of geostatistics, see Journel and Huijbregts (1978), Clark (1979), Knudsen et al. (1978), Rendu (1981), and Kim et al. (1981).

3.8.3 Computer Applications in Exploration

Modern prospecting and exploration have become heavily dependent on the digital computer as a basic tool in all stages of the process. The general trend seems to indicate that this will continue into the future. Consider the following examples:

1. *Geophysical surveys.* Initially, the computer was used to record, store, and analyze data. Later, it was used to solve complex equations used in data analysis, in interpreting the data, and in supplying the geophysicist with confidence limits as well (Van Blaricom, 1992).
2. *Sample and geological mapping.* The computer is now used in every process of mapping a given deposit, including mapping of the topography and the geologic formations, layout of the surface and drillhole samples, three-dimensional analysis of the sample values, and mapping of the zones of mineralization.
3. *Geostatistics.* Ore-reserve estimation by statistical methods would not be possible without the modern computer. For complete analysis, an ore body may be subdivided into tens of thousands of blocks, each of which requires a statistical estimate of grade. The number of computations required can be accomplished only with a sophisticated system of computer software and hardware.
4. *Scheduling and planning.* The determination of grades in a deposit provides the first major element of the decision to proceed with the mining process. The second element requires that the extraction process be planned so that the cash flow can be analyzed to some extent. This will require a computerized plan for the extraction sequence; normally, it also requires three-dimensional mapping of the mineral extraction sequence.
5. *Economic decision making.* This most difficult of all management tasks is made faster and easier if performed by computer. The decision making outlined in Figure 3.1 is normally performed more readily if a computer can be used to gather data and analyze the process. Risk analysis and sensitivity calculations are of increasing concern to the modern decision-maker. Such functions require the use of computers if the process is to be efficient.

3.9 FEASIBILITY ANALYSIS

Prospecting and exploration of a mineral deposit (Stages 1 and 2 in the life of a mine in Table 1.3) culminate in the preparation of a detailed study of the feasibility of proceeding with the development and exploitation of the mine (Stages 3 and 4 in Table 1.3). The resulting analysis and report are primarily economic in nature, but legal, technological, environmental, and sociopolitical aspects are included as well. The key recommendation of a feasibility study is to proceed with, delay, study further, or terminate work on the project.

At this point, a substantial sum has been expended to find and determine the magnitude of the deposit. The investigating engineer in charge may be tempted to take an optimistic view and recommend proceeding with the project in hope of bringing the project to fruition. However, realism is imperative at this point, and the engineer must retain an objective viewpoint and not recommend further expenditures unless the investment is a sound one, fully justified by the evidence at hand, and capable of earning some minimum annual return on the investment.

A high degree of expertise and wisdom is required of the individuals responsible for preparing a feasibility report. Doubts and uncertainties surely remain; the data may be impressive but have some gaps or elements of high risk. The managing engineer or geologist and other members of the decision-making team must determine when to seek additional data and when to proceed. When the decision is made to continue, the team prepares a feasibility report, outlining the many aspects of the deposit that will have an impact on the project's potential success. The following list of topics to be covered in the report is based on suggestions provided by Payne (1973) and Jones and Pettijohn (1973):

1. *Preface:* summary, definitions
2. *General information:* location, climate, topography, history, ownership, land status, transportation, etc.
3. *Environmental concerns:* present conditions, standards, necessary protective measures, land reclamation, special studies, permitting
4. *Geologic factors:* deposit setting, origin, structure, mineralogy, and petrography
5. *Mineral reserves:* exploration procedure, findings, calculation of tonnage and grade, extent of grade of by-products in the ore
6. *Mining plan:* development and exploitation
7. *Processing:* metallurgical results, on-site facilities needed
8. *Surface (and underground) plant:* location, construction, layout
9. *Auxiliary and support facilities:* power, water supply, road access, waste disposal, etc.
10. *Staffing:* workforce available (labor and supervisory), housing facilities, transportation, education facilities, other living requirements
11. *Marketing:* economic survey of supply and demand, price history and projection, existence of short-term and long-term contracts, substitutes, strategic reserves, etc.
12. *Cost projections:* estimation of direct, indirect, and overall costs of development and exploitation; costs of processing, transportation, smelting, etc.

13. *Economic evaluation:* valuation of the deposit, classified into type of reserve or resource; calculation of net present value
14. *Risk analysis:* assessment of variability in key variables such as price, labor costs, regulatory requirements, taxes, etc.
15. *Profit projection:* statistical determination of the profit margin, based on the projected variation in grades, prices, costs, etc.

Students in a beginning course in mining engineering would not normally be expected to have the knowledge to prepare such a report now. Completion of this course will provide more understanding of the factors involved. Students should keep in mind, however, that many mining engineering curricula require a senior project that involves an analysis of a mining project. In this type of senior project, a complete feasibility analysis for a mineral deposit is ordinarily required.

Later in the book, methods to calculate cost, gross value, profit, and present worth will be presented. Some simple examples can demonstrate the basic relationship of these terms as used in evaluation activities.

Example 3.1. Estimate the unit profit (in \$/ton or \$/tonne) of mining and processing a copper ore body averaging 0.60% by weight Cu if the selling price of copper in the concentrate is \$0.74/lb (\$1.63/kg), assuming that the overall cost of mining and processing is \$6.80/ton (\$7.50/tonne). Overall recovery, based on metallurgical testing, is expected to be 92% of the copper.

SOLUTION. Calculate the gross value of the ore using equation (3.7).

$$\text{Gross value} = \text{grade} \times \text{recovery} \times \text{price} \times \text{conversion factor} \tag{3.7}$$

For our deposit,

$$\begin{aligned}\text{Gross value} &= (0.0060\ \text{lb/lb})(0.92)(\$0.74/\text{lb})(2000\ \text{lb/ton}) \\ &= \$8.17/\text{ton}\ (\$9.00/\text{tonne})\end{aligned}$$

Profit in this case can be determined using Eq. 3.8, as follows:

$$\text{Profit} = \text{gross value} - \text{cost} \tag{3.8}$$

In our example, the profit is calculated as:

$$\text{Profit} = \$8.17/\text{ton} - \$6.80/\text{ton} = \$1.37/\text{ton}\ (\$1.51/\text{tonne})$$

Example 3.2. Calculate the cutoff grade for the copper deposit of Example 3.1.

SOLUTION. The *cutoff grade* is defined as the grade at which the gross value of the ore mined is exactly equal to the cost of mining. Therefore, the cutoff grade can be calculated using Eq. 3.9:

$$\text{Cutoff grade} = \frac{\text{cost}}{\text{price} \times \text{recovery}} \tag{3.9}$$

Therefore, our calculation yields:

$$\text{Cutoff grade} = \frac{6.80}{(0.74)(0.92)}$$

$$= 9.99 \text{ lb/ton (5.00 kg/tonne) or } 0.50\% \text{ Cu by weight}$$

Feasibility studies do not always proceed smoothly and produce conclusive findings after Stage 2. Decisions are made progressively, and funds are expended strategically to maximize income and minimize risk of losses. Studies of profitability continue throughout the lifetime of a mine. Exploration, development, and exploitation must keep pace with each other for best results. The need for new deposits will drive the exploration program forward unless an extremely poor market is forecast.

3.10 CASE STUDY: DIAMONDS IN THE NORTHWEST TERRITORIES, CANADA

Diamonds, they say, are a girl's best friend. But diamonds also have many industrial uses, and they are not necessarily good friends to the exploration geologist, because they are not easy to find. Diamonds occur in kimberlite pipes, former volcanic channels allowing deep-seated diamonds from the earth's mantle to find their way to the surface. Most kimberlites are barren; thus, location of kimberlites in an area does not provide enough information to locate a diamond source. Prospectors must first locate a pipe, and then determine whether the kimberlite is likely to contain diamonds. Even that does not provide enough information, as the diamond-bearing pipes normally vary in diamond content from one area to the other. Much more in the way of geological analysis and luck may be necessary.

Diamond prospecting has been going on for some time in North America through investigation of kimberlite occurrences. Many kimberlite pipes have been mapped in western states resulting in a significant find located along the Colorado-Wyoming border in 1986. This eventually became the Kelsey Lake diamond mine (Coopersmith, 1997). Similar exploration of kimberlites has occurred in the Northwest Territories of Canada, with geologic showings of diamonds being the driving force. In many of these explorations, geologists depended on the indirect prospecting technique of studying the indicator minerals associated with diamonds. Previous studies had shown that the chemical makeup of these minerals in diamond-bearing pipes was somewhat different from the composition of those in barren pipes. This finding helped to guide the search.

Two exploration geologists who had prospected for diamonds with a major company (that eventually got out of the mineral exploration business) made the actual discovery of the diamond-bearing pipes in the Northwest Territories. They continued to search on their own, assuming that the first diamonds found had been left behind by the glaciers that sheared off the tops of the kimberlite pipes. An exploration program aimed at discovering the source was pursued.

The geologists tracked the indicator minerals up the glacier paths, using geochemical methods. Based on the geochemical evidence, they staked many claims and went into partnership with BHP Minerals to further explore their claims. BHP Minerals financed a detailed geophysical program to locate the pipes. Both electromagnetic surveys from the air and ground-based geophysical surveys pinpointed some of the diamond-bearing pipes. More sophisticated airborne electromagnetic surveys then located additional pipes. Over 100 kimberlite pipes have been identified on BHP Minerals's 850,000 ac (344,000 ha) of claims (McLaughlin, 1998). See Johnson and Penny (1997) for a more complete summary of these exploration activities.

The project, now referred to as the Ekati diamond mine, includes seven kimberlite pipes considered to be of ore quality. Five of the pipes will be mined on the surface, and two will be mined underground. The development cost of the project is about $700 million, with more than 72,000,000 tons (65,000,000 tonnes) of reserves, averaging about 0.99 ct/ton (1.09 ct/tonne). The average value of the diamonds varies from about $30/ct to about $130/ct. Further information about the sociological, environmental, and economic aspects of this fascinating project can be found in summaries by McLaughlin (1998) and Rylatt and Popplewell (1999). This case study illustrates the often-complicated path to a major discovery; it also points out how good geological mapping, geochemistry, geophysics, and perseverance pay off. Luck played a small part in this discovery, but clearly it was primarily good prospecting and exploration practice that brought the diamonds into the reserve category.

3.11 SPECIAL TOPIC: CALCULATION OF AN ORE RESERVE ESTIMATE

The procedure for the calculation of a reserve estimate — tonnage and average grade of ore — for a mineral deposit is demonstrated in the following example. Occurring near the surface, the deposit is iron ore with soil and rock overburden. The deposit is to be mined by the open pit method, so the volume of overburden and the stripping ratio are also to be determined. Although this example is concerned only with the iron ore grade, typically the average percentages of phosphorus, silica, sulfur, and moisture are found as well.

If a complete valuation of the property were required, then an estimation of the mining costs and selling price of the ore would be necessary so that the profit and present worth of the property could be found.

Example 3.3. Calculate the reserve estimate for the iron ore deposit shown in Figure 3.11, determining the tonnage (in long tons, or tonnes) and the average grade of ore, volume of overburden, and stripping ratio. Use the section method of estimation (volume of influence version). The plan view shows the location of the sections and drillholes. Only one cross-sectional view is shown, but calculations from the other sections are provided. Base the ore weight on a tonnage factor of 14.0 ft^2/long ton (0.390 m^3/tonne). Drillhole samples are 5 or 10 ft (1.5 or 3.0 m) in length, as marked on section $1 + 00$. (Note that

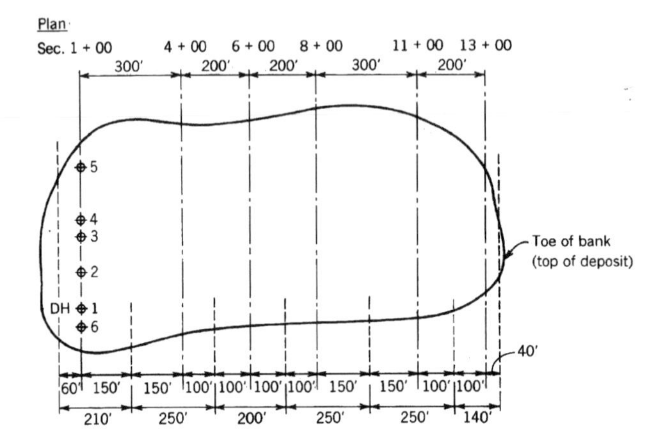

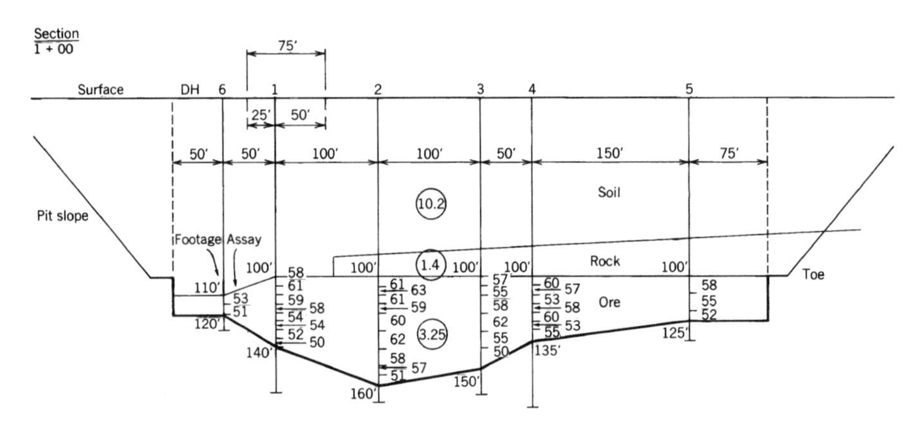

FIGURE 3.11. Plan view and cross section 1 + 00 of iron ore deposit in Example 3.3. Values to the left of the drillholes are depths in ft; those to the right are ore analyses in percent Fe. Circled values are areas measured by planimeter in in.^{2}

because of the amount of detail contained in the accompanying figures and tables, equivalent SI units are omitted. Answers, however, are given in both English and SI units.)

SOLUTION. For convenience, we tabulate all calculations. Width of influence for each drillhole in a cross section is one-half the distance to the adjacent drillholes; for drillholes at the end of the cross section, extend the width of influence on the outside one-half of the distance to the adjacent hole on the inside unless specified otherwise. Determine the width of influence (and hence the volume of influence) for sections in the same manner. Widths for both drillholes and sections are indicated in Figure 3.12. Convert the diagram areas in in.^{2} to ft^2 by multiplying by 6000. Note, however, that the section scales differ vertically and horizontally.

1. Average drillhole and section grades.

Sec. 1 + 00

Hole	Depth D (ft)	% Fe	D × %	Avg. Hole (%)	Hole Width W (ft)	$W \times D$	$W \times D \times$ %	Avg. Sec. (%)
1	5	58	290	$\frac{2230}{40} = 55.8$	25 + 50 = 75	375	21,750	
	5	61	305			375	22,875	
	5	59	295			375	22,125	
	5	58	290			375	21,750	
	5	54	270			375	20,250	
	5	54	270			375	20,250	
	5	52	260			375	19,500	
	5	50	250			375	18,750	
	40		2230			3000	167,250	
2	5	61	305	$\frac{3560}{60} = 59.3$	50 + 50 = 100	500	30,500	
	5	63	315			500	31,500	
	5	61	305			500	30,500	
	5	59	295			500	29,500	
	10	60	600			1000	60,000	
	10	62	620			1000	62,000	
	10	58	580			1000	58,000	
	5	57	285			500	28,550	
	5	51	255			500	25,500	
	60		3560			6000	356,000	

1. Average drillhole and section grades (*Continued*).

Sec. 1 + 00

Hole	Depth D (ft)	% Fe	D × %	Avg. Hole (%)	Hole Width W (ft)	$W \times D$	$W \times D \times \%$	Avg. Sec. (%)
3	5	57	285	$\frac{2810}{50} = 56.2$	50 + 25 = 75	375	21,375	
	5	55	275			375	20,625	
	10	58	580			750	43,500	
	10	62	620			750	46,500	
	10	55	550			750	41,250	
	10	50	500			750	37,500	
	50		2810			3750	210,750	
4	5	60	300	$\frac{1980}{35} = 56.6$	25 + 75 = 100	500	30,000	
	5	57	285			500	28500	
	5	53	265			500	26,500	
	5	58	290			500	29,000	
	5	60	300			500	30,000	
	5	53	265			500	26,500	
	5	55	275			500	27,500	
	35		1980			3500	198,000	
5	10	58	580	$\frac{1390}{25} = 55.6$	75 + 75 = 150	1500	87,000	
	10	55	550			1500	82,500	
	5	52	260			750	39,000	
	25		1390			3750	208,500	
6	5	53	265	$\frac{520}{10} = 52.0$	50 + 25 = 75	375	19,875	
	5	51	255			375	19,125	
	10		520			750	39,000	
Sec.						20,750	1,179,500	$\frac{1,179,500}{20,750} = 56.8$

2. Recapitulation of section and ore body grades and tonnages.

Sec.	Plan Area (in.^{2})	Area A (ft^2)	Section Width l (ft)	Volume $V = A \times l$ (ft^3)	Weight $W = V \div TF$ (tons)	%Fe	$W \times \%$
1 + 00	3.25	19,500	210	4,095,000	292,500	56.8	16,614,000
4 + 00	3.90	23,400	250	5,850,000	417,857	54.3	22,689,635
6 + 00	4.57	27,420	200	5,484,000	391,714	59.1	23,150,297
8 + 00	4.74	28,440	250	7,110,000	507,857	57.0	28,947,849
11 + 00	4.41	26,460	250	6,615,000	472,500	53.7	25,373,250
13 + 00	3.86	23,160	140	3,242,400	231,600	55.9	12,946,440
Mine					2,314,028	56.1	129,721,471

Area conversion factor = (100)(60) = 6000 ft^2/in.^{2}
Tonnage factor = 14.0 ft^3/long ton [1 long ton = 2240 lb (1.016 tonne)]
Total tonnage = 2,314,028 long tons (2,351,161 tonnes)

$$\text{Average grade} = \frac{129,721,471}{2,314,028} = 56.058, \text{ say } 56.1\%$$

3. Recapitulation of section and ore body stripping.

Sec.	Plan. Area (in.^{2}) Soil	Plan. Area (in.^{2}) Rock	Area (ft^2) Soil	Area (ft^2) Rock	Sec. Width (ft)	Volume V (yd^3) Soil	Volume V (yd^3) Rock
1 + 00	10.2	1.4	61,200	8,400	210	476,000	65,333
4 + 00	12.7	3.1	76,200	18,600	250	705,556	172,222
6 + 00	14.2	4.2	85,200	25,200	200	631,111	186,667
8 + 00	15.1	5.1	90,600	30,600	250	838,889	283,333
11 + 00	13.9	4.6	83,400	27,600	250	772,222	255,556
13 + 00	9.8	2.9	58,800	17,400	140	304,889	90,222
Mine						3,728,667	1,053,333

Volume conversion factor = 27 ft^3/yd^3

$$\text{Volume} = \frac{\text{area} \times \text{section width}}{27} \text{ yd}^3$$

Total volume = 3,728,667 + 1,053,333 = 4,782,000 yd^3 (3,656,101 m^3)

4. Stripping ratio SR.

$$SR = \frac{V}{W} = \frac{4{,}782{,}000}{2{,}314{,}028} = 2.066 \quad \text{say } 2.1 \text{ yd}^3/\text{long ton } (1.6 \text{ m}^3/\text{tonne})$$

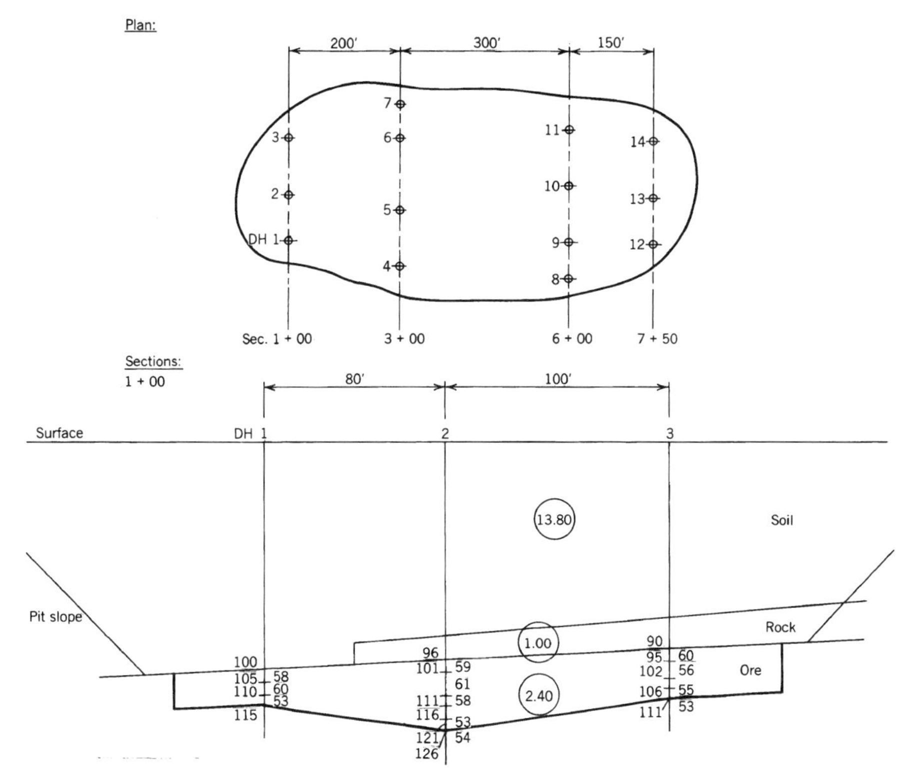

FIGURE 3.12. Plan view and cross section 1 + 00 of iron ore deposit in Problem 3.1.

PROBLEMS

3.1 From the section information and drillhole analyses for the open pit iron ore mine shown in Figures 3.12 and 3.13, calculate the ore reserve estimate of tonnage and average grade. Also calculate stripping yardages (for soil, rock, and total) and the overall stripping ratio. On the diagrams, drillhole footages are listed to the left of the drillhole and % Fe to the right. Circled numbers represent the section diagram areas in in.^{2} for the soil, rock, and ore. Use a tonnage factor of 14 ft^3/long ton (0.390 ft^3/tonne) and assume that the scale of the diagrams in the cross sections is 1″ = 50′. Set up calculations in tabular form, and show all steps. Express grade values to three significant figures, and record drillhole as well as section and mine averages. Note that 1.0 long tons = 1.12 short tons = 2240 lb = 1.016 tonne.

3.2 From section information and drillhole analyses for the open pit copper mine shown in Figures 3.14 and 3.15, calculate the ore reserve estimate of tonnage and average grade. Also calculate stripping yardages (for soil,

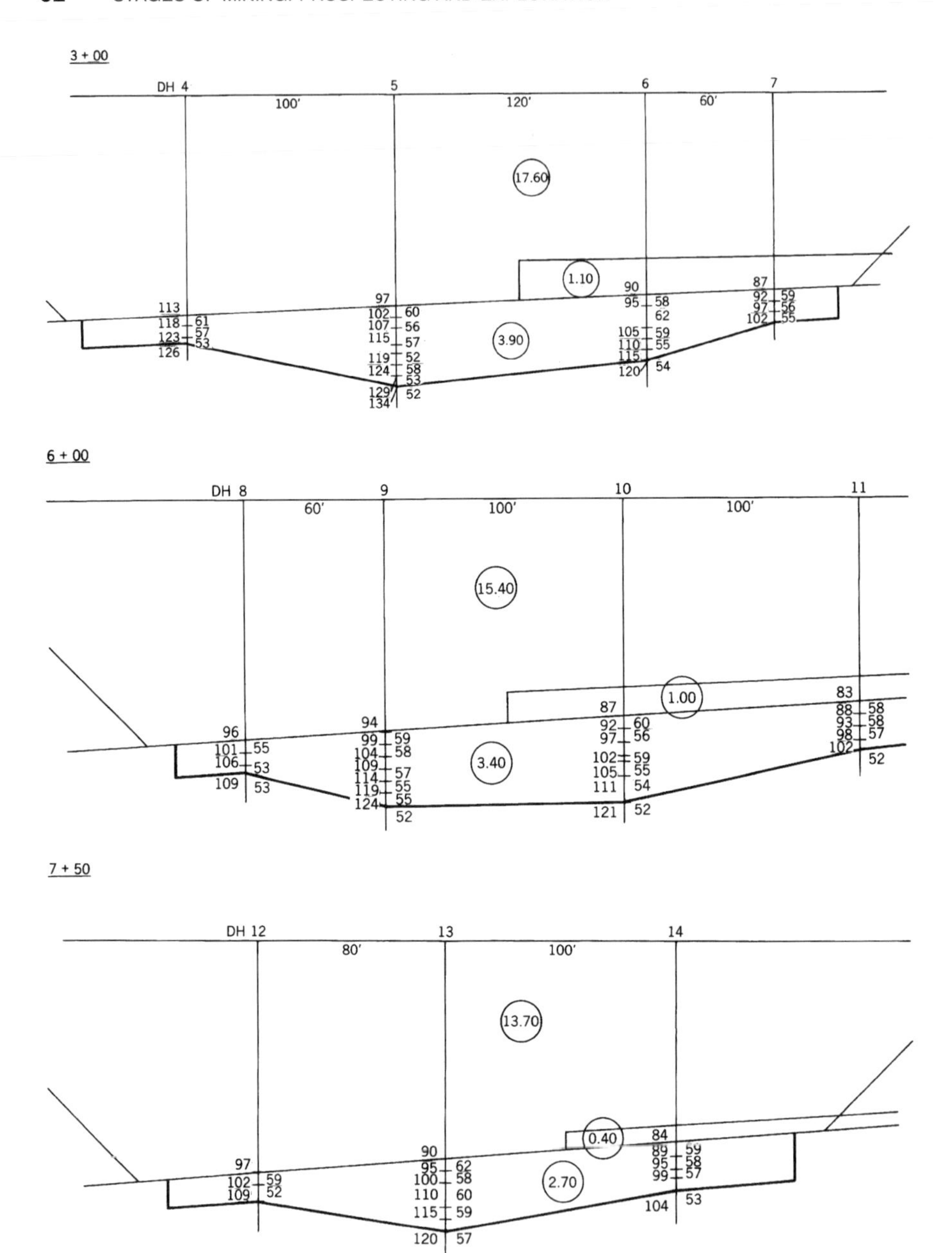

FIGURE 3.13. Cross sections 3 + 00, 6 + 00, and 7 + 50 of iron ore deposit in Problem 3.1.

rock, and total) and the overall stripping ratio. Footages are given to the left of the drillhole and % Cu to the right. Circled numbers represent the section diagram area in in.^{2} for soil, rock, and ore. Assume that the section areas were measured by planimeter on sections with a scale of

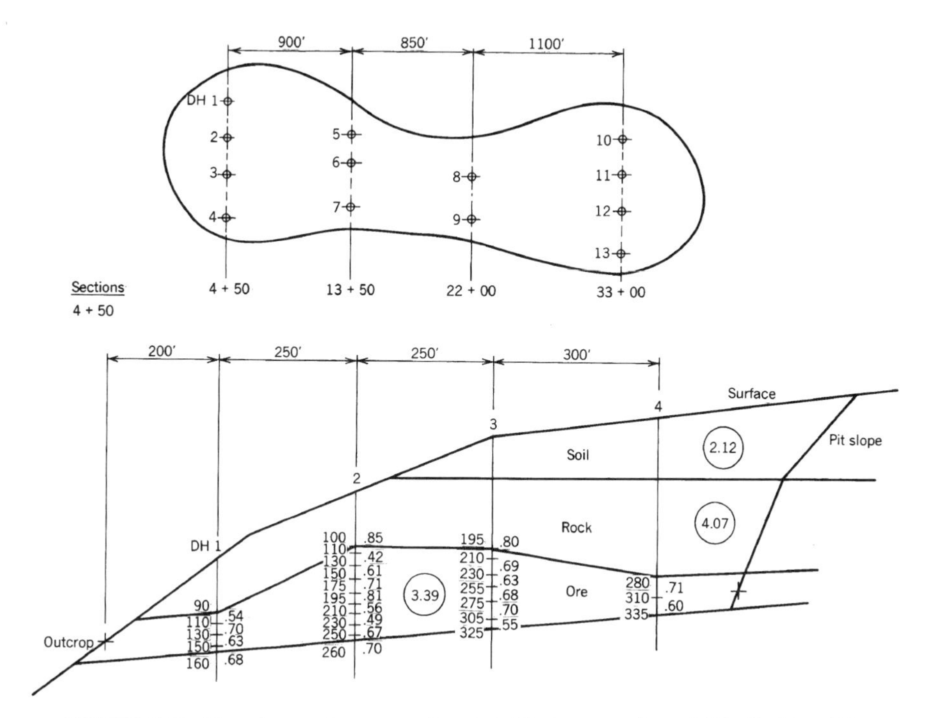

FIGURE 3.14. Plan view and cross section 4 + 50 of copper deposit 9 in Problem 3.2.

1″ = 200′. Use a tonnage factor of 14.2 ft^3/ton (0.443 m^3/tonne). Set up calculations in tabular form, and show all steps. Express grade values to three decimal places, and record drillhole as well as section and mine averages.

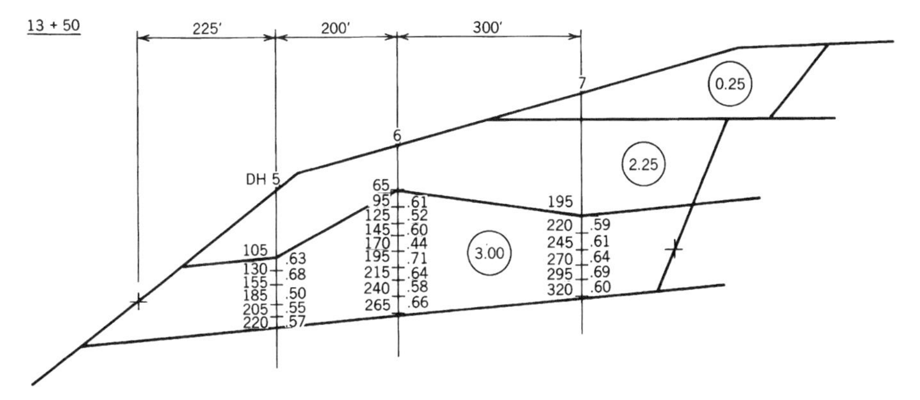

FIGURE 3.15. Cross sections 13+50, 22+00, and 33+00 of copper deposit in Problem 3.2.

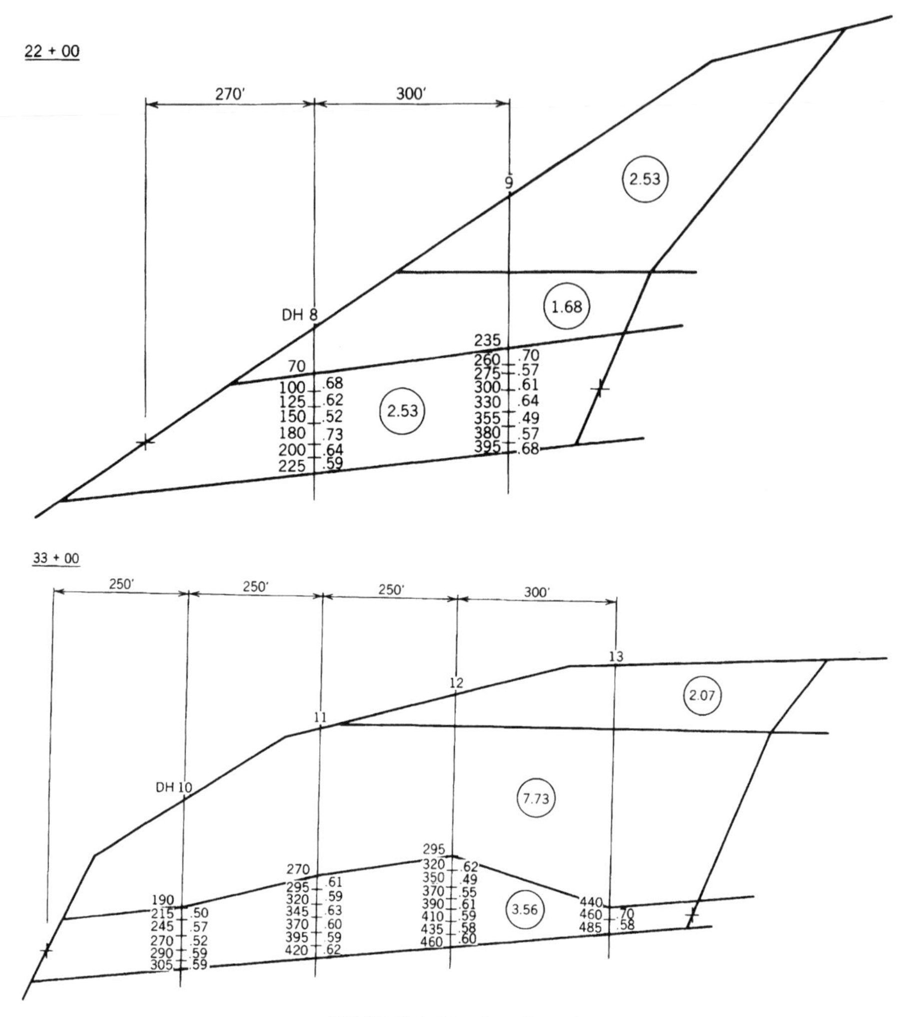

FIGURE 3.15. Continued.

3.3 From drillhole data compiled in the accompanying table, calculate the (a) average seam thickness, (b) average depth of overburden, (c) average analysis (in percent sulfur), and (d) total tons (tonnes) of reserve for the coal seam shown in Figure 3.16. Exploration has been conducted through vertical boreholes, spaced equidistant in a triangular grid. Tonnage factor for coal is 25 ft^3/ton (0.78 m^3/tonne). Note that this is a tabular deposit; statistical averages can be based on areas rather than volumes. (e) Sketch the areas of influence on the plan view, using the polygon method and assuming that the areas of influence for the outside holes run to the outcrop or the boundary of the property. Calculate the overall stripping ratio in feet of overburden removed per foot of coal removed. Borehole data are provided in the following table.

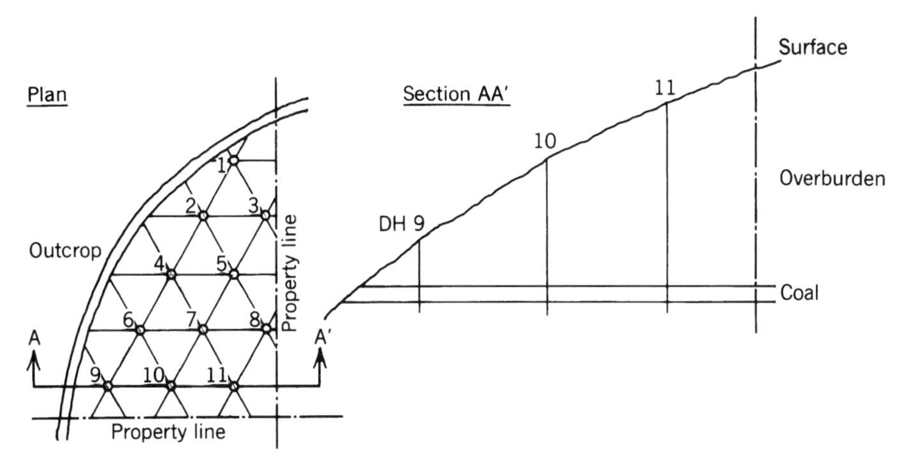

FIGURE 3.16. Plan view and cross section AA′ of coal deposit in Problem 3.3. Other cross sections are similar to the one shown.

Hole	Coal Thickness (ft)	Overburden Thickness (ft)	Analysis (% sulfur)	Area of Influence (10^6 ft^2)
1	3.1	133	0.8	1.56
2	2.8	175	1.6	2.10
3	3.2	283	1.4	1.34
4	3.0	182	1.7	2.32
5	3.1	327	0.9	1.70
6	2.9	156	1.2	2.21
7	3.2	298	1.5	1.73
8	3.2	417	1.4	1.20
9	2.8	149	1.0	2.38
10	3.1	275	1.3	1.98
11	3.0	444	1.2	2.61

3.4 Calculate the cutoff grade of an iron deposit, given the following:

Iron ore price: \$84/long ton (\$82.68/tonne), 100% Fe
Total production cost: \$25/long ton (\$24.61/tonne)
Recovery: 95%
1 long ton = 2240 lb (1.016 tonne)

3.5 An underground mine is producing zinc ore at \$35/ton (\$38.59/tonne), which includes all of its mining and processing costs. If it nets a selling price of \$0.60/lb (\$1.32/kg) of zinc in the concentrate, what is the cutoff grade of the mine (in % zinc) if none of the variables change and recovery of zinc is 88%?

4

STAGES OF MINING: DEVELOPMENT AND EXPLOITATION

4.1 INTRODUCTION

After successful completion of prospecting and exploration—and with a favorable outcome of the mineral evaluation and feasibility studies—the work of mining proper can commence. What was originally an anomalous occurrence of minerals has proved to be an ore deposit (or an economic deposit of coal or stone). With financing in place, the mineral prospect will soon emerge as a producing mine.

The two stages that must now begin involve the actual work of mining the deposit. Development, Stage 3, to a considerable degree precedes exploitation, Stage 4. As in prospecting and exploration, however, we find no sharp demarcation between these stages except that development must get under way first. When exploitation begins, the development of the mine continues but is apt to proceed but one step ahead of exploitation, and concludes only when the last area to be exploited has been properly accessed by the development openings. In most underground mines, development begins a few years before exploitation and ends a few years prior to completion of mining.

The reasons that development and exploitation proceed in sequence but overlap are both fiscal and technological. First, is too expensive to develop—and maintain the openings of— an entire mine without completing some of the exploitation steps to ensure a financial return. Second, the technology may change significantly over time, requiring a different set of development openings. Accordingly, the development proceeds only as fast as the exploitation demands. Just as exploration continues during mining, development occurs throughout exploitation, particularly that closely associated with production openings. The proper coordination of all three of these stages is thus a key task of mine management. Managing these three stages of mining ensures that the mine can be reassessed periodically and that the optimal return is achieved as the economics of the mineral deposit change over its lifetime.

The principle of limiting the amount of development in advance of exploitation is widely practiced in mining. However, severely limiting development may affect the efficiency of the exploitation and constrain the production process. Efficient production scheduling and early exploitation of the ore body require that nearly all major plant facilities be in operation before any ore is produced. It is therefore necessary to allocate a generous amount of capital to ensure that the proper amount of development is completed for optimal operation of the mine. As a general rule, development should be carried out to access the maximum tonnage of ore for a minimum expenditure for development openings. This is especially true where the development openings cost more (in $/ton, or $/tonne) than production openings. Some exceptions to this rule are apparent. For example, development openings in the ore (such as in room-and-pillar mining) will produce a return if the ore is marketed. Thus, there is less economic pressure to minimize such development openings.

Specific practices during both development and exploitation with the various mining methods are discussed in succeeding chapters. This chapter presents the general principles of development and exploitation.

4.2 DEVELOPMENT: GENERAL LOGIC

As indicated in Section 1.4 and Table 1.3, *development* is the work necessary to bring a mine into full, scheduled production. It includes planning, design, construction, and other phases. Development generally proceeds according to the plan adopted during the feasibility study in Stages 1 and 2, but the plan may be altered as more and better information becomes available during successive phases of the project. Most mines that are in existence for more than a decade or so will experience changes in their development plans several times as new technology becomes available, as the economics of their mineral commodity changes, and as the suitability of mining methods to the ore body is better assessed.

From the standpoint of physically opening a mine, the major purpose of development is to provide access to the ore deposit, permitting entry of the miners, equipment, supplies, power, water, and ventilating air, as well as egress for the mineral being mined and the waste produced. Prior to the start of the exploitation phase of mining, development is limited to excavation of the primary or main openings plus other equipment installation and construction required to initiate exploitation. In a surface mine, access to the ore body overlain by waste is gained by stripping the overburden, which is then placed in disposal areas for later reclamation. For an underground mine, the development openings are driven to optimize the operation of the mine during exploitation, and little in the way of ore is recovered during the initial development. Any ore recovered during underground mine development is normally stockpiled until the mine goes into production.

Figure 4.1 gives an example of the many components of operating mines—both surface and underground—that must be developed before a mineral property becomes a source of revenue. During development or early in the exploitation of a mine, all of the facilities shown (or a similar set) must be provided.

4.2.1 Factors in Mine Development

Following a successful exploration program and prior to the development of a mineral property, a variety of factors must be considered. Some of these may have been covered in the feasibility report in summary form. However, they may require additional data and planning before the deposit is mined. Bullock (1982) outlines some of the technical aspects of mine development. Britton and Brasfield (1992) also discuss several of the technical and political variables associated with development of a mineral property. It is important to assess all of the variables associated with mineral development. In today's world, political and environmental variables are particularly important. The following sections discuss some of the concerns that must be addressed.

4.2.1.1 Locational Factors. Ore bodies are located where you find them—and not always in the most advantageous settings. Few are situated in an ideal location, either for supplying the mine or for marketing the product. Geography thus affects the mineral operation significantly and is beyond the control of the mine management. The following are some of the most important variables that are affected by location:

1. Ease of transport of the mineral products and supplies
2. Availability of labor and support services (housing, educational and recreational facilities, health care, etc.)
3. Operational (and psychological) impacts of climate and weather
4. Employees' satisfaction with their lifestyle

Geography particularly affects the mining of those ore deposits that are located in remote arctic regions, in desert locations away from population centers, and in less developed countries. In these areas, mining companies must decide whether to build a company town at the deposit site or to employ fly-in, fly-out (FIFO) workers to exploit the deposit. Company towns are frowned upon in most areas of the world, but in remote regions they may be necessary. Farquharson and Marshall (1996) and Jenkins (1997) have analyzed the process of arriving at the proper decision in the FIFO versus company town choice. A proper analysis clearly must include both economic and human variables. Keeping good employees may also require that remuneration and

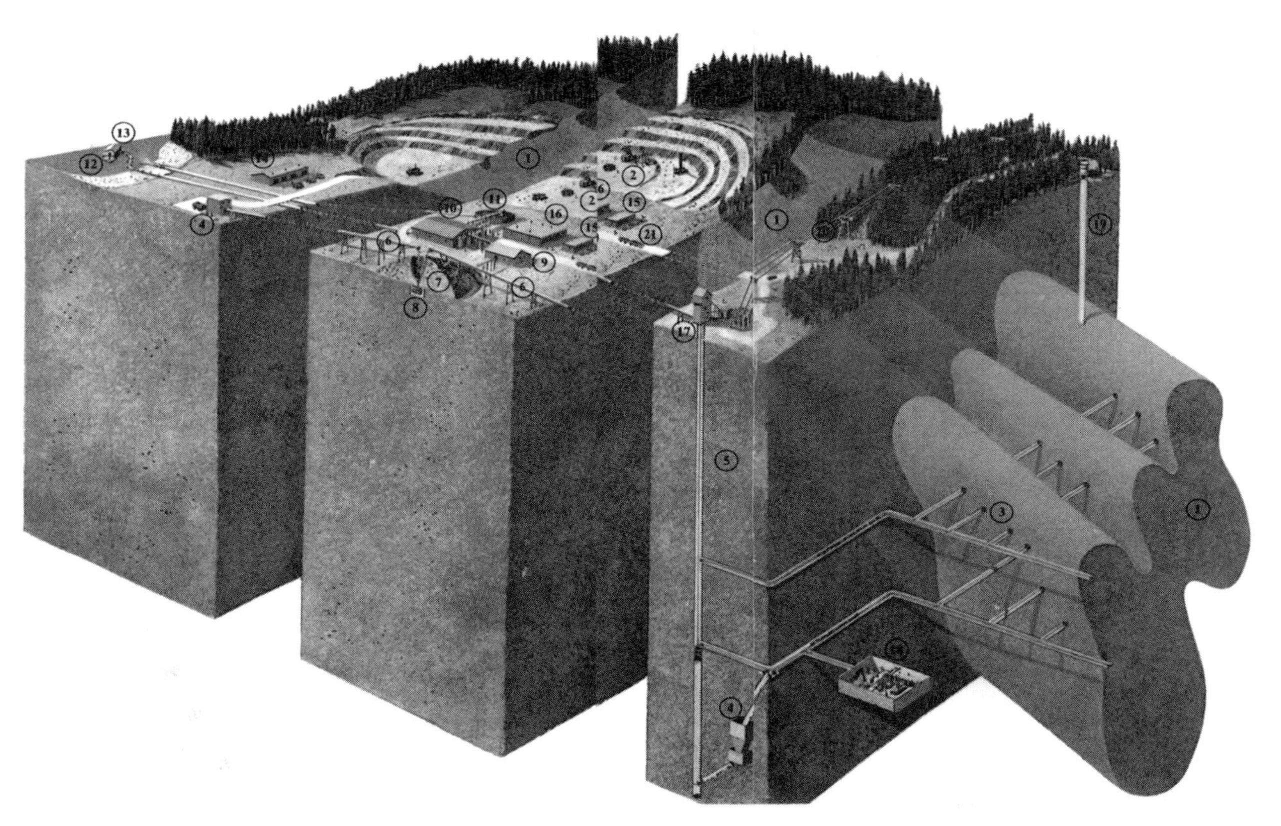

FIGURE 4.1. Idealized diagram of a mine with both surface and underground operations. The surface mine has entered the exploitation stage; the underground mine is under development, and no production workings are shown. (After Placer Development Ltd., 1980. Courtesy of Placer Dome Inc., Vancouver, BC, Canada.)

fringe benefits be well above average for industrial workers in these cases. This becomes an important economic cost that must be considered when the mine location is not otherwise attractive.

4.2.1.2 Natural and Geologic Factors. Mother Nature and the geologic environment govern many key aspects of mine development, especially access openings and surface plant location. The most important factors in this category include the following:

1. Topography
2. Spatial relations (size, shape, attitude, depth, etc.) of the ore body
3. Geologic considerations (mineralogy, petrography, structure, ore body genesis, thermal gradient, presence of water, etc.)
4. Rock mechanics properties (strength, modulus of elasticity, hardness, abrasiveness, etc.)
5. Chemical and metallurgical properties of the ore that affect processing and smelting

Such factors will affect the choice of mining method as well as the decisions that pertain to mine development.

4.2.1.3 Social-Economic-Political-Environmental Factors. Largely out of control of the mining company, a number of variables can exercise a disproportionate influence on both the development and operation of a mine. In addition, their influence may be difficult to predict ahead of time. Among others, these factors include the following:

1. Demographics and occupational skills of the local population
2. Means of financing and marketing (determines the scale and continuity of operation)
3. Political stability of host country (less developed countries constitute a high risk for foreign investors)
4. Environmental legislation
5. Government aids, government restrictions, and taxes on mineral enterprises

These factors govern many critical aspects of both development and exploitation. Mining ventures deemed to be attractive economically have been abandoned or bankrupted by the unforeseen impacts of social, political, governmental, or environmental factors. In seeking new ore deposits, mining companies are often attracted to areas where political and environmental problems can be minimized. *Engineering and Mining Journal* (1999b) discusses U.S. states and Canadian provinces and provides an assessment of their political/environmental attractiveness. Because of strong environmental regulations and public opposition to mining operations in the United States, many mining companies look to foreign deposits for new sources of minerals. This may present the company with other potential problems—for instance, instability of the government, terrorist or criminal activity, and lack of infrastructure. Overcoming the risks of such foreign investments and ensuring that capital investments achieve an economic return become very important. Some thoughts on mitigating risk in overseas minerals development are presented by Murdy and Bhappu (1997) and by the *Engineering and Mining Journal* (1992a).

4.2.2 Sequence of Development

The steps generally carried out during mine development include the following, more or less in the order of accomplishment. Both surface and underground mines follow a similar sequence.

1. Adoption of a feasibility report as a planning document, subject to modification as the project is developed and mined
2. Confirmation of the mining method(s) and general sequence of mining (initial choice of equipment types and size of the workforce may also be made at this point)
3. Arrangement of financing, based on confirmation of ore reserves and cost estimates by independent consultants
4. Acquisition of land and mineral rights as needed
5. Filing of the environmental impact statement, obtaining the mining permit, and posting of bonds subject to both federal and state statutes, as applicable
6. Provision of surface access, transportation, communication, and power supply to the mine site
7. Planning and construction of the surface plant, including all support and service facilities and administrative offices
8. Erection of the mineral processing plant, if required, and ore-handling and shipment facilities, and preparation of stockpiling and waste disposal facilities
9. Acquisition of mining equipment for development and exploitation
10. Construction of the main openings to the ore body in underground mining or advanced stripping in surface mining to provide direct access to the ore zone
11. Construction of the underground facilities (ore passes, crusher stations, transport systems, etc.) or the surface installations (crusher station, water handling system, maintenance garages, etc.) to initiate production

12. Recruitment and training of the labor force and provision of general support services (housing, transportation, educational needs, medical services, consumer goods, etc.), as necessary

Several of these steps may be conducted simultaneously; others may be added because of special needs (e.g., taking on partners, negotiating shipping and sales agreements, etc.), and some may be omitted (e.g., land acquisition if completed during exploration). An overall planning model appears in Figure 4.2; it utilizes three levels of study (conceptual, engineering, and detailed design). To ensure that all development tasks are completed on time, careful scheduling is normally required, as discussed in Section 4.5. Several of the key steps in development are elaborated in subsequent sections.

4.3 DEVELOPMENT: LAND ACQUISITION

Acquiring rights to land and minerals in the United States is a development step that each mining company must undertake. The land can be obtained from certain public lands designated for development or from private sources. On federal lands, the acquiring of mining rights is governed by the Mining Law of 1872, which has been described in some detail in Section 2.2.1 and is further described by Parr (1992). On private land, either the land itself or the mineral rights can be obtained through ordinary real estate transactions. Because the mineral rights and the surface ownership are separable, it is possible and often desirable for a mining company to purchase the mineral rights without the surface. Because of the fact that either public or private land is involved and mining can be conducted on land that is either owned or leased, mining can normally be undertaken under the following four conditions (Parr and Ely, 1973):

1. *Ownership through claims or patents on public land.* The federal government controls about 38% of the land in the United States; this land is known as the public domain. The federal lands on which mineral claims may be made are administered primarily by the Bureau of Land Management (BLM) and are termed *locatable lands.* Excluded from this category are the national parks, wilderness areas, Indian reservations, military installations, and certain other lands where minerals development would not be desirable. Nearly all of the locatable lands are found in the western states. The federal law that permits claiming of federal land for mineral development is applicable primarily to the metallic minerals; fuels and most nonmetallic minerals are excluded (see Sections 2.2.1 and 2.2.2). Following discovery and development of a mineral claim, it may be patented for a nominal fee if its value as an ore deposit is firmly established. The amount of land available for prospecting is decreasing as a result of more public lands being set aside as wilderness areas or placed

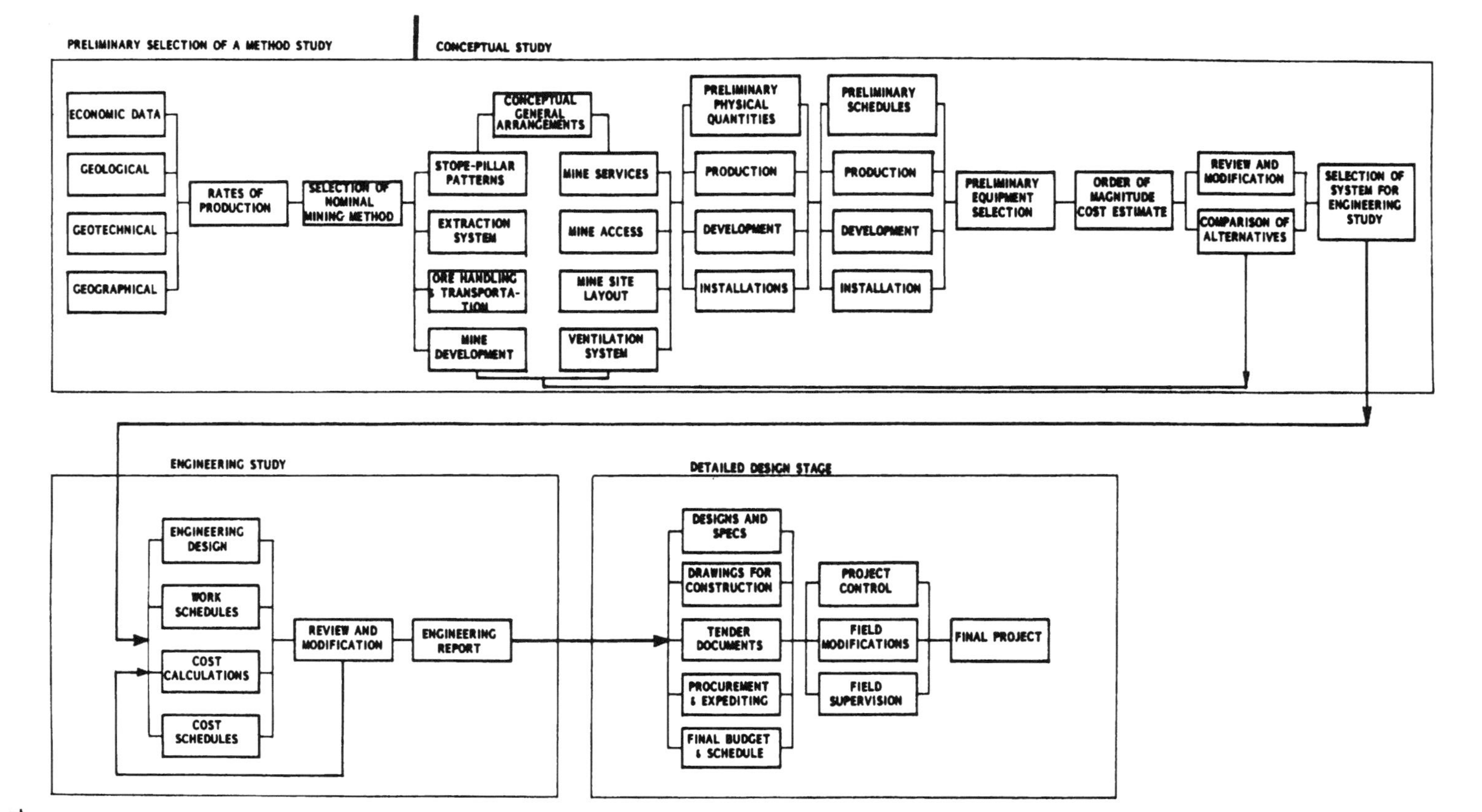

FIGURE 4.2. Planning model and sequence of steps in mine development, showing three levels of study confirming the selection of a mining method. (After Folinsbee and Clarke, 1981. By permission of the Society for Mining, Metallurgy, and Exploration, Inc., Littleton, CO.)

in other nonlocatable categories. The 1964 Wilderness Act established about 9 million ac (3,640,000 ha) as wilderness, but that number has grown to 105 million ac (42,500,000 ha) and many more wilderness areas have been proposed by conservation groups (Arnold, 1999). Although the current wilderness land inventory is only 4.5% of the public lands, much of it is in Alaska and other states where economic mineral deposits are still likely to be located. Thus, it has become increasingly difficult to obtain mineral rights on public lands.

2. *Leasing of public land.* Mining rights may be obtained by bidding on leases for coal, petroleum and natural gas, uranium, and most nonmetallic minerals occurring on federal and many state lands. Payment of a bond and royalties is required; royalties are often in the range of 5% to 15% of the gross value of the minerals extracted.

3. *Ownership of private land.* Outright purchase of lands for mining rights is not as common today as in the past. The reasons are the escalating cost of real estate and the huge capital outlay required for a mining operation of even a nominal size. Ownership of the land is still customary in metal mining, but is decreasing in both coal and nonmetal mining because leasing is becoming more popular.

4. *Leasing of private land.* Mineral rights on private land are increasingly obtained by leases that involve royalties on the minerals as they are mined. Many different leasing agreements can be made, but the normal arrangement is to pay a royalty on a fixed cost per ton or a percentage of the value of the minerals delivered to a buyer. Recent lease agreements outlined by Bourne (1996) indicate that coal leases cost $0.10/ton to $3.00/ton ($0.11/tonne to $3.30/tonne) or 3% to 12.5% of the gross value; metallic leases are generally in the range of $0.25/ton to $25/ton ($0.276/tonne to $27.60/tonne) or 2% to 12.5% of the gross value. Nonmetallic minerals have a wide range of lease costs, depending on the nature of the minerals. One of the high-tonnage commodities in the nonmetallic field is aggregate rock, which leases for $0.05/ton to $1.00/ton ($0.055/tonne to $1.10/tonne) or in the range of 3% to 7% of its value.

Landmen and lawyers are often engaged to acquire the necessary mineral rights for a mining company. The legal aspects of minerals development can be very complicated. A good example is in the recovery of coalbed methane from a coal mine. Many disputes have occurred between coal seam lessees, natural gas lessees, and the surface property owners over ownership of the methane gas in the coal seam. Although these issues have been clarified somewhat in recent court decisions, mining companies must understand all the ownership issues before getting involved in a coal leasing arrangement. McClanahan (1995) has provided a review of applicable legal aspects of this problem. Other complications in the procurement of mineral rights also make legal counsel highly important.

4.4 DEVELOPMENT: ENVIRONMENTAL PROTECTION

No one circumstance of mining has changed the development and exploration activity of a mine more in recent decades than the requirements for environmental protection. Today's regulations require every mining company to adopt modern technology, sometimes at great expense, to protect the environment as much as humanly possible. Thus, every aspect of environmental protection must be carefully planned before the first excavation begins. Mine management must plan how the land is to be restored, how the flora and fauna are to be protected, and how the water and air are to be kept as clean as technology permits. To ensure that such plans have been made, mining permits are withheld until the regulatory agencies are satisfied that environmental laws will be obeyed.

Kaas (1992) has outlined three major Acts that deal with the environmental consequences of mining. The National Environmental Policy Act (NEPA) of 1969 has profound effects on the process of obtaining a mining permit. NEPA requires that an appropriate government agency prepare an environmental impact statement (EIS) for any action it takes (including the issuing of a mining permit) that can have an effect on the human environment. The government agencies involved may have the mining company prepare an environmental assessment (EA), which involves gathering data on the geologic, hydrologic, botanical, zoological, archeological, historical, cultural, sociological, and economic aspects of the proposed mining project. After analysis of these factors, the agency or agencies involved prepare a draft of the EIS. This draft is passed on to other government agencies that may be involved in any way and to the public for comment. Finally, a decision is recorded that outlines the alternatives considered, the environmental consequences of the alternatives, and whether all practical means have been taken to minimize environmental harm. Additional information on this process is provided by Kaas (1992) and Hunt (1992).

The other two major federal Acts that have an effect on mining activities are the Surface Mining Control and Reclamation Act of 1977 and the Resource Conservation and Recovery Act (RCRA) of 1976. The first of these especially affects mining practice and has been described in Section 2.2.5. The RCRA is the federal legislation that deals with solid waste management. It is oriented toward industry in general, but 40% of all solid wastes are generated within the mineral industries. Hence, it is important for the mining industry to keep abreast of the RCRA stipulations on which wastes are hazardous, which are nonhazardous, and the requirements for each category of waste. Additional information on this act can be found in Kaas (1992).

Parr (1992) outlines additional environmental laws that may have significant ramifications for mining companies. The law with the greatest potential impact is the Comprehensive Environmental Response, Compensation, and Liability Act of 1980. This legislation, also known as CERCLA or the Superfund Act, applies to any sites that contain materials mined or deposited in the past that can discharge hazardous substances. CERCLA can be applied to a waste pile or tailings area that was deposited under conditions that were entirely within the law at the time of deposit. In addition, the law can assign the entire cost of cleanup to any one of the responsible parties, either past or present owners, even if they had little or no part in the actual production or deposition of the hazardous material. Clearly, this makes this legislation of tremendous importance in decisions involving old waste and tailings areas. The responsibility of mining companies to protect water quality under the Water Pollution Control Act of 1972 and the Clean Water Act of 1977, to maintain air quality under the Clean Air Act of 1970 and the Clean Air Act Amendments of 1977 and 1990, to protect historical and archeological sites under three different Acts, and to protect endangered species is also summarized by Parr (1992). Detailed description of the application of these laws is beyond the scope of this book.

With so many environmental laws that apply to mining practice, the mining company initiating a new mine must carefully assess the implications of each law on its project. This will enable the planning personnel to integrate the environmental protection into the mine plan so that the overall process of *mining* and *environmental protection* is optimized, rather than either one. Because of the complexity of environmental issues, mining companies often seek help from environmental consultants to plan their operations or to ensure that their development and mining processes are environmentally sound. Development will then begin only after the entire process is planned and the permits obtained.

4.5 DEVELOPMENT: FINANCING AND IMPLEMENTATION

Because mining is considered a high-risk business, it is important that the mining company have a statistically sound ore deposit evaluation, a technologically sound mining plan, and well-established economic feasibility before proceeding to seek financing from any commercial bank or other traditional financial source. The general layout of the mine, the development plan, the extraction sequence, the processing plan, and the location of all facilities will normally be outlined in detail by this time so that financial institutions and investors can evaluate the project.

It is the mining company's decision as to how the project is to be financed. Background information on financing methods for minerals operations is outlined by Brooks and Hursh (1982) and Tinsley (1992). The most common forms of financing are as follows:

1. Loans from commercial banks and private investment sources, such as other corporations, trust funds, insurance companies, or foreign firms, either unsecured or secured by collateral

2. Issues of securities, stocks, and bonds through investment houses or banks (a form of internal financing that avoids borrowing secured by collateral)
3. Leasing of equipment (a method of avoiding capital costs and additional borrowing that results in monthly lease payments over time)
4. Government loans — U.S., foreign, or international — for special projects (e.g., World Bank or International Monetary Fund loans for foreign investment)

Brooks and Hursh (1982) discuss sources of funds for coal mine projects, with about 50% of the financing normally coming from commercial loans, about 30% from internal sources, and the rest from leasing and other sources. Financing arrangements are likely to change from time to time as the perceived risk and the availability of funds change. Commercial banks and similar loan agencies normally have independent consultants evaluate the feasibility of the project and the economics of the development and exploitation phases. When the lending agency is satisfied with the project plan, it may grant the loan so the project can begin.

The development of a mine is ordinarily implemented according to a carefully crafted schedule of activities that will allow the mining company to construct all the necessary elements of the operation. The physical facilities provided to operate the mine are referred to as the *mine plant*. If the mine is an underground operation, the mine plant may be subdivided into surface, shaft, and underground facilities. To provide these facilities in a coordinated manner and to initiate all the development openings in the shortest time and most cost-effective manner, the entire mine development effort is scheduled by utilization of the critical path method (CPM), the project evaluation and review technique (PERT), or a variation of these methods. A CPM diagram for an actual mine and processing plant development (Placer Development Ltd., 1980) is available for study in Hartman (1987, pp. 96–97). Note that the primary objective of the CPM method is to achieve the completion of the development in the shortest possible time so that the exploitation of the deposit can begin to produce a return on the investment. By locating the critical path through the network and ensuring that the activities on the path are not delayed by noncritical activities, the objective can be achieved. Financing agencies may require such a schedule before the financing is granted.

4.6 DEVELOPMENT: TAXATION AND COSTS

4.6.1 Taxation

Taxation of mineral property and mining operations is unique, variable, and complex, both in the United States and elsewhere, and is a cost item that must be properly assessed before mining can proceed. No other cost is so difficult to project with accuracy. In part, the difficulty is traceable to the nature of a mineral deposit, which is considered to be a *wasting* (or depleting) *asset*. This results in a mining company being able to claim a *depletion allowance* applied to its net income and based on the declining value of the unmined mineral content under U.S. tax law. Depletion rates vary with the mineral, from 5% for nonmetallics to 22% for many metallic deposits (Gentry and O'Neil, 1984). Because the depletion allowance reduces the net income, it effectively reduces the income tax a minerals company pays. The economic impact of the minerals depletion allowance on the economics of mining is also discussed by Gentry and O'Neil (1984).

Other taxes that apply to minerals companies are state income tax, taxes on real estate or mineral reserves, severance taxes on the minerals removed, transaction taxes on the sales of minerals, excise taxes (primarily license fees), unemployment taxes, import or export taxes, and so on. Severance taxes are often the most significant of these and are ordinarily collected even if the mining operation does not earn a profit. Gentry and O'Neil (1984) provide a list of states that have severance taxes on minerals. Some of these taxes are significant, up to 15% of the minerals' value or up to 20% of the proceeds of the minerals. Detailed knowledge of the tax laws is thus necessary before a mine is developed.

4.6.2 Cost Estimation

The estimation of development or exploitation costs is an important task in any mine development project. The beginning student may not be fully prepared to perform this task. However, it is important that every student readily gain an appreciation for the relative costs associated with each stage of mining. Although detailed costs are often quite difficult to obtain on specific operations, the technical literature does have average or compiled costs for both development and exploitation methods. The Northwest Mining Association has published several volumes of cost articles from its short courses (Hoskins, 1982, 1986). Many of these articles are useful in providing costs for minerals operations and methods for estimating them. The U.S. Bureau of Mines Staff (1987a, 1987b) has provided a wide-ranging compilation of cost regression equations based on data gathered from a significant number of mining operations. Covering both surface and underground mines and the associated minerals processing, these two books provide the student with an excellent source of data. Additional cost-related data are found in the *SME Mining Engineering Handbook* (O'Hara and Suboleski, 1992; Mutmansky et al., 1992). The first of these chapters provides primarily metal mining information; the second provides data on both coal and metal mining. Finally, the Mining Cost Service of Spokane, Washington, provides the mining industry with cost data covering a wide variety of production, supply, and processing components necessary in a mining operation (Western Mine Engineering, 1998). The advantage of this service is that it is updated annually.

In the initiation of a minerals operation, one of the most important cost estimations is the capital investment required to develop the mine. This value is necessary to secure enough capital to open the mine and initiate revenue from the deposit. It is common practice to base the cost on a unit of capacity—for example, in terms of $/annual ton (tonne) of capacity. The following ranges are commonly used for the capital cost of various mineral operations (mine plant only):

	Coal Mines	
Surface	$40–150/ton/yr	($44–165/tonne/yr)
Underground	$100–200	(110–220)
	Metal Mines	
Surface	$75–200	(83–220)
Underground	$100–300	(110–330)

Obviously, these values provide only a rough idea of the real capital costs of a mine. In an actual mine budgeting analysis, the engineer would provide a more definitive estimate by considering the specific variables of the actual operation and developing a list of equipment and facilities that would be needed to start the mine. Similarly, a list of all openings and related development equipment to develop the mine would be prepared. The capital and development costs would then be compiled from these lists by constructing detailed cost estimates for each element in the mine plan.

4.7 EXPLOITATION: GENERAL STRATEGY

Exploitation, Stage 4 in the life of a mine, is not only the culmination of the three preceding stages, but the end process by which the three previous stages and the fifth stage, reclamation, are economically justified. Without production of ore (or coal or stone) in a substantial amount, there can be no opportunity for a mining venture to succeed, and the mine will be abandoned before it is activated.

Exploitation is the work of recovering mineral from the earth in economic amounts and delivering it to shipping or processing facilities. Although some exploration and development continue during Stage 4, the chief activity is production of minerals that will be marketed. In this production process, a number of extractive unit operations are employed, the primary ones constituting the *production cycle* and the secondary ones the *auxiliary* or support *operations*. We will examine these unit operations in detail in Chapter 5.

The method or methods chosen for exploitation of the mineral deposit defines the fourth stage in the life of the mine. Selection of the method(s) to be used is the key decision to be made in mine development. It must take into account many factors and may have to be refined and changed over time. For example, it may be perfectly logical for a large copper deposit to be mined first by the open pit method, then by block caving methods, and finally by solution mining methods. The initial choice of mining method made in the early stages of planning will require confirmation and refining before exploitation is to begin.

The strategy for conducting the exploitation stage of mining should be clear as mineral production begins. *The cardinal rule of exploitation is to select a mining method that matches the unique characteristics (natural, physical, geologic, social, political, etc.) of the mineral deposit being mined, subject to the requirements of safety, mineral processing, and the environment, to yield the overall lowest cost and return the maximum profit.* In stating this basic objective, safety must be accorded the highest level of concern. It is both socially and economically important that we do so. In addition, we must keep in mind that reducing the cost of mining alone is not our objective; it is the combined cost of mining, processing, and reclaiming the land that is to be minimized. Exploitation must be planned with this goal in mind.

4.8 EXPLOITATION: MINING METHOD

4.8.1 Factors in Selection

There are many factors, both quantitative and qualitative, that must be evaluated in the choice of a mining method. Lists of these factors have been provided by Boshkov and Wright (1973), Morrison and Russell (1973), Folinsbee and Clarke (1981), and Nicholas (1982, 1992b). The following are the primary variables that must be considered:

1. *Spatial characteristics of the deposit.* These factors play a dominant role in the choice of a mining method, because they largely decide the choice between surface and underground mining, affect the production rate, and determine the method of materials handling and the layout of the mine in the ore body.
   a. Size (especially height, thickness, and overall dimensions)
   b. Shape (tabular, lenticular, massive, or irregular)
   c. Attitude (inclination or dip)
   d. Depth (mean and extreme values, stripping ratio)
   e. Regularity of the ore boundaries
   f. Existence of previous mining
2. *Geologic and hydrologic conditions.* Geologic characteristics of the ore and surrounding country rock influence method selection, especially choices between selective and nonselective methods, and ground support requirements for underground mines. Hydrology affects drainage and pumping requirements, both surface and underground. Mineralogy governs solution mining, mineral processing, and smelting requirements.

   a. Mineralogy and petrography (e.g., sulfides vs. oxides in copper)
   b. Chemical composition (primary and secondary minerals)
   c. Deposit structure (folds, faults, discontinuities, intrusions)
   d. Planes of weakness (joints, fractures, shear zones, cleavage in minerals, cleat in coal)
   e. Uniformity of grade
   f. Alteration and weathered zones
   g. Existence of strata gases

3. *Geotechnical (soil and rock mechanics) properties.* The mechanical properties of ore and waste are key factors in selecting the equipment in a surface mine and selecting the class of methods (unsupported, supported, and caving) if underground.
   a. Elastic properties (strength, modulus of elasticity, Poisson's ratio, etc.)
   b. Plastic or viscoelastic behavior (flow, creep)
   c. State of stress (premining, postmining)
   d. Rock mass rating (overall ability of openings to stand unsupported or with support)
   e. Other physical properties affecting competence (specific gravity, voids, porosity, permeability, moisture content, etc.)

4. *Economic considerations.* Ultimately, economics determines whether a mining method should be chosen, because economic factors affect output, investment, cash flow, payback period, and profit.
   a. Reserves (tonnage and grade)
   b. Production rate (output per unit time)
   c. Mine life (total operating period for development and exploitation)
   d. Productivity (tons or tonnes/employee hour)
   e. Comparative mining costs of suitable methods
   f. Comparative capital costs of suitable methods

5. *Technological factors.* The best match between the natural conditions and the mining method is sought. Specific methods may be excluded because of their adverse effects on subsequent operations (e.g., processing, smelting, environmental problems, etc.).
   a. Recovery (proportion of the ore that is extracted)
   b. Dilution (amount of waste that must be produced with the ore)
   c. Flexibility of the method to changing conditions
   d. Selectivity of the method (ability to extract ore and leave waste)
   e. Concentration or dispersion of workings
   f. Ability to mechanize and automate
   g. Capital and labor intensities

6. *Environmental concerns.* The physical, social, political, and economic climate must be considered and will, on occasion, require that a mining method be rejected because of these concerns.
   a. Ground control to maintain integrity of openings
   b. Subsidence, or caving effects at the surface

c. Atmospheric control (ventilation, air quality control, heat and humidity control)
d. Availability of suitable waste disposal areas
e. Workforce (availability, training, living, community conditions)
f. Comparative safety conditions of the suitable mining methods

The selection of a mining method is shifting from an activity that is primarily an art to one that is primarily science. This can be seen in the publications that discuss mining method selection. Boshkov and Wright (1973), Morrison and Russell (1973), and Folinsbee and Clarke (1981) emphasize the subjective variables, whereas Nicholas (1982, 1992b) is predominantly quantitative in approach. The change in emphasis will be discussed in Chapter 14, "Summary of Mining Methods and Their Selection." For now, we can discuss the general procedures of mining method selection.

4.8.2 Guidelines and Procedure

In choosing a mining method, experience with numerous methods will play a major role. The reason is that some of the factors outlined earlier cannot be evaluated on a quantitative basis. Reaching an optimal decision, however, is facilitated and strengthened by the use of quantitative and engineering evaluation, including operations research methods, aided by computerized information gathering and data analysis. It is when quantitative and intuitive methods are in total agreement that the mining method selected will be most successful.

Engineering evaluation in the selection of a mining method is often carried out on three levels (Folinsbee and Clarke, 1981). First, in the *conceptual study*, the physical characteristics and production potentials of a number of mining methods, layouts, and systems are assessed. Next, in the *engineeering study*, the preceding concepts are quantified and compared, resulting in firm designs and costs. Finally, in the *detailed design study*, drawings and specifications for construction for the preferred method are prepared. The result is a final engineering report on which the investment decisions, equipment purchases, and construction schedule are based. During this three-step process, the potential for upgrading any mining method should be considered. The decision makers should consider whether any bigger or better equipment has come on the scene, whether any new variations in the method are being tested to improve output, or whether any new mechanization or automation technologies can enhance the production or cost performance of the chosen mining method. New and better mining methods have evolved through this pprocess.

4.8.3 Classification of Mining Methods

There are a variety of schemes to classify mining methods and aid in selection (Peele, 1941; Young, 1946; Lewis and Clark, 1964), of which the oldest has been most widely copied and may still be the most useful. The basis for classification

TABLE 4.1 Classification of Mining Methods

Locale	Class	Subclass	Method	Commodities	Relative Cost (%)
Surface	Mechanical	—	*Open pit mining	Metal, nonmetal	5%
			Quarrying	Nonmetal	100
			*Open cast (strip) mining	Coal, nonmetal	10
			Auger mining	Coal	5
	Aqueous	Placer	Hydraulicking	Metal, nonmetal	5
			Dredging	Metal, nonmetal	<5
		Solution	Borehole mining	Nonmetal	5
			*Leaching	Metal	10
Underground	Unsupported	—	*Room-and-pillar mining	Coal, nonmetal	20
			Stope-and-pillar mining	Metal, nonmetal	10
			Shrinkage stoping	Metal, nonmetal	45
			*Sublevel stoping	Metal, nonmetal	20
	Supported	—	*Cut-and-fill stoping	Metal	55
			Stull stoping	Metal	70
			Square-set stoping	Metal	100
	Caving	—	*Longwall mining	Coal	15
			Sublevel caving	Metal	15
			*Block caving	Metal	10

*Asterisks indicate the most important and commonly used methods.

in these schemes is some subjective combination of the spatial, geologic, and geotechnical factors we have previously discussed. More recent schemes (Morrison and Russell, 1973; Boshkov and Wright, 1973; Thomas, 1973; Hamrin, 1982; and Nicholas, 1982, 1992b) have added more quantitative elements to the approach, but use the same basic approach as Peele. We will refer to some of these schemes again when we discuss mining method selection in Chapter 14. One of the disconcerting aspects of all of these approaches is the orientation of the scheme toward the underground noncoal methods.

For our purposes, we will devise a classification of mining methods that (1) is generic (that is, applies to both surface and underground and to all types of mineral commodities) but not excessively complicated, (2) includes all current major mining methods, and (3) recognizes the established major class distinctions and relative costs. This classification scheme was introduced in Chapter 1 when exploitation methods were first discussed. The major categories used in our classification are *locale* (surface or underground), *class*, *subclass*, and *method*. Table 4.1 outlines the classification and provides basic information on the typical commodities mined and relative costs.

Features of the various methods are yet to be examined. These include a depiction of the method, sequence of development, cycle of operations, deposit

MINING METHODS NOTEBOOK

Locale:	Class:	Method:

Sketches:

PLAN OR SECTION	SECTION

Sequence of Development:

Cycle of Operations:

Deposit Conditions:

ore strength	deposit size
rock strength	ore grade
deposit shape	ore uniformity
deposit dip	depth

Advantages:	Disadvantages:

Production Rate:	Relative Cost:

Examples:

FIGURE 4.3. Sample form for mining methods notebook page.

conditions, advantages, disadvantages, production rate, relative cost, and examples. A suggested mining methods summary form is provided in Figure 4.3. Surface mining methods will be outlined in detail in Chapters 7 and 8, underground methods in Chapters 10 to 12, and innovative methods in Chapter 13. A summary of all the methods and the procedures for selecting a mining method will appear in Chapter 14. In this text, we will discuss all of the methods listed in Table 4.1. However, the heaviest emphasis will be on the most commonly used methods. Eight of these have been marked with an asterisk in the table. Three are surface methods and five are underground. Together, these methods account for about 90% of U.S. solid minerals production.

4.9 EXPLOITATION: MANAGEMENT AND COST

4.9.1 Mine Management

By the time the exploitation stage of a mine begins, the organization of its management and workforce should be complete. Recruitment of the staff and labor force is normally carried out toward the close of the development stage and is synchronized with the projected opening of the mine. The management structure employed by most mining companies is a *staff-and-line organization* (Boyd, 1973). *Line activities* are those normally considered to be primary functions of the organization such as operations (mining, processing, and smelting), sales, finance, and exploration (Ward and Britton, 1992). *Staff activities* are those functions that support the production activities of the company. Because they are primarily advisory and general support activities, the staff groups may interact extensively with the line managers but report directly to upper management. This management organization keeps the staff activities and line activities from unduly competing with each other for organizational resources and fosters a more cooperative environment.

An example of a staff-and-line management organization for a large mine is shown in Figure 4.4 (Boyd, 1973). Note that the staff groups do not have authority to direct any aspect of the line functions but report to the general manager. In addition to the staff functions shown in the diagram, there may be staff groups to handle public relations, legal issues, employee benefits, and land acquisition. Line managers normally have authority to direct the entire operational structure below them. They usually welcome help from the staff groups but will consider their contributions as advisory in most cases.

The details of mine organization, administration, and operation are best left for an advanced text. For other sources of information on mine management, refer to Britton (1981), Sloan (1983), and Ward and Britton (1992). Students should recognize that mine management is a relatively complex human science and one that must be analyzed carefully in our competitive world. It is also important for students to realize that their future can be in either the staff or the line functions. The two types of positions are somewhat different, but mining engineering graduates fit well in either category.

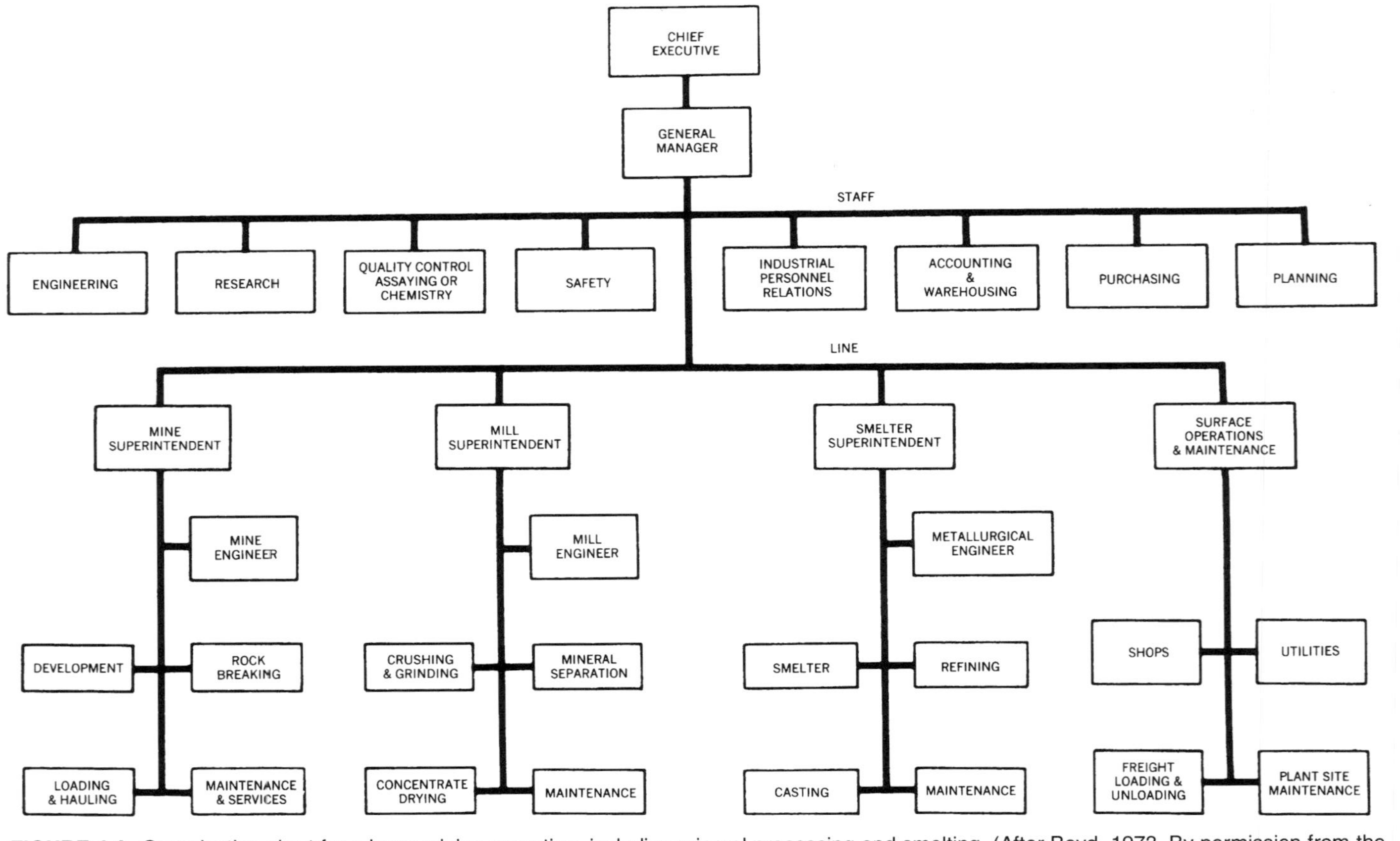

FIGURE 4.4. Organization chart for a large mining operation, including mineral processing and smelting. (After Boyd, 1973. By permission from the Society for Mining, Metallurgy, and Exploration, Inc., Littleton, CO.)

4.9.2 Mining Costs

Building on the discussion of cost estimation introduced during mine development in Section 4.7, we can extend the process to include exploitation costs as well. At this point, it is important to understand more of the nomenclature of costs. The sum of all the costs associated with bringing a mine into production through the stages of prospecting, exploration, development, exploitation, and reclamation is called the *direct mining cost*. If calculated on a gross basis, it is a *total cost*; if on a unit ($/ton, $/tonne) basis, it is a *unit cost*. In addition, *indirect mining cost* is an overhead that usually includes an allowance of 5% to 10% for administration, engineering, and other nonitemized services. To find the *overall mining cost*, it is necessary to sum the two, either on a total or unit basis as follows:

$$\text{Overall mining cost} = \text{direct cost} + \text{indirect cost}$$

If all other costs (processing, smelting, transportation, etc.) are added to this figure, then the *overall production cost* results.

To determine the range of overall mining costs that may exist in a given ore deposit, we can use the data in Table 1.3. Summing all five stages in the mine life, and assuming a 20-year mine life for Stage 4, we obtain the following:

Stage	Unit Cost	Total Cost
1. Prospecting	$0.05–1.00/ton	$0.2–10 million
2. Exploration	$0.20–1.50/ton	$1.0–15 million
3. Development	$0.25–10.00/ton	$10–500 million
4. Exploitation	$2.00–150.00/ton	$100–1000 million
5. Reclamation	$0.20–4.00/ton	$1–20 million
Overall mining costs	$2.70–166.50/ton ($3.00–180/tonne)	$112.2–1545 million

The final figures show the range in the unit costs of mining as well as the range in the total costs. Students should note that detailed cost estimates must be prepared for every mining project to pinpoint the costs.

Because the determination of costs is imperative for the mining engineer, cost estimation examples occur throughout this text. Actual costs are difficult to obtain because companies often consider them proprietary. In addition, costs fluctuate unpredictably over time with price inflation, changes in labor rates, and technological progress. Undated costs used in this text may be taken as current at the time of the writing. To avoid the use of *absolute costs* ($/ton or $/tonne) in our discussions of mining methods—which rapidly become obsolete—it will be more helpful if we employ *relative costs* (in %) for comparisons. Notice that Table 4.1 adopts that standard. To utilize this relative cost standard, the mining cost of the most expensive methods (quarrying and square-set stoping) is established at a 100% relative cost; the costs of all other methods are expressed relative to this standard. Thus, the cheapest mining method is dredging (<5%). The relative costs of some of the methods have changed considerably since the first edition of this text. The labor-intensive methods continue to increase in absolute cost, whereas those methods that have been more intensively mechanized and automated may have decreased in absolute cost. Thus, several of the mining methods show significantly lower relative costs in Table 4.1 than they did in the previous version of this table.

Relative costs should be more stable and useful than absolute costs. However, even relative costs are somewhat limited in value, as many of the mining methods have significant deviations in cost due to geologic conditions, physical dimensions of the deposit, and ground conditions in the production area. Relative costs thus represent average conditions that can be substantially in error for actual mines if converted to absolute costs. They must therefore be used with caution.

4.10 SPECIAL TOPIC: MINING METHODS NOTEBOOK

Probably the surest way to become well acquainted with the various mining methods is to (1) sketch each method, (2) describe how it is developed and operated, and (3) summarize its distinctive conditions and features. Figure 4.3 is provided as a basic summary of each mining method. It is suggested that each student use that form (or a suitable substitute) to write a one-sheet description for each mining method. Information on each method can be assembled from the instructor's lecture, from the text, and from other suitable references. It is the student's job to obtain the information for each mining method, to condense it, and to summarize it neatly. At the conclusion of the course, it is suggested that the forms for all the assigned methods be assembled into a Mining Methods Notebook. Thus, a lasting, well-engineered summary record is produced for future reference.

5

UNIT OPERATIONS OF MINING

5.1 FUNDAMENTAL UNIT OPERATIONS AND CYCLES

In several stages of mining (exploration, development, and exploitation), the activity is often performed in a cyclic manner using a series of fundamental steps to free and transport the mineral being mined. As introduced in Section 1.5, these basic steps are termed the *unit operations of mining*. If they are directly involved in mineral extraction, we call them *production operations*; *auxiliary operations* support production but are usually not directly part of it unless they are essential to worker safety or operating efficiency. Our interest in this chapter is primarily in those unit operations necessary during development and exploitation.

Two basic functions are required in the extraction of minerals: excavation and handling. For mineral resources that are harder to break, the excavation step requires that drilling (rock penetration) and blasting (rock fragmentation) be performed. Where drilling and blasting are employed, we generally call this step *rock breakage*. In softer deposits, the minerals can usually be obtained using mechanical digging devices. We normally call this process *excavation*. Handling, more accurately referred to as materials handling, is ordinarily performed in two steps, loading (often termed excavation) and hauling. If the mine is located underground, then hoisting may also be required. The four most common unit operations—drilling, blasting, loading, and haulage—are discussed as to principle in this chapter. Subsequent chapters will expand on these unit operations and demonstrate how they apply in specific mining methods.

The sequence of unit operations used to accomplish mine development or exploitation is called the *cycle of operations*, that is, the sequence of operations that is repeated over and over to produce the mineral commodity. The most common cycle of operations used for mine production is as follows:

$$\text{Basic production cycle} = \text{drill} + \text{blast} + \text{load} + \text{haul}$$

This cycle of operations is often altered to accommodate the type of equipment and technology best suited to the production process. Because mechanization is constantly being increased in mining operations, the unit operations are often greatly affected by the machinery used. For example, both continuous miners and rapid excavation equipment can replace the unit operations of drilling, blasting, and loading. The basic production cycle would then be simply to excavate and haul. As a second example, we can refer to underground mining operations where support of the surrounding rock becomes an integral part of the cycle of operations. In this case, the unit operations would be drilling, blasting, loading, hauling, and supporting. One of the objectives of mining is to mechanize and automate the unit operations so that they merge into a unitized process of mining. However, truly continuous and automated mining is still some years away. Thus, we must deal with the process of optimizing mining practice by efficient utilization of the unit operations.

5.2 DRILLING AND OTHER ROCK PENETRATION METHODS

5.2.1 Rock Breakage

The freeing or detaching of large masses of harder rock from the parent deposit is termed *rock breakage*. Several historical events contributed to the success of miners in their quest to break rock for mining or construction projects. Some of these are discussed in Section 1.3. The first advance was the discovery in prehistoric times that fire building and water quenching can break rock. This produced thermal stresses to overcome the cohesive strength of the rock. Later it was discovered that black powder, dynamite, ammonium nitrate and fuel oil, and water-based explosives could be used to break rock more efficiently.

Mining today has the same objectives it had in ancient times and employs the same basic elements of the production cycle, breakage and handling. It was with the introduction of blasting into the production cycle that mining became much more efficient in the production of minerals. Today, most progress is being made in the mechanization of certain mineral production operations. Under the right conditions, mechanized rock breakage and haulage without explosives is the most continuous and cost-efficient method of mineral extraction.

In traditional mining practice, two different types of rock breakage operations are performed. In *rock penetration* (drilling, boring, or channeling), a directed hole or kerf is formed, usually mechanically but sometimes hydraulically or thermally, (1) for the placement of explosives for blasting or to produce a free face for blasting, (2) to produce a finished mine opening, or (3) to extract a mineral product of desired size and shape (for example, dimension stone). In contrast, *rock fragmentation* attempts to break rock into manageable sizes, normally by chemical energy in blasting or by mechanical, hydraulic, or other more novel applications of energy.

Although operating on different scales, penetration and fragmentation function by applying energy through similar mechanisms to break rock. The very rapid application of energy is critical in producing rock failure in many of these processes. Thus, we normally apply *dynamic loading* (in preference to *static loading*) to accomplish the penetration or fragmentation. It is ordinarily assumed that the weight W of the rock broken, using any given rock breakage method, is proportional to the energy E consumed in the process:

$$W \propto E \tag{5.1}$$

In drilling and other penetration processes, the rate of advance R, the penetration per unit time, is more important than the amount of rock broken. Hence, the basic relation becomes

$$R \propto dE/dt \tag{5.2}$$

where dE/dt is the time rate of application of the energy to the rock. Because dE/dt is also the power P consumed in the process, we may write Eq. 5.2 in the form

$$R \propto P \tag{5.3}$$

which indicates that — in drilling, cutting, boring, or any other similar process — the penetration rate is directly proportional to the power applied to the process.

In modeling rock breakage processes, we must also understand the characteristics and behavior of rock. We know that the rock parameters (i.e., compressive, shear, and tensile strengths) resist failure induced by the applied loads. Because the strengths decrease in the order stated, it is advantageous to apply the loads to overcome the smallest strength (tensile) and the next smallest strength (shear) before the rock is broken by compressive loads. However, that is normally not easy to do in drilling operations, and the rock is often broken by the application of compressive forces. In an era of increasingly expensive energy, drill designers are paying more attention to Eqs. 5.1 and 5.3 in order to conserve energy.

5.2.2 Principles of Rock Penetration

Rock penetration methods can be classified on several bases. These include the size of hole, method of mounting, and type of power. The scheme outlined in Table 5.1 (Hartman, 1990) is based on the form of attack on the rock and the

TABLE 5.1 Classification of Rock Penetration Methods, Based on Form of Attack[a]

Form of Energy Application	Method	Machine
Practical		
Mechanical (Drilling)	Percussion	
	Drop tool	Churn or cable-tool drill
	Hammer	Rock drill, channeler
	Rotary, drag-bit	
	Blade	Auger or rotary drill, boring
	Stone-set	Diamond drill
	Sawing	Wire-rope, chain, or rotary saw
	Rotary, roller-bit	Rolling-cutter drill, boring
	Rotary-percussion	
	Hammer	Rock drill (independent rotation)
	Rotary	Rolling-cutter drill (superimposed percussion)
Novel		
Thermal	Flame	Jet piercer, jet channeler
	Plasma	Plasma torch
	Hot fluid	Rocket
	Fusion	Subterrene
	Freezing	(Conceptual)
Fluid	Jet	Hydraulic jet, monitor, cannon
	Erosion	Pellet-impact or abrasion drill
	Bursting	Implosion drill
	Cavitation	Cavitating drill
Experimental		
Sonic	Vibration	High-frequency transducer
Chemical	Explosion	Shaped charge, capsule, projectile
	Reaction	Rock “softener,” dissolution
Electrical	Electric arc or current	Electrofrac drill
	Electron beam	Electron gun
	Electromagnetic induction	Spark drill
Light	Laser	Electromagnetic radiation beam
Nuclear	Fission	(Conceptual)
	Fusion	(Conceptual)

[a] Energy applications listed in approximate order of present practicality.

Source: Hartman (1990). By permission from the Society for Mining, Metallurgy, and Exploration, Inc., Littleton, CO.

TABLE 5.2 Classification of Rock Fragmentation Methods Based on Form of Energy Application[a]

Form of Energy Application	Method	Agent or Machine
Chemical	Blasting	High explosive, blasting agent, liquid oxygen (LOX), black powder
Mechanical	Pneumatic	Compressed air or carbon dioxide cylinder
	Ripping	Ripper blade, dozer blade
	Impact	Hydraulic impact hammer, drop ball
Fluid	Mining (soil)	Hydraulicking (monitor)
	Mining (rock)	Hydraulic jet
Electrical	Electric arc or current	Electrofrac machines

[a] Energy applications listed in approximate order of present practicality.

mode of energy application to the rock. This is a general scheme, which includes all forms of penetration, with conventional drilling tools being the most important. However, it contains none of the traditional fragmentation methods, which are outlined in Table 5.2. A discussion of the major categories of penetration methods is in order.

5.2.2.1 Mechanical Attack. The application of mechanical energy to rock masses is the primary method used in traditional drilling. It utilizes either percussive or rotary action, or a combination of both, at the bottom of the hole to penetrate the rock mass. There is a difference in the application of energy between the standard percussion drill and the rotary-percussion drill. The percussion drill employs a drill rod rotational system that indexes each percussive blow of the hammer at a different spot, so as to drill more effectively. However, the rotation does not involve any significant amount of energy application to the rock. The rotary-percussion drill uses an independent rotational system that applies a shearing force to the bottom of the hole by means of the rotation. The rotational energy helps to break additional cuttings from the hole and increases drilling efficiency. In surface mines, roller-bit rotary and large percussion or rotary-percussion drills are in use, with roller-bit rotaries heavily favored for most applications. Underground, percussion drills are used for most applications in hard rock and drag-bit rotaries are used in soft-rock applications.

5.2.2.2 Thermal Attack. Thermal drilling is one of the novel excavation methods that have found practical application in rock penetration. The jet piercer drill, which uses fuel oil and oxygen to produce a flame capable of spalling rock, has been applied in the taconite iron ore range as a practical drilling tool. It has also found some use in the channeling of rock in dimension stone quarries and in a few other hard-rock mining applications. However, it has never been responsible for a large percentage of the minerals produced. The thermal methods have also recently decreased in popularity because of the increased effectiveness of mechanical drilling equipment.

5.2.2.3 Fluid Attack. Using a hydraulic fluid to attack a mineral deposit has been a productive method of mining for more than a century. Its main application has been in placer deposits, where hydraulic monitors have been used. High-pressure water jets have also been applied more recently to the mining of coal, gilsonite, and other consolidated materials. In these cases, however, the action is more fragmentation than penetration. However, it is now well established that fluid attack is capable of drilling rock effectively, both alone and as an assist to a mechanical drill. The applications so far have been confined primarily to specialty cutting and drilling operations.

5.2.2.4 Other Novel Methods. Although other methods of penetration have been applied to rock in experiments, they are nearly all considered to be of scientific, rather than practical, value at the present time. The field of novel drilling techniques has been explored extensively by Maurer (1968, 1980). He continues to pursue the more exotic drilling methods for their practical applications and remains optimistic that some of these methods may be important in the future.

5.2.3 Drilling

Most drilling is employed in mining for the placement and efficient use of explosives. At times, it is used for sampling the mineral deposit during exploration, for the placement of rock bolts, anchors, and cable reinforcements, and for the placement of electrical and communications lines and water pipes in mines. The predominant use of drilling in coal mines today is for the placement of roof bolts. When holes are used for the placement of explosives, the operation is known as *production drilling*. We now look at some of the principles of drill selection and utilization.

5.2.3.1 Operating Components of the System. The drilling system usually consists of four components that work together to penetrate the rock:

1. The *drill*, the mechanical device (and its carriage, if applicable) that converts energy from its original source (electrical, pneumatic, hydraulic, or combustion engine) into rotational and/or percussive energy to penetrate the rock
2. The *drill rod* (also called a steel, stem, or pipe) that transmits energy from the drill to the bit

3. The *bit*, which attacks the rock with rotational and/or percussive action
4. The *circulation fluid* that cleans the hole, controls dust, cools the bit, and (at times) stabilizes the hole

The first three are the physical components of the drilling system, accomplishing the penetration of the rock, and the fourth enhances the efficiency of the drill by removing cuttings from the bottom of the hole.

Much has been done over the last few decades to make drilling tools more efficient by reducing the energy losses in transmission. This has resulted in the use of downhole (in-the-hole) drills of both the large percussion type and the roller-bit rotary (electrodrill and turbodrill) type, although the latter has found application mainly in oil well drilling. The second type replaces mechanical energy with fluid or electrical energy, which usually results in more energy reaching the bit and faster drilling.

5.2.3.2 Mechanics and Performance Factors of Penetration. As indicated previously, there are only two basic ways to attack rock mechanically—percussion and rotation—and all commercial drilling methods to be discussed utilize percussion or rotation principles or combinations of them. A graphical depiction of the different mechanical methods of drilling rock is shown in Figure 5.1. Note that the percussion drill attacks the rock through compression, the rotary (drag-bit) drill by shearing or applying tension to the rock, and the rotary-percussion drill by utilizing a combination of all three methods. More information on the effect of mechanical attack on rock can be found in Maurer (1967) and Clark (1987).

The resistance of rock to penetration by drilling tools is termed its *drilling strength*, an empirical property; it is not equivalent to any of the well-known strength parameters. The drilling strength and the wear on the bits are extremely important in the determination of the drilling cost. Other factors that affect drill performance can be categorized into four groups:

1. *Operating variables.* These affect the four components of the drilling system (drill, rod, bit, and fluid). They are largely controllable and

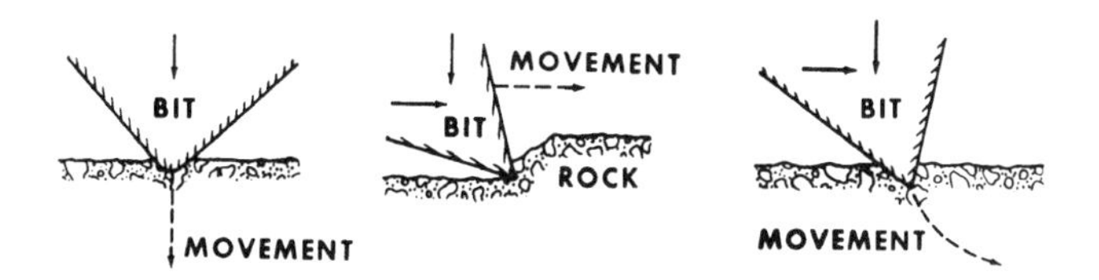

FIGURE 5.1. Types of drilling action in mechanical attack on rock: (*left*) percussion, (*center*) drag-bit, and (*right*) combination (roller-bit and rotary-percussion) drills. (After Maurer, 1967. By permission from the Society for Mining, Metallurgy, and Exploration, Inc., Littleton, CO.)

include factors that affect the tools (drill power, blow energy and frequency, rotary speed, thrust, and rod design) and drillhole cleaning (fluid properties and flow rate).
2. *Drillhole factors.* These include hole size, length, and inclination; they are dictated by outside requirements and thus are largely uncontrollable. Hole diameters in surface mining are typically 6 to 18 in. (150 to 450 mm); underground, they range from 1.5 to 7 in. (40 to 175 mm).
3. *Rock factors.* These consist of properties of the rock, geological conditions surrounding the rock, and the state of stress acting on the drillhole. Often referred to as *drillability factors*, they determine the strength of the rock and limit drill performance. Because these factors are a result of the geologic environment, they are largely uncontrollable.
4. *Service factors.* These variables include labor and supervision, power supply, job site conditions, weather, and so forth. Except for labor and supervision, they are independent factors and cannot often be affected by the drill operator.

Because the operating variables are the most controllable, a complete knowledge of the interaction of the drilling variables and the drilling efficiency is very important. A variety of sources giving information on the effects of drilling parameters on efficiency are available in Clark (1987), Hartman (1990), Lopez Jimeno et al. (1995), and Karanam and Misra (1998). Among the most important parameters to consider are drill power, thrust, torque, rotary speed, blow energy, blows per minute (BPM), and the flow of the hole-flushing fluid. In addition, it is important to choose the proper bit type and rod dimensions to provide a good match with the drill.

5.2.3.3 Drill Selection. In selecting the optimal drilling system, it is necessary to evaluate drill performance. Performance can be measured in a number of ways. The most commonly measured parameters include the following:

1. Process energy and power consumption
2. Penetration rate
3. Bit wear (life)
4. Cost (ownership + operating = overall)

These parameters are all important, but the ultimate measure of performance is cost. If a drill has an excellent penetration rate but is not cost-effective, then an alternative system should be sought. Note also that the overall goal of mining is the minimization of all rock breakage, handling, and processing costs. Drilling affects blasting, which in turn affects excavation and processing costs. Thus, it is important to realize that minimizing overall cost is the primary goal in the choice of the drilling system, and that some drilling efficiency may have to be sacrificed to enhance blasting, excavation, or processing efficiencies.

TABLE 5.3 Application of Drilling and Penetration Methods to Different Types of Rock

Drilling Method	Rock Type/Drillability			
	1 Soft (Shale, Weathered Limestone, Coal)	2 Medium-Hard (Limestone, Weathered Sandstone)	3 Hard (Granite, Chert)	4 Very Hard (Taconite, Quartzite)
Hydraulic jet	×	×		
Rotary, drag-bit	×	×		
Rotary, roller-bit	×	×	×	
Rotary percussion		×	×	
Percussion		×	×	×
Thermal jet piercing			×	×

Source: Hartman (1990). By permission of the Society for Mining, Metallurgy, and Exploration, Inc., Littleton, CO.

In choosing a drilling system, a general qualitative guide such as that in Table 5.3 may be of great help. It allows the decision maker to choose systems that are feasible for the conditions inherent in the drilling situation. *In general, the lowest costs are obtainable in soft rock with rotary drag-bit drilling, in medium and hard rock with rotary roller-bit and rotary-percussion drilling, and in very hard rock with percussion drilling.* It is therefore wise to keep this in mind in choosing among candidate systems. However, it is then important to conduct an engineering evaluation of each candidate system and come up with a performance and cost estimate. The following steps are based on those outlined by Capp (1962):

1. Determine and specify the conditions under which the machine will be utilized, such as the job-related factors (labor, site, weather, etc.) and the geologic variables (rock type and strength, orientation of the geologic materials, existence of faults, etc.). Limit the conditions of use to those that are safe.
2. State the objectives for the rock breakage phases of the production cycle in terms of tonnage, fragmentation, throw, vibrations, and the like. For surface mining, consider the loading and hauling restrictions, pit slope stability, crusher capacity, pit geometry, and other pertinent parameters. Underground, the face area, hole size, hole inclination, round length, and production goal are among the variables that must be considered.
3. Based on the blasting requirements, design the drillhole pattern for surface mining or the round for underground blasting.

4. Determine the drillability factors for the rock anticipated, and identify the drilling methods that appear feasible. Manufacturers may be able to help by performing rock drillability tests and recommending drills and bits.
5. Specify the operating variables for each system under consideration, including the drill, rod, bit, and circulation fluid factors.
6. Estimate performance parameters, including machine availability and costs, and compare. Specify the power source and estimate costs. Major cost items are bits, drill depreciation, labor, maintenance, power, and fluids. Bit life and costs are important but difficult to predict without empirical testing.
7. Select the drilling system that meets all requirements, including safety, and has the lowest overall cost.

The general trends in drills are toward greater efficiency, bigger drill rigs, more hydraulic power, more automation, and increased ability to handle harder rocks as a result of better technology. The most striking change over the last 20 years has been the growth of hydraulic power in percussion and rotary-percussion drills. A very large percentage of all underground percussion drills are now hydraulically powered, and surface drills are also rapidly changing over. Lopez Jimeno et al. (1995) have compared the advantages and disadvantages of hydraulic drills. The primary advantages are lower energy consumption, greater drilling speed, lower rod costs, more flexibility, easier automation, and less noise. The disadvantages are higher capital cost and increased maintenance costs. For most situations, the advantages far outweigh the disadvantages.

The use of hydraulic jets or thermal drills is not a common practice. The availability of better mechanical drills and problems with power supply contribute to their infrequent use in mining. They can be justified on an economic basis if the conditions are not well matched to more conventional drilling methods. However, their economic value must be established before committing capital for their use.

5.2.4 Kerf Cutting and Mechanical Excavation

The development of tungsten carbide cutting elements in 1945 (see Table 1.2) opened up a whole new set of possibilities for extending percussion and rotary drilling principles to the penetration of geologic materials on a larger scale. Some of the tools available combine both penetration and fragmentation principles. These methods have enabled successful cutting of kerfs (channels) and excavation of entire faces for a variety of mining applications. Combined with flame jet tools, wire saws, diamond saws, and conventional mining tools, the technology for cutting kerfs and faces now provides an impressive list of possibilities:

1. Kerf cutting
   a. Coal and soft nonmetallic minerals: chainsaw-type cutting machines with either a fixed cutter bar or a universal (moveable) cutter bar
   b. Dimension stone: channeling machines (percussion or flame jet), wire saws, circular saws with diamond blades
2. Full-face excavating
   a. Underground
      (i) Continuous miners and longwall shearers in coal or soft nonmetallics
      (ii) Boom-type miners (roadheaders) in soft to medium rocks
      (iii) Rapid excavation equipment (tunnel borers, raise borers, and shaft-sinking rigs) for soft to medium-hard rock
   b. Surface
      (i) Rippers for very compact soil, coal, and weathered or soft rock
      (ii) Bucket-wheel and cutting-head excavators for soil or coal
      (iii) Augers and highwall miners for coal
      (iv) Mechanical dredges for placers and soil

Better technology in cutting tools has provided new opportunities for penetrating rock without the use of explosives. The preceding outline lists the variety of applications currently available to the mining industry. The winning of coal with cutting tools is now the standard of the industry, and rapid excavation methods are gaining in their range of applications. This area of rapid excavation is likely to be even more productive in the future.

5.3 BLASTING AND ROCK FRAGMENTATION

5.3.1 Principles of Rock Fragmentation

As discussed previously, *rock fragmentation* is the breakage function carried out on a large scale to fragment masses of rock. In both the mining and construction industries, *blasting* is the predominant fragmentation method employed, but other techniques are becoming more common. Based on distinctions in the way they apply energy to break rock, a classification of rock fragmentation methods is given in Table 5.2. Electrical fracturing of rock has been used sparingly for secondary breakage of boulders in surface mines (Maurer, 1980); the other methods are all in common use. However, only blasting using chemical explosives has widespread use for all consolidated materials in both surface and underground mining. Excluded are excavating machines, which, though they may fragment soil and rock, primarily perform a loading function (Table 5.6), and continuous excavating machines, which mainly produce an opening and are classified as penetration methods (Table 5.1).

5.3.2 Theory of Explosives

5.3.2.1 Nature of Explosives. An *explosive* is any chemical compound, mixture, or device, the primary objective of which is to function by explosion (Hopler, 1998). The decomposition of the explosive is a high-velocity exothermic reaction, accompanied by the liberation of vast amounts of energy and hot gases at tremendous pressure. The process is termed *detonation* if the propagation speed of the reaction through the explosive mass is supersonic. For regulatory purposes, a detonating explosive is a *high explosive* if it can be initiated by a #8 blasting cap (Hopler, 1998). A *blasting agent* is a chemical material that can detonate under the right impulse but meets prescribed criteria for insensitivity to initiation (Hopler, 1998). Blasting agents typically possess a velocity of detonation lower than that of a high explosive. They also require less stringent storage and transportation standards than do high explosives because they are less hazardous during handling and transportation. The decomposition is called *deflagration* and the chemical agent a *low explosive* if the speed of the reaction is subsonic. Black powder (also known as gunpowder) is a low explosive. Detonation may properly be termed an explosion, accompanied by the formation of a shock wave. Deflagration is very rapid burning, but not an explosion and not associated with a shock wave.

Explosives break rock as a result of both (1) the impact of the shock wave and (2) the expanding effect of the high-pressure gas formed during the detonation. The proportion of the breakage attributed to these two factors is dependent on the properties of the explosive and of the rock. Four of the key explosive properties are (1) energy density, (2) bulk density, (3) rate of energy release, and (4) pressure-time history of the gas generated. Important rock properties are (1) density and porosity, (2) strength, (3) energy absorption properties and modulus of elasticity, and (4) rock structure, including jointing, bedding, fractures, alteration, and so forth (Clark, 1968; Hemphill, 1981).

5.3.2.2 Detonation Zone Effects. During an explosion, the chemical reaction of an explosive produces a detonation reaction that propagates through the charge and into the surrounding rock. Figure 5.2 (Dick et al., 1983) illustrates the explosion and detonation pressures when the detonation or reaction zone has reached the midpoint of the explosive mass (borehole charges are typically long, cylindrical columns). The primary reaction occurs in the zone bounded by the shock front and the Chapman-Jourguet (C-J) plane shown in the diagram. This zone is very narrow in a high explosive that has a small *critical diameter* (smallest diameter of a cartridge that can be detonated).

The pressures in Figure 5.2 produce the explosive forces applied to the rock. A comparison is given for a slurry explosive and a slurry blasting agent. The shock (stress) wave moving out from the explosive material creates the initial or *detonation pressure* P_d. It is this pressure that gives the explosive its shattering action in breaking rock. A low explosive generates no shock wave and hence has no P_d. A sustained or *explosion pressure* P_e, also called the

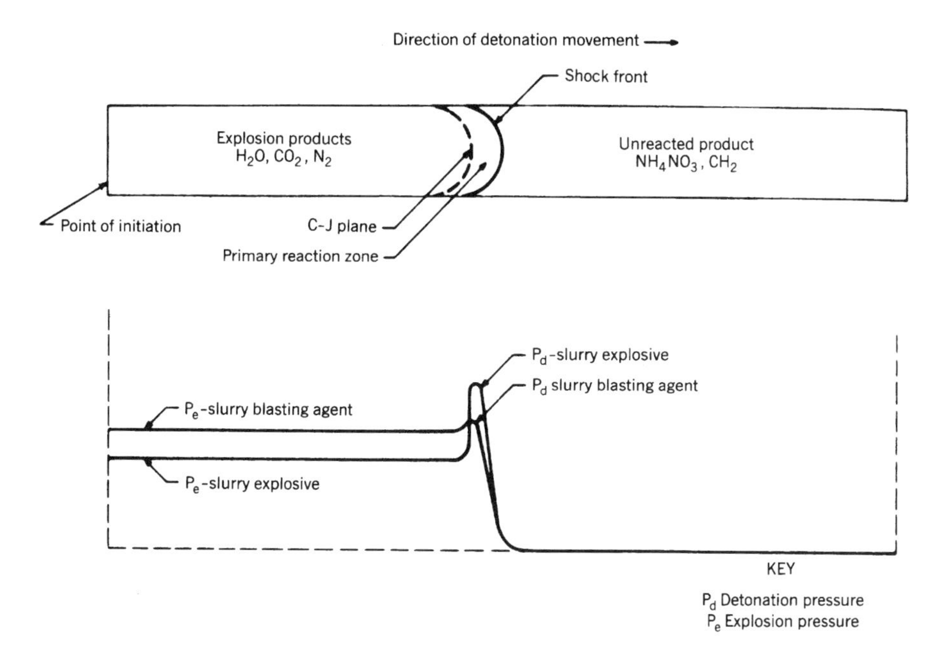

FIGURE 5.2. (*Top*) Propagation of detonation zone through an explosive charge. (*Bottom*) Resulting pressure profiles in rock produced by slurry blasting agent and slurry explosive. (After Dick et al., 1983.)

borehole gas pressure, follows the detonation. Note that the slurry blasting agent has a greater gas pressure and is more capable of heaving the broken rock, whereas the slurry explosive has a higher detonation pressure and is more capable of shattering hard rock.

5.3.2.3 Chemical Reactions of Explosives. Explosives consist of *oxidizers* and *fuels* mixed together in a proper proportion to produce the desired violent chemical reaction. Relatively inexpensive oxidizers and fuels are generally used to produce a cost-efficient mixture. Most ingredients used consist of the elements oxygen, nitrogen, hydrogen, and carbon, plus certain metallic elements (aluminum, magnesium, sodium, calcium, etc.). The primary criterion of efficient energy release is the oxygen balance. *Zero oxygen balance* is the point at which an explosive has sufficient oxygen to completely oxidize all the contained fuels but no excess oxygen to react with the contained nitrogen (Clark, 1968, 1987).

There are two reasons for the careful balancing of the oxygen-fuel mixture: (1) The energy output is optimized, and (2) the formation of toxic chemical gases such as oxides of nitrogen, carbon monoxide, methane, and others is minimized. An equation can be written for the oxygen balance in an explosive reaction if the starting and ending products are known. For example, if only oxygen, hydrogen, and carbon are involved, and the reaction forms oxygen, carbon dioxide, and water, the proper equation is

$$\text{OB} = \text{original oxygen content} - \text{oxygen formed into carbon dioxide} - \text{oxygen formed into water} \quad (5.4)$$

This is normally written in the following form:

$$\text{OB} = \text{O}_0 - 2\text{C}_0 - 0.5\text{H}_0 \quad (5.5)$$

where OB represents the oxygen balance (g-atoms/kg) and O_0, C_0, and H_0 represent the number of g-atoms/kg of the oxygen, carbon, and hydrogen (respectively) in the original explosive mixture. Note that the coefficients of C_0 and H_0 in Eq. (5.5) represent the number of atoms of oxygen required to satisfy one atom of carbon or hydrogen in the reaction. The equation becomes slightly more complicated if metals are present in the original mixture. For example, assume that some aluminum, calcium, and sodium are present in the explosive, and that they form Al_2O_3, CaO, and Na_2O during the reaction. Then the oxygen balance equation becomes

$$\text{OB} = \text{O}_0 - 2\text{C}_0 - 0.5\text{H}_0 - 0.667\text{Al}_0 - \text{Ca}_0 - 0.5\text{Na}_0 \quad (5.6)$$

The energy released during the reaction is obtained by calculating the difference in the heats of formation of the ingredients and the products. The consequences of departing from zero oxygen balance can be illustrated by the evaluation of three different ammonium nitrate (AN)–fuel oil (FO) mixtures (ANFO) (Dick et al., 1983). In these idealized equations, fuel oil is represented approximately by the formula CH_2, and the energy release is expressed in kcal/kg.

1. 94.5% AN plus 5.5% FO (oxygen-balanced)

   $$3\text{NH}_4\text{NO}_3 + \text{CH}_2 \rightarrow 7\text{H}_2\text{O} + \text{CO}_2 + 3\text{N}_2 + 930 \text{ kcal/kg}$$

2. 92.0% AN plus 8.0% FO (fuel excess)

   $$2\text{NH}_4\text{NO}_3 + \text{CH}_2 \rightarrow 5 \text{ H}_2\text{O} + \text{CO} + 2\text{N}_2 + 810 \text{ kcal/kg}$$

3. 96.6% AN plus 3.4% FO (fuel shortage)

   $$5\text{NH}_4\text{NO}_3 + \text{CH}_2 \rightarrow 11\text{H}_2\text{O} + \text{CO}_2 + 4\text{N}_2 + 2\text{NO} + 600 \text{ kcal/kg}$$

Note that Hopler (1998) indicates that the optimal percentage of fuel oil is 5.7%. Table 5.4 provides the elemental content of many explosive ingredients to use in oxygen balance equations. Additional information on the content of explosives can be found in Cook (1974) and Clark (1968, 1987).

TABLE 5.4 Composition of Explosive Ingredients

Name	Mol. Wt.	Formula	Composition (g-atoms/kg)				Heat of Formation	
			C	H	N	O	kcal/mole	kcal/kg
Nitroglycerin	227.09	$C_3H_5(ONO_2)_3$	13.21	22.02	13.21	39.63	82.66	364.00
Ethylene glycol dinitrate	152.97	$C_2H_4(NO_3)_2$	13.15	26.30	13.15	39.46	56.00	367.00
Nitrocellulose								
11.05% N_2			23.90	31.90	35.70	7.90		754.00
11.64% N_2			23.20	30.30	35.90	8.30		699.00
12.20% N_2			22.50	28.70	36.20	8.70		664.00
12.81% N_2			21.80	27.20	36.50	9.10		605.00
13.45% N_2			21.00	25.40	36.70	9.60		558.00
14.12% N_2			20.20	23.60	37.00	10.10		500.00
Trinitrotoluene (2-4-6)	227.13	$C_6H_2CH_3(NO_2)_3$	30.82	22.01	13.21	26.42	13.0	57.20
Dinitrotoluene	182.13	$C_7N_2O_4H_6$	38.43	32.94	10.98	21.96	6.900	38.00
Lead azide	291.3	$Pb(N_3)_2$	Pb = 3.430	—	20.60		−107.0 to −112.0	−364.0 to −386.0
Mercury fulminate	284.65	$Hg(CNO)_2$	7.026	Hg = 3.513	7.026	7.026	−65.46	−230.00
S:G pulp			41.70	63.00	—	21.40		1050.00
X pulp			40.50	58.50	—	28.00		1000.00
Paraffin			71.00	148.00	—	—		500.00
Cellulose			37.10	61.80	—	30.90	2270	1400.00
Ammonium nitrate	80.05	NH_4NO_3	—	49.97	24.98	37.48	87.93	1098.00
Sodium nitrate	85.01	$NaNO_3$	—	Na = 11.76	11.76	35.30	112.45	1323.00
Calcium carbonate	100.09	$CaCO_3$	9.999	Ca = 9.999	—	30.00	287.93	2876.00
Water	18.00	H_2O					57.80	

Source: Clark (1968). By permission of the Society for Mining, Metallurgy, and Exploration, Inc., Littleton, CO.

5.3.3 Properties of Explosives

5.3.3.1 Classification Schemes. Several methods of classifying explosives are commonly used. First, government agencies have designated classes for various explosive mixtures. The current classes are as follows (Hopler, 1998):

1. Division 1.1 or 1.2 (formerly called Class A) explosives possess detonation or other maximum hazard properties and must be handled with utmost care. Dynamite, nitroglycerine, and blasting caps are all in this category.
2. Division 1.3 (formerly Class B) explosives present a flammability hazard with little chance of detonation.
3. Division 1.4 (formerly Class C) explosive mixtures contain some Division 1.1, 1.2, or 1.3 explosive components, but in restricted quantities.
4. Division 1.5 (blasting agents) includes substances that have a mass explosion hazard but are so insensitive that they present very little probability of initiation or transition from deflagration to detonation during transit.

Explosives are also often categorized in a more general manner as low explosives, high explosives, and blasting agents (see Section 5.3.2.1). None of these classes, however, addresses one other special type of explosive. That category is the *permissible*, an explosive formulated to present a very minimal fire or a gas ignition hazard in coal mining. Permissibles are normally fuel-deficient and contain chemical additives to limit the flame temperature and duration. The Mine Safety and Health Administration (MSHA) normally publishes a list of explosives with the appropriate characteristics (Hopler, 1998, pp. 725–727).

5.3.3.2 Ingredients. As mentioned in Section 5.3.2.3, the principal components of an explosive are fuels and oxidizers. The most common fuels used are fuel oil, carbon, aluminum, and trinitrotoluene (TNT). Oxidizers include ammonium nitrate (AN), sodium nitrate, and calcium carbonate. Most commercial explosives today use AN as the base and the primary oxidizer. Other chemical substances may be added to affect the strength, sensitivity, water resistance, stability, or other important parameters of the explosive. These include *sensitizers* (nitroglycerine, TNT, nitrostarch, aluminum, etc.), *energizers* (aluminum or other metal powders), and *miscellaneous agents* (water, thickeners, gelatinizers, emulsifiers, stabilizers, flame retardants, dyes, etc.).

The development of slurry explosives has had a great effect on the content and properties of commercial explosives. These allow the creation of water gels (solid particles of chemicals permanently suspended in a liquid, water) and emulsions (water dispersed permanently through a solid chemical mass) for blasting purposes. Slurries are among the most common types of explosives used worldwide. They can be formulated in a variety of strengths and have much more water resistance than ANFO. Another category of explosive that has significant amounts of added ingredients is the permissible. These explosives are formulated with additives to reduce the hazards associated with igniting coal or gas in coal mines.

U.S. sales of explosives for 1995 (Kramer, 1999) show the following breakdown based on weight figures:

Type	Percentage
Ammonium nitrate and ANFO	85%
Water gels and emulsions	14%
Other high explosives	1%
Permissibles	0.1%

The use of dynamite has nearly disappeared from the scene as most manufacturers have switched from nitroglycerine to AN as their primary base for commercial explosives. It should be noted that mining applications utilize about 90% of all the explosives manufactured in the United States.

5.3.3.3 Blasting Properties of Explosives. Explosives are characterized by various properties that determine how they will function in blasting under field conditions. Although these are not necessarily the fundamental properties that govern the behavior of explosives, they represent the practical aspects of an explosive that govern its ability to perform under various conditions. The following are the most important (Atlas Powder Company, 1987; Hopler, 1998):

1. *Velocity of detonation* is the speed at which a detonation wave travels through a column of the explosive. Commercial explosives have a velocity of detonation of 8000 to 26,000 ft/s (2450 to 7925 m/s), with the higher velocities associated with greater energy release and the ability to shatter hard rock. High-velocity explosives have the ability to shatter rock, but the extra energy they provide is wasted on soft, plastic, or fractured rock.
2. *Detonation pressure* is normally measured in kilobar units and represents the pressure exerted as the detonation wave travels through an area (see Section 5.3.2.2, where detonation pressure is designated P_d). High detonation pressures are required in boosters used to initiate blasting agents.
3. *Borehole* (explosion) *pressure* is the gas pressure resulting from the explosion. Designated as P_e in Section 5.3.2.2, the borehole pressure is normally in the 10 to 60 kilobar (1000 to 6000 kPa) range. Borehole pressure is a good measure of the ability of an explosive to displace rock.
4. *Density* is normally expressed as the specific weight of the explosive. It is an important property in many granular explosives because the strength and density are positively correlated.

5. *Energy output* or *strength* provides a measure of the total energy released by the explosive during its reaction. The relative weight strength (RWS) and the relative bulk strength (RBS) are often used to express the energy content. RWS is defined as the percentage of the energy available as compared with that in an equal weight of standard ANFO, and RBS is the percentage of the energy available as compared with an equal volume of standard ANFO at a given density. Table 5.5 provides these strength values for some common explosives.
6. *Cap sensitivity* is a measure of the ease of initiation of an explosive when subjected to a shock wave from a detonator. Blasting agents have very low cap sensitivity and are relatively safe to handle, whereas most high explosives have a high sensitivity and can be detonated with a small initiator.
7. *Gap sensitivity* (also called sensitiveness) is the ability of one explosive charge to detonate another with an air gap between the two. Gap sensitivity must be controlled in industrial explosives so that one borehole's charge does not detonate another charge prematurely.
8. *Water resistance* is the ability to withstand exposure to water without losing sensitivity or efficiency. Water resistance is minimal in ANFO, higher in slurries, and maximum in some of the high explosives like dynamite.
9. *Fume class* represents a measure of the amount of toxic fumes produced by the blast. The Institute of Manufacturers of Explosives (IME) classification designates explosives that produce less than 0.16 ft^3 (0.005 m^3) of

TABLE 5.5. Important Properties of Explosives

Explosive	Specific Gravity or Density (g/cm^3)	Relative Weight Strength (ANFO = 100)	Relative Bulk Strength (ANFO = 100)
ANFO	0.85	100	100
ANFO (dense)	1.10	100	130
15% Al/ANFO	0.85	135	135
15% Al/ANFO (dense)	1.10	135	175
Pelletized TNT	1.00	90	106
1% Al/NCN slurry	1.35	86	136
20% TNT slurry	1.48	87	151
49% Dynamite	1.44	82	139
25% TNT/15% Al slurry	1.60	140	264
95% Dynamite	1.40	138	193

Source: Modified after Bucyrus-Erie Co., 1976. By permission of Bucyrus International Inc., South Milwaukee, WI.

toxic gases per 1.25 in. by 8 in. cartridge as Fume Class 1; 0.16 to 0.33 ft^3 (0.005 to 0.009 m^3) as Fume Class 2, and 0.33 to 0.67 ft^3 (0.009 to 0.019 m^3) as Fume Class 3. The fume class is important if the explosive is to be used in underground mines.

5.3.3.4 Initiation Systems. To properly trigger a blast, whether underground or on the surface, specially manufactured initiation components provide safety and optimally control the explosive action. A variety of initiation systems, both traditional and modern, are in use. To understand the systems, one must recognize that three major elements are required. These elements and the options for supplying them are outlined here:

1. Initial energy source (to initiate the reaction)
   (a) Match and ignitor cord (to initiate blasting fuse)
   (b) Blasting cap (to initiate detonating or low-energy cord)
   (c) Electric current source (to power an electrical blasting circuit)
   (d) Low-energy firing cap (to initiate a shock tube reaction)
2. Distribution network (for distributing energy to the various blastholes)
   (a) Blasting fuse (an old component no longer in common use)
   (b) Electrical circuit (to carry current to electrical blasting caps)
   (c) Detonating cord (a cord containing 15 to 60 grains/ft or 3.19 to 12.76 g/m of high explosive)
   (d) Low-energy detonating cord (a cord containing 2.4 to 6 grains/ft or 0.51 to 1.28 g/m of high explosive)
   (e) Shock tube (a small-diameter plastic tube with a very low loading of an explosive dust)
   (f) Gas tube (a small-diameter plastic tube that is explosive only after gas is pumped into the tube)
3. Detonator (the device that causes the detonation in each blasthole)
   (a) Blasting cap (fuse or electric variety)
   (b) Cartridge explosive
   (c) Booster (also called a primer) that detonates a blasting agent

More information on the properties of initiation system components can be found in Dick et al. (1983), *Engineering and Mining Journal* (1992b), and Hopler (1998). It should be noted that blasting fuse is no longer in common use. Electrical circuits, detonating cord, and shock tube systems are all utilized, with the shock tube system becoming more prevalent because of its relative safety. In choosing initiation systems, safety should be the prime concern, and cost, noise, and ease of hookup the secondary variables.

5.3.4 Selection of Explosives

The selection of an explosive and a blasting system is a relatively complicated task. As with a drilling system, the objective is not to optimize the blasting alone but the overall process of excavation. Therefore, the blasting operation must consider the cost of drilling and the cost of excavating the broken muck and hauling it to its final destination. Clearly, the explosive chosen for use will affect the drilling; blasting agents will require larger drillholes than explosives with higher strength. In addition, the cost of the excavation unit operation in many mines will be affected by the blasting. A significant portion of the excavation cost will be avoided if explosives are used to cast the overburden. This method of blast casting makes perfect sense if the cost of the extra blasting products is less than the cost of the avoided excavation. The blasting effectiveness will also determine the cost of comminution in some mines. This must be considered in the choice of an explosive. To deal with this problem, Dick et al. (1983) suggest consideration of the following selection criteria:

1. *Explosive costs.* The cost of individual explosive products is a vital factor in their selection. Plainly, this is the primary reason for the infrequent use of dynamite in today's world and the growth of AN-based explosives.
2. *Charge diameter.* The borehole diameter and the critical diameter of the explosive limit the charge diameter. Changing the drilling tools can change the hole diameter, but the critical diameter of the explosive is a property of the explosive. This will have a bearing on what explosive agent can be chosen.
3. *Rock blastability.* Rock conditions and the geologic environment must be considered and evaluated. Hard, dense, brittle, plastic, soft, or variable rock may necessitate a unique blasting system.
4. *Water conditions.* Wet overburden requires the use of a water-resistant explosive or water-repellent containers.
5. *Fume release.* If fumes are substantial, adequate ventilation must be provided in underground mines. Explosives with high fume levels are banned from underground mines.
6. *Other conditions.* These include ambient temperature (freezing temperatures can cause insensitivity), propagating ground (that may cause unintended detonations between holes), storage and transportation requirements (less stringent for blasting agents), and the existence of potentially explosive atmospheres (may require permissible explosives).

5.4 LOADING AND EXCAVATION

5.4.1 Materials Handling

The unit operations involved in excavating or moving bulk materials during mining are termed *materials handling*. In cyclic operations, the two principal operations are loading and haulage, with hoisting an optional third where essentially vertical transport is accomplished. Much of the productivity increase in recent mining history has been accomplished by eliminating unit operations to make the overall process more efficient. In continuous mining, where the excavator performs the breaking and handling functions, cutting, drilling, and blasting are eliminated. Extraction and loading are performed in a single function (excavation). In a machine that combines loading and hauling, such as the load-haul-dump (LHD) device, materials handling is conducted in a single operation.

Another way of making the process of mining more efficient is to increase the size of the equipment so that productivity is increased and labor cost is decreased per unit of material produced. For example, the scale of surface mining equipment has increased significantly over the last few decades. Upper limits have risen to 360 tons (325 tonnes) of capacity for off-highway haulage trucks, bucket capacities of 220 yd^3 (170 m^3) for draglines, 180 yd^3 (140 m^3) for overburden shovels, and 80 yd^3 (61 m^3) for electric loading shovels. These capacities have increased steadily, except for those of draglines and overburden shovels, which have suffered from a declining market and some perceptions of having reached their optimum size. The reasons for the gigantic scale of surface mining equipment are found in its high productivity and low unit operating cost. In part, this is increasingly due to the better computer and automation technology currently being implemented in this type of equipment. In this section, our attention will be on principles, reserving applications for later discussion.

5.4.2 Principles of Loading and Excavation

The extraction and elevation of minerals, either broken or in place, is termed *loading* or *excavating*. These terms are not synonymous, but they are used almost interchangeably. *Excavation* implies some action of extracting from the solid; *loading* suggests scooping and elevating material without any extraction. However, the term *loading* is often used to indicate that the material is placed in a haulage device. Blasted overburden in a surface coal mine is said to be excavated if it is cast onto a spoil pile, and loaded if it is placed in a haulage vehicle.

Loading or excavation is one of the primary unit operations performed in any mining operation. Many types of equipment are employed for this unit operation; these are classified in Table 5.6. Bases for the classification are the locale of the mining (surface or underground) and continuity of operation (cyclic or continuous). Common examples of individual machines are given for each category. Many types of equipment are listed; each machine has characteristics that distinguish it and help earmark it for selection. In each category, certain equipment types dominate the usage because their characteristics are better matched to the designated task. This will become evident when mining methods are discussed.

TABLE 5.6. Classification of Loading-Excavating Methods and Equipment

Operation	Category or Method	Machine (Application)
Surface		
Cyclic	Shovel	Power shovel, front-end loader, hydraulic excavator, backhoe (mining ore, stripping overburden)
	Dragline	Crawler, walking (stripping overburden)
	Dozer	Rubber-tired, crawler (blade)
	Scraper	Rubber-tired, crawler
	Blasting	Explosives stripping (overburden)
Continuous	Mechanical excavator	Bucket wheel (BWE) (overburden), cutting-head (soil, coal)
	Highwall mining	Auger, highwall miner (coal)
	Hydraulicking	Monitor or giant (placer)
	Dredging	Bucket ladder, hydraulic (placer)
Underground		
Cyclic	Loader	Overhead, gathering arm, shovel, front-end
	Shaft mucker	Clamshell, orange peel, cactus grab
	Self-loading transport	Load-haul-dump (LHD)
	Slusher	Rope-drawn scraper (metal ore)
Continuous	Continuous miner	Milling type, drum, ripper, borer, auger, plow, shearer (coal or nonmetallic)
	Boring machine	Tunnel-boring machine (TBM), roadheader, raise borer, shaft borer (soft rock)

Most loaders or excavators are required to work in three *working zones* and with a variety of constraints (Martin et al., 1982). These zones—digging, maneuvering/transport, and dumping—and some of their constraints are listed in Figure 5.3. The situation illustrated occurs primarily in stripping overburden or loading ore with a boom-type excavator (power shovel, hydraulic shovel, dragline, or bucket wheel) or with a front-end loader. The major advantages and disadvantages of shovels, draglines, and bucket-wheel excavators for stripping purposes are provided in Table 5.7. Note that wheel excavators have nearly disappeared in North America but are used in Europe with more frequency. A similar comparison of loading (rope) shovels, hydraulic shovels, and front-end loaders for use in open pits is provided in Table 5.8. Loading shovels are still widely preferred, but improvements in hydraulic shovels and front-end loaders in terms of equipment lifetime and maintenance costs have made them more popular than in the past.

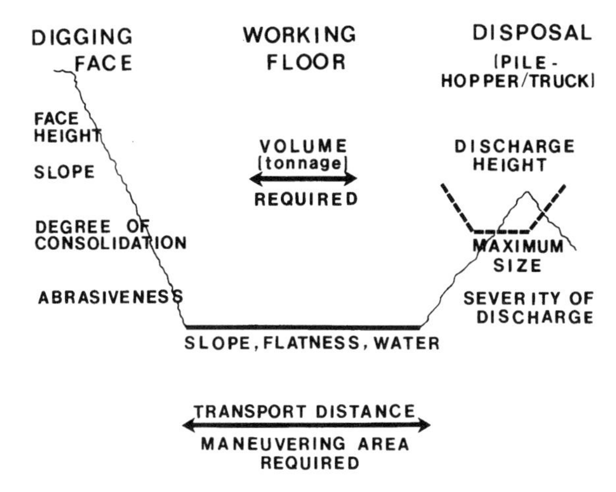

FIGURE 5.3. Working zones and constraints in loading or excavating in a surface mine. Conditions: shovel, dragline, or wheel excavator. Zones: digging, maneuvering-transport, and dumping. (After Martin et al., 1982. By permission from Martin Consultants, Inc., Golden, CO.)

5.4.3 Selection of Equipment

A relatively large number of factors must be considered in the selection of the prime excavators that are used in a mining operation (Martin et al., 1982; Atkinson, 1992a). The following list of variables was drawn up considering surface mining applications, but the factors are applicable to underground equipment selection as well. The factors to be considered can be logically grouped into four categories:

1. *Performance factors.* These are related to machine productivity and include cycle speed, available breakout force, digging range, bucket capacity, travel speed, and reliability.
2. *Design factors.* The design variables apply to the quality and effectiveness of detail design, including the sophistication of human-machine interaction for operators and maintenance personnel, the level of technology employed, and the types of control and power available.
3. *Support factors.* These variables involve the service and maintenance aspects of machine operation. Sometimes overlooked in the past, these factors are becoming more important as computer diagnostics and monitoring systems allow maintenance personnel to better maintain and repair the equipment.
4. *Cost factors.* Probably the most important category in the selection process, the costs of purchasing and operating the equipment can be estimated by standard estimating procedures. Publications by Western

TABLE 5.7. Comparison of Features of Shovel, Dragline, and Bucket-Wheel Excavators

Machine	Advantages	Disadvantages
Shovel	1. Lower capital cost per yd^3 (m^3) of bucket capacity, although when boom length or machine weight is considered, the capital costs are roughly equivalent. 2. Digs poor blasts and tougher materials better. 3. Can handle partings well.	1. More coal damage can result in lower coal recovery. 2. Susceptible to spoil slides and pit flooding. 3. Cannot easily handle spoil having poor stability. 4. Cannot dig deep box cuts easily. 5. Reduced cover depth capability compared with a dragline of comparable cost. 6. Difficult to move.
Dragline	1. Flexible operation; easy to move. 2. Large digging depth capability. 3. Can handle and stack overburden having poor stability. 4. Completely safe from spoil pile slides or pit flooding during normal operation. 5. High percentage of coal recovery; less coal damage. 6. Will dig a deeper box cut. 7. Low maintenance cost. 8. Can handle partings well. 9. Is not affected by an uneven or rolling coal seam top surface. 10. Can move in any direction.	1. Requires bench preparation. 2. Does not dig poor blasts well. 3. Higher capital cost per yd^3 (m^3) of bucket capacity, although when boom length or machine weight is considered, capital costs are roughly equivalent.
Bucket wheel	1. Continuous operation; no swinging necessary. 2. Long discharge range. 3. Can be operated on a highwall bench or on the coal seam. 4. Can easily handle spoil with poor stacking characteristic and poor stability. 5. Can extend range of shovel or dragline when operated in tandem. 6. Can facilitate land reclamation as it dumps surface material back on top of the spoil pile.	1. Will not dig hard materials. 2. Some surface preparation required. 3. Lower availability. 4. Large maintenance crew required. 5. High capital cost compared with output. 6. Can be susceptible to spoil slides and flooding. 7. Can cause coal damage with resulting lower coal recovery. 8. Poor mobility.

Source: Bucyrus-Erie Co., 1976 (By permission from Bucyrus International Inc., South Milwaukee, WI.)

TABLE 5.8. Comparison of Open Pit Loading Equipment

Machine	Advantages	Disadvantages
Loading (rope) shovel	1. Proven in hard, dense rock. 2. Low operating cost. 3. Less sensitive to poor maintenance. 4. Operator fatigue not a serious problem. 5. Low ground pressures. 6. Long lifetime.	1. Lack of mobility. 2. High capital cost. 3. Poor cleanup capability. 4. Will not travel on steep grades. 5. Affected by obsolescence.
Hydraulic shovel	1. Excellent breakout threat. 2. Can mine selectively. 3. High fill factor. 4. Good cleanup capability. 5. Good load placement. 6. Short assembly time.	1. Relatively short lifetime. 2. Relatively high maintenance. 3. High initial cost.
Front-end loader	1. Excellent mobility. 2. Very versatile. 3. Lower capital cost. 4. Can operate on moderate grades.	1. Unsuitable for hard, dense rocks. 2. High tire and operating costs. 3. Digging area must be kept clean. 4. High operator fatigue. 5. Short lifetime.

Sources: Martin et al. (1982); Atkinson (1992a) By permission from Martin Consultants Inc., Golden, CO, and the Society for Mining, Metallurgy, and Exploration, Inc., Littleton, CO.

Mine Engineering (1998, 1999) are very helpful in this regard. The customary basis is to use unit costs, estimating *overall costs* as the sum of *ownership* and *operating costs* computed on a \$/hr basis and converted to \$/ton (\$/tonne) or \$/yd^3 (\$/m^3).

The data in Table 5.9 may be useful in estimating the physical limits and costs related to surface mining equipment. Also of interest in the selection of surface mining equipment will be the information provided by Martin et al. (1982) and Atkinson (1992a). Ratings and information on surface mining equipment are available from a number of literature sources, but the data on underground mining equipment are somewhat more limited. In recent years, mergers, takeovers, and bankruptcies have caused major changes in the manufacturing of mining equipment. It is therefore important to seek out the latest data from manufacturers when selecting equipment.

TABLE 5.9. Estimating Parameters for Surface Excavators

	Capacity yd³ (m³)	Est. Weight, lb/yd³ (kg/m³)	Est. Power, hp/yd³ (kW/m³)	Est. Life, hr	Est. Price, $/yd³ ($/m³)
Rubber-tired scraper	25–54 (19–41)	3100 (1840)	14 (14)	12,000	22,000 (28,800)
Front-end loader	3.5–36 (2.7–28)	13,000 (7,700)	52 (51)	12,000	85,000 (111,200)
Hydraulic excavator	4–67 (3–51)	30,000 (17,800)	72 (70)	30,000	140,000 (183,000)
Electric power shovel	6–80 (4.6–61)	50,000 (29,700)	95 (93)	75,000	190,000 (249,000)
Walking dragline	9–180 (6.9–138)	105,000 (62,300)	75 (73)	100,000	320,000 (419,000)
Bucket-wheel excavator	0.1–5.2 (0.1–4)	—	—	30,000	—

Sources: Compiled from data in Martin et al. (1982) and Western Mine Engineering (1999). Used with permission of Martin Consultants, Inc., Golden, CO, and Western Mine Engineering, Inc., Spokane, WA.

5.5 HAULAGE AND HOISTING

5.5.1 Principles of Haulage and Hoisting

Bulk materials in mining are transported by *haulage* (primarily horizontal) and *hoisting* (primarily vertical) systems. The equipment used to perform these operations is outlined in Table 5.10, classified on the same basis as excavating equipment. Information is provided on the normal range of haul distances and the gradeability, both normal and maximum, of the equipment. The primary function of all the equipment is haulage, but some (front-end loader, dozer, rubber-tired scraper, slusher, and LHD) are self-loading. The only true hoisting machines are the skip and the cage; however, pneumatic and hydraulic conveyors and high-angle conveyors can perform vertical transport as well. Although there are many choices of equipment, a few provide most of the haulage service in mining operations. Haulage trucks and conveyors are prevalent in surface mines. In underground mines, rail, trucks, shuttle cars, LHDs, and conveyors are all widely used.

Like excavators, haulage units also perform in working zones. For the most widely used haulage machine, the truck, and similar equipment, there are four working zones: loading, traveling loaded, dumping, and traveling empty. In Figure 5.4, these are illustrated in the lower diagram. Because waiting lines can develop at the loading and dumping points, it is important that these zones be

TABLE 5.10. Classification of Haulage and Hoisting Methods and Equipment

Operation	Method	Haul Distance	Gradeability (degrees)	
			Avg.	Max.
Surface				
Cyclic	Rail (train)	Unlimited	2	3
	Truck, trailer	1–10 mi (0.6–16 km)	3	6
	Truck, solid-body	0.2–5 mi (0.3–8 km)	8	12
	Scraper (rubber-tired)	500–5000 ft (150–1500 m)	12	15
	Front-end loader	< 1000 ft (300 m)	8	12
	Dozer	< 500 ft (150 m)	15	20
	Skip	< 8000 ft vert. (2400 m)	Unlimited	
	Aerial tramway	0.5–5 mi (0.8–8 km)	5	20
Continuous	Belt conveyor	0.2–10 mi (0.3–16 km)	17	20
	High-angle conveyor (HAC)	< 1 mi (1.6 km)	40	90
	Hydraulic conveyor (pipeline)	Unlimited	Unlimited	
Underground				
Cyclic	Rail (train)	Unlimited	2	3
	Truck, shuttle car	500–5000 ft (150–1500 m)	8	12
	Slusher (scraper)	100–300 ft (30–90 m)	25	30
	LHD	300–2000 ft (90–600 m)	8	12
	Skip, cage	< 8000 ft vert.	Unlimited	
Continuous	Conveyor (belt, chain and flight, monorail)	0.2–5 mi (0.3–8 km)	17	20
	Hydraulic conveyor	Unlimited	Unlimited	
	Pneumatic conveyor	Unlimited	Unlimited	

Source: Modified after Hartman (1987).

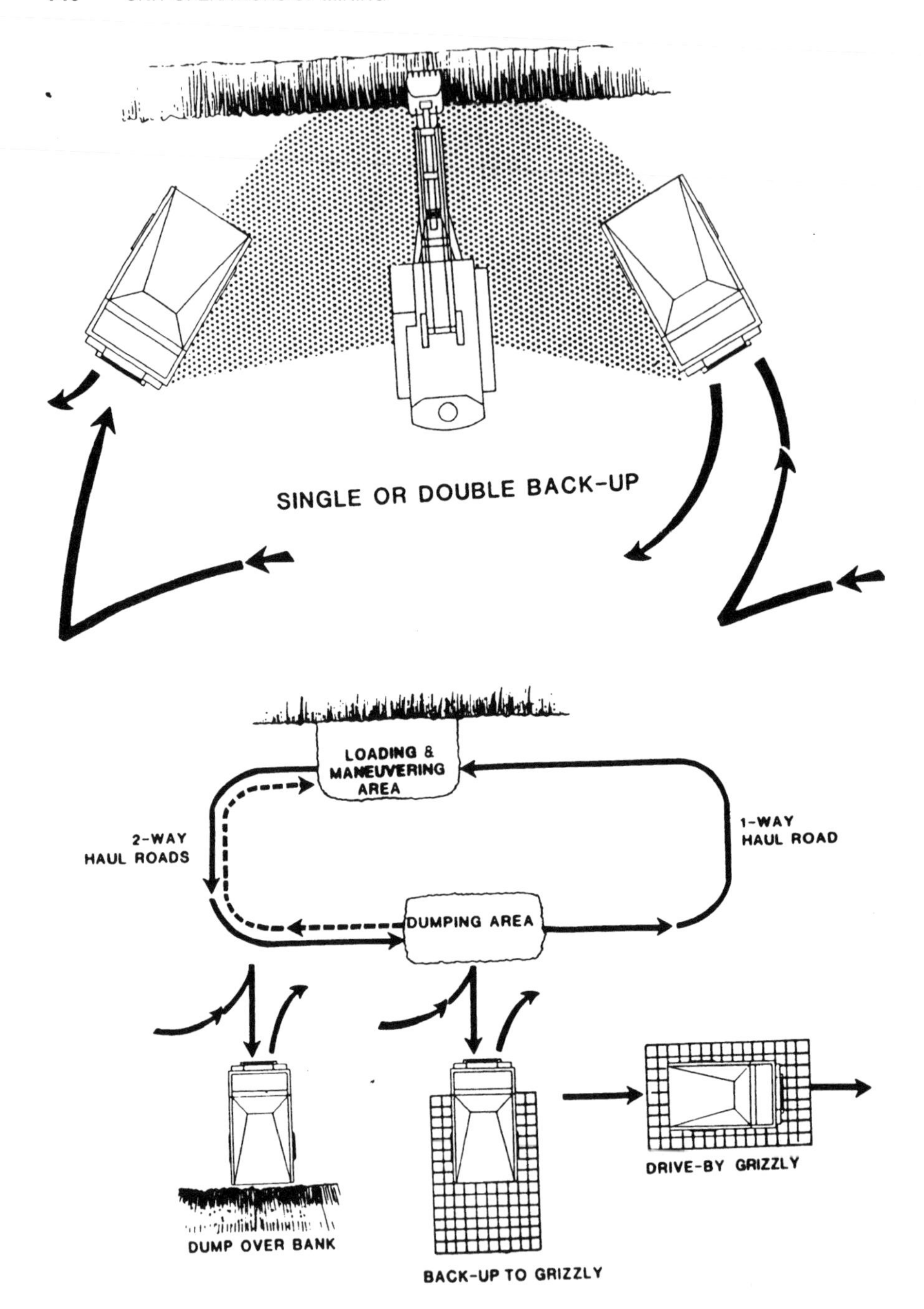

FIGURE 5.4. Working zones in haulage in a surface mine. Conditions: truck haulage, shovel loading, grizzly or bank dumping. Zones: loading, traveling loaded, dumping, and traveling empty. (After Martin et al., 1982. By permission from Martin Consultants, Inc., Golden, CO.)

TABLE 5.11 Comparison of Features of Principal Haulage Units

Machine	Advantages	Disadvantages
Dozer	1. Flexible 2. Good gradeability 3. Negotiates rough terrain	1. Limited to short haul 2. Discontinuous 3. Low output, slow
Truck	1. Flexible and maneuverable 2. Handles coarse, blocky rock 3. Moderate gradeability	1. Requires good haul roads 2. Slowed by bad weather 3. High operating cost
Scraper (rubber-tired)	1. Flexible and maneuverable 2. Good gradeability	1. May require push loading 2. Limited to soil, small fragments 3. High operating cost
Rail	1. High output, low cost 2. Unlimited haul distance 3. Handles coarse, blocky rock	1. Track maintenance costly 2. Poor gradeability 3. High investment cost
Belt conveyor	1. High output, continuous 2. Very good gradeability 3. Low operating cost	1. Inflexible 2. Limited to small or crushed rock 3. High investment cost

Sources: Modified from Pfleider, 1973; Martin et al., 1982. By permission of Martin Consultants, Inc., Golden, CO, and the Society for Mining, Metallurgy, and Exploration, Inc., Littleton, CO.

carefully evaluated in every haulage system where they occur (Martin et al., 1982). Note in the upper diagram of Figure 5.4 that either a single location or a double location can be used for trucks being loaded. *Double-spotting*, or using two loading stations, can greatly improve productivity where it can be designed into the system. In the dumping zone, a drive-by grizzly is preferable to a backup grizzly. However, if a backup dump point is used, more than a single dumping location should be designed into the dump point. Different working zones characterize other haulage machines, and they all require careful analysis to ensure that the system operates efficiently.

Table 5.11 summarizes some of the main features of haulage equipment. Most of these equipment types are used both on the surface and underground. However, the equipment used underground is often more compact and has less capacity than equipment used for surface mines.

5.5.2 Selection of Haulage and Hoisting Equipment

The factors to be considered in selecting a haulage system are similar to those concerning excavators. For a discussion of some of the unique aspects of surface-mine haulage system design, see Martin et al. (1982), Kennedy (1990), and Sweigard (1992a). For underground equipment, the discussion by Hustrulid (1982, pp. 1169–1266) gives details of the various choices available and their capabilities. Additional insights and selection criteria can be found in Lineberry and Paolini (1992).

The selection of hoists and related equipment is another important decision in materials handling. Hoists are a rather specialized area of materials handling, and many hoists are custom-designed for particular mining applications. A more complete discussion of the components of a hoisting system and the calculations necessary to design such a system are detailed in Chapter 9.

5.6 AUXILIARY OPERATIONS

Auxiliary operations consist of all activities supportive of but not contributing directly to the production of coal, ore, or stone. Many of the auxiliary unit operations are scheduled prior to or after the production cycle so as to support but not interfere with production operations. Others, like ground control and ventilation, are performed as an integral part of the production cycle if they are essential to health and safety or to efficient operation.

Because unit operations generate no income, there is a tendency in mining organizations to assign them a staff function and a low priority. However, mine managers must ensure that these tasks receive proper attention and that technological advances and efficient operation are part of the auxiliary operations. There are good economic reasons for doing so. Recalling our discussion of mine administration (Section 4.9), approximately three support personnel are required in auxiliary operations for every two in production. Thus, 60% of all personnel are involved in the support of the production operations. Accordingly, the auxiliary operations must also be managed in a proper manner to conduct a mining operation that is efficient overall.

Many of the auxiliary operations common to mining operations are listed in Table 5.12. Most are classified as supportive of the exploitation function, but unit operations associated with development and reclamation are included as well. Note that health and safety is an extremely important auxiliary operation that should not take a back seat to the production goals of the organization. It is important that this function be given proper emphasis in the management scheme. Also important to the public image of the mining company are those functions associated with the environment. The public image of a mining company is heavily dependent on what is visible. It is therefore essential to reclaim the land properly after mining and to let the public know of the diligence exercised in environmental protection.

Given their importance, auxiliary operations must be studied and optimized in every mining operation. The most important of these tasks, health and safety and environmental control, are treated individually in other chapters. Additional discussions of auxiliary and supportive tasks in mining can be found in other resources. Of particular interest are the discussions of unit operations in coal mining (Stefanko and Bise, 1983) and of unit operations that pertain to a more general class of mining operations (Mutmansky and Bise, 1992).

TABLE 5.12 Classification of Mine Auxiliary Operations

Function	Surface Operation	Underground Operation
Exploitation		
Health and safety	Dust control[a]	Gas and dust control[a]
	Noise abatement	Ventilation and air-conditioning[a]
	Prevention of spontaneous combustion[a]	Noise abatement
	Disease prevention	Disease prevention
Environmental control	Air and water protection	Groundwater protection
	Waste disposal[a]	Subsidence control
Ground control	Slope stability	Roof control[a]
	Soil erosion control	Controlled caving
Power supply and distribution	Power distribution (electrical)	Power distribution (electricity, compressed air)
Water and flood control	Pumping, drainage	Pumping, drainage
Waste disposal	Storage, dumping	Backfilling, hoist to surface[a]
Materiel supply	Storage, delivery of supplies	Storage, delivery of supplies
Maintenance and repair	Shop facilities	Shop facilities
Lighting	Portable floods	Stationary, portable lights
Communications	Radio, telephone	Radio, telephone
Construction	Haul roads, etc.	Haulageways, etc.
Personnel transport	Personnel trucks, buses	Man cages, trips, cars
Development (in support of production)		
Site preparation	Clearing land, access, etc.	—
Topsoil removal	Stripping, stockpiling[a]	—
Surface reclamation	Replacement, grading, revegetation	—
Reclamation (simultaneous with production)		
Topographic work	Regrading	—
	Restoring soil	—
Revegetation	Planting grasses	—
	Planting trees	—
Erosion control	Establishing sediment basins	—
	Riprapping streams	—
Monitoring	Water quality	Water quality
	Air quality	Air quality

[a] May be incorporated in production cycle.

Source: Modified after Hartman (1987).

5.7 CYCLES AND SYSTEMS

Having studied unit operations of mining, it is desirable to summarize the importance of the cycles of unit operations and how they contribute to the overall efficiency of the system in which they are utilized. A typical cycle of unit operations for metal and many nonmetal mines consists of the following: (1) drilling, (2) blasting, (3) loading, and (4) hauling. This cycle is performed over and over to accomplish the winning of the ore. Note that each of the unit operations cannot begin its handling of the mineral product before the previous unit operation has completed its work. From a systems engineering standpoint, cyclic production processes are troublesome because the overall production is limited to the production of the slowest unit operation. Also note that the overall availability of the production system is a product of the availabilities of the individual unit operation subsystems unless standby equipment is maintained. It is therefore important that cyclic unit operations be designed with care to achieve the most efficient operation.

Another manner of optimizing the unit operations in mining is to eliminate those that may be superfluous. Several good examples are in evidence. At one time, the cycle of operations in underground coal mining was (1) undercut the face, (2) drill, (3) blast, (4) load, (5) haul, and (6) support the roof. When the continuous miner was developed, the cycle of operations was greatly simplified. The continuous miner eliminated the tasks of undercutting, drilling, blasting, and loading by substituting the unit operation that might be called *excavating the face*. Obviously, that greatly reduced the complexity of the production process and made it more efficient. In today's world, the coal industry is seeking to simplify the process even further by simultaneously attempting to excavate the face with a continuous miner while providing ground control using an integral roof bolter. A second example that is practiced in hard-rock mining is rapid excavation, which is the process of excavating the face of a drift, a raise, or a shaft using a boring device. This allows the unit operations of drilling, blasting, and loading to be replaced by a single unit operation, excavation.

In assessing the importance of unit operations in the overall operation of a given mining operation, a mining engineer cannot overemphasize their value. Unit operations are the basic components of the mining process. Proper equipment choices must be made, and the unit operations must be carefully designed. If the unit operations are designed to be productive in their own right and in the way they interact with other unit operations, then the overall mining process will be efficient.

5.8 SPECIAL TOPIC: CHEMICAL DESIGN OF EXPLOSIVES

The oxygen balance of explosives can be accomplished by means of the physical-chemical properties presented in Section 5.3. Information on this process can be found in the discussions by Clark (1968, 1987).

Example 5.1. A mixture of chemicals has been produced, containing the following:

Nitroglycerine (NG)	18%
Trinitroglycerine (TNT)	3%
Ammonium nitrate (AN)	55%
Sodium nitrate	10%
S:G pulp	12%
Calcium carbonate	2%
Total	100%

(a) Calculate the oxygen balance of the mixture.
(b) If the explosive is not balanced, what change in composition would you recommend to improve the oxygen balance (OB)?

SOLUTION. (a) Determine the oxygen balance using data from Table 5.4 and Eq. 5.4. This is best performed in a tabular form, as follows:

		H_0	N_0	O_0	C_0	Ca	Na
NG	18%	3.964	2.378	7.133	2.378	0.0	0.0
TNT	3%	0.660	0.396	0.793	0.925	0.0	0.0
AN	55%	27.484	13.739	20.614	0.0	0.0	0.0
$NaNO_3$	10%	0.0	1.176	3.530	0.0	0.0	0.0
S:G pulp	12%	7.560	0.0	2.568	5.004	0.0	0.0
$CaCO_3$	2%	0.0	0.0	0.600	0.200	0.200	0.0
Totals		39.668	17.689	35.238	8.507	0.200	1.176

$$OB = 35.238 - 2(8.507) - 0.5(39.668) - 0.200 - 0.5\,(1.176) = -2.398 \text{ g-atom/kg}$$

(b) Because the explosive is slightly oxygen-deficient (fuel-rich), reduce the fuel content (NG, TNT) or increase the oxidant (AN).

PROBLEMS

5.1 Consider the following seven blasting materials (or combinations of materials):

1. Ammonium nitrate, NH_4NO_3
2. Ammonium nitrate and lampblack, $NH_4NO_3 + C$
3. Ammonium nitrate (96.6%) and fuel oil (3.4%), fuel-shortage $NH_4NO_3 + CH_2$
4. Ammonium nitrate (94.5%) and fuel oil (5.5%), oxygen-balanced $NH_4NO_3 + CH_2$
5. Ammonium nitrate (92%) and fuel oil (8%), fuel-excess $NH_4NO_3 + CH_2$

6. Metallized ammonium nitrate, $NH_4NO_3 + Al$
7. Metallized (9.9%) ammonium nitrate (87.6%) and fuel oil (2.5%), $NH_4NO_3 + CH_2 + Al$
   (a) Write simplified, balanced chemical equations of detonation for each of the seven materials or combinations of materials, expressing the exothermic heat of reaction in kcal/kg. The references by Clark (1968, pp. 341–346), Dick (1973, pp. 11:78–80), Dick et al. (1983, pp. 3–4), and Clark (1987, pp. 341–363) may be of help.
   (b) Compare the heats of reaction, ranking the level of explosives from greatest to least. Account for the variation. Can the higher energy release of certain explosives be utilized in blasting? How? Are there limits?

5.2 Calculate the oxygen balance (OB) for 1 kg of a high explosive with the following composition:

NG	54%
TNT	9%
AN	19%
Sodium nitrate	4%
S:G pulp	12%
Calcium carbonate	2%
Total	100%

Is this explosive in balance? What changes in composition would you recommend to improve the OB?

5.3 Calculate the OB for 1 kg of a high explosive with the following composition:

NG	9%
TNT	2%
AN	63%
Sodium nitrate	12%
S:G pulp	12%
Calcium carbonate	2%
Total	100%

Is this mixture in balance with regard to oxygen? What change in composition would you suggest to improve the OB?

6

SURFACE MINE DEVELOPMENT

6.1 THE NATURE OF SURFACE MINING

Surface mines come in many different forms and sizes. However, most of our surface-mined mineral products come from large mines that practice *large-scale* (mass production) *methods*. The sheer magnitude of the volume or tonnage of material broken and handled in surface mining is staggering. Table 6.1 outlines some of the statistics on production of ore and waste in surface and underground mining in the United States for the year 1997. Note that the average tonnage of waste required to produce a ton of ore is about 15 for surface coal and about 2.6 overall for surface mining, but only 0.1 for coal and 0.07 overall for underground mining. Because of the waste associated with surface mining, the cost to mine a ton of material in surface mines must be much lower than that for underground mines.

The proportions of ore and total material tonnage produced in surface and underground mines are provided in Table 6.2. The tonnage of ore and coal being mined on the surface keeps increasing. When the first edition of this book appeared in 1987, the proportion of ore and coal mined in surface mines in the United States was 85%; it is now more than 88%. Note also that the percentage of all material mined in surface mines is greater than 96% on a tonnage basis. This heavy emphasis on surface mining is due to two factors: the increasing difficulty of finding deposits that can be economically mined underground and the ever-increasing efficiency of mining in surface mines. This is true even though the cost of reclamation for a surface mine is increasing steadily. The success of surface mining operations is thus highly dependent on efficient drilling, blasting, loading, and haulage.

Table 6.1 Estimates of Ore and Waste Production for Different Mineral-Commodity Classes in Surface and Underground Mining in the United States, 1997

	Surface			Underground			All Mining		
Commodity	Ore	Waste	Total	Ore	Waste	Total	Ore	Waste	Total
				(million tons)					
Metals	1290	1,863	3,153	64	3	67	1,354	1,866	3,220
Nonmetals	2778	449	3,227	123	0	123	2,901	449	3,350
Coal	669	10,303*	10,972	421	45*	466	1,090	10,348	11,438
Total	4737	12,615	17,352	608	48	656	5,345	12,663	18,008

Sources: Moore (1997) and Energy Information Administration (2000).

* These values based on ore/waste ratios reported by Energy Information Administration (1983).

Table 6.2 Proportions of Ore and Coal Tonnage and Total Tonnage from Surface and Underground Mines

	Ore Tonnage		Total Tonnage	
	Surface	Underground	Surface	Underground
Metals	95.3%	4.7%	97.9%	2.1%
Nonmetals	95.8%	4.3%	96.3%	3.7%
Coal	61.4%	38.6%	95.9%	4.1%
Total	88.6%	11.3%	96.4%	3.6%

Sources: Energy Information Administration (1983, 2000); Moore (1997).

Mine planning and development are crucial steps in the operation of a surface mine. Certain factors (see Section 4.2) may require special attention in preparation for surface mining. Of the location factors, climate is of more critical concern in surface operations than in underground mines. Today, harsh climates at high altitudes or in northern latitudes rarely preclude surface mining, but they can be detrimental to efficient and cost-effective operation. Among the natural and geologic factors, terrain, depth, spatial characteristics of the deposit, and presence of water are very important variables. The environmental concerns and the costs of overcoming environmental problems rank among the most important considerations in the planning and development of a surface mine.

The steps in the sequence of mine development are enumerated in Section 4.2; three of these are unique to surface mining:

1. Initiation of a land reclamation plan as part of the environmental impact statement (EIS)
2. Provision of topsoil stockpiles and waste disposal dumps
3. Performing advanced stripping of overburden to gain access to the deposit

Other steps in mine development must be carried out as well, but these three steps require significant resource allocation and planning for a surface mine. Figure 6.1 shows the mine development schedule for a typical mine (Petty, 1981). Note that the environmental studies and the permitting process are a significant part of the overall mine development, and that environmental concerns are continuous throughout the development process. Thus, the first of the aforementioned three steps in development will normally be a major consideration in the development of the mine. Land reclamation, waste disposal, and advanced stripping are scheduled during Stage 3 of Figure 6.1. (Notice that it is customary terminology to say we *mine* ore, stone, or rock but

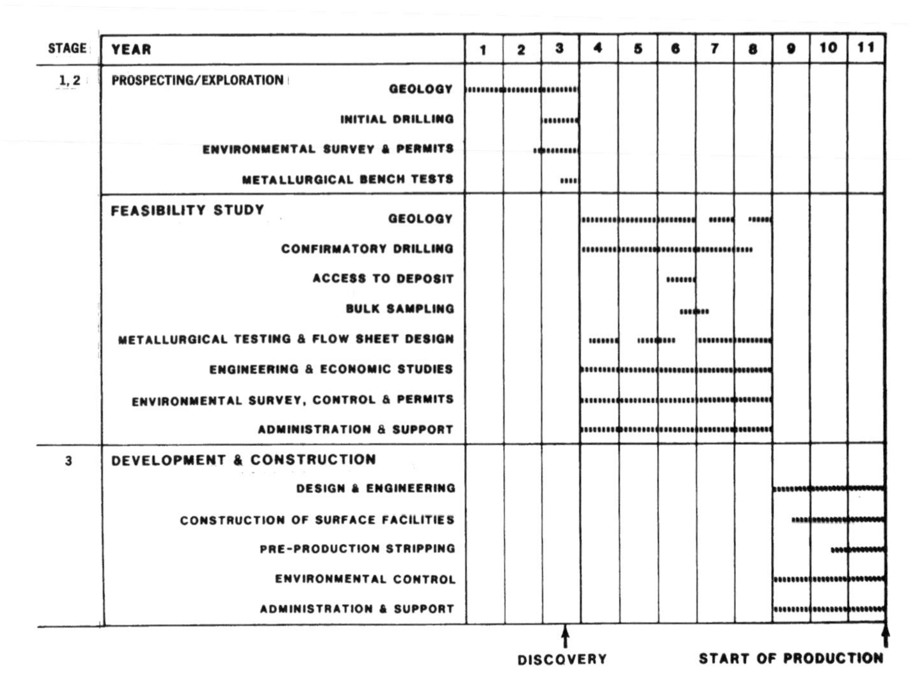

FIGURE 6.1. Scheduling diagram for surface metal mine. (After Petty, 1981. By permission from American Institute of Professional Geologists, Colorado Section, Golden, CO.)

that we *strip* overburden or waste.) We will consider the three tasks, as well as plant layout, in more detail in the sections that follow.

6.1.1 Land Reclamation

The federal Surface Mining Control and Reclamation Act of 1977 requires that land disturbed by mining be reclaimed and restored to its premining condition or better. Some of the many requirements are summarized in Section 2.2.5. During development, the first steps are taken to ensure that the EIS filed by the company is fully implemented. The company must post a sizable bond to cover the expected cost of reclaiming the land. The bond on each section of land is released only after the land has been reclaimed and suitable vegetation has been established. Provisions of the law, which are primarily applicable to surface coal mines, are quite stringent and costly. Restoration of land to "approximate original contour" following mining is a requirement that necessitates careful planning, surveying, and mapping. Preserving surface drainage may require stream relocation or diversion. Careful control of the runoff from the mine is also necessary to prevent stream siltation. Maintaining wildlife and fisheries is of special concern on lands to be surface mined. Provisions to protect fish and game resources and to provide acceptable habitat must be initiated early in the development process and carefully maintained throughout the life of the mine. Finally, archaeological sites, both known or uncovered during mining, must be protected. State or international codes may also enter significantly into the planning picture. For provisions of a typical state mining code, see Atkinson (1983) and the Illinois Department of Mines and Minerals (1985).

6.1.2 Topsoil Stockpiles and Waste Disposal

During the development stage of any surface mine, topsoil stockpiles and waste disposal areas are located. Separate areas may be required for topsoil, subsurface soil, rock, low-grade or leachable ore, and tailings. Separating these materials enhances the opportunities to better utilize the materials in the extraction of valuable components and in the reclamation of the mined area. Site selection must ensure convenient disposal and retrieval but must avoid interference with production and its related auxiliary operations. Advanced planning is necessary to ensure that the materials in storage never conflict with mining activities.

6.1.3 Advanced Stripping

The geometry of the mineral deposit and the overburden and the planned production rate dictate the minimum amount of stripping that must be done to maintain the desired rate of ore or coal extraction. Economic considerations would normally suggest that the stripping be performed only as needed because the expenses associated with stripping are not matched by an economic return (see Section 4.1). However, Fourie and Dohm (1992) outline three general plans for stripping in a surface mine: the *increasing stripping ratio method*, the *decreasing stripping ratio method*, and the *constant stripping ratio method*. The increasing stripping ratio method strictly follows the rule of stripping only as much overburden as required for production; this procedure is optimal in terms of the cash flow if other variables do not significantly enter the picture. The other two methods attempt to level out the stripping requirements and spread them more evenly over time. In a deposit that does not outcrop, advanced stripping is required before ore production can begin. In addition, it is a general rule that a certain amount of ore should always be available for mining so that production scheduling is not constrained by the lack of available ore. A rule of thumb, common in truck-shovel operations, is to maintain at least a 30-day supply of broken ore available for the loading equipment.

A major consideration in stripping decisions may be the climate. Severe cold weather may favor stripping in the summer months when the ground is thawed, with mining being conducted during the winter. In the iron ranges of the Lake Superior district and in regions of Alaska, shipping may be restricted by ice formation on the waterways. Thus, stripping may be emphasized in winter and mining during the summer.

A final consideration may be the decision as to whether the stripping should be performed in-house or by a contractor. Large mining companies generally prefer to do their own stripping, but a smaller company may find it expeditious to contract it out. Contract stripping is often more expensive, but it may be quicker. In addition, the mining company is relieved of the capital expense of purchasing stripping equipment.

6.1.4 Plant Layout

Some of the factors that must be considered in selecting the mine plant size and layout were identified in Section 4.2. For a surface mine, the task is complicated by the special considerations just discussed: land reclamation, topsoil stockpiling, waste disposal, and overburden stripping. These must be carefully planned to minimize the cost of mining. In addition, the support activities will become an important additional mine plant layout task. The plant layout of the Black Thunder mine in Wyoming, among the largest coal mines in the world, provides an interesting case study and is shown in Figure 6.2. The pit, waste dumps, and topsoil stockpiles are off the diagram to the lower right. The coal haul road and crusher, mineral processing plant, slot storage facility, and clean coal storage facility are all connected by single-flight belt conveyors for efficient materials handling. The storage silo is on a rail loop that facilitates unit train loading. The maintenance shops, administration building, change house, warehouse, fuel storage, and other necessary facilities are then arranged in close proximity to the processing facilities. When laid out in a logical manner, the physical plant enhances the ability to mine and contributes to the efficiency of the operation.

6.2 PIT PLANNING AND DESIGN

Open pit mining is a method of operating a surface mine that is simple in concept but complex in its cost and efficiency requirements. Recall the earlier discussion in Section 6.1 about the amount of waste mined; it is evident that open pit mining must be carefully planned and executed to keep unit costs to a minimum. Accordingly, the average open pit mine is heavily engineered even though it is simple in configuration. There are a number of factors that must be considered in the initial planning. The following are based on information provided by Atkinson (1983):

1. *Natural and geologic factors:* geologic conditions, ore types and grades, hydrologic conditions, topography, metallurgical characteristics, climate, and environmental variables of the site
2. *Economic factors:* ore grade, ore tonnage, stripping ratio, cutoff grade, operating cost, investment cost, desired profit margin, production rate, processing and/or smelting costs, and market conditions

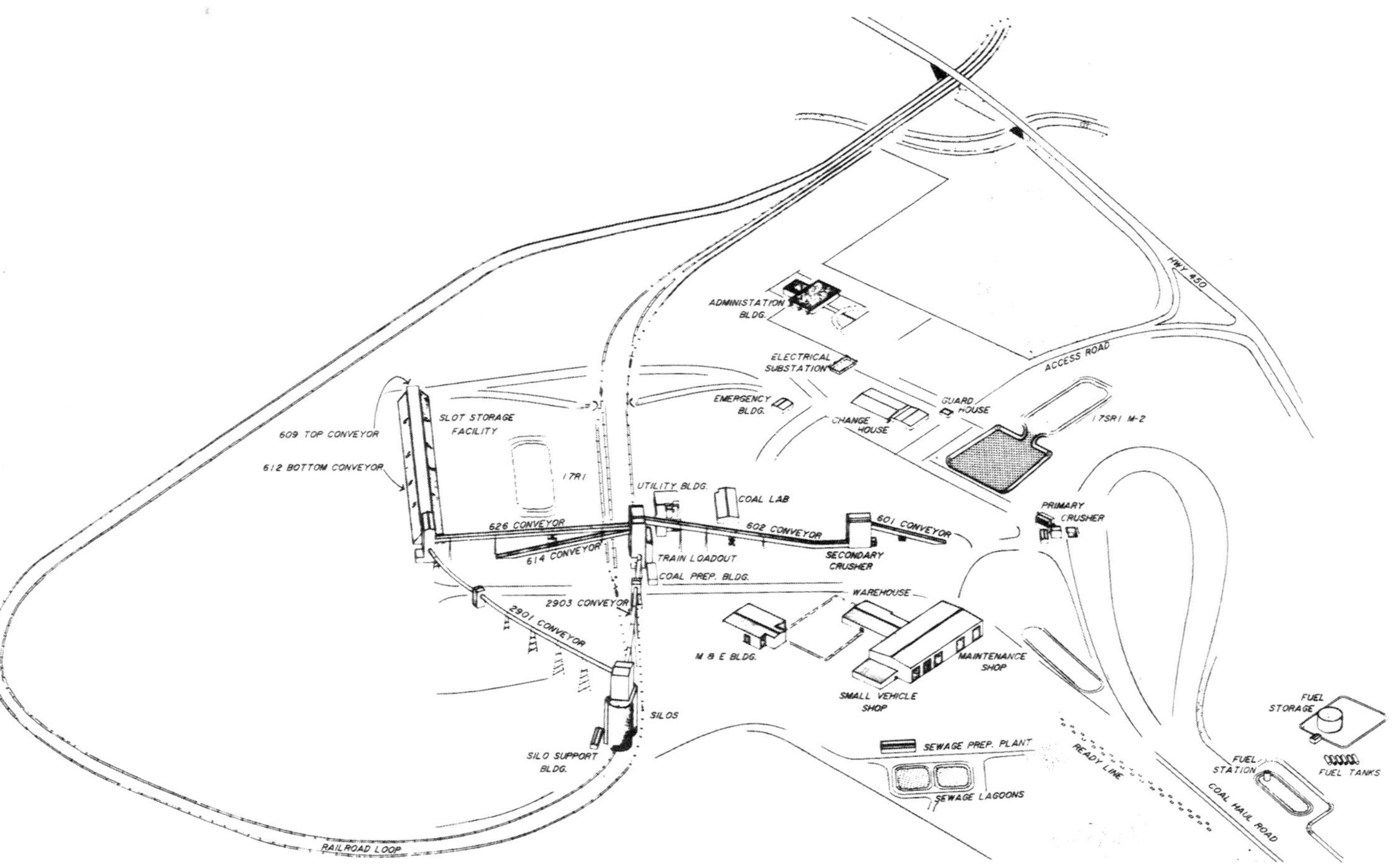

FIGURE 6.2. Plant layout of Black Thunder surface coal mine. (By permission from Thunder Basin Coal Co., Wright, WY.)

3. *Technological factors:* equipment, pit slope, bench height, road grade, property lines, transportation options, and pit limits

The pit planning team will most likely strive to optimize the pit design in respect to the technological factors. Most of the other factors are beyond their control and become part of the constraints.

The overall plan for the pit is then studied, including both the overall pit limit and the sequence of extraction. Many variables must be considered in this exercise so that the ore is brought into production as early as possible and the sequence is conducted without disrupting production or cash flow. Mathieson (1982) makes the point that the initial cash flow is very important, as the income generated during the first five or ten years of exploitation is more apt to make or break the mine than the economics of the long-term mine plan. In this regard, he lists a number of objectives that apply to pit planning in most open pit operations:

1. Mine the ore body so that the production cost per lb (kg) of metal is a minimum (i.e., mine the "next best ore" to generate income as early as possible).
2. Maintain proper operating parameters (adequate bench width and haul roads).
3. Maintain sufficient exposure of ore to overcome miscalculations or delays in drilling and blasting.
4. Defer stripping as long as possible without constraining equipment, manpower, or the production schedule.
5. Follow a logical and achievable start-up schedule (for training, equipment procurement and deployment, etc.) that minimizes the risk of delays in the initial cash flow.
6. Maximize pit slopes, while maintaining reasonably low likelihood of slope failure (provide safe berms, employ good rock mechanics, implement good slope monitoring systems, etc.).
7. Examine the economic merits of various production rates and cutoff grades.
8. Subject the favored choice of method, equipment, and pit sequence to exhaustive contingency planning before proceeding with development.

To accomplish these goals, the mine planning department may analyze the overall economics of the deposit and its extraction using several different alternatives. Most companies divide this task into the following three methods: (1) long-range mine planning, (2) short-range mine planning, and (3) production scheduling. The following sections outline these functions.

6.2.1 Long-Range Mine Planning

There are no formal definitions for long-range mine planning, but common practice is to apply this term to the general extraction plan for a mine with emphasis on the entire life of the mine or a major portion thereof. To accomplish this task, the mine is normally evaluated by dividing the deposit into relatively large geometric blocks and assigning values to each block based on the estimated ore grade within it. The possible extraction sequences are then analyzed to provide an estimate of the overall pit limits and the gross sequence of exploitation. A long-term mine plan is subject to change over the lifetime of the mine as the market and technological conditions change. However, it forms the basic plan of attack on the deposit and is important to the overall economic success of the mine. As a result, the long-range plan should be updated at regular intervals.

Ordinarily, the long-range mine plan will result from the evaluation of many possible pit limits, using the ore grades determined from the exploration drillholes. The economic determination of the *maximum allowable stripping ratio* (SR_{max}) is generally used to determine the pit limits (Pana and Davey, 1973). This ratio, determined solely by economics, establishes the ultimate boundary of the pit where breakeven occurs, that is, where the profit margin is zero. Mathematically, it is computed as follows:

$$SR_{max} = \frac{\text{value of ore} - \text{production cost}}{\text{stripping cost}} \tag{6.1}$$

The units in this equation are normally \$/ton (\$/tonne) in the numerator and $\$/yd^3$ ($\$/m^3$) in the denominator. The SR_{max} value is then provided in yd^3/ton ($m^3/tonne$). The numerator is often called the *stripping allowance* because it represents the amount of income that can be applied to the stripping function per ton (tonne) of ore mined without creating an economic loss.

In today's world, a long-range plan is normally produced by computer software that carefully evaluates all the blocks of ore in the deposit and determines which contribute to the profitability of the mine. Figure 6.3 shows a three-dimensional representation of an ultimate pit that may have been determined in a long-range mining plan. However, because the market price of the minerals and the costs of exploitation change over time, the ultimate pit limit determined in a long-range plan is dynamic in nature. It is therefore necessary to update a long-range plan periodically to guide the overall mining process. Note also that surface mining profitability of a given deposit does not automatically mean that the deposit should be mined by surface mining methods. It may be that underground methods are appropriate for the deposit if stripping ratios are high. In addition, it may be economically optimal to operate an open pit for the upper part of the ore body and an underground mine for deeper portions of the ore zone. Nilsson (1992) provides insights on methods of analyzing the surface-versus-underground decision.

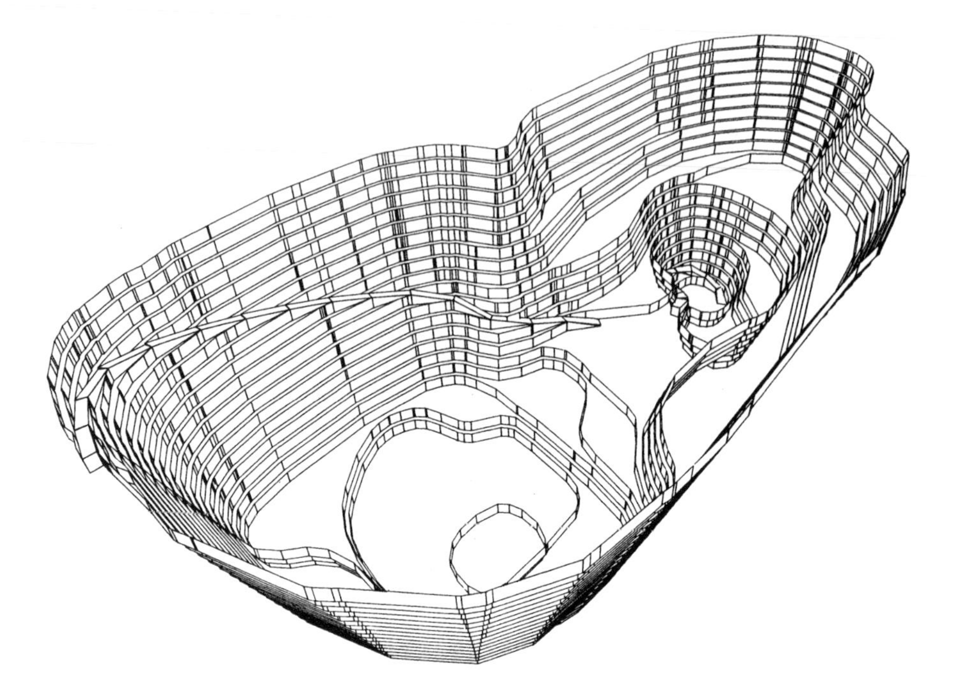

FIGURE 6.3. Computer simulation of pit boundaries of surface mine, three-dimensional depiction. (By permission from ACS Inc., Dallas, TX.)

6.2.2 Short-Range Mine Planning

Once a long-range surface mining plan has been completed, it is the usual practice to develop a series of short-range mining plans to better guide the mining process (Pana and Davey, 1973). A short-term plan ordinarily outlines a sequence of mining that dictates the blocks of ore and overburden to be mined a few months to 10 or more years in advance. In this plan, more attention is paid to better defining the ore and waste blocks as to grade and metallurgical characteristics. Smaller blocks of ore are normally utilized in this stage, and additional drillholes may be required to better estimate the block grades. Computerized grade determination is the norm; geostatistics may play an important part in the estimation of the block grades.

In determining a short-term mining plan, several of the objectives (Mathieson, 19982) listed in Section 6.2 will be key to the process. Of particular interest will be maintaining cash flow, stripping sufficiently ahead to keep new production faces available, and maintaining a constant ore grade to the processing facilities. The sequencing of the ore blocks is important, and the ability to analyze this problem may be quite complex. Operations research techniques such as mathematical programming, dynamic programming, critical path scheduling, and other methods may be used to aid in the establishment of a short-term mine plan.

6.2.3 Production Scheduling

Although there is no universally accepted definition of how a production schedule differs from a short-range mine plan, the term *production scheduling* is generally used to mean the assignment of production equipment to blocks in the pit on an hour-to-hour or shift-to-shift basis. Normally, the production schedule is drawn up for periods of less than a month or so, with emphasis on what is to be accomplished in the next few shifts. The production plan must work within the constraints of the short-range mine plan and is altered daily, or more often, to accommodate changes in the availability of the equipment and the new blocks of ore and waste that are prepared for mining (Kim, 1979).

Production scheduling is now more useful than ever before because of the ability of mining companies to segregate mined material within a given block into various products to be hauled to different destinations. This is particularly true in metalliferous mining but is practiced in other mine types as well. Scheduling at this level often requires improving the grade estimates by analysis of the cuttings from the production drillholes. This allows a mine production plan to divide the material broken in a blast into several categories. For example, the loader operator can identify ore, leach, and waste material and load them into separate haulage vehicles. In addition, the muck pile can be divided into subcategories to be processed in different metallurgical recovery systems. Production scheduling becomes a powerful tool when used in this manner.

6.3 EQUIPMENT AND SYSTEM SELECTION

The general principles of equipment and system selection were set forth in Section 5.4.3. Now we apply them to the operating circumstances of the surface mine being contemplated. It is not surprising that the selection of equipment is nearly synonymous with specifying the stripping method, because materials handling lies at the heart of surface mining.

In selecting a particular method of removing ore, coal, or waste, the ultimate aim is to excavate the material at the lowest possible cost, subject to the limitations imposed by other elements of the mining system. Many factors are involved; these differ in coal, metal, and nonmetal mining. Primary considerations are the size and shape of the ore body, distribution of mineral values within the deposit, and consolidation and compaction of the overburden. Additional variables of interest are the presence of geologic structures (faults, folds, shear zones, water-bearing formations, etc.) and alteration products (which may render haul roads impassable or affect mineral processing), production rate and mine life, horizontal haul and vertical hoist distances, and usefulness of stripping equipment for subsequent mining or reclamation.

The choices of equipment and systems to mechanically strip ore, coal, or stone deposits are relatively many, but not all will be considered for all mining operations (Martin et al., 1982; Atkinson, 1983). Three rock-breakage and six materials-handling systems are outlined here, followed by comments concerning their primary usefulness.

1. Rock breakage
   (a) No breakage necessary (material: typical soils)
   (b) Ripping with a dozer (material: stiff soil or weaker rock formations)
   (c) Drilling and blasting (material: medium to hard rock formations)
2. Materials handling
   (a) Dragline (direct casting)
   (b) Power shovel or front-end loader and trucks
   (c) Dozer and front-end loader
   (d) Dozer and rubber-tired scraper
   (e) Shovel or loader plus hopper, crusher, and belt conveyor
   (f) Bucket-wheel excavator and belt conveyor

The choice of a rock-breakage system is largely determined by the properties of the geologic materials to be mined. Rock-breakage methods increase in cost from first to third in the list. It is therefore the normal practice to choose the first method in the list that applies to the materials to be broken. When drilling and blasting are to be performed, it is necessary to choose a basic drill type (see Section 5.2.3.3) and a blasting agent (see Section 5.3.4). Figure 6.4 illustrates some of the choices available for drilling and blasting. Roller-bit rotary drills and ammonium nitrate–fuel oil (ANFO) are the most popular choices in surface mines.

In choosing a materials handling system, the pit geometry will play a major part in the type and size of the equipment that may be chosen. Figure 6.5 shows six large excavators for use in surface operations. Two of them, the dragline and the stripping (overburden) shovel, are direct casting devices that excavate the material and cast (dump) the material in its final resting place. The remaining excavators all require that the material be hauled to its final destination. Today, the stripping shovel and the bucket-wheel excavator have just about disappeared from use in North America. Table 6.3 (Martin et al., 1982) outlines selection guidelines for six materials-handling methods that are commonly used today. Note that deposit type, pit depth, and production rate play a big part in the choice of a system.

Haulage may also be a significant part of the selection process in choosing equipment. Five of the primary choices are illustrated in Figure 6.6. In the five haulage devices shown, only the scraper (Figure 6.6c) is also an excavator. The remaining devices must be loaded by a separate piece of equipment. Note that rail haulage is used sparingly today, primarily in the iron mining industry. The

FIGURE 6.4. Unit operations and equipment for surface mining: drilling and blasting. (a) Small-hole, percussion trac drill. (By permission from Gardner-Denver Co., Cooper Industries, Houston, TX.) (b) Large-hole, roller-bit, rotary drill. (By permission from Marion Div., Dresser Industries, Marion, OH.) (c) Bulk explosives truck (p. 166). (By permission from Orica USA, Watkins, CO.)

belt conveyor is gaining rapidly as a haulage device in some areas of mining, because it is low in cost and is now available in designs that can be easily moved and can turn horizontal curves. Belt conveyors ordinarily require that a crusher be used to prepare the material for haulage.

FIGURE 6.4. *Continued.*

6.4 STRIPPING RATIOS AND PIT LIMITS

6.4.1 Maximum versus Overall Stripping Ratio

It is by calculating stripping ratios that we are able to locate pit limits and to express volumes of overburden to be moved per unit weight of ore, coal, or stone recovered. The basic concepts used here to define stripping ratios were first introduced to the senior author of this book by E. P. Pfleider and are outlined in his book (Pfleider, 1968) by Soderberg and Rausch (1968). Generally, two stripping ratios are used, which are expressed in units of yd^3/ton (m^3/ton):

1. Maximum allowable stripping ratio SR_{max}
   = volume of overburden/weight of ore at the economic pit limit
   $= v/w$
2. Overall stripping ratio SR_0
   = volume of overburden/weight of ore for entire ore body or cross section
   $= V/W$

SR_{max} establishes the pit limit except in the unusual circumstance where (1) the

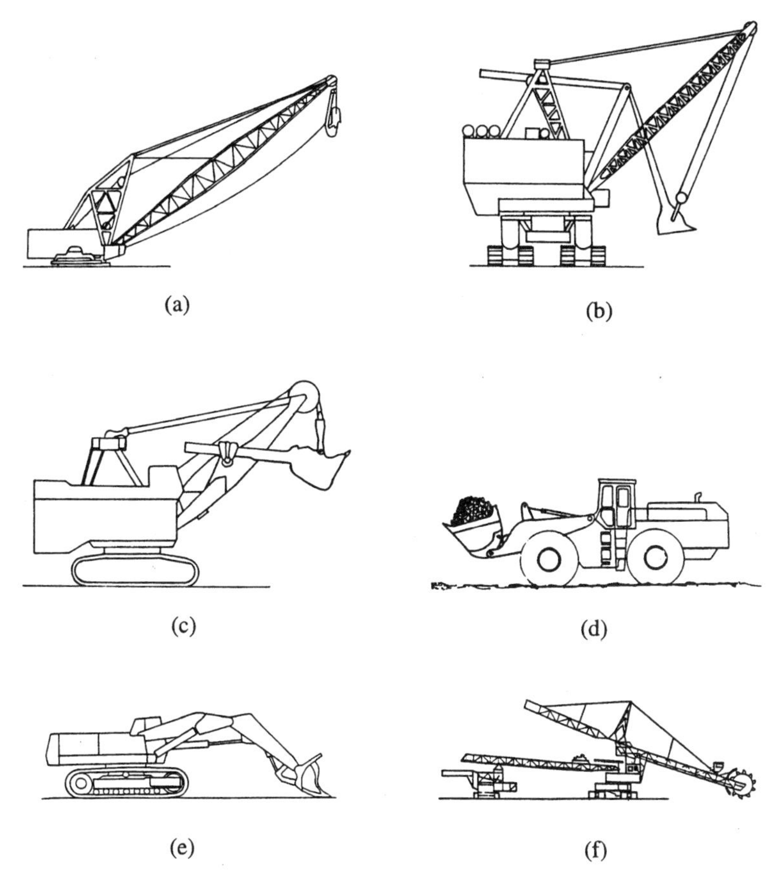

FIGURE 6.5. Large excavators for surface mining: (a) walking dragline, (b) stripping shovel, (c) loading shovel, (d) front-end loader, (e) hydraulic shovel, and (f) bucket-wheel excavator. *Source:* Martin et al. (1982). (By permission of Martin Consultants, Inc. Golden, CO.)

surface is flat and (2) the deposit is flat, tabular, and of constant thickness. In that case, the calculation of SR_{max} is valid, but the stripping ratio of the deposit never reaches that value, and the pit limit is at the boundary of the property. Note that SR_0 is an actual numerical ratio of yd^3/ton (m^3/tonne), but in this book the SR_{max} value is expressed in equivalent yd^3/ton (m^3/tonne) to aid in economic analysis (see Section 6.4.2).

In Section 6.2.1, we defined the maximum stripping ratio. Although this is a physical quantity, it is determined solely by economics. The overall stripping ratio has a more direct physical significance. Because of the economic basis for SR_{max}, it can be used to locate pit limits for ore bodies of varying dimensions and orientations and applies well to overburden materials that differ in diggability and other physical properties.

FIGURE 6.6. Unit operations and equipment for surface mining: rubber-tired haulage. (a) Diesel-electric rear-dump truck. (By permission from Terex Corporation, Westport, CT.) (b) Bottom-dump tractor-trailer. (By permission from Terex Corporation, Westport, CT.) (c) Wheeled scraper. (By permission from Terex Corporation, Westport, CT.) (d) Side-dump car. (By permission from Difco, Inc., Findlay, OH.) (e) Overland belt conveyor. (Photo courtesy of Continental Conveyor & Equipment Company, Winfield, AL.)

(c)

(d)

FIGURE 6.6. *Continued.*

FIGURE 6.6. *Continued.*

6.4.2 Equivalent Yards

The construction industry has often employed the term *equivalent yards*. Equivalent yardage is a unitless multiplier used to differentiate overburden materials with different diggabilities. A cubic yard (or cubic meter) of a standard material from a mine or district is assigned an equivalent yard value of 1.0. Other materials are then assigned an equivalent yard value or ratio that reflects their excavation cost in $/yd^3 ($/m^3) as compared with that for the standard material. The following are examples of possible standards and typical costs:

1. *Lake Superior iron ranges* (*loaded and hauled*)
   Glacial till: $0.25–0.50/yd^3 ($0.33–0.65/m^3)
2. *Eastern U.S. coal fields* (*cast overburden*)
   Soil or decomposed rock: $0.10–0.30/yd^3 ($0.13–0.39/m^3)
3. *Western U.S. porphyry copper deposits* (*blasted, loaded, hauled*)
   Quartz monzonite porphyry: $0.50–1.00/yd^3 ($0.65–1.31/m^3)

The equivalent yardage value e of a material is calculated with reference to one of the these standards, which are assigned a value of unity ($e = 1.0$). For

Table 6.3 Selection Guidelines for Materials Handling in Surface Mining

	Dozer–Front-end Loader	Dozer–Scraper	Dragline (direct casting)	Excavator–Truck	Excavator–Hopper–Crusher–Conveyor	Wheel Excavator–Conveyor
Maximum production	Medium	Medium	High	High	High	High
Production rate	Medium	Low	High	Medium	Medium	High
Pit life	Short	Short	Long	Medium	Long	Long
Pit depth	Medium	Flat and shallow	Medium	Deep	Deep	Medium
Deposit	Unconsolidated	Unconsolidated	Consolidated	Consolidated	Consolidated	Uniform, no large boulders
Preparation (if required)	Ripping	Ripping	Drill and blast	Drill and blast	Drill and blast	Drill and blast
System complexity	Low	Medium	Low	Medium	High	High
Operational flexibility	High	Medium	Low	High	Low	Low
Blending capability	High	High	Low	Medium	Low	Low
Selective placement (disposal)	Good	Excellent	Poor	Good	Medium	Medium
Wet weather impact	High	High	Low	Medium	Low	Low
Scheduling requirements	Low	High	Low	High	Medium	Medium
System availability	Medium	Medium	High	Medium	Low	Low
Support equipment	Low	Low	Medium	Medium	High	High
Ease of start-up	Simple	Simple	Moderate	Simple	Complex	Complex
Investment	Low	Low	Medium	Medium	High	High

Source: Martin et al. (1982). By permission of Martin Consultants, Inc., Golden, CO.

example, if the standard material is a soil with an equivalent yardage value of 1.0, then a sandstone with an equivalent yard value of 2.0 will have an excavation cost in $/yd^3 ($/m^3) that is twice that of the soil. A table of typical yardage ratings follows.

Material	Rating
Dredged mud, water	0.5
Loose sand	0.7
Common soil (sand, loam, till)	1.0
Hard soil (clay, hardpan)	1.5
Shaley rock	1.5–2.5
Sandstone, limestone	2.0–3.0
Hard taconite	3.0–5.0

The concept of equivalent yardage is useful in dealing with a variety of stripping materials, as discussed in the following sections.

6.4.3 Relationships for Maximum Stripping Ratio and Pit Limit

The physical and economic relationships of the maximum allowable stripping ratio at the pit limit enables us to develop a mathematical expression to locate the pit limit. We can do this using simplified geometric representations of the pit. In Figure 6.7a, an inclined mineral deposit intersecting a horizontal surface is presented in cross-sectional view. The deposit thickness is t, its dip is α, its tonnage factor is TF, and its inclined length to the pit limit is m. The pit slope is β, its inclined length is l, its vertical height is h, and the horizontal distance from the outcrop to the pit limit is d. Note that d is measured to the *crest* of the bank and m to the *toe*. For purposes of simplicity, we assume the thickness of the cross section to be 1.0 ft (see Figure 6.7b). The overall volume of the overburden is V, and the overall weight of the ore is W.

We can establish a geometric relation for SR_{max} if we assume that the cost of removing the cross-hatched quadrilateral of overburden in Figure 6.7a will equal the net value of the ore that is uncovered. The volume of overburden removed is designated v, and the tonnage of ore recovered is w. Mathematically, we can write for the overburden

$$v = \frac{1.0 \times b \times l}{27} \tag{6.2}$$

where b and l are in ft (m) and v is expressed in units of yd^3 (m^3). Note that the constant 27 must be dropped if meters are used as the unit in the equation. For the ore,

$$w = \frac{1.0 \times 1.0 \times t}{TF} \tag{6.3}$$

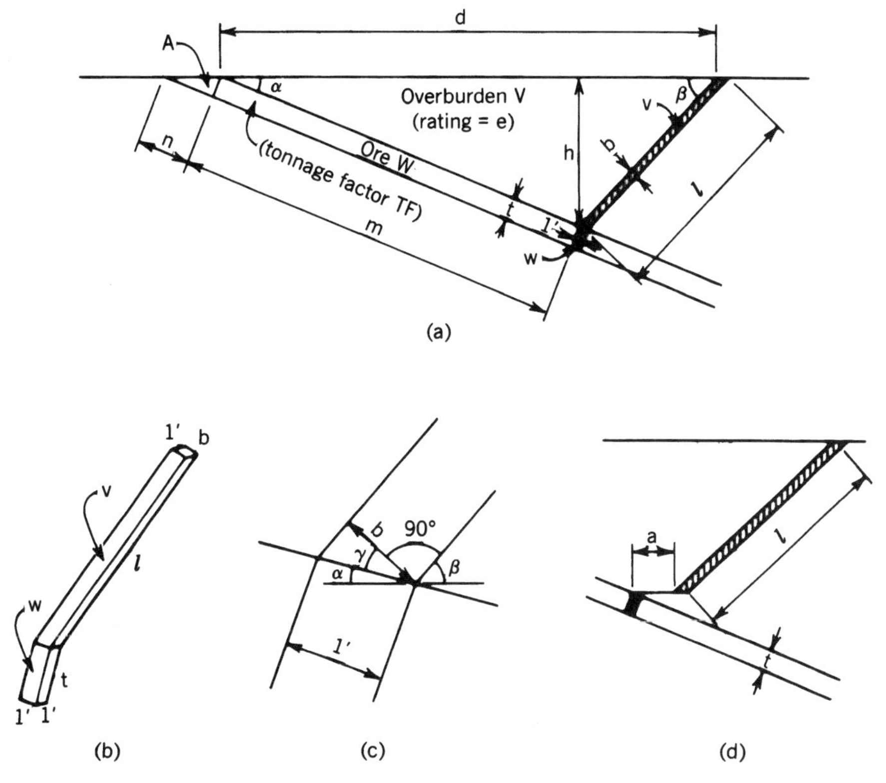

FIGURE 6.7. Geometric relation of pit parameters and SR_{max} at pit limit. (a) Cross section through ore body and overburden. (b) Detail at pit limit. (c) Detail at ore-overburden intersection. (d) Effect of berm at pit limit. Unit dimension indicated by 1′.

where t is in ft (m), TF is the tonnage factor in ft^3/ton (m^3/tonne), and w is expressed in tons (tonnes). The ratio of the two is the maximum allowable stripping ratio,

$$SR_{\text{max}} = \frac{v}{w} = \frac{ebl/27}{t/TF} \tag{6.4}$$

Again, omit the 27 when using SI units. Insertion of the equivalent yardage e in the equation will permit us to use the equation when different materials are present in the overburden. Finally, it is convenient to determine a geometric expression for b. Referring to Figure 6.7c, we see that

$$\alpha + \beta + \gamma + 90^\circ = 180^\circ$$

and therefore

$$b = 1.0 \times \cos \gamma = \cos(90^\circ - \alpha - \beta) \tag{6.5}$$

Because SR_{max} is calculated using the economic relationship in Eq. 6.1, we are in a position to determine l, the inclined slope at the economic pit limit. Rewriting Eq. 6.4, we obtain

$$l = \frac{27t \times SR_{max}}{eb \times TF} \tag{6.6}$$

By trigonometry, we can find the vertical height of pit slope h in ft (m) as

$$h = l \sin \beta \tag{6.7}$$

Assuming a berm width a as a safety feature (see Figure 6.7d), the horizontal distance from the outcrop m in ft (m) is

$$d = a + \frac{h}{\tan \alpha} + \frac{h}{\tan \beta} \tag{6.8}$$

Finally, the inclined length of ore from the outcrop m in ft (m) is

$$m = \frac{h}{\sin \alpha} \tag{6.9}$$

There are several variations of the previous deposit geometry that affect the determination of SR_{max} and the pit limits. Two of them are shown in Figure 6.8. In both, two different overburden formations occur, with equivalent yardage ratings of e_1 and e_2, respectively. In Figure 6.8a, the surface is horizontal and the deposit dips. In Figure 6.8b, the surface is inclined and the deposit is flat. In writing a relation for maximum stripping ratio, we modify Eq. 6.4 as follows:

$$SR_{max} = \frac{v}{w} = \frac{(e_1 b_1 l_1 + e_2 b_2 l_2)/27}{t/TF} \tag{6.10}$$

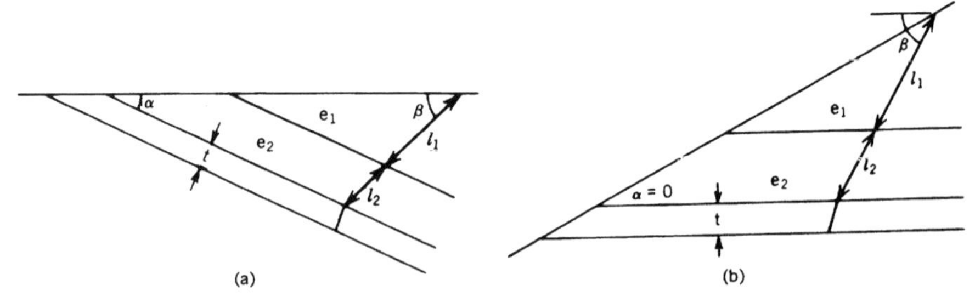

FIGURE 6.8. Variations in deposit geometry and overburden composition. (a) Typical dipping tabular noncoal deposit with different overburden formations e_1 and e_2. (b) Typical flat tabular coal deposit with different overburden formations e_1 and e_2.

If the bank angle is constant in different overburdens, then $b_1 = b_2$. After calculating a numerical value for l_2 using trigonometry, the equation can be solved for l_1, the only remaining unknown. Then the inclined pit slope $l = l_1 + l_2$, and values of h, d, and m can be found using Eqs. 6.7, 6.8, and 6.9, modified trigonometrically if the surface slopes. One caution in calculating the value of b using Eq. 6.5: both α and β should be measured from the horizontal. An error in b will result otherwise.

If there are two variables in the deposit geometry (i.e., both t and l vary), then Eq. 6.6 must be solved by trial and error. In this case, a graphical solution may be quicker than an algebraic one.

6.4.4 Determination of Overall Stripping Ratio

In calculating the potential return on a surface mining venture, the overall stripping ratio is more important than the maximum stripping ratio. We must therefore also calculate the overall stripping ratio for a given cross section or for the entire mine. In our case, we may now want to determine SR_o for a cross section where we have already calculated SR_{max}.

Referring to the general case in Figure 6.7, the total volume of overburden V in yd^3 is found trigonometrically as

$$V = \frac{1.0 \times \frac{1}{2} \times h \times d}{27} \tag{6.11}$$

Similarly, the total weight of ore W in tons (tonnes) is

$$W = \frac{1.0 \times m \times t}{TF} \tag{6.12}$$

Therefore, the resulting expression for overall stripping ratio in yd^3/ton is

$$SR_o = \frac{V}{W} = \frac{\frac{1}{2}hd/27}{mt/TF} \tag{6.13}$$

Note that Eqs. 6.11 through 6.13 must be altered slightly to produce answers in m^3/tonne. Eq. 6.13 ignores the area A in Figure 6.7a. To correct the volume of ore removed, that area must be calculated trigonometrically and added to W.

The configuration of surface mines used here is simplified to illustrate the stripping ratio principles. In practice, the cross sections will not normally consist of simple geometric figures. Therefore, most of the cross sections must be analyzed using area determinations that are more sophisticated than those demonstrated here. For many years, engineers performed their volume calculations using polar planimeters (area measurement devices). Today they have software programs and digitizing equipment that perform the same function mathematically in less time and with more reliable results. These calculations are now routinely performed by computer systems that determine pit limits, calculate reserves, perform pit planning, and estimate grades and costs.

6.5 SPECIAL TOPIC: CALCULATION OF STRIPPING RATIOS AND PIT LIMITS

Using the basic equations developed in this chapter, we can now apply the principles to calculate the maximum and overall stripping ratios and the pit limit for ore, coal, and stone deposits. The following example outlines the procedure for a dipping mineral deposit of modest value.

Example 6.1. The following data are given for a mineral deposit occurring under conditions similar to those in Figure 6.7:

Value of ore = \$4.80/ton (\$5.29/tonne)
Costs (excluding stripping) = \$3.30/ton (\$3.64/tonne)
Stripping cost (for overburden of $e = 1$) = \$0.20/ton (\$0.26/tonne)
Berm dimension $a = 0$ ft (0 m)
Dip of deposit $\alpha = 20^\circ$
Pit slope $\beta = 60^\circ$
Deposit thickness $t = 50$ ft (15.2 m)
Equivalent yardage $e = 2.0$
Tonnage factor $TF = 15$ ft^3/ton (0.47 m^3/tonne)

(a) Calculate SR_{max}; (b) locate the pit limit for the deposit in terms of h; and (c) calculate SR_o.

SOLUTION. (a) Use Eqs. 6.1, 6.5, 6.6, and 6.7 to find SR_{max} and the pit limit:

Stripping allowance (for breakeven or zero profit)

$$= \text{value} - \text{cost} = 4.80 - 3.30 = \$1.50/\text{ton}$$

(b) $b = \cos(90^\circ - 20^\circ - 60^\circ) = \cos 10^\circ = 0.9848$

$$l = \frac{27(50)(7.5)}{(2)(0.9848)(15)} = 343 \text{ ft}$$

$$h = 343 \sin 60^\circ = 343\,(0.8660) = 297 \text{ ft}$$

(c) Use Eqs. 6.8, 6.9, and 6.13 to find SR_o:

$$d = 0 + \frac{297}{\tan 20^\circ} + \frac{297}{\tan 60^\circ} = 816 + 171 = 987\,\text{ft}$$

$$m = \frac{297}{\sin 20^\circ} = 868\ \text{ft}$$

$$n = \frac{50}{\tan 20^\circ} = 137\ \text{ft}$$

$$A = \frac{1}{2}\,(50)(137) = 3427\ \text{ft}^2$$

$$SR_o = \frac{\frac{1}{2}(297)987)/27}{[3427 + (868)(50)]/15} = 1.739 \text{ or about } 1.7\ \text{yd}^3/\text{ton}\ (1.5\ \text{m}^3/\text{tonne})$$

If a berm had been used in this problem, there would be no change in SR_{max} or h, but SR_o would increase.

PROBLEMS

6.1 Calculate the maximum allowable stripping ratio, locate the pit limit (specify h = vertical depth to the deposit and d = horizontal distance along the surface from the outcrop), and calculate the overall stripping ratio (actual yd^3/ton or m^3/tonne) in mining the ore body shown in Figure 6.9 by surface methods. Allow a 30 ft (9.1 m) berm and a pit slope of 45°. Cost of excavating an equivalent yard of overburden is \$0.35/yd^3 (\$0.46/m^3), the stripping allowance is \$2.10/ton (\$2.31/tonne) of ore uncovered, and the ore tonnage factor is 16 ft^3/ton (0.50 m^3/tonne). Work the problem first in English units, then in SI units, and check the results to be certain they are compatible.

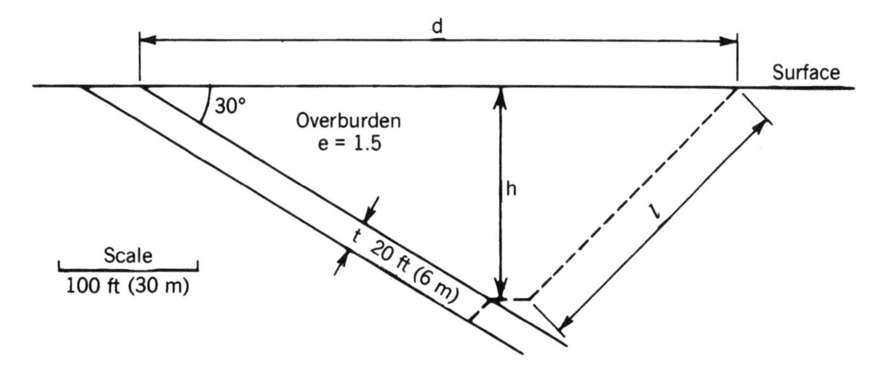

FIGURE 6.9. Cross section of surface mine in Problem 6.1.

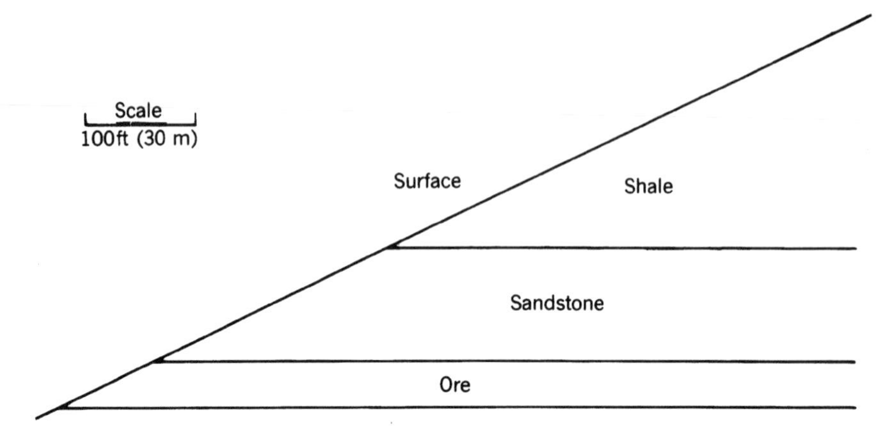

FIGURE 6.10. Cross section of surface mine in Problem 6.2.

6.2 Calculate the maximum stripping ratio and the overall stripping ratio for the open pit shown in Figure 6.10, given the following conditions:

Stripping allowance	\$1.50/ton (\$1.65/tonne) of ore
Stripping cost ($e = 1$)	\$0.25/yd^3 (\$0.33/m^3) of overburden
Pit slope	65°
Berm, at top of ore	30 ft (9.1 m)
Surface inclination	25°
Dip of bed	0°
Ore thickness	40 ft (12.2 m)
Sandstone thickness	100 ft (30.5 m)
Tonnage factor	12 ft^3/ton (0.37 m^3/tonne)

Equivalent yards, overburden:

Shale	1.5
Sandstone	2.5

Locate and draw the economic pit limit to scale for the cross section shown.

6.3 Locate and sketch the pit limit (expressed as the vertical depth of overburden at the maximum stripping ratio) for the coal deposit shown in Figure 6.11, assuming that a berm is not used. The following conditions are given:

Value of coal	\$24/ton (\$26/tonne)
Production cost	\$12/ton (\$13/tonne)
Stripping cost	\$0.60/yd^3 (\$0.78/m^3)
Soil thickness	80 ft (24.4 m)

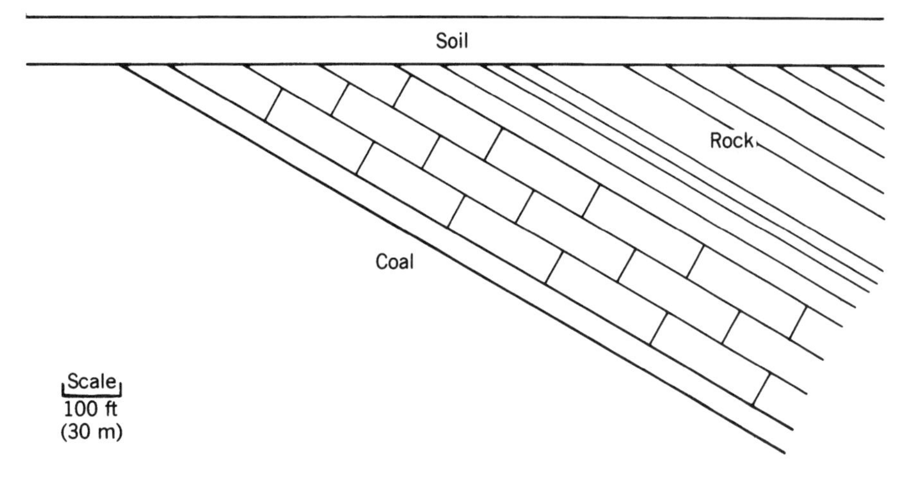

FIGURE 6.11. Cross section of surface mine in Problem 6.3.

Pit slope	56°
Dip of the seam	33.7°
Coal thickness	40 ft (12.2 m)

Equivalent yards, overburden:

Soil	0.5
Rock	2.5
Tonnage factor	24.5 ft^3/ton (0.765 m^3/tonne)

Calculate the maximum and overall stripping ratios.

6.4 Locate the pit limit (t, l, h, and d) by trial and error for the ore body shown in Figure 6.12 under the following conditions:

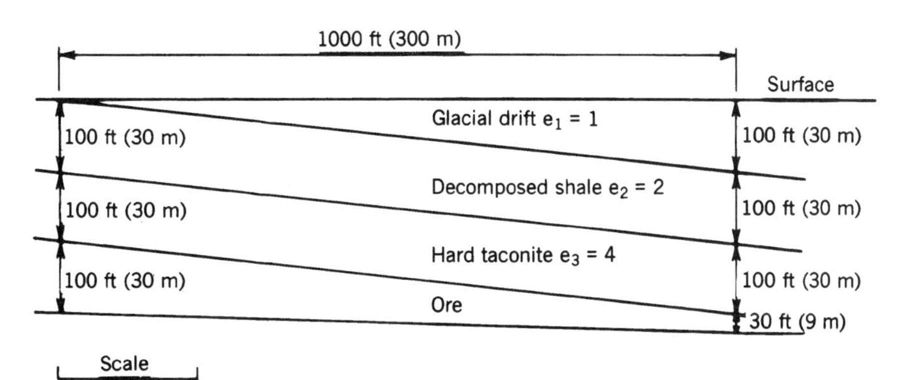

FIGURE 6.12. Cross section of surface mine in Problem 6.4.

Maximum stripping ratio	6.0 yd^3/ton (5.1 m^3/tonne)
Pit slope	35°
Tonnage factor	13.5 ft^3/ton (0.421 m^3/tonne)

Give answers to the nearest 5 ft (1.5 m), and sketch the pit limit to scale.

6.5 A vertical kimberlite pipe, cylindrical in shape with a diameter of 300 ft (91.4 m), is to be exploited by open pit mining. The pit slopes in the kimberlite and the surrounding overburden will be 45°. If the kimberlite is worth $20/ton ($22.05/tonne) and the rock costs $4/ton ($4.41/tonne) to remove, what is the depth of the pit at the maximum stripping ratio? Assume that the pit limit is an inverted cone and that no berms are left in the ore or the rock. Does the cross-sectional approach work in this case? If not, why not?

7

SURFACE MINING: MECHANICAL EXTRACTION METHODS

7.1 CLASSIFICATION OF METHODS

Exploitation in which mining of ore, coal, or stone is carried out at the surface with essentially no exposure of miners underground is referred to as *surface mining*. Although occasional openings may be placed below the surface and limited underground development is occasionally required, this type of mining is essentially surface-based.

In Table 4.1, the surface mining methods were divided into two classes. In this chapter, we turn our attention to the *mechanical extraction* methods. The four methods in this class are as follows:

1. Open pit mining
2. Quarrying
3. Open cast (strip) mining
4. Auger or highwall mining

These methods are responsible for more than 90% of the surface mine production in the United States and the bulk of the nation's total tonnage of coal, ore, and stone. Two of these methods—open pit and open cast mining—rank as the most important surface methods and are among the eight most important of all the methods.

The terminology of surface mining applied to the mining of stone should be clarified at this point. The term *quarry* is used in the mineral industry and the general literature to mean a mine that produces stone or aggregate. The stone industry even uses the term to describe an underground operation that mines stone. However, this term does not describe a mining method, only the product mined. Most surface quarries use the open pit mining method; most underground quarries utilize the room-and-pillar or stope-and-pillar mining method. In this book, we define *quarrying* as the mining method associated with the production of intact blocks of rock called *dimension stone*, typically for architectural or decorative use.

Applicable to many near-surface deposits, the open pit and open cast mining methods employ a conventional mining cycle of operations to extract mineral. Rock breakage is normally accomplished by drilling and blasting, followed by the materials handling operations of loading and haulage. Quarrying and augering are more specialized and less frequently used methods in which breakage is by alternative means. Explosives are not normally used for breakage, although they can be used in quarrying to free large blocks of stone. The variations of these two methods are outlined in the sections that follow. The student should prepare a sheet on each method for his or her Mining Methods Notebook; see Scction 4.10.

7.2 OPEN PIT MINING

Open pit mining is the process of mining any near-surface deposit by means of a surface pit excavated using one or more horizontal benches. Open pit mines are often used in mining metallic or nonmetallic deposits and more sparingly in coal and other bedded deposits. Both the overburden (if present) and the ore are typically removed in benches that vary from 30 ft (9 m) to 100 ft (30 m) in height. By adding additional benches, a pit of any depth can be extracted. A thick deposit requires many benches and may resemble an inverted cone, with the higher benches being larger than the lower benches (see Figure 7.1). A single bench may suffice if the deposit and overburden are relatively thin (50–150 ft or 15–45 m), which is typical of some U.S. coal and nonmetallic ores.

The purpose of the benches is to control the depth of the blastholes, the slope of the pit walls, and the dangers of highwall faces. The benches also provide enough length of face to allow sustained, uninterrupted production. After advanced stripping uncovers the deposit, stripping and mining are coordinated so that ore revenues will reimburse waste costs, while at the same time long-range objectives are being met.

Individual benches are designed to accommodate the materials-handling equipment utilized. The reach of the excavator limits the height of the bench; a power shovel can trim a higher bank than a front-end loader or a hydraulic excavator. The width must be sufficient to contain most of the flyrock from a bench blast and provide maneuvering room for excavator and haulage units. The slope of the bench and of the pit itself is the maximum dictated by rock or soil mechanics concerns (see Section 7.7). Common practice is as follows:

FIGURE 7.1. Bingham Canyon Mine near Salt Lake City, UT. Photo courtesy of Kennecott Utah Copper Corporation.

	Bench Dimensions		
Mineral	Height, ft (m)	Width, ft (m)	Slope
Copper	40–60 (12–18)	80–125 (24–38)	50°–60°
Iron	30–45 (9–14)	60–100 (18–30)	60°–70°
Nonmetallics	40–100 (12–30)	60–150 (18–45)	50°–60°
Coal (Western U.S.)	50–75 (15–23)	50–100 (15–30)	60°–70°

Open pit mining is a large-scale method in terms of production rate, responsible for more than 60% of all surface output. It continues to produce more mineral resources at lower cost to enable the mining of ever-decreasing grades of most metallic deposits. It permits the utilization of highly mechanized, mass production equipment that is capital intensive but labor conserving (Martin et al., 1982). Figure 7.2 shows some of the variations of open pit mining. Note that it can be used on flat-lying seams (Figure 7.2a) and that it is often used where several seams exist such as in some iron or coal measures. It also applies to a number of other deposit types, as shown in Figures 7.2b through 7.2e.

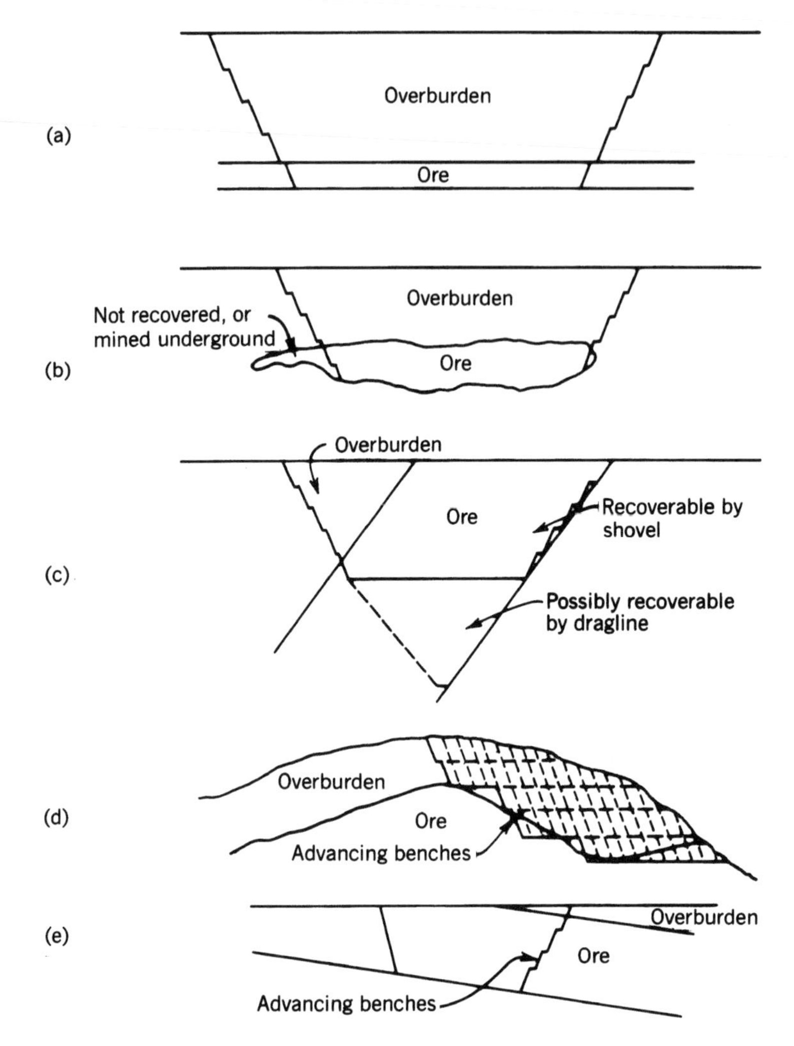

FIGURE 7.2. Variations of open pit mining. (a) Flat-lying seam or bed, flat terrain. *Example:* iron, taconite. (b) Massive deposit, flat terrain. *Example:* iron. (c) Pitching seam or bed, flat terrain. *Example:* anthracite. (d) Massive deposit, high relief. *Example:* copper. (e) Thick-bedded deposits, little overburden. *Example:* nonmetallics, western U.S. coal.

7.2.1 Sequence of Development

The procedure for laying out a surface mine was detailed in Chapter 6. The sections that have special applicability to open pit development are Sections 6.1 and 6.2. They should be reviewed now. Note that open pit mining normally involves the haulage of moderate to large amounts of waste and ore out of the pit for relatively long distances at steep grades. These haulage requirements will affect the layout of the pit, the selection of equipment, and the prescribed production rate. Because ore grades are typically low, and in most commodities, decreasing steadily, ratios must be held to modest levels (usually 1 to 5 yd^3/ton, or 0.8 to 4 m^3/tonne). Thus, most open pit mines are less than 1000 ft (300 m) in depth unless they are associated with a deposit of better than average grade, stripping ratio, or size (Crawford and Hustrulid, 1979; Atkinson, 1983).

Major steps in the development of an open pit mine are as follows: After all permits are obtained, the land is cleared. Surface buildings are located and constructed as described in Section 4.2. Particularly important is the location of waste dumps and, if reclamation of the surface is required, topsoil stockpiles. Ore storage, processing, and storage facilities are located, keeping the ultimate pit limit and outside access in mind. Equipment is selected and acquired as needed. Advanced stripping of overburden then commences to allow exploitation of the ore to begin on schedule. Stripping and mining are later conducted in a carefully coordinated manner, in keeping with short-range and long-range mining plans.

Establishment of the first bench and each succeeding bench in waste or ore is a critical operation. The initial entry in a bench is referred to as the *box cut* (or drop cut), a wedge-shaped volume of rock that must be removed to establish a new bench face. Drillholes are placed in parallel rows in descending order of depth so that when the box cut is blasted and excavated, there is a ramp of negotiable grade from the upper to the lower bench. The many considerations in executing a box cut are outlined in detail by Hustrulid and Kuchta (1995).

A major concern in pit development is the proper design of the benches and the haul roads. The excavator reach and other dimensions of the equipment used generally determine the height and width of the working benches. Pits are often developed using a *working slope*, a slope angle with a relatively high factor of safety so that the slopes are stable during the exploitation of the deposit. When the pit is in its final stage, the slopes may be steepened to achieve a lower stripping ratio. The maximum slope of the pit is then called the *ultimate slope*. Particularly when the ultimate slope is being used, the benches must be designed as *catch benches* so that any material that is dislodged from the pit walls is caught on the benches. Call (1986) has outlined recommendations for the design of these benches.

The design of haul roads is important for both safety and efficiency. The best-managed pits will have proper haul road width, banking, safety berms, and curve designs to allow the trucks to operate with the greatest safety and a low rolling resistance. Haulage trucks in many pits travel on the left side of the road to allow the driver to gauge the location of the vehicle with respect to the berm. This decreases the probability of misjudging the location of the berm and accidentally driving off the road. The many aspects of proper haul road design are discussed by Kaufman and Ault (1977), Atkinson (1992b), and Hustrulid and Kuchta (1995).

7.2.2 Cycle of Operations

The requirements for exploitation of an ore body include the following: stripping overburden, mining the valuable minerals, and the auxiliary operations that will enable the operation to proceed in a safe and efficient manner. Each of these is examined separately.

7.2.2.1 Stripping Overburden. *Stripping* is the term applied to removing overlying material to expose the deposit and excavating overburden within the confines of the pit after the ore is exposed. The nature of the overburden determines the cycle of operations. Some softer materials may not require breakage. More consolidated rocks will require breakage by explosives. Materials-handling equipment is then selected to satisfy the operating conditions. Systems of equipment were compared in Table 6.3. Additional information on the selection of equipment can be found in Martin et al. (1982) and Atkinson (1983, 1992a).

The following are alternative methods for performing each of the steps in the cycle of operations:

Drilling: auger (weak rock), roller-bit rotary (average rock), percussion (hard rock)

Blasting: ammonium nitrate–fuel oil (ANFO) or slurry, loaded by bulk explosive trucks or by hand, firing by electrical caps or detonating cord

Excavation: power shovel, hydraulic shovel, front-end loader, dozer, scraper (soil), bucket-wheel excavator (soil)

Haulage: truck, belt conveyor, dozer, scraper (soil)

7.2.2.2 Mining Ore, Coal, or Stone. In open pit stripping and mining, equipment and cycles of operations can often be very similar or even identical—the determining factor is the difference or similarity of the ore and waste. If they are very much alike, then it is to the mine operator's advantage to employ the same equipment and cycle. The primary reason is that equipment can be interchanged at times when breakdowns occur or when unexpected production or stripping demands occur. The mining cycle of operations and the equipment used generally consist of the following:

Drilling: roller-bit rotary (average rock), percussion or rotary-percussion (hard rock)

Blasting: ammonium nitrate and fuel oil (ANFO) or slurry (*alternative:* soft rock can be ripped, coal can be loaded directly), loading and firing similar to the processes used in stripping

Excavation: power shovel, hydraulic shovel, front-end loader, dragline, scraper (soil-like ores)

Haulage: truck, belt conveyor, rail

Hoisting (very steep pits): high-angle conveyor, skip hoist, hydraulic pipeline

7.2.2.3 Auxiliary Operations. Auxiliary operations are similar for stripping and mining. They are outlined in Table 5.12 for the general case. For open pit mining, the auxiliary operations that will be of concern are slope stability, dust control, pumping and drainage, waste disposal, maintenance of equipment and haul roads, and personnel transport. Environmental control must also be emphasized throughout the lifetime of the mine. Reclamation is now an auxiliary operation that is pursued during exploitation to ease any environmental problems and minimize the expense of reclaiming after mining is terminated.

7.2.3 Conditions

The natural, spatial, and geologic conditions associated with successful open pit operations are indicated in the following list, which has been compiled from information in a variety of sources. It should be remembered that these represent the most logical circumstances under which the mining method known as open pit mining can be applied.

1. *Ore strength:* any
2. *Rock strength:* any
3. *Deposit shape:* any, but prefer deposits parallel to the surface
4. *Deposit dip:* any, prefer deposits with low dip
5. *Deposit size:* large or thick
6. *Ore grade:* can be very low if other conditions are favorable
7. *Ore uniformity:* prefer uniform ore, but blending can be easily implemented in most operations
8. *Depth:* shallow to intermediate (limited by the economic strip ratio)

The interaction between these deposit conditions can alter the variables that apply to a mining method. For example, a deposit of high grade and large extent can be mined to a much greater depth. Note also that ore strength and rock strength will be referred to in general terms. For more specific strength values, readers may use the data provided in Table 14.2.

7.2.4 Characteristics

The characteristics that normally apply to any mining method affect the choice of the method. These are summarized in the following lists of advantages and disadvantages This information has been compiled from discussions in Pfleider (1973), Crawford and Hustrulid (1979), and Hustrulid and Kuchta (1995).

Advantages

1. High productivity (U.S. averages for copper and iron mines have been in the range of 100 to 400 tons or 90 to 360 tonnes per employee-shift, including both ore and waste).
2. Lowest cost of the broadly used methods (relative cost about 5%).
3. High production rate (in most mines, production can be increased by increasing the number of excavation units).
4. Low labor requirement; can be relatively unskilled labor for the most part.
5. Relatively flexible; can vary output if demand changes.
6. Ideal for large equipment, which permits high productivity.
7. Fairly low rock breakage cost; superior to underground mining.
8. Simple development and access; minimal openings required, although advanced stripping may be considerable.
9. Little support normally required; proper design and maintenance of benches can provide stability.
10. Good recovery (approaches 100%); moderate to low dilution.
11. Favorable health and safety factors; no underground hazards.

Disadvantages

1. Limited by depth to about 1000 ft (300 m) because of the technological limits of equipment; deposit beyond pit limits must be mined underground or left in place.
2. Limited by stripping ratio (range of 1 to 5 yd^3/ton or 0.8 to 4 m^3/tonne) because of economics.
3. High capital investment associated with large equipment.
4. Surface may require extensive reclamation, an expense added to the production cost.
5. Requires large deposit and large equipment to achieve lowest cost.
6. Weather detrimental; can impede or prohibit operations.
7. Slope stability is critical; proper design and maintenance of benches plus good drainage essential.
8. Must provide waste disposal; provision of dump area and proper dump design is essential.
9. Pit may fill with water after mining; water may be polluted.

7.2.5 Applications

7.2.5.1 Open pit mining has been applied to many types of mineral deposits. In the United States the method has been extensively applied to copper, gold, and iron deposits. However, it has also been applied to coal, uranium, bauxite, molybdenum, and most nonmetallics. Open pit mining is also the most common method used in surface limestone and other surface stone quarries.

7.2.5.2 Case Study: Bingham Canyon Mine. Kennecott Utah Copper Corporation's Bingham Canyon copper mine is located about 30 miles (42 km) southwest of Salt Lake City, Utah. It was the world's first open pit mine and is still operating today. The history of Bingham Canyon began with underground mining in the 1800s. As the high-grade copper ores were being depleted, a young engineer, Daniel C. Jackling, proposed in 1898 the mining of lower-grade copper mineralization by excavation of a large open pit. In 1906, the Bingham Canyon open pit mine was initiated.

Bingham Canyon proved that the open pit concept was economical, and, as a result, the mine grew into the largest human-made excavation of its type on earth. The pit is shown in Figure 7.1. The production system eventually grew into a shovel-rail system for ore production and a shovel-truck combination for stripping waste. In 1980 the mine was using 150 ton (136 tonne) trucks and was producing 370,000 tons (336,000 tonnes) of ore and waste per day (Voynick, 1998). However, low copper prices pushed the mine into unprofitable operation, and in 1985 its operators closed it in order to rethink and redesign the mine and the mill complex.

After new labor agreements and a new mine plan were in place, the mine reopened in 1987. The new layout, shown in Figure 7.3, was instituted to modernize the operations, cut down on labor and transportation costs, and provide a lower overall cost of producing copper. Gone were the trains, replaced by a shovel-truck-crusher-conveyor system to deliver the ore to the new Copperton Concentrator Plant. Hydraulic pipelines were established to move concentrate to the smelter, tailings to the waste impoundment, and water to the mine. In addition to its primary target (copper), gold, silver, and molybdenum are processed. The approximate size of the workforce was reduced from 2,500 to 475 after the new plan was in place. This economic rejuvenation added about 25 years to the mine's life as an open pit mine (Tygesen, 1992). Recently, the total production of ore and waste was back up to 320,000 tons (290,000 tonnes) per day. Carter (1990) has listed details concerning the pit and its logistics:

- Longest horizontal dimension of pit: 2.5 mi (4.0 km)
- Pit depth: 0.5 mi (0.8 km)
- Drills: roller-bit rotary drilling of 12 in. (0.3 m) holes
- Shovel size: 34 yd^3 (26 m^3)
- Truck size: 190 tons (172 tonnes)
- In-pit crusher: 10 in. (0.25 m) gap; 1000 hp (745 kW); 10,000 tons/hr (9070 tonnes/hr)
- Belt conveyor: 72 in. (1.8 m) wide, 5 mi (8 km) long, 16,500 hp (12,300 kW)

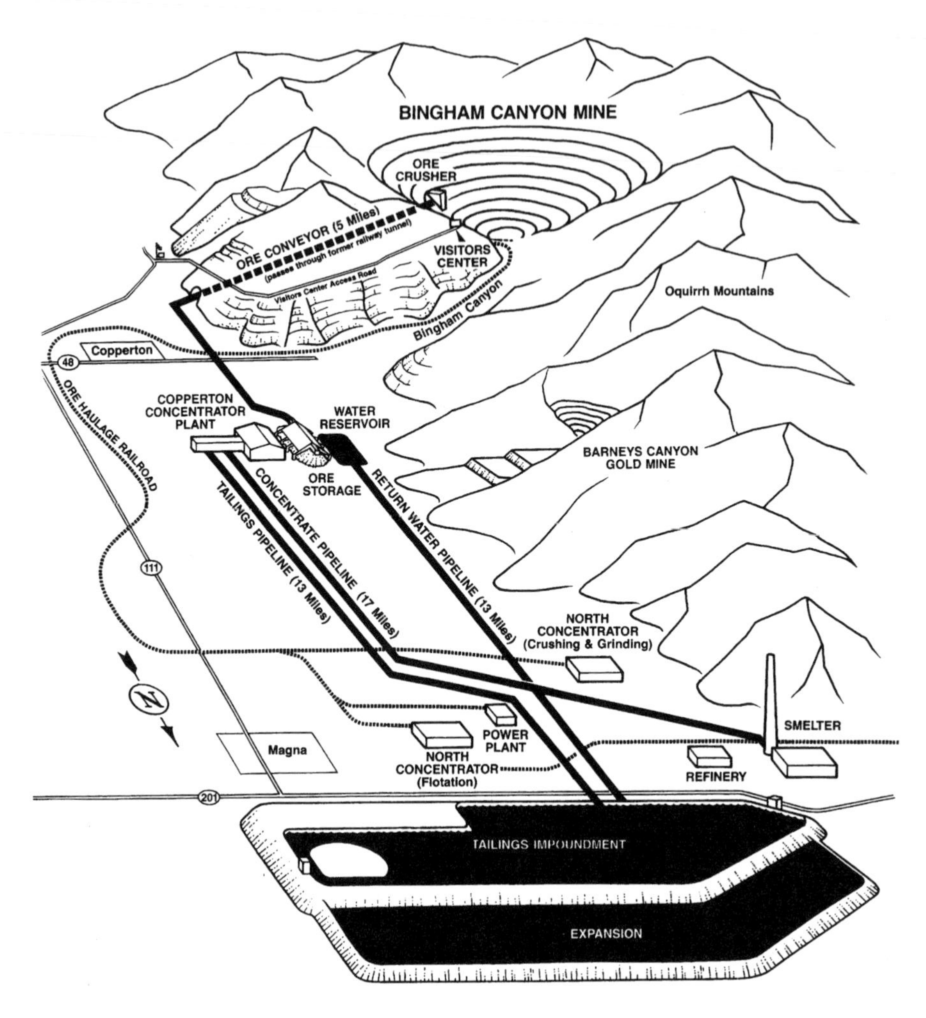

FIGURE 7.3. Layout of the mine, concentrator, smelter, refinery, and related facilities of the Bingham Canyon Mine. Illustration courtesy of Kennecott Utah Copper Corporation.

Concentrator storage: 350,000 tons (317,000 tonnes)
Metals recovered: Cu, Mo, Au, Ag
Strip ratio: 0.83/1.0
Average grade: 0.6% Cu
Cutoff grade: 0.23% Cu

The Bingham Canyon operation also continues to extract some copper from very low grade ores by leaching it with sulfuric acid and processing the pregnant liquor in 13 precipitation cones in which shredded iron material is sacrificed to collect the copper (Tygesen, 1992). This process provides additional copper that would otherwise be wasted. Bingham Canyon is a worthwhile case study. It has been operating for nearly a century, proving the value of open pit mining to the mining industry and to consumers.

7.3 QUARRYING

Dimension stone *quarrying* produces rectangular blocks of rock that are roughly sized and shaped. Surface quarries resemble open pits, but the benches (called faces) are lower and generally vertical (see Figure 7.4). The highwalls of a quarry are often of imposing height and steepness, some attaining a vertical dimension approaching 1,000 ft (300 m). Although the term *quarry* is sometimes applied to any surface mine extracting a stone product, it is used here to mean the mining method employed to produce intact blocks of dimension stone. Thus, crushed limestone may be mined in a quarry, but it is produced by the mining method known as open pit mining. By the same logic, dimension stone is extracted from a quarry using the mining method we call quarrying.

Dimension stone is used for architectural building blocks, stone monuments, decorative building slabs, flagstone, curbing, roofing, and miscellaneous other uses. However, the dimension stone industry is receding in importance. The tonnage produced dropped about 30% over a period of 15 years to about 1,080,000 tons (980,000 tonnes) in 1998 (Antonides, 1999). The value of the stone was somewhat less than \$200/ton (\$220/tonne), making this mining method one of the most expensive. The reason is that it is a highly selective, small-scale, method with low productivity. Quarrying and square-set stoping are the most costly mining methods in use today, which accounts for their being used only where other methods do not apply.

7.3.1 Sequence of Development

The properties that make a deposit of rock commercially viable as dimension stone are largely physical and mineralogical: color and appearance, competence, uniformity, strength, and freedom from cracks, flaws, and other defects (Singleton, 1980). Therefore, prospecting, exploration, and development are often carried out in quite a different manner than for other mineral commodities.

Development begins with clearing the land and developing the processing and necessary support facilities adjacent to the quarry site. The processing plant is ordinarily used to cut and polish the rock. However, sculpting and polishing may be performed elsewhere if required. Because blocks are cut roughly to size in the quarry, all mining and processing operations are geared to handle individual blocks of stone. Waste is often substantial in the cutting of the stone to more exact dimensions. Therefore, a disposal area must be prepared for the waste produced in processing. Any overburden or weathered

FIGURE 7.4. Dimension-stone quarry. (By permission from Imerys, Roswell, GA.)

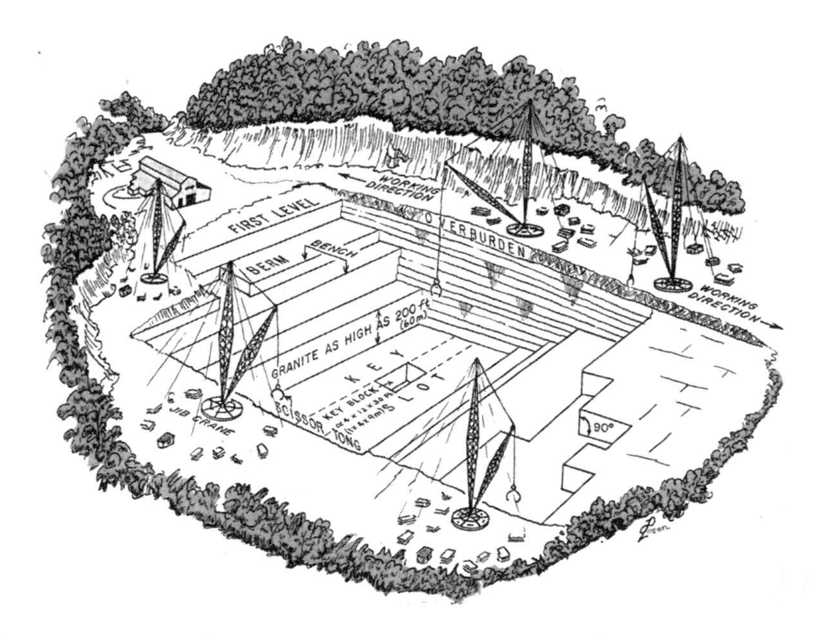

FIGURE 7.5. Diagram of operations in dimension-stone quarry. (By permission from West Virginia Geological and Economic Survey, Morgantown, WV.)

rock that must be removed (usually minor) is then stripped off the desired rock mass, using any means that applies to the type of material to be moved.

Opening the first face is accomplished by forming a cut across the quarry width, orienting the cut with due regard to the bedding planes, joints, and other geologic variables (Morrison and Russell, 1973; Power, 1975). The cut is established by cutting or channeling a key block, often about 4 ft (1.2 m) wide, 12 ft (3.6 m) deep, and up to 30 ft (9.1 m) long (see Figure 7.5). Once the block is removed, the key slot can be cut across the quarry, which establishes the first cut. Now there are two free faces to facilitate breakage and blocks of the desired size are more readily cut and removed. When the level is advanced sufficiently, a new working level can be initiated by removing another key block and key row.

7.3.2 Cycle of Operations

7.3.2.1 Stripping Overburden. Because the overburden associated with a dimension stone deposit is generally not of great thickness, simple stripping methods suffice. Quarries may contract this operation to another company, as their equipment is not the type normally used for stripping purposes. Stripping methods parallel those in open pit mining; the cycle of operations normally consists of the following:

Drilling: auger (weak rock or soil), roller-bit rotary (average rock); percussion drills (very hard rock)

Blasting: ANFO (alternative: rip if soil or very weak rock)

Excavation: dragline, scraper, or monitor (if soil); front-end loader (if rock)

Haulage: truck, scraper, or cast by dragline

7.3.2.2 Quarrying Stone. Because of the highly specialized nature of dimension stone recovery, a customized cycle of unit operations is normally employed. The method being used in Figures 7.4 and 7.5 is oriented toward freeing manageable sized blocks in the pit and hauling or hoisting them to the surface. A general description of the methods used is found in Spielvogel (1978). Traditional methods include freeing the blocks by wedging with *plug and feathers* or cutting the blocks free by using a *wire saw*. After the blocks are freed, they are hauled or hoisted to the surface. Alternative methods to free the blocks employ circular diamond saws, diamond chain saws, channeling machines, and flame-jet or water-jet channeling technology. Milling or finishing operations take place in the processing plant (Kostner, 1976).

Recently, methods of freeing larger blocks of stone in the pit have been utilized. These methods, developed in the 1970s in Finland (*World Mining Equipment*, 1997), utilize precise drilling and toppling of blocks in the pit to break and handle the blocks of dimension stone more efficiently. The initial blocks are as much as 140,000 ft^3 (4000 m^3) in volume. A block is first drilled with a precise row of holes along its back and bottom. The sides are then freed by channeling or diamond wire sawing. A controlled blast using plastic cartridge explosives is utilized to free the block from the surrounding rock (Singh, 1988). The large block is next cut into smaller blocks, normally using a controlled drill and blast method. These blocks are then toppled one by one onto a bed of sand or crushed rock, where they are cut into slabs by wedging or wire sawing. The slabs are then loaded and hauled with the use of front-end loaders. Figure 7.6 shows a similar method used in a Maryland quarry (Meade, 1992). This operation uses wire sawing to provide the initial freed blocks. Toppling and further processing are then performed via the methods discussed earlier.

The cycle of operations in quarrying consists of the following:

Cutting: diamond circular, chain, or wire saws; in-line percussion drills; flamejet or water-jet channeler (hard rock)

Wedging (freeing): drill and broach; wedge, plug and feathers; controlled blasting

Excavation/hoisting: crane, stiffleg derrick; front-end loader

Haulage: rail, truck, front-end loader

7.3.2.3 Auxiliary Operations. Quarrying requires relatively simple ancillary operations, similar to those for other surface mining methods. These

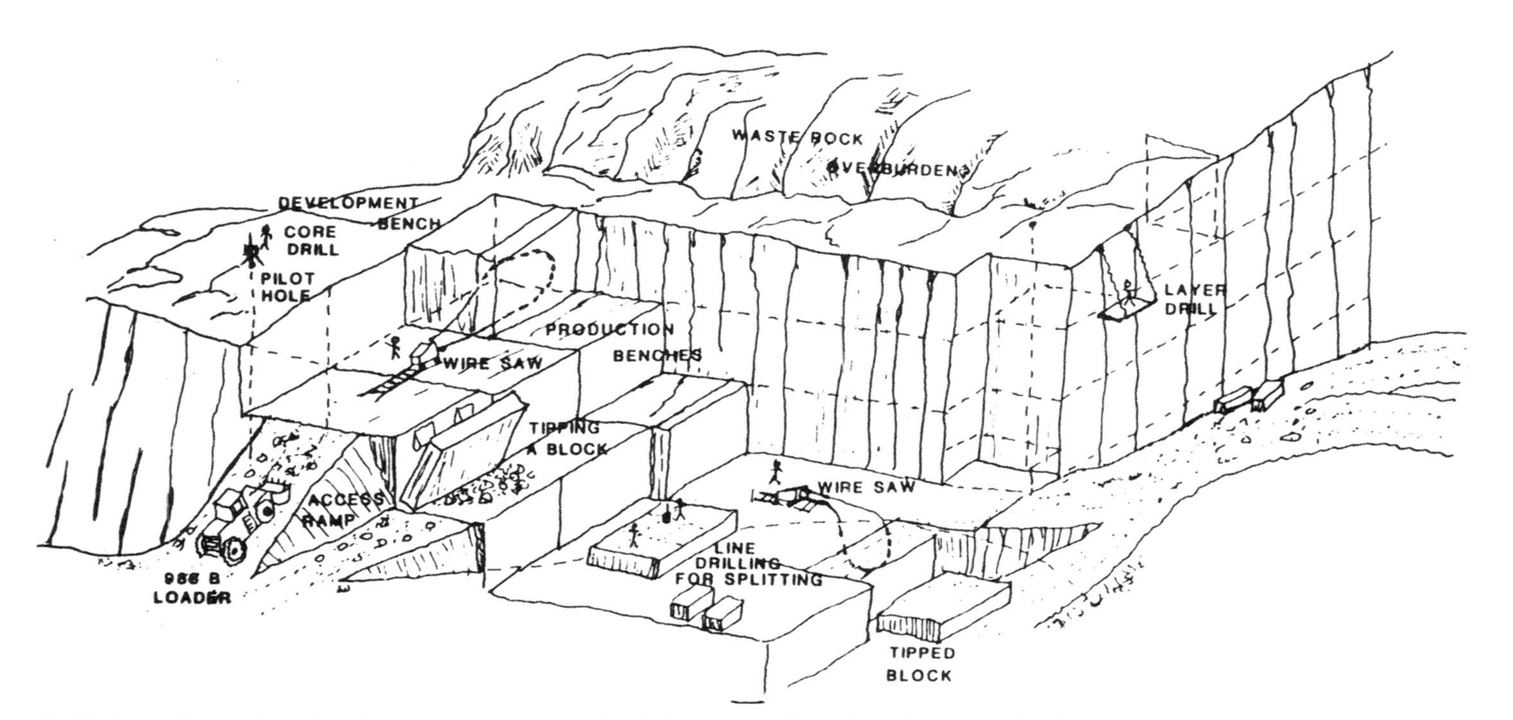

FIGURE 7.6. Method of toppling dimension-stone blocks, Friendsville Quarry, Friendsville, MD (Meade, 1992). By permission from the Society for Mining, Metallurgy, and Exploration, Inc., Littleton, CO.

include power supply, equipment maintenance, pit drainage, waste disposal, material supply, and reclamation.

7.3.3 Conditions

The mineral and related conditions amenable to dimension-stone quarrying are listed below. They have been compiled from information found in Bowles (1958), Barton (1968), Morrison and Russell (1973), and Singleton (1980).

1. *Ore strength:* structurally sound; free of unwanted defects
2. *Rock strength:* any
3. *Deposit shape:* thick-bedded or massive; large in lateral extent
4. *Deposit dip:* any, if thick
5. *Deposit size:* large, thick
6. *Ore grade:* high in physical and visual qualities
7. *Ore uniformity:* uniform
8. *Depth:* shallow to intermediate

7.3.4 Characteristics

The following information is condensed from that provided by Morrison and Russell (1973) and Power (1975).

Advantages

1. Low capital cost; mechanization not extensive.
2. Suited to some small deposits.
3. Easily accessible; hoisting may complicate moving stone, supplies, and workers.
4. Stable walls and benches; generally no bank support required.
5. High selectivity; can disregard or discard low-quality stone.
6. Good safety; little chance of slope failure.

Disadvantages

1. Somewhat limited by depth; usually less than 300 ft (90 m); can be up to 1,000 ft (300 m).
2. Low productivity; high labor cost.
3. Highest mining cost because of low productivity (relative cost = 100%).
4. Low production rate.
5. Relatively skilled labor required.
6. Inflexible; cannot easily change the mining plan at depth.
7. Mechanization is limited by the nature of the method.

8. Complicated and costly rock breakage because of inability to use the full power of explosives.
9. Waste can be 60% to 90% .

7.3.5 Applications

Quarrying for dimension stone is limited to the production of specific geologic materials that have either structural or aesthetic architectural properties. Examples include granite (Vermont, Wisconsin, Massachusetts, Pennsylvania), Bedford limestone (Indiana), marble (Georgia, Colorado, and Tennessee), and slate (Pennsylvania). Nearly all these mines are small surface operations, but there are some variations. Dimension stone companies seldom go underground because of the high cost of doing so. However, Colorado Yule marble, which was used on the Lincoln Memorial, is mined today in an underground mine in Marble, Colorado (McGee, 2000). In an underground dimension stone quarry, similar benching methods for extracting the blocks are used, with the openings being arranged like those in a large room-and-pillar or stope-and-pillar mine.

7.4 OPEN CAST (STRIP) MINING

Open cast (strip) mining is a surface exploitation method, used mainly for coal and other bedded deposits, which resembles open pit mining. However, it differs in one unique respect: The overburden is not transported to waste dumps for disposal but cast directly into adjacent mined-out panels. *Casting* is the term commonly used in mining to indicate this process of excavation and dumping into a final location. Materials handling thus consists of excavation and casting generally combined in one unit operation and performed by a single machine. Sometimes overburden is loaded into conveyances and deposited in mined-out areas; at other times, the stripping is done partly by casting and partly by haulage. It is casting in the pit, however, that makes this method distinctive, achieving for it the highest productivity and often a low cost as well. Open cast mining is classed as a large-scale mining method and is one of the most popular surface methods. In U.S. coal mining operations, more than 50% of all the tonnage is produced by open cast mining. The emphasis on stripping overburden in open cast mining has led to the use of a colloquial term for the method, *strip mining*. However, *open cast mining* is a more descriptive generic term for the method and will be used here.

It is not just the replacement of haulage with casting that makes open cast mining attractive. The depositing of *spoil* (overburden) in mined-out areas means that the mining activity is concentrated in a relatively small area and that reclamation can immediately follow mining. This is shown in Figure 7.7 (both top and bottom), where the reclamation is performed as soon as possible after mining. Another advantage is that the pit is kept open a relatively short time, permitting a steeper slope to the overburden bank (called a *highwall*).

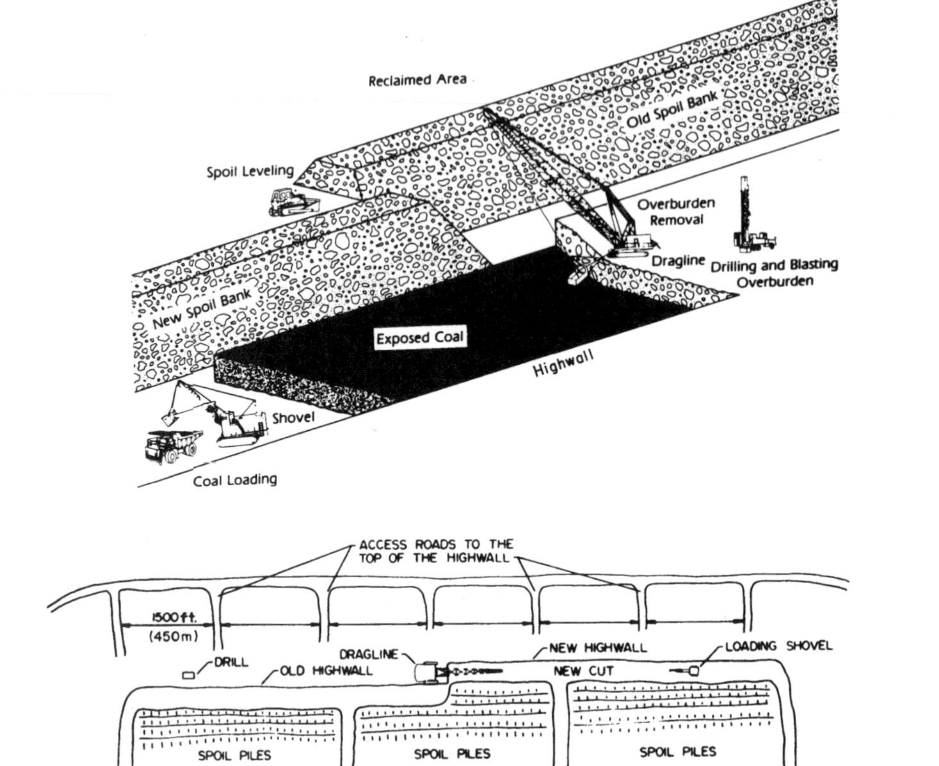

FIGURE 7.7. Open cast mine in single-seam area mining. (*Top*) Nomenclature and main components of stripping and mining. (After U.S. Department of Energy, 1982). (*Bottom*) Plan view of pit, showing access roads to highwall and cut. Road interval on the highwall is limited to the maximum length of power cable for the dragline that can be handled readily. (After Bucyrus-Erie Co., 1976. By permission from Bucyrus International Inc., South Milwaukee, WI.)

Typical dimensions in an open cast mine are 100 to 200 ft (30 to 60 m) for the height of the highwall, 70 to 150 ft (23 to 45 m) for the width of the cut, 60° to 70° for the slope of the highwall, and 35° to 50° for the slope of the spoil pile (Bucyrus-Erie Company, 1976).

The key to productivity in the open cast method is the output of the stripping excavator. By utilizing the largest land machines in the world, the number of active faces in the mine is reduced and the productivity is enhanced. However, a corresponding disadvantage is that a single excavator is responsible for the entire production, and major breakdowns can have dire consequences. Today, increasing the size of the stripping machine is no longer a major priority in open cast mining. More versatility and reliability are often sought instead.

Unlike open pit mining, the open cast method does not normally employ the same equipment for both stripping overburden and mining coal or mineral. Casting overburden requires specialized boom-type excavators, whereas mining is carried out with conventional loading and haulage equipment. Further, differences in the overburden (soil or blasted rock) and the mineral mined (usually coal, blasted or not blasted) require different equipment.

7.4.1 Variation in Open Cast Mining

Open cast mining has developed into a versatile method with variation abounding to match the mining depth, slope of the original topography, and types of equipment available. The two major variations of open cast mining are *area mining* and *contour mining*. Area mining (illustrated in Figure 7.8) is carried out on relatively flat terrain with flat-lying seams; mining cuts are made in straight, parallel panels, advancing across the property. It is common in the U.S. West and Midwest. Contour mining is conducted in hilly or mountainous terrain, with cuts placed on the contours of the topography. The mining proceeds around the hills and mountains, as shown in Figure 7.9, extracting the coal seam to a depth fixed by the economic strip ratio. Contour stripping is commonly practiced in the Appalachian coal fields. Note that the configuration of the pit in Figure 7.9 allows the disturbed land to be limited to about 1000 ft (300 m) along the outcrop, with reclamation following close behind the coal recovery operation.

Other variations of open cast mining are often utilized as well. Skelly and Loy, Inc. (1979) illustrate several methods of box-cut and block-cut mining in its surface mining manual. These methods emphasize mining the overburden in rectangular blocks rather than in continuous strips, often utilizing dozers or scrapers in the mining process. Another variation is mountaintop mining, in which the overburden from a coal seam near the top of a mountain is removed and placed in valley fills. This is allowed under the regulations of the Surface Mining Control and Reclamation Act (SMCRA) if the land is reclaimed for a higher use than would result otherwise. However, the method is under attack because of its potential for stream degradation.

Multiple-seam mining is a commonly practiced method of removing more than one seam of coal from a pit in a carefully sequenced series of steps (McDonald, 1992). Figure 7.10 illustrates the process of removing two coal seams using a single dragline. The initial cut exposing the upper coal seam is removed, as shown in Figure 7.10 (top). After the upper seam is recovered, the dragline is moved to the spoil pile to remove the overburden from the lower

FIGURE 7.8. Area mining method using a dragline (Skelly and Loy, 1979). By permission from Skelly and Loy, Harrisburg, PA.

seam, as shown in Figure 7.10 (bottom). Similar operations can be performed using two excavators (two draglines, shovel and dragline, shovel and bucket-wheel excavator, etc.). Examples of these methods are discussed by Skelly and Loy, Inc. (1979).

Explosives casting (also called cast blasting) is a method of open cast mining in which the energy generated by blasting is utilized to move a portion of the

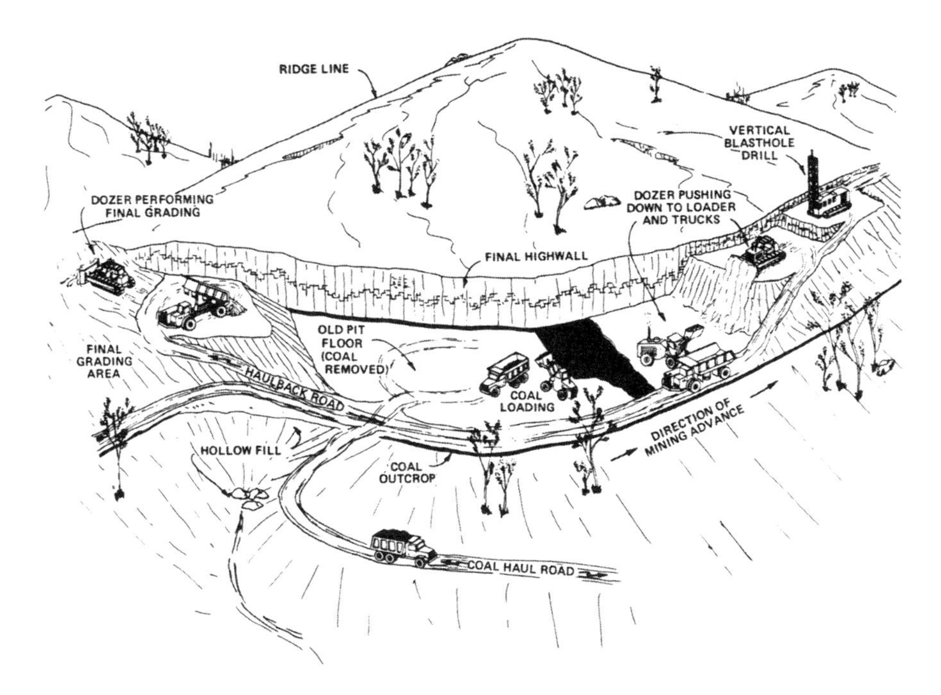

FIGURE 7.9. Contour mining via the haulback method (Mathtech Inc., 1976).

overburden into the area where it would normally be cast by the excavator (Chironis, 1980; Bauer et al., 1983; Atkinson, 1992a). Under the proper circumstances, explosives casting can be less expensive for casting than an excavator. With proper blast design, 40% to 60% of the overburden can be cast by explosives into the area designated for spoiling, with overall savings in the stripping costs.

7.4.2 Sequence of Development

The factors and sequence of development outlined in general for mine development in Section 4.2.2 are applicable to open cast mining. The most important steps are (2) confirmation of the mining method, (5) writing an environmental assessment and an EIS (if required), (7) locating the surface plant, and (9) selection of stripping and mining equipment. In connection with step 5, the provisions of SMCRA must be carefully followed, because open cast mining of coal is specifically covered in that Act.

In developing a large open cast mine in a flat area or in one of low relief, the surface plant is located at or near the center of the reserve. A central location ensures minimum haul distance and ready access to all parts of the reserve. If the deposit outcrops, however, it may be expeditious to locate the plant adjacent to the outcrop, but on barren land so as to avoid interference

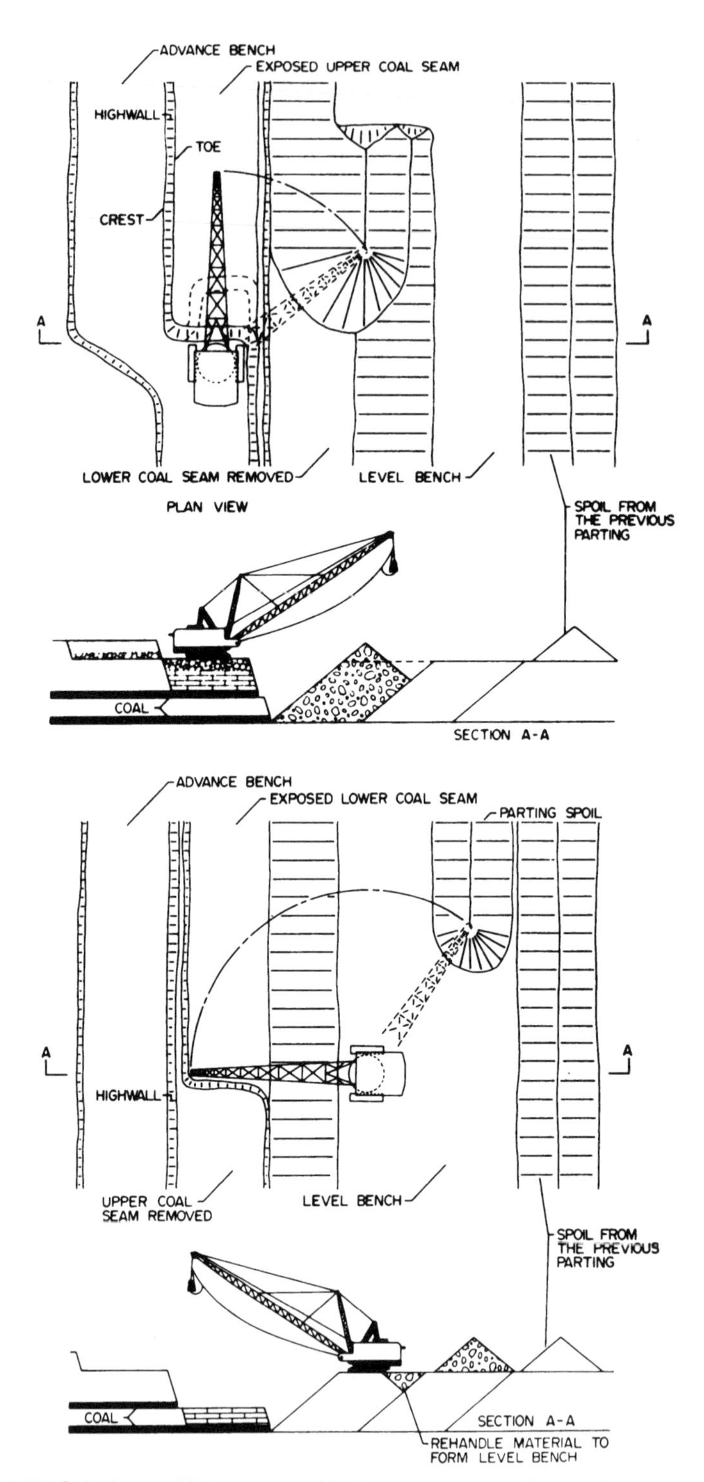

FIGURE 7.10. Stripping multiple seams with the open cast method, using one excavating machine. (*Top*) Casting overburden on first pass. (*Bottom*) Casting parting on second pass. (After Bucyrus-Erie Co., 1976. By permission from Bucyrus International Inc., South Milwaukee, WI.)

with mining. The type of surface transportation (truck, rail, or water) may influence the location of the plant significantly.

By its very nature, open cast mining depends heavily on the selection of equipment. Reviewing the options in Table 6.3, the planning group must consider those equipment types that can be applied under the conditions that exist in the area to be mined. At one time, draglines, overburden shovels, and bucket-wheel excavators were the primary choices. Today, hydraulic shovels, loading shovels with trucks, front-end loaders, and scrapers may also be considered. The mix of equipment used has changed substantially.

Consider the choice between overburden shovels and draglines. A cost comparison between the two in terms of $/yd^3 ($/m^3) is shown in Figure 7.11. Given that these costs are indicative of the two machines, it seems reasonable that overburden shovels would be chosen over draglines for most

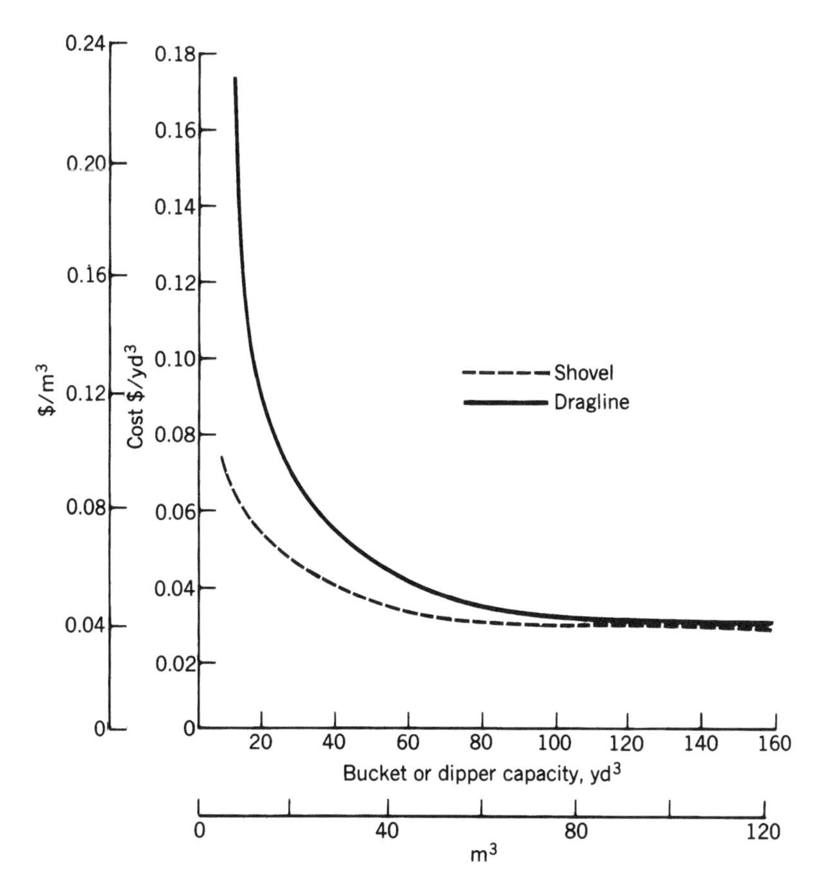

FIGURE 7.11. Comparison of unit operating costs (per yd^3, or m^3, of material stripped) for various bucket capacities. (After Pfleider, 1973. By permission from the Society for Mining, Metallurgy, and Exploration, Inc., Littleton, CO.)

open cast mines. The exact opposite is true. However, the reasons are independent of the costs of moving a fixed volume of overburden. Today the dragline is more popular because of its versatility in the mine. Draglines have the ability to work on the spoil pile, can be used to excavate ramps and other pit features, and can reclaim the spoil pile area. None of these capabilities is available in the shovel. The curves do suggest one reason for the halt in increasing dragline sizes. Two excavators of the 100 yd^3 (75 m^3) category can produce as cheaply as one dragline in the 200 yd^3 (150 m^3) range. In addition, two draglines are more versatile and result in less downtime costs than a single larger dragline.

Figure 7.12 provides additional cost information on open cast mining of lignite using various equipment types. The dragline and bucket-wheel excavator are the most cost-efficient equipment types, but the wheel excavator is not heavily used in the United States because of the lack of ideal overburden conditions. Truck/shovel and scraper operations show the highest costs and therefore are avoided when a dragline can perform the job. However, they find use when the conditions are not suited to a dragline or bucket-wheel excavator.

The major activities in development of an open cast mine start with clearing of the land. A surface plant is then located and constructed. Because of the importance of reclamation in this method, special attention is given to the location and maintenance of topsoil stockpiles. Environmental and restoration procedures are worked out with care so that they proceed in a logical and expedient manner along with mining. Coal dumping, storage, processing (if any), and transport facilities are located logically with respect to the mining. After the equipment is selected, initial pit development takes place. As in open pit mining, the first cut may be a drop cut. This cut is the most difficult; some haulage of the overburden may be required, and progress may be relatively slow. Because a highwall is normally maintained after the first cut is completed, the succeeding cuts are more efficient. Simultaneous development and exploitation can then proceed in a normal fashion.

7.4.3 Cycle of Operations

7.4.3.1 Stripping Overburden. The cycle of operations during stripping in open cast mining is determined largely by the nature of the overburden. Soil and decomposed rock can sometimes be excavated without prior breakage, stiff soil or weak rock can be ripped prior to excavation, and hard rock requires drilling and blasting. The drilling method is also a function of the overburden material. Auger drills can be used in soil or soft rock, roller-bit rotary for intermediate rock, and percussion or rotary-percussion for hard rock. Drilling is normally done using a pattern of vertical holes that terminates a few feet or about a meter above the coal to avoid breaking the coal seam too severely. Ordinarily, the cycle of operations in stripping consists of the following:

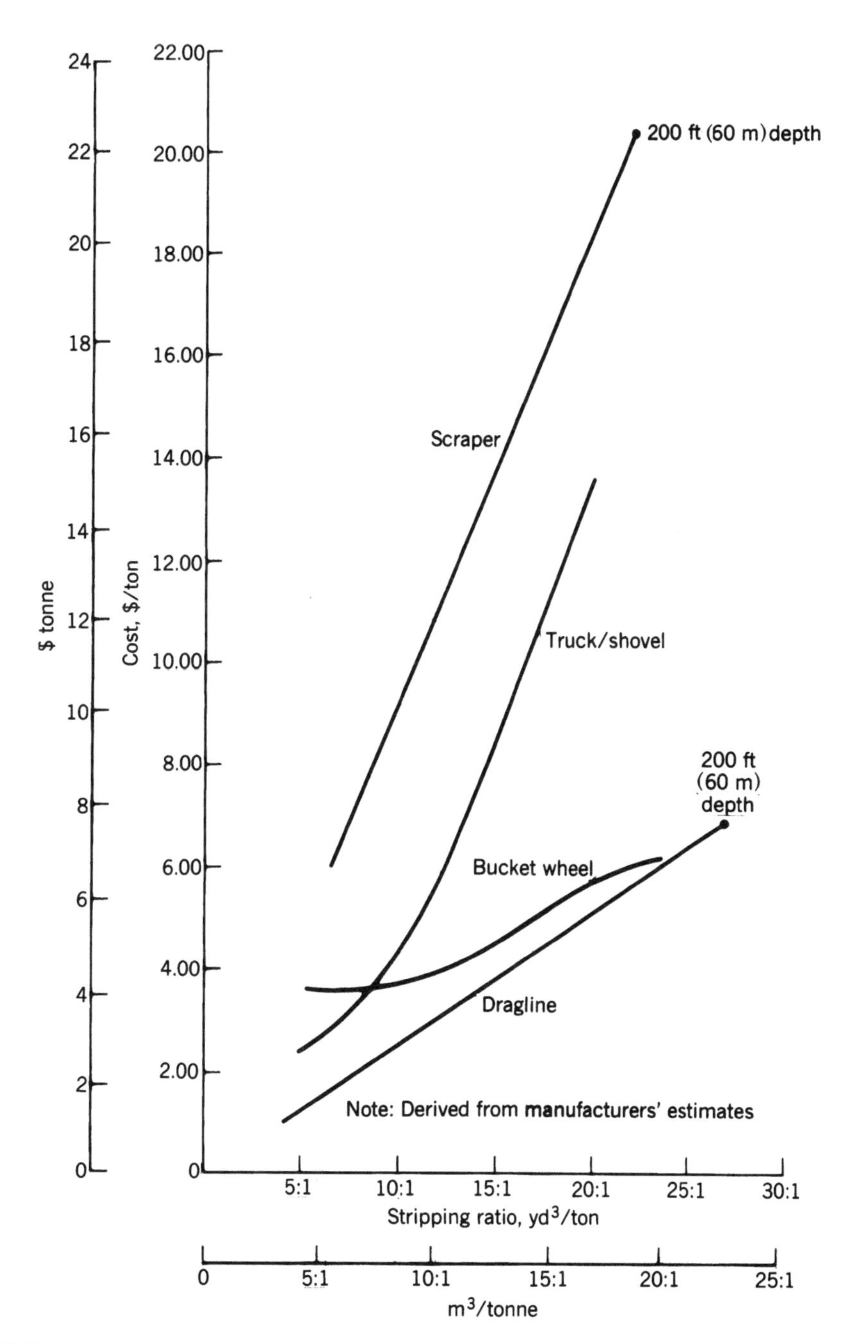

FIGURE 7.12. Comparison of unit operating costs (per ton of mineral produced) for four excavating systems for varying stripping ratios. Conditions: hypothetical Gulf Coast lignite mine, annual production 5 million tons/yr (4.5 million tonnes/yr). (After Kahle and Mosely, 1983. By permission from the Society for Mining, Metallurgy, and Exploration, Inc., Littleton, CO.)

Drilling: auger drill (soft rock), roller-bit rotary (average rock), percussion or rotary-percussion (hard rock)

Blasting: ANFO, ammonium nitrate (AN) gel or slurry; loading by machine (bulk) or hand (bagged explosives); firing by electric caps or by detonating cord

Excavation: dragline, overburden shovel, hydraulic shovel, bucket-wheel excavator (soil), dozer, scraper (soil), cast blasting

Haulage (if needed): truck, scraper (soil), conveyor

7.4.3.2 Mining Ore or Coal. The nature of the coal or ore also determines the cycle used in this operation. The process may involve direct loading, ripping and loading, or drilling and blasting. Many coal seams can be excavated directly without prior preparation. A power shovel or front-end loader can be used in these cases. In phosphate operations, the phosphate rock is sometimes excavated out of the pit with the use of a dragline. After the material is placed on the highwall, hydraulic monitors flush the ore into slurry form to be hauled away via a hydraulic pipeline. The mining cycle typically consists of the following:

Cleaning: Rotary brush or dozer cleans top of the seam.

Drilling: Small auger or percussion drill used where needed.

Blasting: ANFO (ripping with a dozer is an alternative).

Excavation: Front-end loader, power shovel, continuous miner (designed for surface mines).

Haulage: Truck, tractor-trailer, belt conveyor, hydraulic conveyor, rail.

7.4.3.3 Auxiliary Operations. In open cast mining, auxiliary operations include reclamation, slope stability, haul road construction and maintenance, equipment maintenance, drainage and pumping, communications, power distribution, dust control, and safety. Reclamation may be the most important; sizeable bonding costs can be averted by expediently reclaiming the disturbed area.

7.4.4 Conditions

The deposit conditions amenable to open cast mining vary significantly with the type of mineral extracted and the geologic conditions that exist in the deposit area. The following are typical conditions and have been compiled from information in Pfleider (1968, 1973), Bucyrus-Erie Company (1976), and Skelly and Loy, Inc. (1979):

1. *Ore strength:* any
2. *Rock strength:* any

3. *Deposit shape:* tabular, bedded
4. *Deposit dip:* any, prefer horizontal or low dip
5. *Deposit size:* prefer continuous deposit, large in lateral extent
6. *Ore grade:* can be low if other conditions are favorable
7. *Ore uniformity:* prefer uniformity
8. *Depth:* shallow to moderate to control stripping ratio

7.4.5 Characteristics

The application of open cast mining has numerous advantages and disadvantages. The following information was gathered from Pfleider (1973), Bucyrus-Erie Company (1976), and Kahle and Moseley (1983).

Advantages

1. Highest productivity of any coal mining method, with an average of about 9.85 tons per employee-hour (8.94 tonnes per employee-hour), more than 2.5 times that of underground mining (Energy Information Administration, 2000).
2. Lowest cost per ton (tonne) of the coal mining methods (large equipment reduces unit costs, relative cost 10%).
3. High production rate.
4. Early production (modest development requirements allow rapid exploitation).
5. Low labor intensity, as compared with underground mining.
6. Relatively flexible, can increase production by expanding operations.
7. Suitable for large equipment, permits high productivity.
8. Low blasting costs; large bench faces permit efficiency, provide multiple free faces for blasting.
9. Simple development and access.
10. Highwall support measures seldom necessary.
11. Good recovery (can approach 100%); low dilution.
12. Normally eliminates haulage of overburden.
13. Good health and safety factors.

Disadvantages

1. Economic limits of the method and technological limits of the equipment impose depth limits (generally about 300 ft or 90 m).
2. Economics impose limits on stripping ratios; typical strip ratios from 1.5 to 22 yd^3/ton or 1.3 to 19 m^3/tonne (Mutmansky et al., 1992).
3. Surface is damaged; extensive environmental reclamation required; environmental expense is substantial.

4. Public image of open cast mining is decidedly negative; mining company must promote positive public relations.
5. Some excavators require skilled operators.
6. Requires large equipment to realize lowest cost; small deposits mined with small equipment.
7. Weather can impede operations.
8. Requires careful sequencing of operations, especially during stripping.
9. Slopes must be monitored and maintained.
10. Surface runoff must be controlled; can damage streams if not properly managed.

7.4.6 Applications

In the United States open cast mining is used extensively for surface mining of coal. It is also used for producing anthracite coal (Pennsylvania), bentonite (Arizona), lignite (Texas, North Dakota), phosphate (Florida), tar sands (Canada), and uranium (Wyoming).

Students looking for good case studies of open cast mines will find a number of interesting descriptions in the *SME Mining Engineering Handbook* (Hartman, 1992). Variations of open cast mining discussed include methods for Appalachian contour stripping (Tussey, 1992), Texas lignite (Rand, 1992), Florida phosphate (Olson, 1992), and a number of western multiseam coal mines (Bricker, 1992; Anderson and Kirk, 1992; McDonald, 1992; Stubblefield and Fish, 1992). The Athabasca oil sands mining operations in Alberta, Canada, provide other interesting case studies of open cast mining. Descriptions of those operations can be found in McKee (1992) and Phelps (1999).

7.5 AUGER MINING

Auger mining is a mining method that recovers coal or other minerals from under the highwall when the ultimate stripping ratio has been achieved in open cast mining operations. An augering machine or a continuous mining device that bores parallel holes or entries into the highwall extracts the coal. While the equipment goes underground, the crew remains on the surface and operates the equipment remotely; hence, we classify the method as a surface mining procedure. The method is used primarily in Appalachian contour mines, where the long highwalls permit recovery of a significant additional tonnage from the coal seams.

Two variations of auger mining are in current practice. The first uses traditional augering equipment developed in the 1940s to drill holes in the coal seam from the pit bottom. A typical augering setup is shown in Figure 7.13. Augering machines employ drag-bit drilling heads of 2 to 8 ft (0.6 to 2.4 m) in

FIGURE 7.13. Auger working in an Appalachian coal mine. (Volkwein and Ulery, 1993).

diameter to place circular holes in the coal. Generally, holes of 200 to 300 ft (60 to 90 m) in length can be placed under the highwall. To improve recovery and productivity, augers with multiple heads are used, especially in thin seams. Still, the typical augering operation extracts only 40% to 65% of the coal in place (Ford, Bacon, and Davis, Inc., 1975; Blakely, 1975). Examples of auger utilization can be found in McCarter and Smolnikar (1992).

The second variation of auger mining is often called highwall mining. *Highwall mining* is the practice of sending a small, remotely controlled continuous miner into the highwall to mine parallel entries of about 10 ft (3 m) in width and the height of the coal seam. A highwall miner is shown in Figure 7.14. Small pillars are left between the entries, the width depending on the support needed. The depth of the holes can be up to 1000 ft (300 m) but may be held to lesser amounts to overcome logistics problems. Recovery is about 60% in most operations. A summary of some equipment capabilities is available in Chadwick (1993).

On a purely economic basis, the use of auger mining is quite attractive. The extraction of additional coal from the highwall can be done at a much lower cost than in the primary stripping operation. However, the mine must have the highwall length, bench width, roof conditions, dip, and reserves to make the auger or highwall mining operation feasible. The augering operation is not generally possible in mines other than open cast mines.

FIGURE 7.14. ADDCAR highwall mining system. Photo courtesy of Mining Technologies Inc., Ashland KY. Used with permission.

7.5.1 Sequence of Development

Developing and preparing a site for auger or highwall mining is a relatively simple task if coordinated with an ongoing stripping operation. Practically no additional excavation is required to extract the additional coal. However, the highwall must be trimmed of loose rock for safe operation, and the bench must be cleared to a width of 50 to 75 ft (15 to 25 m). Storage areas must be provided for the coal extracted and for the auger flights or the pusher/conveyor units for the highwall miner. A haul road must be maintained for the coal haulage trucks.

7.5.2 Cycle of Operations

Generally, no stripping is required for an augering operation unless the extraction is performed at the outcrop. The mining is normally initiated at the ultimate pit limit highwall in a contour mining operation. In this case, it follows the open cast mining equipment around the contour and provides additional production. The augering will delay the ability of the mine to backfill the pit, but augering operations can be coordinated with other operations in the open cast mine to minimize this delay. The cycle of operations during production is simplified and conducted as follows:

Excavation: augering machine (one, two, or three heads) or highwall mining system

Haulage (for augering): auger flights, conveyor or loader, trucks

Haulage (for highwall miner): conveyor units arranged in a train, conveyor or loader, trucks

Auxiliary operations are simple; providing electrical power and maintaining haul roads are the primary concerns.

7.5.3 Conditions

The following have been compiled from information in Ford, Bacon, and Davis, Inc. (1975), Skelly and Loy, Inc. (1979), and Chadwick (1993).

1. *Ore strength:* any
2. *Rock strength:* any
3. *Deposit shape:* tabular, bedded
4. *Deposit dip:* low, nearly horizontal
5. *Deposit size:* small to large
6. *Ore grade:* can be low
7. *Ore uniformity:* uniform in thickness, free of partings
8. *Depth:* shallow to moderate

7.5.4 Characteristics

The following information was obtained from Blakely (1975), Ford, Bacon, and Davis, Inc. (1975), and Chadwick (1993).

Advantages

1. High productivity (25 to 500 tons or 22 to 450 tonnes per employee-shift).
2. Low mining cost; lowest of any coal mining method (relative cost about 5%).
3. Intermediate production rate (100 to 2500 tons or 90 to 2200 tonnes/shift).
4. Little development required when used in conjunction with an open cast mine.
5. Low labor requirement.

6. Low capital investment for an augering machine; highwall mining systems can be contracted on a royalty basis.
7. Surface topography is preserved, no separate reclamation costs incurred.
8. Recovers coal that would otherwise be lost.
9. Good health and safety characteristics; explosion hazards can exist.

Disadvantages

1. Application is limited to certain conditions associated with open cast mining; method rarely employed alone.
2. Low coal recovery (40% to 65% in most cases).
3. Methane explosions a hazard, can be mitigated (Volkwein and Ulery, 1993).
4. Production capability dependent on a single extraction unit.

7.5.5 Applications

Augering has been used since the 1940s in Appalachian surface mines. Generally, about 2% of U.S. coal production is derived from this source. The method is most prevalent in Pennsylvania, West Virginia, and Kentucky, with occasional use elsewhere. Highwall mining has been around since the Joy Pushbutton Miner was developed in the 1950s. That machine had technological problems that made it unsuccessful. Today's highwall miners have been much more reliable and productive. They have been used in the Appalachian coal fields, in New Mexico, and in Australia (Chadwick, 1993).

7.6 SPECIAL TOPIC A: SURFACE MINE BLASTING

Although blasting theory and practice are covered later in the discussion of underground mining (Section 11.6) and the general topic of explosives has already been discussed (Section 5.3), surface blasting is a sufficiently unique topic that deserves its own coverage. In this section, we note common blasting practices and explore the design of blasting patterns in surface mining.

7.6.1 Blasting Practice

Since the advent of ammonium-nitrate-based explosives and blasting agents and the demise of nitroglycerine-based high explosives in mining, blasting practices have been standardized substantially, particularly in surface mining. Moreover, the provision of several free faces with the existence of benches in virtually all methods of surface mining further narrows the practice of surface mine blasting. The major elements of surface blasting practice consist of the following:

1. *Inclination of the borehole:* Generally vertical; may be inclined to improve efficiency.
2. *Subdrilling:* Depth of hole extends below the intended floor (provided to ensure complete breakage to floor; increases in back rows of holes).
3. *Rows of holes:* Usually multiple (two, three, or more).
4. *Hole pattern:* Square, rectangular, or triangular.
5. *Explosives:* ANFO (bulk or packaged); AN slurry or gel.
6. *Hole loading:* Performed by machine (bulk) or hand (packaged explosives); explosives confined by stemming devices or drill cuttings; may deck explosives (load in layers with inert material between layers) to better distribute explosive energy.
7. *Hole detonation:* Double line of detonating cord is used to set off primers in each hole; cord detonated by blasting caps fired electrically; millisecond delays provide delays between holes and rows to improve fragmentation, reduce noise, and minimize seismic effects (ground vibrations).
8. *Secondary breakage:* Additional breakage required for boulders performed by mudcapping, blockholing, dropballing, or hydraulic impact hammer.

7.6.2 Design of a Blasting Pattern

The design of blasting patterns in surface mining is based on theoretical concepts, but is modified by experience and empirical considerations. Two important factors are often considered to evaluate the drilling and blasting operations. The first is called the *drilling factor*, the length of hole drilled per ton (tonne) of material blasted. The second is the *powder factor*, the weight of explosives per ton (tonne) of material blasted. These two factors have a wide range of values, depending on the strength of the material to be blasted and the number of planes of weakness distributed through the rock. Hartman (1987) reports that the drilling factor in surface mines ranges from 0.02 to 0.04 ft/ton (6.7 to 13.4 mm/tonne) and the powder factor varies from 0.05 to 1.0 lb/ton (0.03 to 0.5 kg/tonne). In any given situation, a mining operation will seek means of reducing these two factors to the lowest possible values consistent with good blasting results.

Several aspects of the drilling pattern used in surface mines are shown in Figure 7.15. Note in Figure 7.15a that two primers are used to ensure proper detonation. In Figure 7.15b, a plan view of a triangular pattern of holes is shown for a surface mine blast with two free faces. The distance between rows is known as the *burden*, which is shown as 21 to 24 ft (6.4 to 7.3 m) in the figure. The distance between the individual holes in any row is known as the *spacing*. The spacing in the triangular system of holes is 26 to 28 ft (7.9 to 8.5 m). In Figure 7.15c, a square pattern of holes is shown with the burden (3.0 m or 9.8 ft) equal to the spacing.

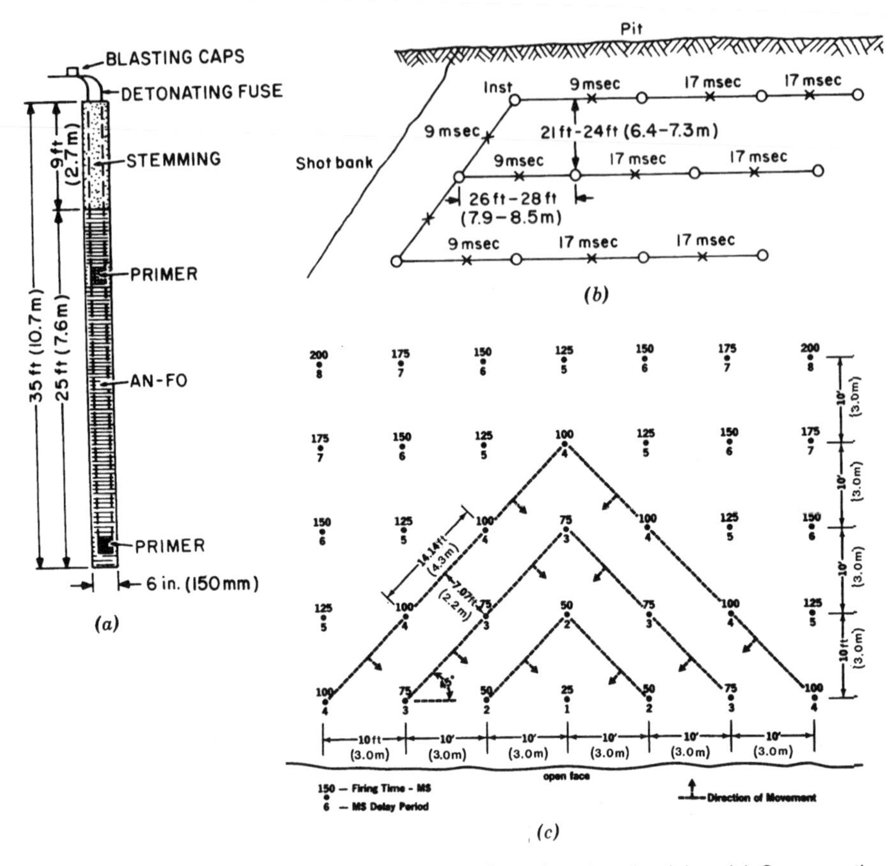

FIGURE 7.15. Hole loading and blasting patterns in surface bench mining. (a) Cross section through drillhole showing loading and charging of explosives. (b) Plan view of diagonal blasthole pattern; breaking to two vertical faces with millisecond delays. (After Stefanko and Bise, 1983. By permission from the Society for Mining, Metallurgy, and Exploration, Inc., Littleton, CO.) (c) Plan view of echelon blasthole pattern; breaking to single vertical free face with millisecond delays. (After E.I. duPont de Nemours & Co., 1977. By permission from the International Society of Explosives Engineers, Cleveland, OH.)

In setting up the basic design of a blast in an open pit mine, the following parameters must be determined:

1. *Bank height and slope:* Based on excavator and slope stability considerations.
2. *Hole diameter:* 6 to 18 in. (150 to 460 mm).
3. *Hole angle:* Normally vertical; can be sloped at up to 30° from the vertical; horizontal holes not normally used.
4. *Hole depth:* Bank height plus subdrilling length; subdrilling length is commonly 0.2 to 0.5 times the burden (Atlas Powder Company, 1987; Hustrulid, 1999).

5. *Burden:* Perpendicular distance from the nearest free face to the center of the charge in ft (m); generally ranges from 15 to 30 ft (4.6 to 9.1 m) in surface mines; recommended values can be found in Atlas Powder Company (1987) and Hustrulid (1999).
6. *Explosives loading density:* Weight of explosives per length of drillhole in lb/ft (kg/m).
7. *Height of charge in drillhole:* Usually continuous but will stem the upper part of the hole, determined in ft or m.
8. *Weight of charge per drillhole:* In lb (kg).
9. *Hole spacing:* Distance between the holes in a row; generally ranges from 15 to 30 ft (4.6 to 9.1 m) in surface mines; recommended values can be found in Atlas Powder Company (1987) and Hustrulid (1999).
10. *Number of holes required:* Depends on the area to be blasted.
11. *Weight of rock (ore, coal) broken:* Tons (tonnes).
12. *Drilling factor:* Length of drillhole per unit weight of rock broken, in ft/ton (m/tonne).
13. *Powder factor:* Weight of explosives per unit weight of rock broken, in lb/ton (kg/tonne).

Blasting theory will be discussed further in Chapter 11; theoretical considerations are important in the general application of explosives to the breakage of rock. The determination of most of the aforementioned parameters will be done by empirical methods, that is, by experimenting with them in the mine and determining the best set of parameters for the given geology and material characteristics. A basic design example follows.

Example 7.1. Provide the design parameters for the surface mine blast shown in Figure 7.16, given the following data:

Copper ore specific weight w	2.0 tons/yd^3 (2.4 tonnes/m^3)
Bank length L	200 ft (61 m)
Explosive type	AN slurry
Loading density r	70 lb/ft (104 kg/m)
Powder factor PF	0.5 lb/ton (0.25 kg/tonne)
Height of charge in hole k	35 ft (11 m)
Hole diameter	12 in. (305 mm)

SOLUTION

Bank area $A = bh = (40)(50) = 2000$ ft^2 (186 m^3)

$$\text{Bank volume } V = \frac{LA}{27} = \frac{(200)(2000)}{27} = 14{,}815\,\text{yd}^3\ (11{,}325\ \text{m}^3)$$

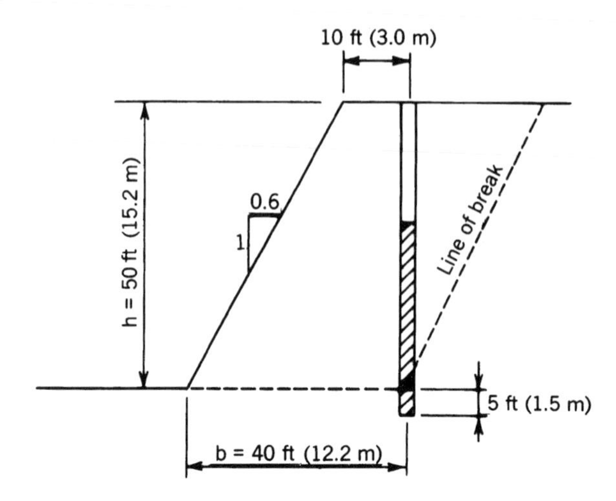

FIGURE 7.16. Design of surface blast, showing bank cross section and charged drillhole. See Example 7.1.

Bank weight $W = wV = (2.0)(14{,}815) = 29{,}630$ tons (26,880 tonnes)

Weight of charge/drillhole $c = rk = (70)(35) = 2450$ lb (1110 kg)

$$\text{Weight of rock broken/drillhole } m = \frac{c}{PF} = \frac{2540}{0.5} = 4900 \text{ tons (4445 tonnes)}$$

$$\text{Number of holes needed } n = \frac{W}{m} = \frac{29{,}630}{4900} = 6.05 \text{ (call it 6)}$$

$$\text{Hole spacing } s = \frac{L}{n} = \frac{200}{6} = 33 \text{ ft (10 m)}$$

7.7 SPECIAL TOPIC B: SLOPE STABILITY

In the engineering of surface mines, there are many occasions, such as when the design of slopes is necessary, that the stability of the soil and rock is important. The design of pit slopes and mine highwalls comes to mind first, but the stability of soil and rock is also important in the construction of buildings and roads, the use of waste dumps and tailings ponds, and the design of the final reclaimed landscape. In each of these cases, the properties of the geologic material must be analyzed to ensure its stability.

In this section, we look at the types of problems that occur in soils and poorly consolidated and broken rock materials that behave like soil. Typically, rock in jointed or fractured masses may resemble soil and may be susceptible to analysis by soil mechanics. Similarly, in underground mining, backfill, caved ground, and soft running ground are often more like soil than rock. Hence, soil mechanics has widespread application in mining engineering. We shall examine only one common usage — the design of a slope in surface mining.

7.7.1 Principles of Soil Mechanics

Soil mechanics is the study of the properties of soils and their behavior in relation to the design, construction, and performance of engineering works. It is also applicable to unconsolidated materials deposited during human activity such as fills, wastes, tailings, dumps, stockpiles, and materials in bins, which broadens its utility in mine engineering. Soil mechanics is a companion field to rock mechanics, and the two have much in common in the design of geotechnical structures.

Soil is defined as the uncemented aggregate of mineral grains and decayed organic matter along with the liquid and gas that occupy the empty spaces between the solid particles (Das, 1990). For our purposes, we will also consider unconsolidated human-made materials as soil and analyze them in the same manner. In visualizing the components of a soil, the three phases of the soil can be depicted in two ways, as shown in Figure 7.17. The upper diagram shows the soil in its natural form. The lower diagram illustrates the soil separated into its three phases. Consider that the natural soil has a volume V, a grain volume V_g, and a void volume V_v. Assume also that the volume of air is V_a and the volume of water is V_w. If the specific weight of the unit volume of soil is W, the weight of the grains is W_g, and the weight of the waters is W_w (the weight of the air W_a is normally ignored), we can derive the following

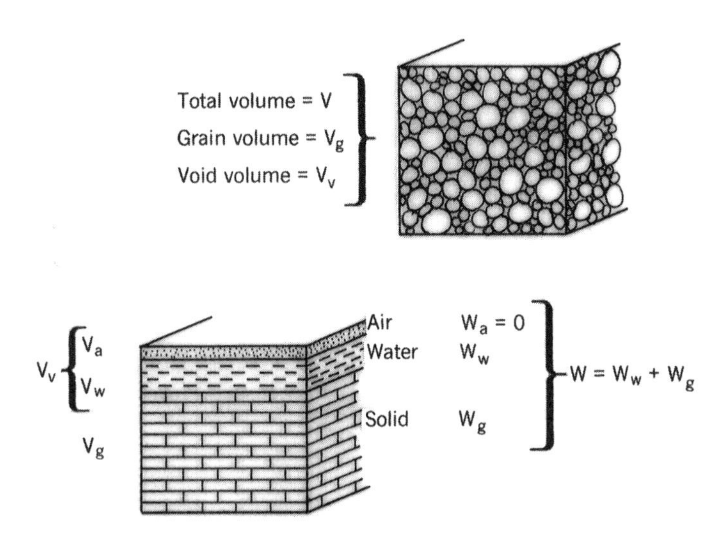

FIGURE 7.17. Soil represented as a three-phase system, consisting of solid (grains), liquid (water), and gas (air). (*Top*) Natural composite sample. (*Bottom*) Sample stratified by phase.

parameters:

$$\text{Porosity } n = \frac{V_v}{V} \times 100\% \tag{7.1}$$

$$\text{Moisture content } m = \frac{W_w}{W_g} \text{ (expressed as a decimal)} \tag{7.2}$$

$$\text{Specific weight } w = \frac{W}{V} \text{ in lb /ft}^3 \text{ (kg/m}^3\text{)} \tag{7.3}$$

Another common measure of w in surface mining is *bank* or *solid measure*, in units of tons/yd^3 (m^3/tonne). This is normally utilized for the description of soil in situ.

Soil properties are usually grouped into two categories. *Index properties* are identification properties used to classify soils. *Mechanical properties* are physical properties that determine soil behavior.

The most important soil index property is *grain size*. Grain size distribution is measured by sieving, sedimentation, or a commercial particle size analyzer. Once determined, the size distribution is plotted as a frequency graph (Figure 7.18a), as a cumulative undersize graph (Figure 7.18b), or as a straight line on log probability paper (Figure 7.18c). The size distribution can help to classify the soil. Soil classification is performed with the use of several standard procedures, as outlined by Das (1983) and Terzaghi et al. (1996). The commonly used methods are the Public Roads Administration (PRA) system, adopted many years ago and modified over time, and the United Soil Classification System, put together by the U.S. Army Corps of Engineers and the Bureau of Reclamation (U.S. Army Corps of Engineers, 1953). The classification chart in Figure 7.19 is based on the PRA classification. It considers three prominent size fractions of the soil, consisting of very fine particles (*clay*), fine particles (*silt*), and coarse particles (*sand*). A fourth category is sometimes employed for the very coarse particles (*gravel*). The upper limits are 2 μm for clay, 60 μm for silt, and 2 mm for sand. Gravel is considered to be material above 2 mm, and particles below 0.2 μm are

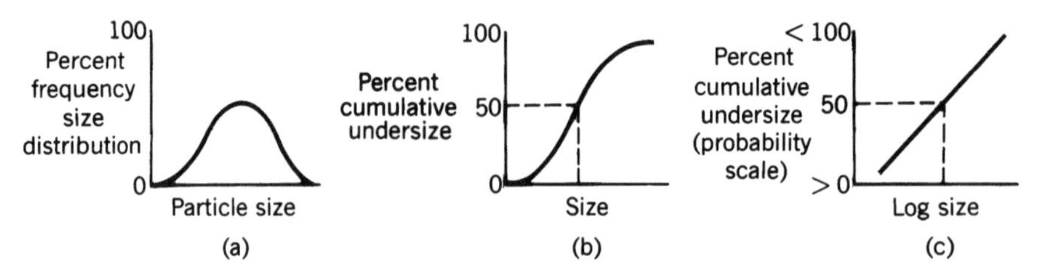

FIGURE 7.18. Grain size distribution of soil, plotted as (left) frequency graph, (*center*) cumulative-undersize graph, and (*right*) cumulative-undersize graph on log-probability paper.

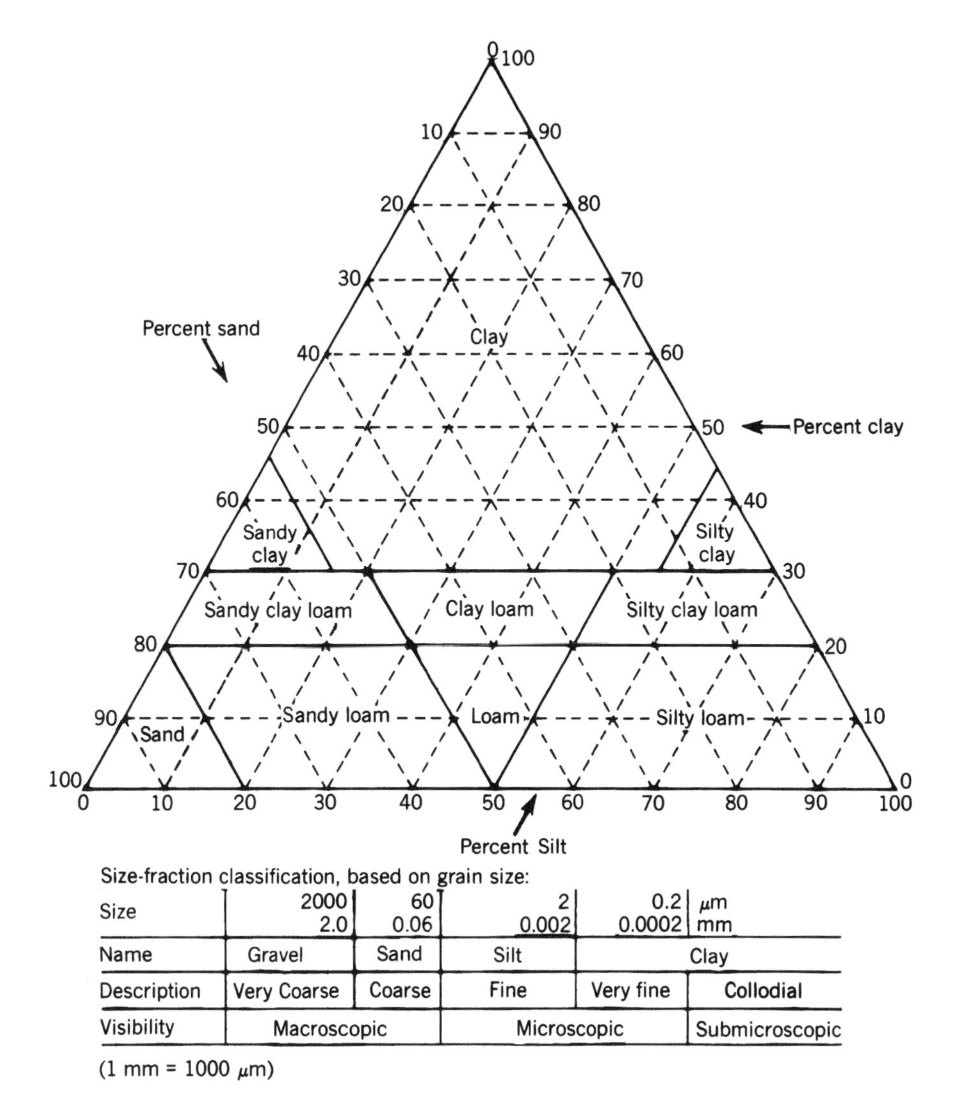

Size	2000 μm / 2.0 mm	60 μm / 0.06 mm	2 μm / 0.002 mm	0.2 μm / 0.0002 mm	μm / mm
Name	Gravel	Sand	Silt	Clay	
Description	Very Coarse	Coarse	Fine	Very fine	Collodial
Visibility	Macroscopic		Microscopic		Submicroscopic

(1 mm = 1000 μm)

FIGURE 7.19. Soil classification chart, based on grain size distribution. (Modified after U.S. Public Road Administration chart.)

considered to be colloidal, but these two fractions are not used in the classification system. Once the fractions are determined, they are plotted on the PRA chart and the soil type is read. The following is an example of the process.

Example 7.2. The grain size distribution of a given soil is plotted in Figure 7.20. Classify the soil with the aid of the PRA chart.

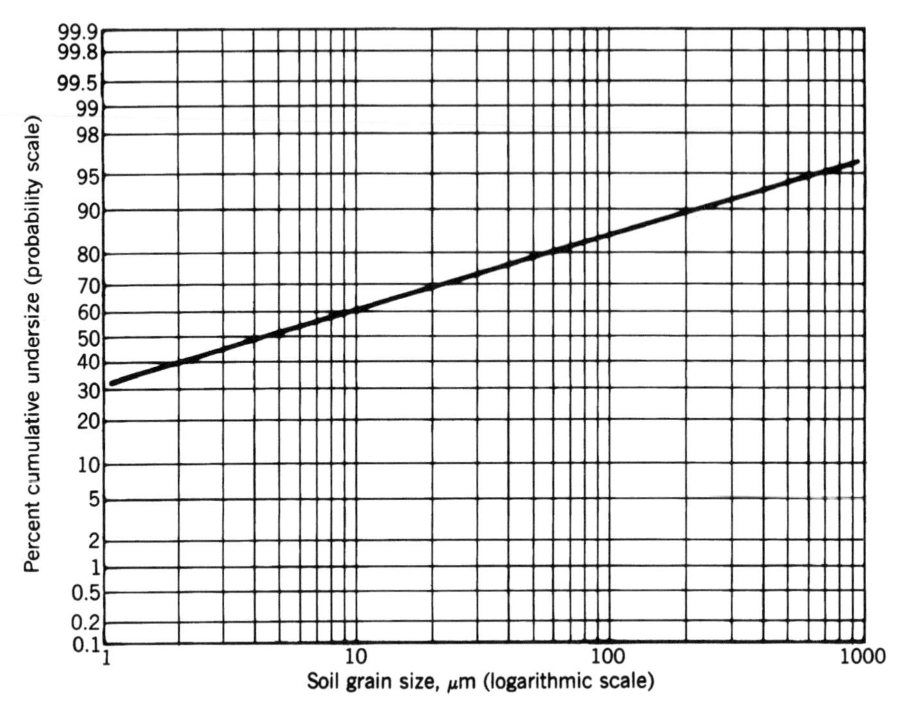

FIGURE 7.20. Grain size distribution of soil plotted on log-probability paper as percentage cumulative undersize. See Example 7.2.

SOLUTION. Read the cumulative undersize at 60 μm as 80% and at 2 μm as 40%. The distribution of the three sizes in the soil is

Sand	$100 - 80 = 20\%$
Silt	$80 - 40 = 40\%$
Clay	$40 - 0 = 40\%$

Plotting the distribution on Figure 7.19 identifies the soil as a clay.

Much information can be gleaned from knowledge of the soil type. Certain behavioral and engineering properties are associated with sandy, silty, and clayey soils. It is the fine fraction of a soil, however—the clay fraction—that is critical and governs its behavior.

Of the mechanical properties of soil, the most important is *strength*, the allowable working stress. Compressive, tensile, and shear strengths are the most commonly used. In soils, the most important is shear strength, expressed in the following manner (Sherman, 1973):

$$\text{Shear strength} = \text{cohesion} + \text{friction}$$

$$s = c + p \tan \phi = c + wh \tan \phi \tag{7.4}$$

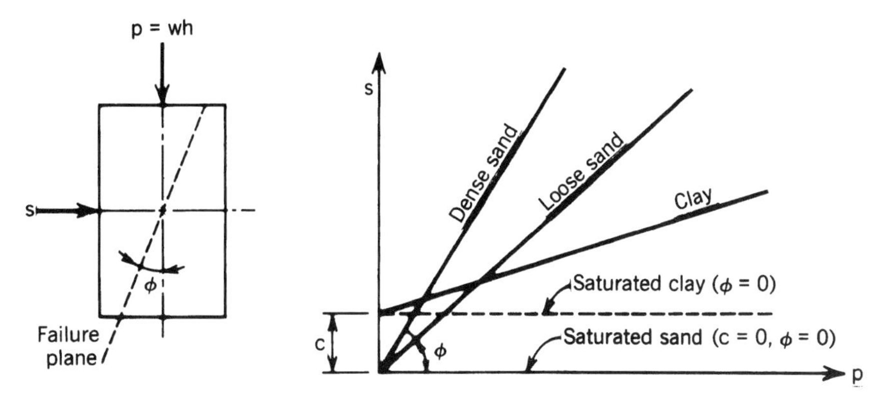

FIGURE 7.21. Relation of shear strength to normal stress in unconfined soil.

where s is the shear strength in lb/ft^2 (kPa), c is cohesion in lb/ft^2 (kPa), p is normal stress in lb/ft^2 (kPa), w is specific weight in lb/ft^3 (kg/m^3), h is depth in ft (m), and ϕ is angle of internal friction (angle of repose) in degrees. The diagrams in Figure 7.21 indicate the differences in shear strength for different soil types as a function of the normal stress. Most sands have low cohesion and high friction between particles, whereas clays have high cohesion and low friction. A wet soil usually possesses both higher cohesion and a larger angle of internal friction; if saturated, however, both may fall to zero.

We can now offer some general observations about soil behavior. A *sandy soil* is usually stable and permeable and can be kept drained to prevent saturation and a loss of shear strength. *Quicksand* is a dangerous subtype of soil exhibiting spontaneous liquefaction; it consists of rounded grains and is unstable when wet. A *clayey soil* has an affinity for water, is cohesive and plastic, may swell when wet, and can be treacherous when saturated. A *silty soil* (loam) demonstrates properties intermediate between sands and clays. In designing a soil-supported structure—bank slope, building, or road—the strength properties of the dry (or moist) soil are utilized, and provisions are made to keep the soil well drained. Soil drainage is a subfield of soil mechanics, one of great importance in surface mining, but its exploration is beyond the scope of this book.

7.7.2 Analysis of Slope Stability

A *slide* is a term given to the sudden or gradual failure of the soil mass in a slope, causing the slope mass to move outward and downward. Because they can be catastrophic in deep surface mines, engineers must learn how to prevent slides in slopes of soil and similar materials. The design of such slopes is often referred to as *slope stability*.

There are many events that may trigger a slide in a mine slope: (1) maintaining too steep a slope, (2) excavating or undercutting in the toe area,

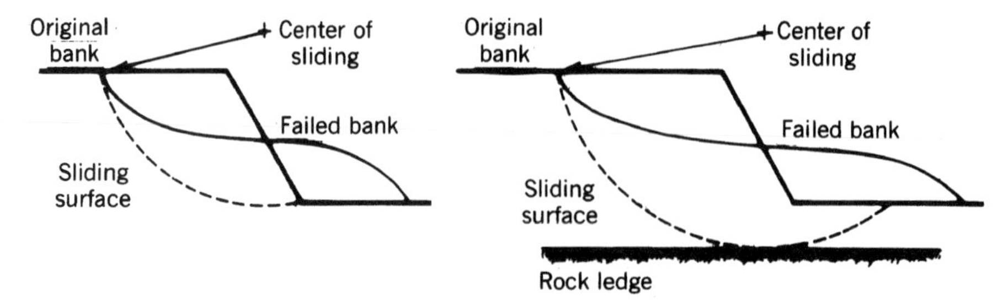

FIGURE 7.22. Types of bank failure. (*Left*) Slope failure. (*Right*) Base failure.

(3) a rise in the pore water pressure, (4) excessive shock or vibration, such as blasting or earthquake, (5) existence of old surfaces of slides, cracks, seams, or rock layers, and (6) existence of discontinuities such as bedding, cleavage, faults, and fractures. Two major types of failures are often recognized, depending on the location of the failure surface (see Figure 7.22). A slope failure is one that occurs between the crest and the toe of a bank. A base failure occurs at a plane of weakness (such as those in causal factors 5 and 6, mentioned earlier) that is below the toe of a bank. There are analytical procedures that can be used to overcome each of these failures.

Table 7.1 outlines some slope angles that have been recommended over the years as safe values to start with in many mining situations involving soils and

Table 7.1 Recommended Pit Slopes[a]

A. Soils	
Clean, loose sand	$1\frac{1}{2}:1$ (34°)
Sand and clay	$1\frac{1}{3}:1$ (37°)
Wet sand	$2\frac{1}{2}:1$ (22°)
Gravel	$1\frac{1}{3}:1$ (37°)
Dry clay	$1:1$ (45°)
Wet clay	$3\frac{1}{2}:1$ (16°)
Loose rock	$1:1$ (45°)
Glacial till	$3:1$ (20°)
Mud	$1:1$ (45°)
B. Rock	
Solid country rock	$\frac{1}{3}$ or $\frac{1}{2}:1$ (72°–63°)
Solid iron ore, bench	$\frac{1}{2}$ or $\frac{2}{3}:1$ (63°–56°)
Solid iron ore, total bank	$1:1$ (45°)
Solid copper ore, bench	$\frac{3}{5}:1$ (60°)
Solid copper ore, total bank	$1\frac{1}{2}:1$ (34°)

[a] Values expressed as slope ratio, horizontal to vertical, and slope angle (in degrees).

rocks. These values are dependent on many variables and will vary from mine to mine. When failures do occur, changes in the design may be required. In most cases, failures can be prevented by good analysis. Analytical methods for two types of soils are provided in the following paragraphs.

7.7.2.1 Cohesionless Soil. In dry or drained sandy or silty soil, a stable bank can be designed based on the requirement that the bank angle be less than the angle of internal friction of the soil (Sherman, 1973). Usually the design is based on the factor of safety (FS) that is calculated using the angle of the slope β and the angle of internal friction ϕ. The factor of safety then becomes:

$$FS = \frac{\tan\ \phi}{\tan\ \beta} \tag{7.5}$$

Numerically, values of FS range from 1.0 to 5.0, with a commonly accepted design limit of at least 1.25 for nonpermanent slopes.

7.7.2.2 Cohesive Soil. In a bank consisting of a cohesive soil like clay or silty clay, the slope is also stable if the slope angle $\beta < \phi$, but the soil will possess some cohesion that will allow us to increase the bank angle to an even steeper value. The method used for analysis is known in soil mechanics as the *slip circle technique* (Sherman, 1973; Cording and Cepeda-Diaz, 1992). It is based on the assumption that the rotational angle of sliding during failure is circular (see dotted sliding failure surface lines in Figure 7.22). This permits a simple analytic solution. The mathematical premise for analysis is the summing of the moments of all forces acting on the bank, including the gravitational force on the soil mass and shear stress along the supposed failure surface. If the moments are unbalanced in favor of movement, then the bank fails. Otherwise, the bank remains intact. The factor of safety can be calculated in this case as the ratio of the moments resisting failure to those causing failure.

The forces acting on the bank and certain key dimensions are represented diagrammatically in Figure 7.23 (top), showing a bank undergoing a base failure. Assuming that failure occurs along a rotational arc of sliding, the center of rotation O is located by trial and error, the objective being to place O in the worst possible position for the bank geometry given. The radius of the arc is r, and the length of the sliding surface L; the depth of the plane of weakness along which failure occurs is h (although an average height $\bar{h}$ is generally assumed in the calculation of s, because the actual depth varies from 0 to h). We represent the bank weight by two components W_1 and W_2, each acting at a respective lever arm (l_1 and l_2) and creating moments ($W_1 l_1$ and $W_2 l_2$). (An alternative is to use a single combined weight and moment arm.) The two moments act in opposite directions around O, $W_1 l_1$ being a sliding moment and $W_2 l_2$ a resisting moment. In addition, there is another resisting moment,

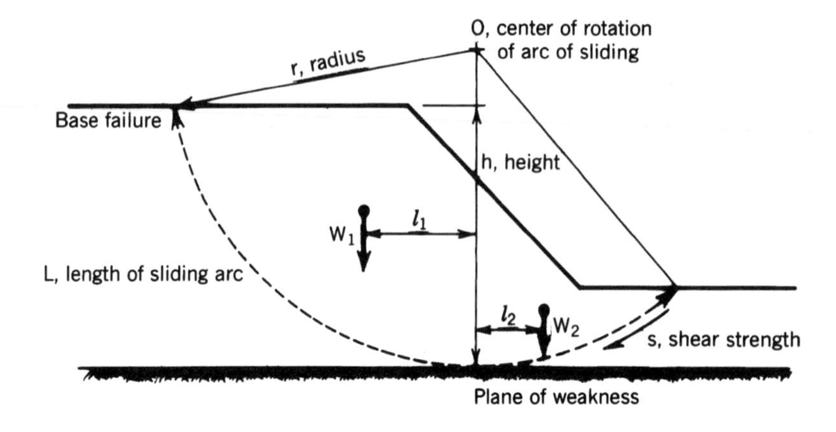

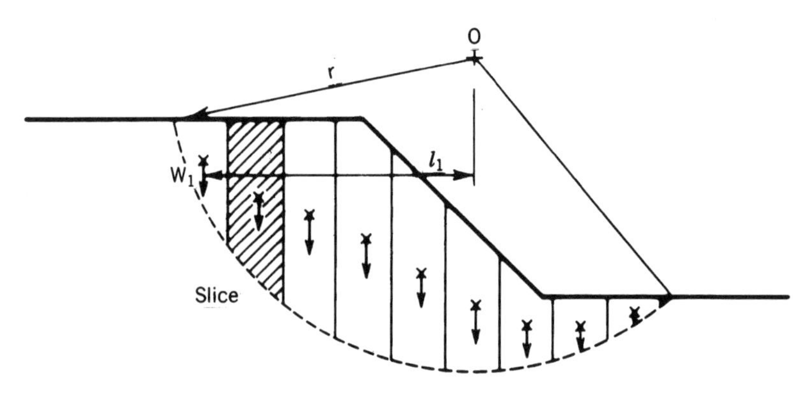

FIGURE 7.23. Bank failure analysis by slip circle analysis. (*Top*) Analysis by opposing bank weights. (*Bottom*) Analysis by method of slices.

the shear strength s acting along the length of the sliding L at moment arm length r. Summing the three moments enables us to determine whether the bank is stable:

$$W_1 l_1 \lesssim W_2 l_2 + sLr \tag{7.6}$$

The factor of safety is

$$FS = \frac{W_2 l_2 + sLr}{W_1 l_1} \tag{7.7}$$

And we again usually assume stability if $FS \geqslant 1.25$. For greater accuracy, the method of slices is used (see Figure 7.23 bottom). This method uses vertical segments through the projected slide area to analyze the stability.

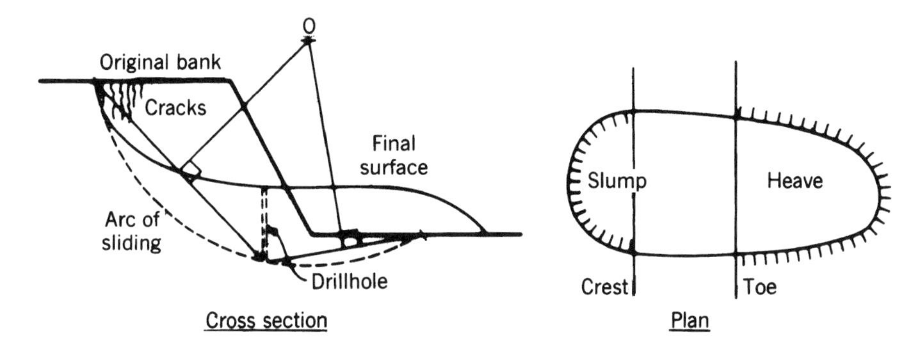

FIGURE 7.24. Field analysis of bank failure with drillhole to determine location of sliding surface and center of arc of sliding.

Details of this method are provided by Cording and Cepeda-Diaz (1992).

The biggest problem of analysis in the past has been location of the rotational center O. This can be accomplished by using a physical model of the bank and experimenting with the model. It can also be performed by calculation methods. Today, the rotational center can be located quite readily using a computer to perform the computations.

In the field, slope stability experts look for evidence of unstable slopes as a means of overcoming failures and detecting incipient slides. A sure sign of a pending slide is the appearance of tension cracks in the crest of the bank some distance from the crest (Figure 7.24). Preventive measures can often be taken at this time by correcting the geometry of the slope or reducing the opportunity for water to affect its stability. In addition, instrumentation can be installed in slopes to detect initial evidence of a slide before a major movement of earth occurs.

The analytical procedure for analyzing the stability of a slope is demonstrated by a numerical example.

***Example** 7.3.* Analyze the soil bank shown in Figure 7.25. The worst-case center of rotation is shown. The bank consists of a sand with properties as follows:

$$w = 120 \text{ lb/ft}^3 \text{ (1922 kg/m}^3\text{)}$$

$$c = 115 \text{ lb/ft}^2 \text{ (5.51 kPa)}$$

$$\phi = 30^\circ$$

Analyze for two cases: (1) saturated and (2) dry.

SOLUTION. (1) Assume that the soil is saturated. Then $c = 0$, $\phi = 0$, and $s = 0$. Therefore, the moment $sLr = 0$.

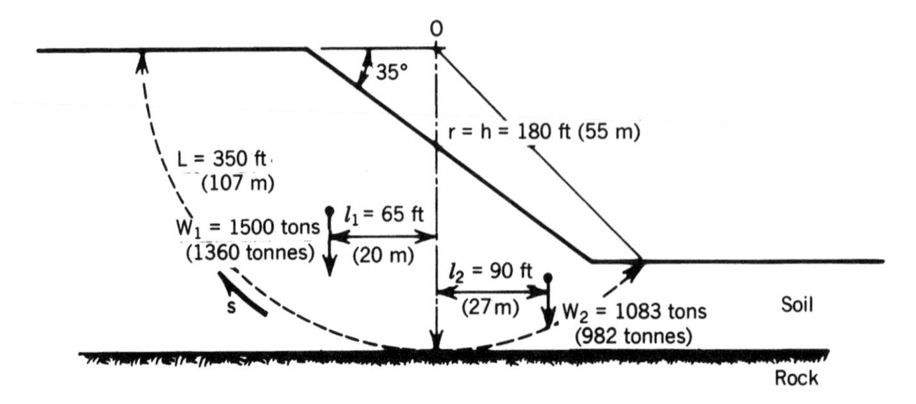

FIGURE 7.25. Analyzing a base failure by the slip circle technique. See Example 7.3.

Writing the moment equation (7.6) per unit of length along the bank in units of ft-tons (m-tonnes):

$$W_1 l_1 \leqslant W_2 l_2 + sLr$$

$$(1500)(65) \leqslant (1083)(90) + 0$$

$$97{,}500 = 97{,}500 \text{ ft-tons } (26{,}960 \text{ m-tonnes})$$

From Eq. 7.7,

$$FS = \frac{97{,}500}{97{,}500} = 1.00 < 1.25$$

This indicates that the slope is unstable if it is saturated.

(2) Assume that the soil is dry (or only modestly moist). Calculate the shear strength by Eq. 7.4:

$$\bar{h} = h/2 = 180/2 = 90 \text{ ft } (27 \text{ m})$$

$$s = c + w\bar{h} \tan \phi = 115 + (120)(90) \tan 30^\circ$$

$$= 6350 \text{ lb/ft}^2 \ (304 \text{ kPa}) = 3.175 \text{ tons/ft}^2 \ (31.0 \text{ tonnes/m}^2)$$

$$sLr = (3.175)(350)(180) = 200{,}020 \text{ ft-tons } (55{,}310 \text{ m-tonnes})$$

Summing by moments using Eq. 7.6,

$$97{,}500 \leqslant 97{,}400 + 200{,}020 \leqslant 297{,}500 \text{ ft-tons } (82{,}\ 260 \text{ m-tonnes})$$

Using Eq. 7.7, the factor of safety is

$$FS = \frac{297{,}500}{97{,}500} = 3.05 > 1.25$$

which is safe by a considerable margin. This example shows the value of keeping water out of this soil to maintain its stability, an important principle to be used in soil stability problems.

Today, more sophisticated and accurate methods of analysis are possible with computer and numerical techniques. Although the preceeding example is rather simple, actual slopes may contain a number of soil types with different properties. The more complicated the slope, the more helpful it will be to have a computerized method of analysis like the finite element method. More can be learned about the finite element method in Naylor and Pande (1981) and Brady and Brown (1993).

PROBLEMS

7.1 Particle size analyses are obtained in index-property tests of five soils samples, and the resulting cumulative size distributions are plotted on the log-probability graph in Figure 7.26. Using the PRA classification chart, identify and name each soil. Read the graph to the nearest 5%.

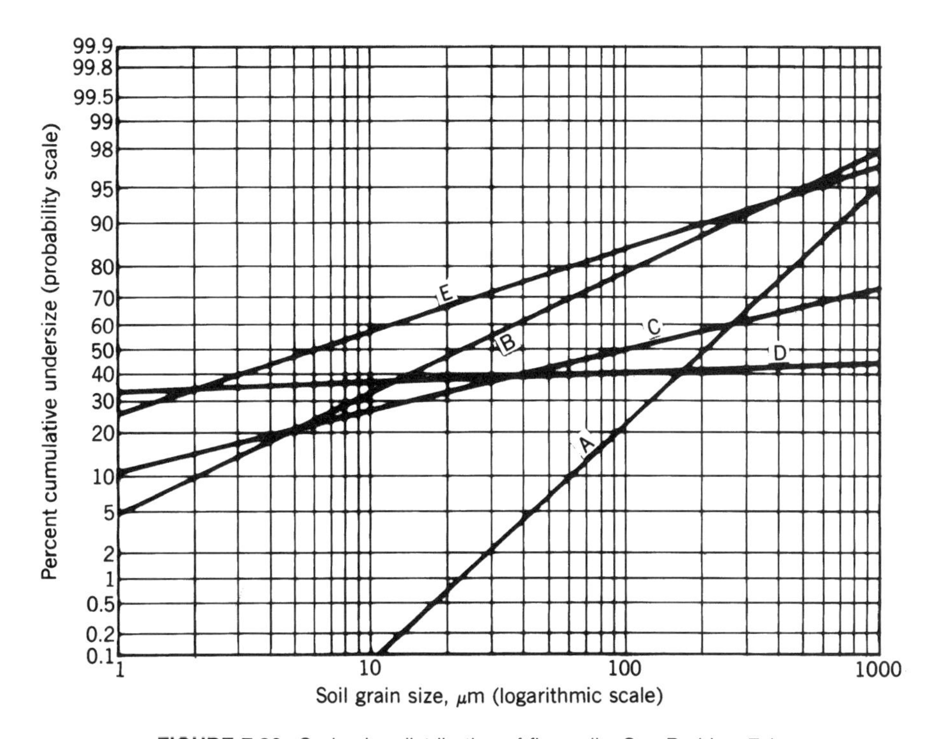

FIGURE 7.26. Grain size distribution of five soils. See Problem 7.1.

7.2 The following information has been determined in the stability analysis of a slope. All of the mass has been combined.

Shear strength	1200 ft^2 (57.5 kPa)
Length of arc sliding	290 ft (88.4 m)
Total weight of bank mass failing	826 tons (746 tonnes)
Lever arm	26.3 ft (8.0 m)
Radius of arc	120 ft (36.6 m)

Is the bank safe, or will it fail? If safe, what is the factor of safety?

7.3 An open pit bank in soil has been analyzed under dry conditions and the following data obtained using a combined moment:

Factor of safety	1.3
Overturning moment	6154 ft-tons (1702 m-tonnes)
Bank height	52 ft (15.8 m)
Radius of arc of sliding	60 ft (18.3 m)
Length of arc	124 ft (37.8 m)
Bank angle	44°
Soil cohesion	80 lb/ft^2 (3.8 kPa)
Soil friction angle	20°
Soil friction weight	162 lb/ft^2 (2595 kg/m^3)

Determine the factor of safety of the bank during the rainy season when the soil is saturated. Is the bank safe, or will it fail?

7.4 Investigate by the slip circle method the stability of the bank shown in Figure 7.27, assuming that a base failure is most likely to happen and the

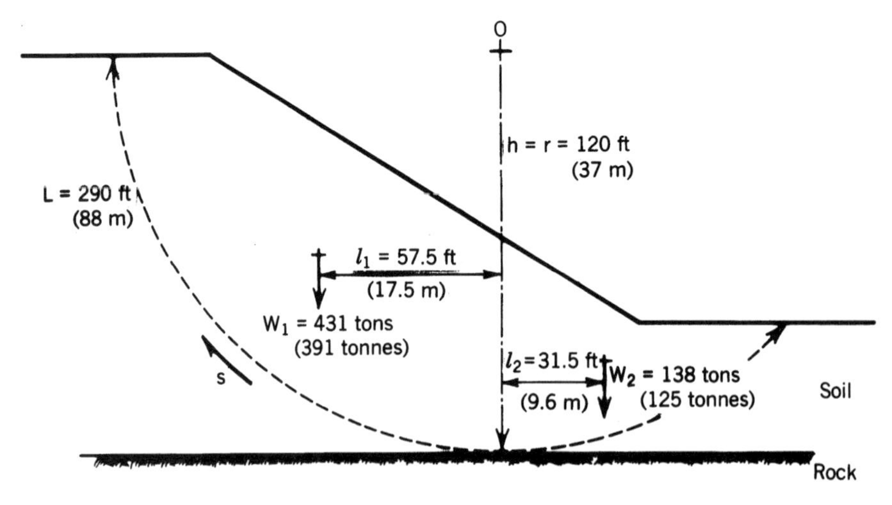

FIGURE 7.27. Analysis of possible base failure of bank. See Problem 7.4.

center of sliding is as shown. Analyze (a) dry soil and (b) saturated soil. Soil properties and bank dimensions are as follows:

Soil specific weight	105 lb/ft^3 (1682 kg/m^3)
Soil cohesion	112 lb/ft^2 (5.36 kPa)
Soil internal friction angle	15°
Pit slope	32°
Bank height	80 ft (24 m)
Depth to rock ledge	120 ft (36 m)

Calculate the factor of safety. Is the slope safe?

7.5 Investigate by the slip circle technique the stability of the slope shown in Figure 7.28, assuming that a slope failure is the most likely to occur. The center of rotation shown is considered to be the worst case. Analyze (a) dry and (b) saturated. Soil properties and bank dimensions are as follows:

Soil specific weight	169 lb/ft^3 (2707 kg/m^3)
Soil cohesion	215 lb/ft^2 (10.29 kPa)
Soil internal friction angle	25°
Pit slope	40°
Bank height	60 ft (18 m)
Depth to rock ledge	80 ft (24 m)

Calculate the factor of safety. Is the slope safe?

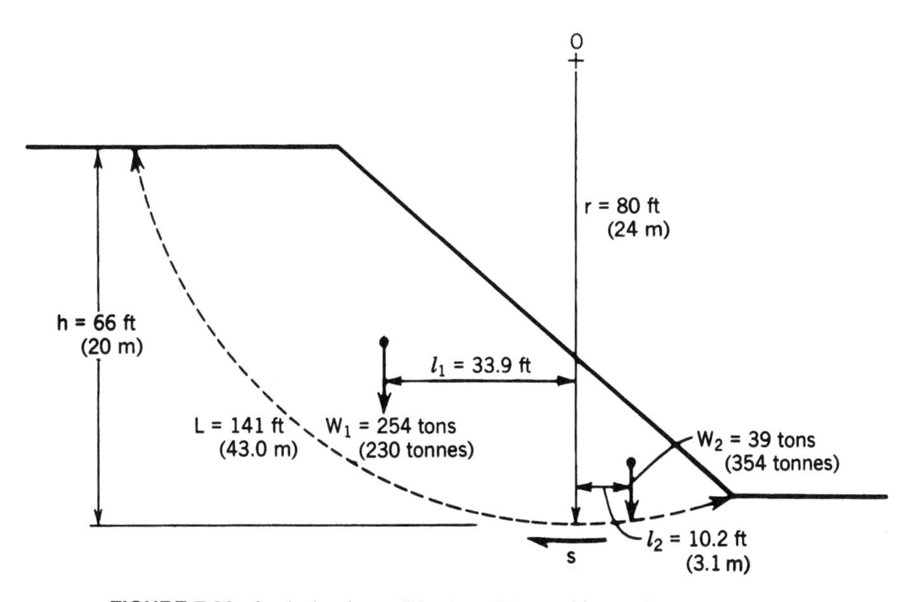

FIGURE 7.28. Analysis of possible slope failure of bank. See Problem 7.5.

7.6 Consider the blasting pattern for the surface mine in Example 7.1 and perform the following:

(a) Modify the design of the blasting pattern by increasing the powder factor to 0.6. What is the effect on the hole spacing and number of holes required?
(b) Design a blasting pattern to conserve explosives (reduce the powder factor) in Example 7.1 by modifying the hole burden, depth, or spacing.

8

SURFACE MINING: AQUEOUS EXTRACTION METHODS

8.1 CLASSIFICATION OF METHODS

In addition to mechanical cxtraction, there is a second category of surface mining methods. *Aqueous extraction* includes all the methods that employ water or a liquid solvent to recover minerals from the earth. Both hydraulic action of a liquid (normally water) and solution chemical attack on the minerals are commonly utilized. The methods in this category are used less often than the mechanical methods, generally accounting for about 10% of surface mineral production. However, in certain minerals, the variety of methods and the percentage of minerals mined by aqueous methods are increasing significantly. Most aqueous mining methods are comparatively inexpensive, with relative costs of less than 5% (see Table 4.1).

The aqueous extraction class is made up of two subclasses. *Placer mining*, the first subclass, is the recovery of heavy minerals from alluvial or placer deposits using water to excavate, transport, and/or concentrate the mineral. *Solution mining* is another subclass, employed for the recovery of soluble or fusible minerals or those that can be easily recovered in slurry form. Solution mining is customarily applied using ordinary water, heated water, or liquid solvents to extract the minerals. The two classes, with their subclasses and variations, are as follows:

A. Placer mining
   1. Hydraulicking
   2. Dredging
      a. Shallow-water dredging
      b. Deep-sea dredging

B. Solution mining
   1. Borehole extraction
      a. Frasch method
      b. Multiple-well method
   2. Leaching procedures
      a. Dump/heap leaching
      b. In situ leaching
   3. Evaporite/evaporation methods

These methods are relatively limited in application, but the leaching methods have recently been used more frequently, particularly in gold and copper mining. These methods are quite unique in terms of their sequences of development and cycles of operation.

8.2 PLACER MINING: HYDRAULICKING

A placer deposit is any concentration of minerals that has been redeposited in unconsolidated form by the action of a fluid. We can classify such deposits by agent of deposition as alluvial (continental detrital), eolian (wind), marine, or glacial. Most placers are sand and gravel deposits deposited by water. Additional variations of placer types can be found in McLean et al. (1992). These are the sources of gold, diamonds, tin, titanium, platinum, tungsten, chromite, magnetite, coal, phosphates, and sand and gravel.

The properties that make a particular placer minable by aqueous methods and of interest to a mining company are as follows (Daily, 1968a):

1. Material amenable to disintegration by the action of water under pressure (or mechanical dislodgement and hydraulic transport)
2. Adequate water supply available at the required head
3. Adequate space for waste disposal
4. Difference in density, or a similar property between ore minerals and gangue to allow efficient mineral processing
5. Natural gradient amenable to hydraulic transport of mineral (if hydraulicking)
6. Ability to comply with environmental regulations

The two principal methods of placer mining in common use are hydraulicking and dredging. In *hydraulicking*, a high-pressure stream of water is directed against a bank to undercut and cave it (Figure 8.1). As the material disintegrates, the loosened particles (mineral, sand, and gravel) are slurried in water and washed into a *sluice*, either a natural trough in the ground or a metal or wooden box, where it is transported by gravity to a *riffle box* or other more elaborate concentrating device.

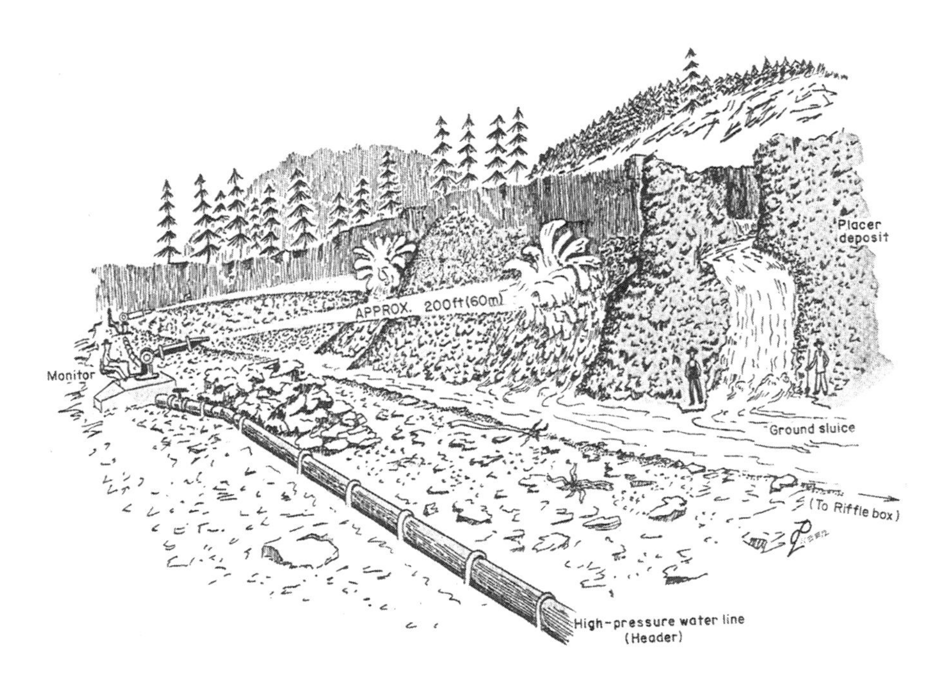

FIGURE 8.1. Placer mining by hydraulicking. (By permission from the West Virginia Geological and Economic Survey, Morgantown, WV.)

Water is applied to the bank by a hydraulic monitor or giant, a nozzle designed to direct hydraulic energy at the bank. The monitors are connected to a supply line or header to provide a continuous flow of high-pressure fluid. One design of monitor, called an Intelligiant (see Figure 8.2), utilizes curved sections of pipe to balance all the reactive forces in the water delivery system, easing the anchoring and control of the monitor nozzle during use. In some mining operations, the giants are mounted on mobile carriers and are programmed to cycle back and forth across the bank. Design specifications for monitors are as follows:

Nozzle diameter:	1.5–6.0 in. (40–150 mm)
Pressure:	45–200 lb/in.^{2} (300–1400 kPa)
Volume flow rate:	1200–8000 gal/min (77–510 L/sec)
Water jet velocity:	
Sand	30 ft/min (0.15 m/sec)
Gravel	300 ft/min (1.5 m/sec)
Boulders	600 ft/min (3.0 m/sec)

8.2.1 Sequence of Development

Development of a hydraulicking operation will not proceed unless an adequate supply of water (preferably upstream), a suitable waste disposal area (prefer-

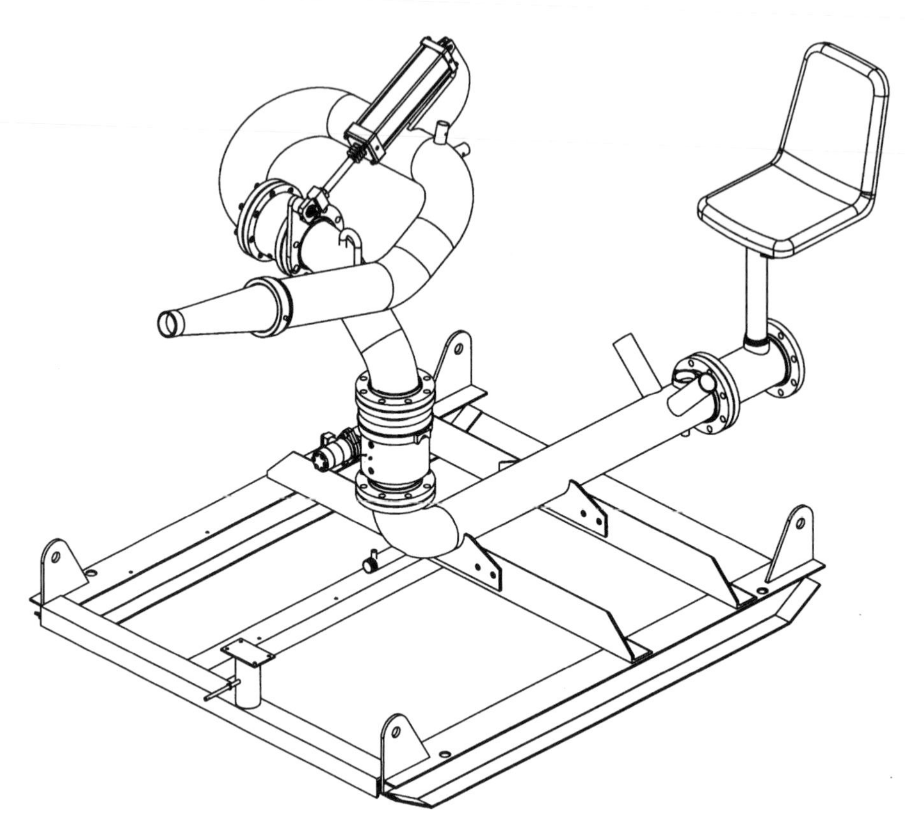

FIGURE 8.2. The Stang Intelligiant monitor mounted on a skid. (Diagram courtesy of Stang Industrial Products, Huntington Beach, CA.)

ably downstream), and control of water quality are possible. In addition, the method must be accepted by the local regulatory agencies. Hydraulicking disturbs the land and water significantly and is therefore often outlawed.

Stripping is not often required for hydraulicking. When necessary, it is ordinarily accomplished using monitors or any other convenient stripping equipment. The development also requires a pump station or a holding basin for water with sufficient head to operate the monitors. A large-diameter header is laid to connect the basin with the monitors. Probably the most critical step in development is providing the resources necessary for environmental protection. Suitable provisions to safeguard water quality and to reclaim the surface must be made before mining begins.

8.2.2 Cycle of Operations

Because excavation and materials handling are both accomplished hydraulically, the production cycle is an abbreviated one. No prior breakage or transport is normally required. The simplicity of the cycle and multipurpose use of the water helps to account for the attractiveness of this method in terms of productivity and cost. Auxiliary operations are negligible, and reclamation is often incorporated in the production cycle.

8.2.3 Conditions

The following conditions are summarized from discussions in Daily (1968), Morrison and Russell (1973), and McLean et al. (1992):

1. *Ore strength:* Prefer heavy minerals in unconsolidated soil or gravel with minimum boulders.
2. *Rock strength:* Unconsolidated.
3. *Deposit shape:* Placer type, tabular, bank or bench.
4. *Deposit dip:* Nearly flat (2–6% grade).
5. *Deposit size:* Small to intermediate with thickness of 15–200 ft (5–60 m).
6. *Ore grade:* Can be very low.
7. *Ore uniformity:* Fairly uniform.
8. *Depth:* Very shallow, little overburden.
9. *Other:* Requires large quantities of water.

8.2.4 Characteristics

The following advantages and disadvantages of hydraulicking are based on descriptions in Daily (1968), Morrison and Russell (1973), and Cook (1983).

Advantages

1. Fairly high productivity of 100–300 yd^3 (75–230 m^3) per employee-shift
2. Low mining cost (relative cost about 5%).
3. Intermediate production rate.
4. Low capital cost; simple equipment and cycle.
5. Can automate some operations.
6. Can operate a mine with a few workers.

Disadvantages

1. Environmental damage is severe unless elaborate protection is implemented.
2. Extensive water requirements.
3. Limited to unconsolidated deposits that disintegrate under hydraulic attack.
4. Cutting action is difficult to control.

8.2.5 Applications

Because of environmental concerns, hydraulicking is little used today in the United States, except in Alaska. Cook (1983) provides descriptions and photos of a number of hydraulicking operations in the Alaska gold fields. Hydraulic mining is still used in other parts of the world, however, primarily for gold and tin. Occasionally, it is used for stripping the overburden from a deposit (Li, 1976; Thomas, 1978). It also is employed in some Florida phosphate mines to slurry phosphate pebble for transport to a processing plant. In this application, the hydraulic monitor is utilized after the phosphate pebble is mined with a dragline and placed on the highwall (Hoppe, 1976). McLean et al. (1992) also describe three locations where hydraulic giants recover material from tailings piles.

8.3 PLACER MINING: DREDGING

Dredging is the underwater excavation of a placer deposit, usually carried out from a floating vessel called a *dredge*, which may incorporate processing and waste disposal facilities. The dredge has been around for a long time, with the first known use in the Netherlands in 1565 (Macdonald, 1983). The body of water used for dredging may be natural or human-made. Depending on the size of the dredge and the size of deposit, 200 to 2000 gal/min (13 to 125 l/sec) of water may be required for mining, processing, and waste disposal (Daily, 1968a; Macdonald, 1983). Dredging methods are often subdivided into shallow-water methods and deep-sea or marine mining methods. We consider only the shallow-water methods here; deep-sea mining is covered in Chapter 13.

Shallow-water dredges are often classified by method of excavation and materials transport (Turner, 1996; Herbich, 1992). *Mechanical dredges* are those that mechanically excavate and transport the mineral. They include dipper, bucket, and ladder dredges, with several variations in the bucket dredge category (Herbich, 1992). *Hydraulic dredges* (also called suction dredges) are designed to transport the mineral in slurry form, using water as the transport medium. Herbich (1992) also lists four subcategories of hydraulic dredges: hopper, sidecasting, pipeline, and agitation dredges, with four variations of pipeline dredges. Many of the categories of dredges are used strictly for river and harbor dredging applications and are not covered here. The primary types of dredges employed in minerals recovery are (1) bucket-line dredges, (2) cutter-head suction dredges, and (3) cutter-wheel suction dredges.

Bucket-line dredges are mechanical dredges that were used heavily in the past for extraction of gold from placers up to 160 ft (50 m) in depth. The buckets are continuously moved around the bucket ladder, excavating and elevating the placer material into the processing plant located aboard the dredge. As shown in Figure 8.3, the bucket ladder is controlled with the use of a large crane, and the waste material is discarded from the back of the dredge by means of a stacking conveyor. The dredge is moved around the body of water using a pair of spuds located at the back of the barge and wire ropes

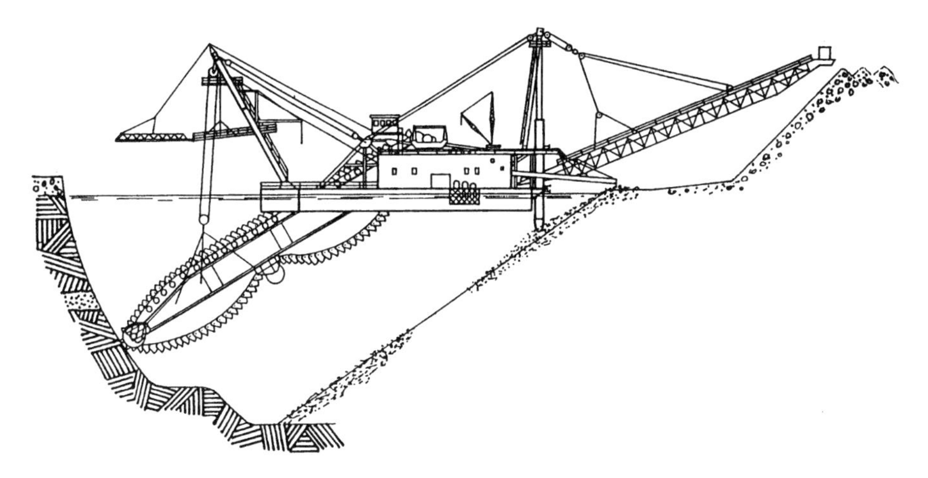

FIGURE 8.3. Typical bucket-line dredge. *Source:* Brooks (1991). (By permission of the Institute of Mining and Metallurgy, London.)

anchored to the banks. Often, the water body is a pond specifically created to float the dredge. In this situation, the dredge carries its pond with it by advancing into the forward bank and filling behind. Bucket-line dredges can also be applied to shallow marine deposits and have been used for gold, tin, and diamonds in such environments.

Suction dredges (also called hydraulic dredges), mentioned earlier are dependent on slurry transport of the placer from the point of digging to the dredge and then to the final destination. Ordinarily, no processing is performed on the dredge; it is typically conducted on land. A cutter-head suction dredge is shown in Figure 8.4. The cutter head frees the placer material, which is then hydraulically moved into the pipeline on which the cutter head is mounted. Digging depths of up to 60 ft (18 m) are common (Herbich, 1992). A bucket-wheel dredge is similar in concept, but a vertical bucket wheel supplies the cutting action. A typical bucket-wheel dredge is shown in Figure 8.5. The digging depths (Bray et al., 1997) normally reach as much as 115 ft (35 m). Details on digging bucket and cutter-head designs can be found in Herbich (1992) and Bray et al. (1997).

Like draglines in open cast mines, dredges can be behemoths that move vast volumes of material. The bucket-line dredge is the best example. The output of such a dredge is estimated from its design specifications in the following example.

Example 8.1. Find the low to high output of a bucket-line dredge (in yd^3/day or m^3/day) if the specifications are as follows:

Bucket capacity	10 ft^3 (0.28 m^3)
Bucket-line speed	22 buckets/min
Shifts	3/day, 22.5 hr total
Bucket fill factor	60% to 87%

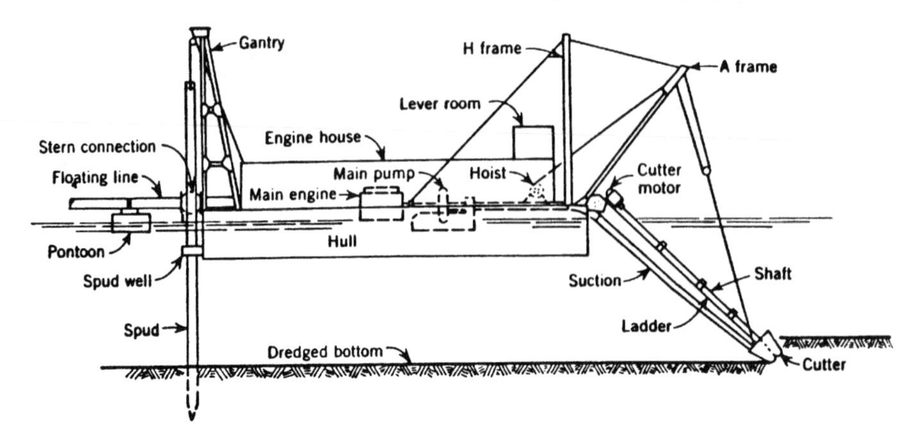

FIGURE 8.4. General layout (simplified) of a cutter-head suction dredge. (From *Hydraulic Dredging* by John Huston. Copyright © 1970 by Cornell Maritime Press, Inc. Used by permission.)

SOLUTION. Using the lowest and highest values of the bucket fill factor, calculate the following:

$$\text{Low output} = \frac{(10)(22)(60 \text{ min/hr})(22.5)(0.60)}{27 \text{ ft}^3/\text{yd}^3} = 6600 \text{ yd}^3/\text{day } (5050 \text{ m}^3/\text{day})$$

$$\text{High output} = \frac{(10)(22)(60)(22.5)(0.87)}{27} = 9600 \text{ yd}^3/\text{day } (7340 \text{ m}^3/\text{day})$$

Using this information, one can provide a rule of thumb that the output of a bucket-line dredge is 500 to 1000 yd^3/day per ft^3 (13,500 to 27,000 m^3/day per m^3) of bucket capacity. A large dredge may excavate 9 million yd^3 (7 million m^3) of placer material a year.

Bucket-line dredging is not only capable of producing large volumes of material, it is also highly productive, perhaps the most productive of all methods of mining. However, the capital costs of the equipment are quite high, ranging from $9 million for a bucket-line dredge with 10 ft^3 (0.28 m^3) buckets to $50 million for a dredge with 30 ft^3 (0.85 m^3) buckets (McLean et al. 1992). The productivity that results when using these dredges will overcome the capital costs if a suitable deposit can be located. McLean et al. (1992) estimate costs of $0.25 to $1.20/yd^3 ($0.37 to $1.57/m^3) for mineral dredging. Costs for shallow-water marine mining applications are found in Cruickshank (1992).

8.3.1 Sequence of Development

Development for placer mining is substantially the same, whether it is performed by hydraulicking or dredging. Again, provision of an adequate water supply is imperative. In dredging, a pond must be created by damming a

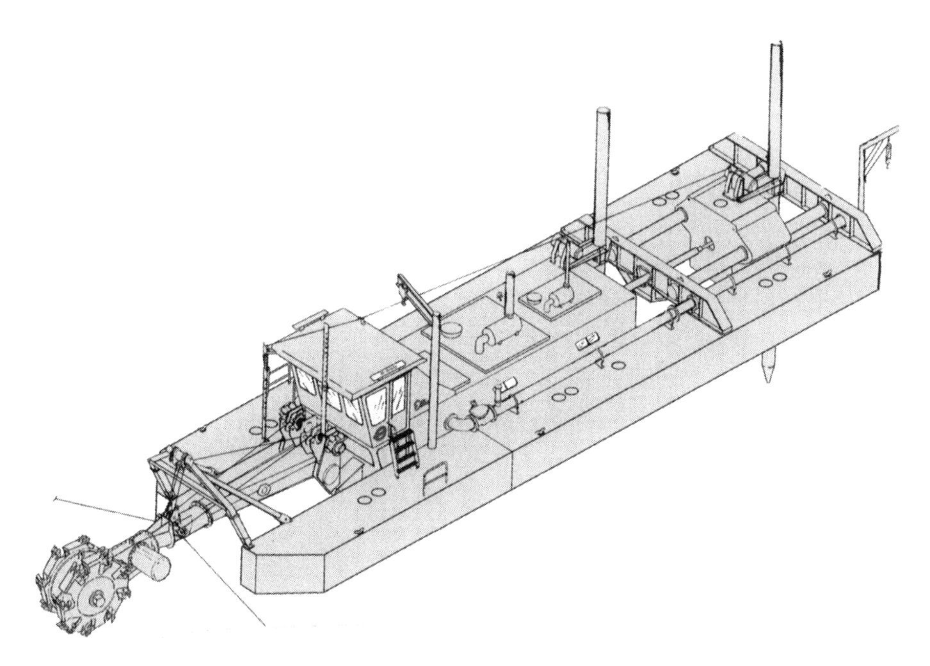

FIGURE 8.5. General diagram showing layout of a bucket-wheel hydraulic dredge. (Photo courtesy of Ellicott International, worldwide dredge manufacturer.)

stream, using a pumping system, or using the existing groundwater if it exists at a sufficiently high level. In addition, a suitable waste disposal and reclamation plan must be provided. Overburden removal is minimal in most cases and is done conventionally unless the overburden and placer are mined together (Pfleider, 1973).

8.3.2 Cycle of Operations

Because the dredge is a continuous mining device, no breakage is required and materials handling is conducted with very little interruption. Water aids in excavation, although much of the digging is accomplished mechanically. Mineral processing is often performed using wet gravity separation on board the dredge. The processing of the placer material on board allows the waste to be dumped directly into the pond behind the dredge, filling the pond as the dredge moves forward. Reclamation is often conducted as an integral part of the operating cycle.

8.3.3 Conditions

The types of materials processed by dredging methods include soil overburdens that are removed via dredge, river gravels that are processed for their mineral content or for sand and gravel, alluvial fans left behind by geologic erosion, and ocean beach deposits containing valuable minerals. Although these deposits are quite diverse, they have much in common as to the conditions that must be present. The following conditions are derived from descriptions in Daily (1968a, 1968b), Huston (1970), Turner (1996), and McLean et al. (1992):

1. *Ore strength:* unconsolidated deposit of soil, gravel, or similar material; some boulders permissible, depending on dredge type; prefer valuable mineral to be heavier than waste
2. *Rock strength:* unconsolidated
3. *Deposit shape:* placer, tabular, bank or bench
4. *Deposit dip:* preferably flat (maximum of 2% to 6% grade)
5. *Deposit size:* intermediate to large (thickness 25 to 200 ft, or 8 to 60 m)
6. *Ore grade:* can be very low
7. *Ore uniformity:* fairly uniform
8. *Depth:* very shallow, little overburden
9. *Other:* moderate quantities of water required (200 to 2000 gal/min, or 13 to 125 l/sec)

8.3.4 Characteristics

The advantages and disadvantages of dredging are well known. The following are outlined by Morrison and Russell (1973), Pfleider (1973), Macdonald (1983), and McLean et al. (1992). These characteristics apply to shallow-water dredging only.

Advantages

1. Most productive of all mining methods (250 to 400 yd^3, or 190 to 300 m^3, of gravel per employee-shift).
2. Lowest mining cost (relative cost $<5\%$).
3. High production rate (maximum of 9 million yd^3 or 7 million m^3 a year).
4. Low labor requirements (crew of 2 to 30 people).
5. Good recovery (up to 90%).
6. Continuous operation with no breakage required.

Disadvantages

1. Environmental damage can be severe; environmental protection must be exercised; outlawed in some states.
2. Moderate water requirements (600 to 800 gal/yd^3, or 3000 to 4000 l/m^3, of material mined).
3. Limited to unconsolidated deposits that disintegrate under hydraulic or mechanical attack.

4. High capital investment for large dredges.
5. Inflexible and unselective; limited to placer-type deposits.

8.3.5 Applications and Variations

Dredging has applications to harbor and river dredging to enhance the transportation potential of such waterways. As a mining method, however, it is not used extensively in the United States ($<2\%$ of surface mineral production). Currently, two large bucket-line dredges are being used in Yuba County, California, to recover gold from the large alluvial fan deposits located along the Yuba River (Lewis, 1984; Wolf, 1999). One of these dredges also recovers sand and gravel from a 7000 acre (2700 ha) site. Other gold dredges are operating in Alaska (Garnett, 1997). Dredges have also been used in the United States for removing overburden from iron, coal, and phosphate operations, for beach-type titanium deposits, for the recovery of sand and gravel, and for recovery of tailings.

Dredging is widely used in other parts of the world for producing gold, tin, titanium, diamonds, and some of the heavy-sand minerals. The most common applications are found in South America, Southeast Asia, Australia, and the Far North (Daily, 1968b; Macdonald, 1983; Lewis, 1984; McLean, 1992; and Foster, 1994). McLean (1992) presents case studies on Malaysian tin, Peruvian gold, and Australian titanium.

Students should note that some placers are mined without using water in *dry mining*, sometimes called dry-land dredging (McLean et al. 1992; Foster, 1994). This method uses conventional surface mining equipment and may be applied to deposits that are too small for dredging, in conditions that are not suitable to a dredge, and to materials that are better processed dry. It should also be noted that dredging methods are available for mining of deep-sea minerals. These will be covered in Chapter 13.

8.4 SOLUTION MINING

As conventional ore production methods have become more costly and difficult, the mining industry has increasingly turned to the category of methods known as solution mining. For our purposes, *solution mining* is defined as the subclass of aqueous surface mining methods in which minerals are recovered by leaching, dissolution, melting, or slurrying processes. It should be noted that some of these methods have been developed for underground use (for example, to recover remaining values in a block caving operation); however, the methods are operated similarly to a surface-based method. Therefore, all solution mining methods will be covered in this chapter.

Solution mining in this country was first used in 1922 and in the early days was applied primarily to evaporites, sulfur, and copper. However, it has now grown into a significant part of the mineral industry, as shown in Table 8.1. Note that more than a quarter of our gold, silver, copper, uranium, salt, magnesium, sulfur, and lithium are produced with the use of solution mining methods. Clearly, these methods have been among the fastest growing in the last decade or so.

It is difficult to categorize the methods of solution mining because of the different kinds of deposits (in situ minerals, previously mined materials, evaporites, and lake deposits) and the variety of methods of attack (dissolution with water alone, dissolution with chemicals, evaporation of water containing salts, slurrying, and melting). However, the following can be considered distinct categories:

1. Borehole extraction systems
2. Leaching methods
3. Evaporite/evaporative procedures

This classification scheme is discussed in the following sections, even though there is clearly some overlap among the three methods. Further discussion of solution mining systems can be found in Schlitt (1992) and Bartlett (1998).

8.4.1 Borehole Extraction

A number of solution mining methods utilize wells to access mineral values located under the earth. These methods are usually called *borehole mining* or *borehole extraction.* The wells are generally used to inject water and/or a lixiviant into the mineral deposit to effect the mineral extraction. The methods of attack are melting, leaching, dissolution, and slurrying the valuable mineral for recovery purposes. Possibilities include the following:

1. Melting of sulfur (Frasch process)
2. In situ leaching of uranium, copper, gold, and silver
3. Dissolution of salt, potash, and trona (to be discussed in Section 8.4.3)
4. Slurrying of phosphate, kaolin, oil sands, coal, gilsonite, and uranium (technologically feasible but not widely utilized because of economics)

Melting, dissolution, and slurrying are normally conducted using single-well procedures in which each well is a producing entity. A good example is the Frasch process illustrated in Figure 8.6 (right). In this single-well method, the wellbore contains three concentric pipes. The outer pipe is used to inject superheated water into the deposit to melt the sulfur. The inner pipe allows compressed air to flow to the bottom of the well, where it flows upward through the annulus between the inner and middle pipes, taking the melted sulfur with it. The flow is produced primarily by the difference in densities between the water and the aerated sulfur. This is often referred to as an airlift system.

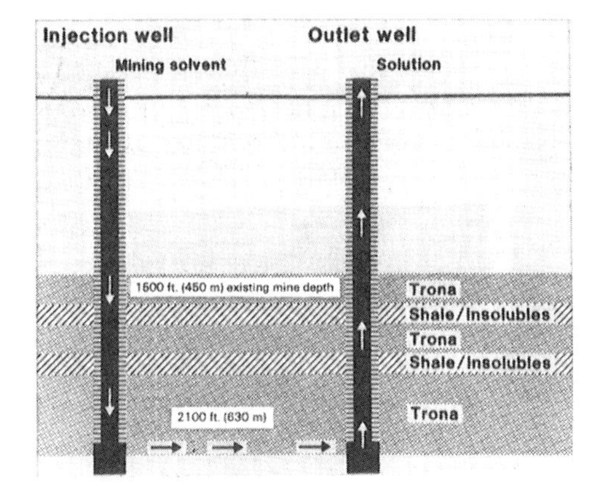

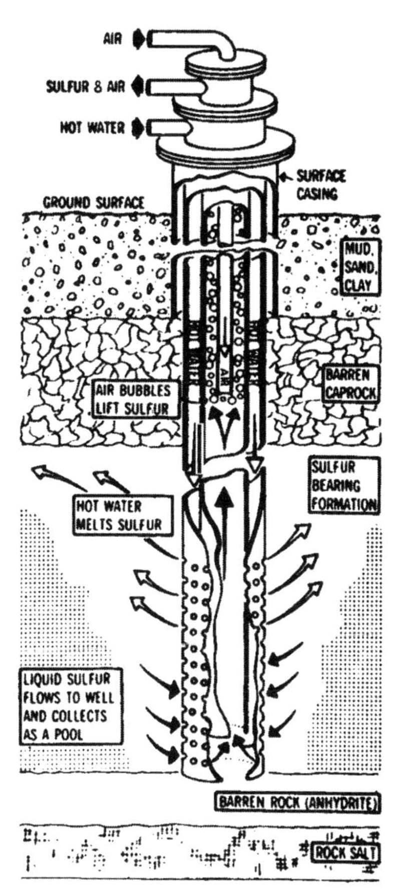

FIGURE 8.6. Solution mining by boreholes. (*Left*) Multiple wells for trona. (After American Mining Congress, 1981; Kostick, 1982; by permission from National Mining Association, Washington, DC.) (*Right*) Concentric well for sulfur, Frasch process. (After Marsden and Lucas, 1973. By permission from the Society for Mining, Metallurgy, and Exploration, Inc., Littleton, CO.)

Sulfur wells require the following three properties for economic operation (Shock, 1992): (1) a large deposit of >5% sulfur, (2) an adequate and inexpensive supply of water, and (3) a low-cost source of fuel to heat the water. Most sulfur produced in the United States comes from the Gulf states of Texas and Louisiana, where sulfur-bearing salt domes and generous supplies of natural gas coexist. Both onshore and offshore mines are operated in this region. In the Frasch process, superheated water at 320°F (160°C) is pumped into the deposit. The wells are spaced 100 to 1000 ft (30 to 300 m) apart. The process uses a large amount of water, ranging from 1000 to 12,000 gal/ton (4200 to 50,000 l/tonne) to produce the sulfur. Hence, there is need for a suitable supply of both water and fuel.

Other commodities can be produced efficiently only by using multiple-well methods. In these methods, several wells work together to permit flow through the deposit (see Figure 8.6, left). Some of these wells are injection wells in which a fluid (normally water and a lixiviant) is introduced into the ore zone; others are recovery wells through which the pregnant solution is extracted. The wells are often arranged in logical geometric patterns to facilitate fluid recovery. Three patterns are shown in Figure 8.7. In the left diagram, a line of injection wells is surrounded by recovery wells to provide proper flow through the deposit. In the five-spot and seven-spot systems, each injection well is surrounded by a number of recovery wells. Note also that the well field is often surrounded by monitor wells to ensure that the chemical agents used do not escape into surrounding sediments where they may have environmental consequences. A monitor well is shown in Figure 8.8, where a line-drive well field is illustrated as it might look in a uranium solution-mining system. Ahlness et al. (1992) show a more sophisticated system of monitor wells that can detect the flow into the formations above and below the seam being mined.

The requirements for a multiple-well solution mine are that the ore horizon be relatively permeable or easily fractured to enhance the flow through the deposit and that the valuable mineral be amenable to chemical dissolution. Both explosive and hydraulic fracturing can be used to enhance the permeability and flow between the wells. Figure 8.8 (Bartlett, 1998) illustrates the flow as it might occur in a permeable layer that contains uranium minerals. Note that the overlying and underlying sediments are impermeable formations and that monitor wells are used to ensure that the lixiviant does not escape into surrounding formations. This type of solution mining is termed flooded leaching. *Flooded leaching* is the method used when the deposit is subjected to a lixiviant without air being present. It is used primarily in buried deposits of copper and uranium where the grade is too low for underground mining and the depth too great for surface mining.

8.4.1.1 Cycle of Operations. Well methods use similar cycles, whether they are single-well or multiple-well methods. The differences are primarily the fluids introduced into the deposit, the method of attack on the deposit, and the method of recovering the valuable mineral. The cycle can be divided into three steps:

1. Preparing the solution by heating or adding the proper lixiviant
2. Pumping the solution into the deposit to bring the valuable mineral into the solution (by melting, dissolution, or slurrying)
3. Raising the pregnant solution (or slurry or molten material) to the surface

Because of the various possibilities, each of these steps can vary significantly. However, the overall process follows the same three steps.

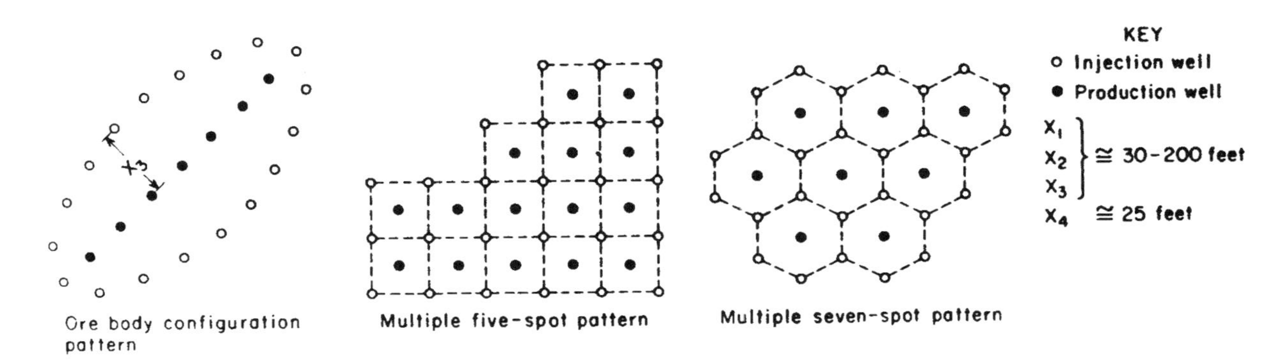

FIGURE 8.7. Well field patterns (Ahlness et al., 1992. By permission from the Society for Mining, Metallurgy, and Exploration, Inc., Littleton, CO).

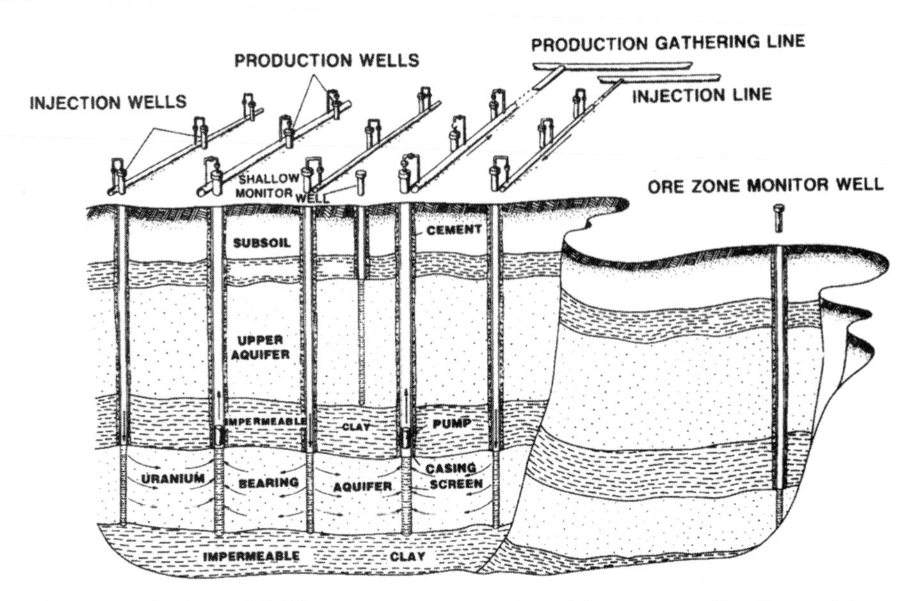

FIGURE 8.8. Typical well field layout for uranium solution mining from a confined horizontal ore deposit. *Source:* Bartlett (1998). This figure is under copyright ownership of OPA (Overseas Publishers Association) N.V. By permission of Taylor & Francis Ltd., Reading, United Kingdom.

8.4.1.2 Conditions. The following information on well methods is summarized from White (1975), Kostick (1982), Ahlness et al. (1992), and Bartlett (1998):

1. *Ore strength:* reasonably competent but porous and permeable (alternatives: must dissolve, melt, or slurry in water)
2. *Rock strength:* surrounding rock must be competent and impervious
3. *Deposit shape:* any, but prefer tabular deposit with large extent
4. *Deposit dip:* any, preferably flat or low
5. *Deposit size:* moderate to large, thickness >50 ft (15 m)
6. *Ore grade:* intermediate (sulfur $>5\%$)
7. *Ore uniformity:* variable to uniform
8. *Depth:* intermediate to high (sulfur generally 200 to 2500 ft or 60 to 750 m; typically less than 10,000 ft or 3000 m for other commodities)
9. *Other:* moderate to large quantities required

Note that well methods compete with surface and underground mining methods in many cases. Thus, a deposit may require evaluation for surface mining, underground mining, and solution mining before a borehole mining method is adopted.

The growth of solution mining is evidence of its advantages for some deposits. The following advantages are compiled from information in Bhappu (1982), Hrabik (1986), Shock (1992), Ahlness et al. (1992), and Bartlett (1998).

Advantages

1. High productivity and moderate production cost.
2. Low capital cost as compared with other methods.
3. Low mining cost (average relative cost: 5%).
4. Applicable to deep and low-grade deposits.
5. Reduced development time and cost.
6. Continuous operation.
7. Breakage not required.
8. Disturbs less surface area.

Note that well fields are normally operated with a controlled number of wells being utilized at any time. New wells are brought on-line as older ones become less productive and are taken out of the system. Using this procedure, the output from the well field is kept relatively constant over time. Note also that the relative cost will vary considerably, depending on the value of the ore. Clearly, the ability to spend more on a solution mining method depends on the value of the ore.

Disadvantages

1. Limited to deposits that dissolve, melt, or slurry in water.
2. Moderate water requirements.
3. Testing often proves difficult.
4. Frasch process requires significant amounts of energy.
5. Unselective: Control of the extraction zone is difficult.
6. Recovery normally low to moderate.
7. Possible groundwater contamination.

Although the well methods of solution mining may be of increasing importance in the future, environmental concerns will likely temper their growth. Thus, planning and control of environmental protection is of utmost importance in these methods.

8.4.1.3 Applications. A variety of well methods have been practiced in the minerals industry, both single-well and multiple-well types. The Frasch process is the only means of recovering sulfur directly involving the mining process; all other sulfur is recovered as a byproduct of oil and gas refining or from metal smelters (Ober, 2000b). The Frasch process has been used in the areas surrounding the Gulf of Mexico, both onshore and offshore. More information on these wells can be found in Donner and Wornat (1973), Ackerman (1992), Shock (1992), and Bartlett (1998). Single-well methods have also been used for evaporite production. These applications are discussed in Section 8.4.3. In single-well salt operations, a secondary result is produced—a salt cavern. Such cavities are useful for sequestering toxic materials and can be used for storage of oil and gas products (Richner, 1992).

Multiple-well methods have been applied in the United States, primarily in mining uranium and copper. Both can be brought into solution using either sulfuric acid or ammonium carbonate as the lixiviant, provided that the deposit is sufficiently permeable. Chemical reactions for the processes can be found in Schlitt (1992) and Bartlett (1998). Uranium is ordinarily mined by this method in Wyoming and Texas, and copper has been mined in the copper-producing states of the southwest. Additional information on these mines can be found in Ahlness et al. (1992).

8.4.2 Leaching Methods

Leaching is the chemical extraction of metals or minerals from the confines of a deposit or from material already mined (Schlitt, 1982). The concept has been around for a considerable length of time, with the first large-scale operation reported in the mid-eighteenth century. Two variations of this method are recognized: percolation leaching and flooded leaching (Bartlett, 1998). The processes are basically chemical but may be bacteriological as well (certain bacteria act as catalysts to speed up some of the reactions in leaching sulfide minerals). If the extraction is carried out on mineral in place, it is termed *in situ leaching*. In situ leaching can be performed with the use of boreholes (as described in Section 8.4.1) or by methods that apply to ore broken in place (described in the following paragraphs). If it is performed in previously mined dumps, tailings, or slag piles, it is called *heap* (or dump) *leaching* and is accomplished by percolation of the lixiviant through the broken ore mass. Another variation, performed in vats or tanks, is *vat leaching*. This method uses the flooded leaching procedure.

Heap leaching has evolved from the early practice of leaching additional copper out of waste dumps, using sulfuric acid, and extracting the copper ions taken into solution. This was a secondary mining method often employed as a result of the opportunity that presented itself when the primary mining process was completed. Today this practice has evolved significantly, with leaching being the only method of recovery of mineral values from some ore bodies. It has found increasing use in the recovery of gold and silver over the last few decades and is also used on copper and uranium. Appropriate lixiviants for these minerals are listed in Table 8.1.

Leaching in designed heaps, with asphalt, impervious soil layers, or geomembranes used to control the liquids, is the most common method of leaching for gold and silver. A heap may be designed in a number of ways to

Table 8.1 Metals and Minerals Commonly Mined by Solution Mining Methods in the United States

Metal or Mineral	Approximate Primary Production	Dissolution Agent/Method
Gold	35%	NaCN
Silver	25%	NaCN
Copper	30%	H_2SO_4; $(NH_4)_2CO_3$
Uranium	75%	H_2SO_4; $(NH_4)_2CO_3$
Common salt	50%	Water
Potash	20%	Water
Trona	20%	Water
Boron	20%	HCl
Magnesium	85%	Seawater, lake brine processing
Sulfur	35%	Hot water (melting)
Lithium	100%	Lake brine processing

Sources: Bartlett (1998), Aplan (1999), and Ober (2000b). Information from Bartlett (1998) is under copyright ownership of OPA (Overseas Publishers Association) N.V. Permission to use this information was granted by Taylor & Francis Ltd., Reading, United Kingdom.

facilitate the solution mining process. Hutchinson and Ellison (1992) outline three methods that are commonly used: (1) the reusable pad method, (2) the expanding pad method, and (3) the valley leach (multiple-lift) method. A schematic for a generic leaching operation is shown in Figure 8.9. Note that the ore is mined, crushed, dumped into a heap of specific height, and leached for a period of time. The same general procedure is used for all three leaching methods, but the physical layout of the heap or dump changes to suit each method. Note that successful operation requires a pad that will reliably contain the liquids and a plan for maintaining water balance in the system.

Details on the parameters of the heap leaching system can be found in Hutchinson and Ellison (1992) and Bartlett (1998). The heaps are constructed with the use of trucks, front-end loaders, or conveyor stackers, with suitable measures taken to preserve the permeability of the heaps. Depending on the design, the thickness of the dump will be anywhere from 10 ft (3 m) to hundreds of feet (100 m) if the heaps are multiple-layered in design. The reusable-pad heaps are ordinarily leached for 30 to 90 days, but thicker dumps are normally leached for longer periods. Solutions are applied to the surface of the heap with sprinklers, wobblers, or drip emitters so that the lixiviant is evenly distributed through the heap. Solution is introduced at the rate of 0.003 to 0.008 gal/min/ft^2 (0.002 to 0.005 l/sec/m^2).

The control of liquids used in leaching operations is crucial to sound environmental operation of the leaching system. The heap must be placed on a pad that keeps the liquids circulating in the system and out of the

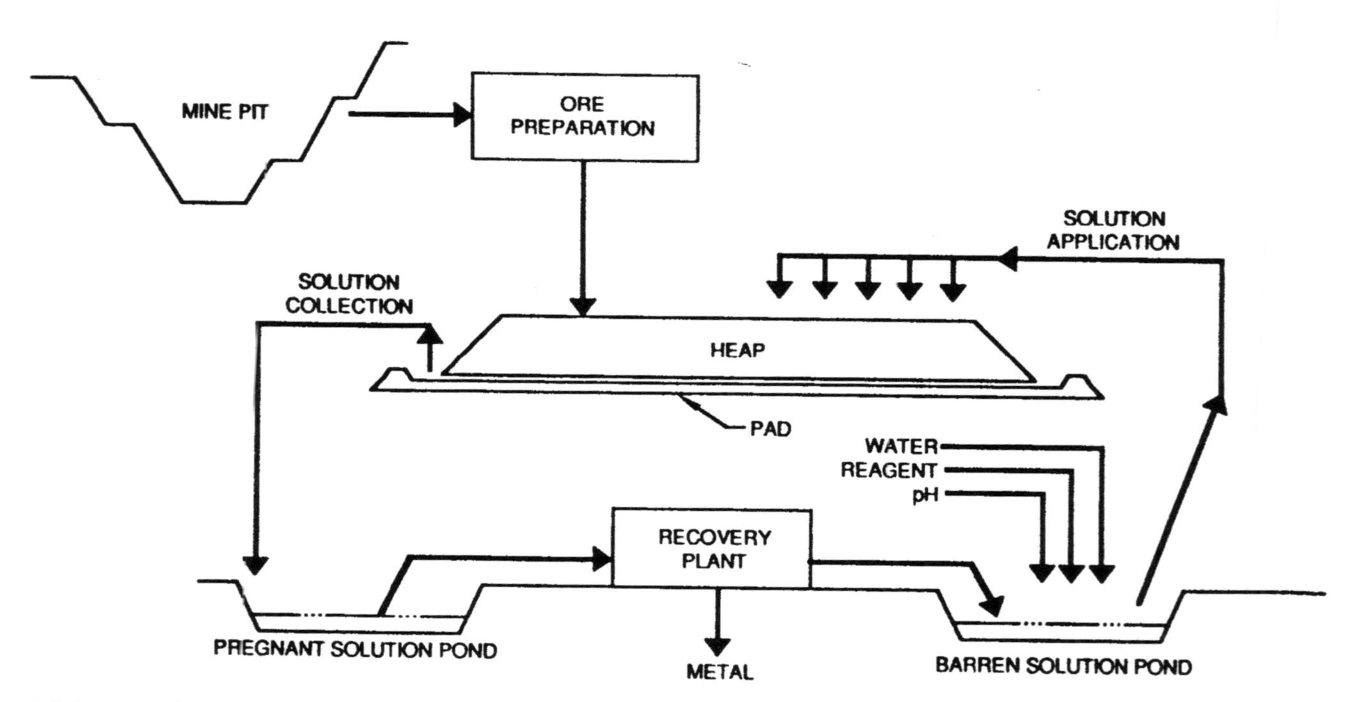

FIGURE 8.9. Generic heap leach operation for recovering metal from a mined ore. Reprinted with permission from *Mine Waste Management* (Hutchirson and Ellison, 1992). Copyright Lewis Publishers, an imprint of CRC Press, Boca Raton, Florida.

surrounding geologic materials. Asphalt pads or pads constructed with geomembrane liners are commonly used. Figure 8.10 shows a number of alternatives for leaching systems. The geomembranes are often made of high-density polyethylene, polyvinyl chloride, and very low density polyethylene. The various properties of these liners are outlined in Hutchinson and Ellison (1992) and Bartlett (1998). Note that drainage layers are often used to detect leakage. An important design consideration in heap leach design is the fluctuations in liquid in the system as a result of rainfall and evaporation. In a continuously gaining system, it is necessary to treat and dispose of the extra water. This may be quite expensive. Typical gold leaching operations are performed in climates where the evaporation can maintain a balance with the precipitation. Bartlett (1998) shows a diagram of the seasonal variation in water volume for a leaching system. If the system is not continuously gaining,

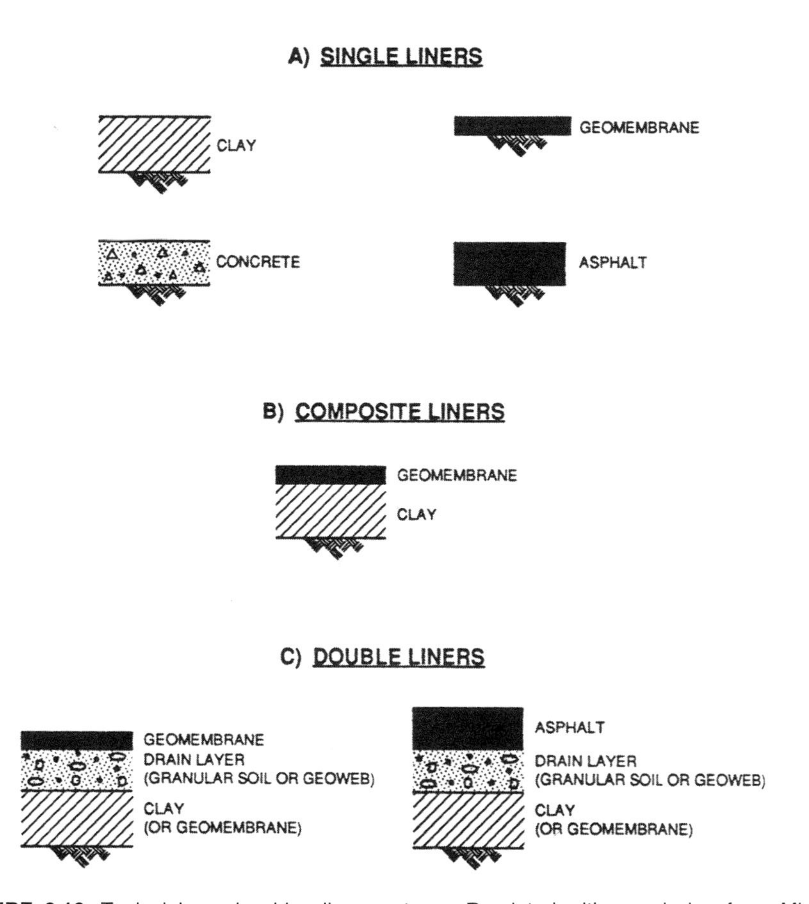

FIGURE 8.10. Typical heap leaching liner systems. Reprinted with permission from *Mine Waste Management* (Hutchinson and Ellison, 1992). Copyright Lewis Publishers, an imprint of CRC Press, Boca Raton, Florida.

then the critical design parameter is the pond capacity. Bartlett (1998) also identifies the usefulness of having an overflow pond. This allows more variation in the water volume and provides for maintenance work on the ponds.

In situ leaching attacks the mineral values in the same way as heap leaching if percolation leaching is used. In this case, however, the ore body must be prepared for leaching in a different way. Three types of leaching operations are used in in situ percolation: (1) leaching of the marginal remaining ore in an underground caving operation (see Figure 2.5), (2) leaching of a marginal near-surface deposit after fragmentation of the ore body using coyote tunnels, and (3) leaching of open pit ore after blasting a portion of the final highwall. A summary of some of these types of leaching systems is presented by Bhappu (1982) and Bartlett (1998).

A number of different remediation procedures are used after leaching to ensure that chemicals do not become an environmental hazard (Mudder and Miller, 1998). The active leaching chemicals can be neutralized in the case of acids or destroyed by bacteria in the case of cyanide. The use of bacteria or other naturally occurring materials to overcome environmental problems is called *bioremediation* (Bartlett, 1998). The use of biological systems for environmental protection is an area of constant research and one that is likely to grow in the future.

8.4.2.1 Sequence of Development. The development steps for a leaching system are similar for both surface and in situ systems. The descriptions given here emphasize the steps for heap leaching on the surface. The following steps are based on descriptions in Hutchinson and Ellison (1992) and Bartlett (1998):

1. Remove vegetation from the area of the leach pad.
2. Create a slope of about 5% to facilitate pad drainage.
3. Construct an impervious pad and proper pad berms to control leaching fluids.
4. Construct a pregnant solution pond and a barren solution pond (if necessary).
5. Mine, crush, and remove fines from the ore material to be leached.
6. Spread material in lifts of 10 ft (3 m) or more (alternative: use a single lift and remove after leaching), using trucks, front-end loaders, or stacking conveyors.
7. Rip the surface of the heap (if necessary) to enhance permeability.
8. Install the irrigation system and initiate leaching.

Steps 4 through 8 apply to in situ leaching operations as well as heap leaches.

8.4.2.2 Cycle of Operations. The cycle of operations is similar in most leaching operations. It normally consists of the following steps:

1. Preparation of the material for leaching in vat, heap, dump, or in situ
2. Application of the solvent
3. Percolation of the solvent through the material
4. Collection of the pregnant solution
5. Processing to recover mineral and regenerate solvent

Additional steps may be required if the ore is difficult to leach, as with refractory gold ore. In this case, roasting or pressure oxidation of the ore may be required (*Engineeering and Mining Journal*, 1993).

8.4.2.3 Conditions. The conditions for surface heap or dump leaching are rather obvious. Accordingly, the parameters specified below apply primarily to in situ operations. Additional information can be found in Schlitt (1982) and Bartlett (1998):

1. *Ore strength:* Permeable or rubblized material.
2. *Rock strength:* Can be weak but must be impervious to fluid transport.
3. *Deposit shape:* Massive or large vein.
4. *Deposit dip:* Steep, if a vein.
5. *Deposit size:* Any, prefer large.
6. *Ore grade:* Can be very low.
7. *Ore uniformity:* Variable, minerals should be accessible to the leach solution.
8. *Depth:* Depends on the type of leach; typically less than 1000 ft (300 m).
9. *Other:* Moderate to large amounts of water required.

8.4.2.4 Characteristics. The use of leaching in the recovery of minerals is growing. However, both the economic and the environmental ramifications must be considered in choosing a leaching method. The advantages and disadvantages outlined here are derived from discussions in Schlitt (1992), Marcus (1997b), and Bartlett (1998).

Advantages

1. Low mining cost (relative cost averages about 5%).
2. Low labor requirements.
3. Applicable to low-grade deposits.
4. Reduced development time and cost.
5. Can supplement primary mining.
6. Good health and safety factors.
7. Biological science can enhance reactions and help with environmental protection.

Disadvantages

1. Limited to soluble minerals that can be chemically leached.
2. Moderate water requirements.
3. Large land areas required for leach dumps.
4. Can be unselective.
5. Recovery may be low in some cases.
6. Possible environmental hazards, especially groundwater contamination.
7. Water balance must be controlled by proper design.
8. Lixiviants may be dangerous to birds; netting or other protection may be required over the ponds.

8.4.2.5 Applications and Variations. The use of leaching technology has been growing steadily in the extraction of gold, silver, and copper on the surface and in uranium in situ. There are many examples in the literature cited earlier. In addition, specific descriptions of leaching applications are provided here. For copper oxide leaching in multiple-lift leach pads after open pit mining, the article by Jenkins (1994) should be consulted. This can be contrasted with the reusable-pad method employed in Chile; it has been described by Bernard (1995) and by Martin et al. (1995). Publications on gold leaching using open pit mining and multiple-lift heaps (Carter, 1997; Marcus, 1997b) may also be of interest. The following case study describes a successful gold leaching operation.

Case Study: Round Mountain Mine, Nye County, Central Nevada

Round Mountain Gold Corporation operates the Round Mountain Mine using open pit mining and multiple-lift leaching (Spickelmier, 1993). The ore is quite variable in grade, resulting in two different leaching pads being utilized. The south pad allows processing of the higher-grade ore after the ore is crushed to minus 0.75 in. (19 mm). The dedicated pad is used for low-grade ore with as little as 0.006 oz/ton (0.2 g/tonne) of gold. The material sent to the dedicated pad is run-of-mine material that would be sent to the waste dump otherwise. Spickelmier (1993) has provided the following information:

Ore type: Au and Ag ore, both oxidized and nonoxidized, varying in grade
Minimum grade: 0.006 oz/ton (0.2 g/tonne) of gold
Mining method: open pit, primarily using shovels and trucks
Production: 45,000 tons/day (40,800 tonnes/day)
Pad type: asphalt with internal rubberized layer and 5% slope
Heap type: multiple-lift with 35 to 50 ft (10.7 to 15.2 m) lifts and 300 ft (91 m) maximum height

Heap placement: linear stacker conveyor system

Leach solution: water plus 0.5 lb/ton (0.21 kg/tonne) of NaCN

Solution application: 0.004 gal/ft^2/min (0.16 l/m^2/min) sprayed through wobblers

Processing: Au and Ag recovered in carbon adsorption columns, then stripped and sent to electrorefining

Total cost: <$5/ton of ore leached

Several additional facts about this mine may be of interest. First, the solution sent to the south dump in winter is warmed by 182°F (82°C) water obtained on the site from geothermal wells. Second, because of the remote location, the mining company has made low-cost homesites available to the workers. The workers own their own manufactured homes that can be moved if they take employment elsewhere.

8.4.3 Evaporite/Evaporation Operations

An *evaporite* is a sedimentary deposit composed primarily of minerals produced from a saline solution by evaporation in a closed basin. Typical minerals that fall into this category include halite (NaCl), potash, and trona. These minerals can be recovered either by conventional mining practice or by solution mining. When extracted by solution mining, the recovery of the valuable minerals is often accomplished by evaporation of the water from brines in solar ponds. In addition to the evaporites mentioned earlier, salts of lithium, magnesium, and boron are commonly recovered from the Great Salt Lake, from Searles Lake in California, and from similar saline sources around the world.

The operation of mineral recovery systems usually requires the use of solar energy. Saline solutions are pumped into large, shallow ponds to allow the water to evaporate. A warm, dry climate and level land are typically required at these mines. The calculations required to provide adequate evaporation for a given amount of solution are outlined in Bartlett (1998). Evaporation conditions, volume of solution to be processed, and the expected rainfall in the area are often the major parameters of concern in this type of mineral extraction.

8.4.3.1 Sequence of Development. The extraction of an evaporite normally requires that the mineral be located in solution in an inland sea or put into solution using an in situ mining method. Processing of brine from an existing lake simplifies the effort by using natural processes to bring the mineral into solution. The sequence of development given here assumes a lake brine operation. Most of the information in the following sections is derived from Richner (1992) and Bartlett (1998). The usual development steps are as follows:

1. Locate or produce large, nearly flat areas.
2. Lay down a layer of fine-grained material like sand or clay as a bed for the geomembrane or other containment layer.
3. Ensure the integrity of the containment layer.
4. Initiate the flow of solution into the ponds.

8.4.3.2 Cycle of Operations. The cycle of operations in this type of mining is relatively simple; the usual steps in the process are as follows:

1. Pump the solution into the pond, matching the inflow rate with the net evaporation rate.
2. Allow solar energy to concentrate and crystallize the minerals.
3. When a sufficient layer of minerals has been precipitated, allow the sun to evaporate all remaining water.
4. Harvest the mineral, ordinarily leaving a layer of salts to protect the lining of the pond.
5. Reinitiate the flow of solution into the pond.

8.4.3.3 Conditions. The conditions required for this method are quite different from those used for other methods:

1. *Ore:* brines or evaporites that may be turned into brines
2. *Rock strength:* medium to strong if the salts are extracted in situ
3. *Deposit type:* large underground evaporite or mineral content in brines
4. *Deposit dip:* not important
5. *Ore grade:* variable
6. *Ore uniformity:* variable
7. *Ore depth:* variable; prefer natural surface brines

8.4.3.4 Characteristics

Advantages

1. Low cost, particularly with lake brines.
2. Natural concentration often available.
3. Uses free solar energy.
4. Minerals relatively easily found.
5. Reduced development time and cost.
6. High recovery.
7. Good health and safety factors.
8. Low labor costs.

Disadvantages

1. Specialized method; applies primarily to brines.
2. Moderate water requirements.
3. Requires large land areas for recovery.
4. Possible environmental consequences.

8.4.3.5 Applications and Variations. Solution mining for a common salt such as NaCl, potash, and trona has been conducted using both in situ methods and surface methods. Very large percentages of the U.S. production of lithium, boron, and magnesium are normally derived from solution mining methods. Lithium and boron compounds are ordinarily obtained from lake brines, and magnesium is extracted from lake brines, well brines, and seawater.

8.5 SPECIAL TOPIC: SELECTION OF MATERIALS HANDLING EQUIPMENT

The proper choice of surface mining equipment, as required during the development stage, is one of the most critical decisions in surface mining (Sections 5.4 and 5.5). Technological feasibility, economic suitability, and safety are key criteria that must be satisfied. Because it is central to all mine operations, the selection of equipment for materials handling and its two unit operations (excavation and haulage) is one of the first tasks undertaken. Choosing equipment for other operations logically follows and complements choosing equipment for materials handling. Because open pit mining always requires both loading and haulage in the production cycle for stripping as well as mining, we choose it as the typical surface mining method for which to demonstrate the equipment selection procedure. The logic is also applicable to coal loading and haulage operations in open cast mining, as well as to various underground methods in which the loading and haulage steps occur.

8.5.1 Excavator

The machine we choose as a typical excavator to demonstrate the selection procedure is the power shovel (often referred to as a rope shovel because its bucket is lifted by wire ropes). The power shovel is a cyclic loader, so designated because the machine cycles back and forth from the muck pile to the haulage device with its loads. It competes with other cyclic loaders such as the hydraulic shovel, the front-end loader, and the dragline. The power shovel is still the workhorse of the open pit industry, although hydraulic shovels are now becoming much more common. The selection procedure that follows is based largely on material in Atkinson (1992a).

1. *Idealized output.* Power shovels are normally nominally rated in yd^3/hr (m^3/hr), based on idealized conditions. The output is a function of bucket size, type of material, working time, difficulty of digging, type of haulage unit, and working conditions. The units used are customarily *bank (solid) measure*, the volume of material in place. In dealing with haulage units, however, *loose (broken) measure*, the volume of material after excavation, is employed. If actual values cannot be obtained, the following approximate measures of specific weight may be used:

Measure	Soil tons/yd³ (tonnes/m³)	Rock tons/yd³ (tonnes/m³)	Coal tons/yd³ (tonnes/m³)
Bank (solid)	1.5 (1.8)	2.0 (2.4)	1.1 (1.3)
Loose (broken)	1.3 (1.5)	1.3 (1.5)	0.7 (0.8)

Table 8.2 provides the cycle times in seconds that would be expected under different digging conditions. If the time obtained from Table 8.2 is designated t_c, then the number of cycles per hour C is

$$C = \frac{3600}{t_c} \tag{8.1}$$

Using the cycles/hour and the bucket capacity, a person can calculate the idealized output for the machine if it works for a given time period. This idealized output must be tempered, based on the operating factors of the operation.

2. *Operating factors.* The output calculated using idealized data must be modified for "real-world" conditions. Corrections should be made for three factors idealized in the table: working time, swing angle, and working conditions.

Working time. The actual time that a shovel operator will be loading in a given shift is a function of the operator changeout system, number of shovel moves required per shift, meal arrangements, human limitations, and other variables. Among other things, the average worker may be operating the equipment only seven hours of an eight-hour shift. Excavation engineers have long provided a derating factor in acceptance of the fact that only a portion of the operating time is productively utilized. The following time factors are normally assumed:

Availability Conditions	Actual Time
Favorable	55 min/hr
Average	50 min/hr
Poor	45 min/hr

Table 8.2 Loading Shovel Cycle Times (sec)

B_c		Digging Conditions			
yd^3	m^3	E	M	M-H	H
4	3	18	23	28	32
5	4	20	25	29	33
6	5	21	26	30	34
7	5.5	21	26	30	34
8	6	22	27	31	35
10	8	23	28	32	36
12	9	24	29	32	37
15	11.5	26	30	33	38
20	15	27	32	35	40
25	19	29	34	37	42
45	35	30	36	40	45

E: Easy digging, loose, free-running material; e.g., sand, small gravel.
M: Medium digging, partially consolidated materials; e.g., clayey gravel, packed earth, clay, anthracite, etc.
M-H: Medium-hard digging; e.g., well blasted limestones, heavy wet clay, weaker ores, gravel with large boulders, etc.
H: Hard digging—materials that require heavy blasting and tough plastic clays; e.g., granite, strong limestone, taconite, strong ores, etc.

Unless otherwise specified, average availability is assumed, and 50 min/hr and 7 hr/shift are used. To better utilize this information, we define the unitless production factor P as:

$$P = \frac{\text{Actual min/hr}}{\text{60 min/hr}} \tag{8.2}$$

Swing angle. Most shovels operating on a bench will have a bench geometry resulting in a swing angle of about 90°. Table 8.2 assumes a 90° swing and must be corrected if the angle of swing differs from the assumed value. Atkinson (1992a) suggests that the following values of swing factor S be used:

Angle of swing, degrees	45	60	75	90	120	150	180
Swing factor S	1.2	1.1	1.05	1.00	0.91	0.84	0.77

Working conditions. Working conditions include the condition of the bench, the type of material being excavated, and the degree of fragmentation of the muck. These conditions primarily affect the ability of the shovel operator to fill the bucket in an expedient manner. The *fill factor* F (Atkinson (1992a) terms this variable the *fillability*) is the portion of the bucket volume actually utilized during normal operation (Caterpillar, Inc., 1997). This value varies from about 0.75 for difficult-to-dig material to 1.10 for mixed soil and rock. Nominal values that can be used are condensed from tables in Atkinson (1992a) and (Caterpillar, Inc., 1997), as follows:

Material	Factor
Rock-soil mixtures	1.10
Easy digging material (sand, small gravel)	1.00
Medium digging material (coal, light clay, wet soil, soft ores)	0.90
Medium-hard digging material (iron ore, phosphate, copper ore, hard limestone)	0.85
Hard digging material (blocky iron ore, sandstone, basalt, heavy clay)	0.80

Note that the condition of the muck affects both the cycle times of the shovel and the bucket fill factor, emphasizing the need for good fragmentation. To complete the information needed to calculate the size of the shovel, the dipper factor is normally calculated to determine the relationship between material in the solid and material in the bucket. This can be done using the following relationship:

$$B_f = \frac{F(\text{weight/loose unit volume})}{\text{weight/bank unit volume}} \tag{8.3}$$

where B_f is the dipper factor in yd^3 in the bank per yd^3 in the bucket (m^3/m^3) and F is the unitless bucket fill factor.

The bucket capacity of the shovel can now be calculated, assuming that the various aforementioned factors can be measured or estimated. The calculation, adapted from Atkinson (1992a) is as follows:

$$B_c = \frac{Q}{CPSB_f} \tag{8.4}$$

where B_c is the bucket capacity in yd^3 (m^3), Q is the production required in bank yd^3/hr (m^3/hr), C is the number of cycles per hr, P is the production factor, S is the swing factor, and B_f is the dipper factor. Note that the propel time factor defined in Atkinson (1992a) is assumed to be part of the production factor in this treatment and that an ideal bank height is assumed. The following example illustrates the selection procedure for a power shovel.

Example 8.2. The following data are given for a copper mine using shovel-truck production equipment:

Required production: 15,000 tons/shift
Availability: 50 min/hr
Bank (solid) density: 2.0 tons/yd^3
Swing angle: 120°
Shift operating time: 7 hr
Muck: medium digging
Loose density: 1.30 tons/yd^3

What shovel size should be chosen to perform the production?

SOLUTION. The size of the shovel is calculated using Eqs. 8.1 to 8.4. To initiate the data collection, it will be necessary to estimate the size of the shovel so that a value of t_c can be obtained. Starting with a 25 yd^3 shovel, we read $t_c = 34$ sec. The calculations proceed as follows:

$$Q = \frac{(15{,}000 \text{ tons/shift})}{(2.0 \text{ tons/yd}^3)(7 \text{ hr/shift})} = 1071 \text{ yd}^3/\text{hr}$$

$$C = \frac{3600}{34} = 105.9 \text{ cycles/hr} \tag{8.1}$$

$$P = \frac{50}{60} = 0.833 \tag{8.2}$$

Read $S = 0.91$ and $F = 0.90$

$$B_f = \frac{0.90(1.30)}{2.00} = 0.585 \tag{8.3}$$

$$B_c = \frac{1071}{(105.9)(0.833)(0.91)(0.585)} = 22.8 \text{ yd}^3 \ (17.4 \text{ m}^3) \tag{8.4}$$

We can assume that a 25 yd^3 (19 m^3) shovel is available and use that in our subsequent calculations. Note that the cycle time we assumed at the beginning of the problem is appropriate for this shovel.

8.5.2 Haulage Units

The off-highway haulage truck is the most commonly used haulage device in open pit mines, and it is chosen here for that reason. As with the power shovel, ample data are available to help in the choice of units. Two important tasks must be undertaken if haulage trucks are to be used properly. First, the trucks must be matched in size to the excavator. Second, the number of trucks must be matched to the haulage layout so that the system produces in a near-optimal manner.

8.5.2.1 Haulage Truck Size. Haulage truck capacity (*live load*) is normally measured on the basis of weight rather than volume so as to prevent overloading. A volume measure is normally also provided for trucks, generally expressed in struck (leveled) capacity of loose material. Truck capacities range from about 22 tons (20 tonnes) to about 350 tons (317 tonnes). Common practice is to use the smallest third of the range in underground operations, the middle of the range in small surface mines, and the largest third in large open pit mines. To size a truck for use with a particular excavator, a number of variables must be considered:

1. *Optimal number of swings.* The number of swings used to fill a haulage truck has been a question that mining engineers have considered ever since trucks were first used in mining. The general rule of thumb seems to vary around the country, with three to four passes normally used in large western open pit mines and four to six passes in the Appalachian coal fields (Mace, 2000). These rules apply primarily to power shovels, hydraulic shovels, and front-end loaders. Where hydraulic backhoes are used, the number of passes is generally somewhat higher.
2. *Working time.* The same availability estimates are made for trucks as for shovels: 50 min/hr and 7 hr/shift are used unless other conditions are specified.
3. *Fill factor.* The ability to fill the bucket of the excavator will affect the number of passes needed to fill a haulage truck. The fill factors used in determining the shovel size are used here as well.

In determining the number of passes needed to fill a truck, it is important to guard against overfilling the truck for two reasons: (1) Too much weight can potentially damage the truck and (2) too much volume will result in spills in the loading area and along the haul roads. The example that follows illustrates the calculation of truck size to optimally match the previously chosen shovel.

Example 8.3. If the 25 yd^3 shovel chosen in the previous example is used in an open pit mine, what is the normal range of truck size that can be used if:

(a) Three to four swings are considered to be a good match
(b) Four to six swings are considered to be the proper match

SOLUTION. For both of these rules of thumb, it is necessary to know how much weight is present in the bucket if it is filled to its normal fill factor. For medium rock, the fill factor F is 0.90. The tonnage moved with each swing of the bucket is then

$$\begin{aligned}\text{Weight/bucket} &= (0.90)(25\ \text{yd}^3/\text{bucket})(1.3\ \text{tons/yd}^3)\\ &= 29.25\ \text{tons (26.5 tonnes)}\end{aligned}$$

(a) Three swings will move about 88 tons into the truck.

Four swings will move about 117 tons into the truck.
The choices seem to be either a 100 ton truck or a 120 ton truck.

(b) Four swings will provide about 117 tons of load.
Six swings will provide about 146 tons of load.
For this rule, either a 120 ton (109 tonne) truck or a 150 ton (135 tonne) truck seems appropriate.

For this shovel-truck situation, we will choose a 120 ton (109 tonne) truck. Note that not every manufacturer may have a 120 ton (109 tonne) model. This may force the decision maker to consider other sizes. We will assume here that the 120 ton (109 tonne) truck is available and that we have confidence in the manufacturer.

8.5.2.2 Number of Haulage Units Required. Having chosen the size of truck required to service the excavator, we still have the task of choosing how many trucks are necessary to adequately process the design tonnage from a production face without the trucks waiting in queues at the excavator or the dump point. To accomplish this job, it is essential to establish the cycle time of the haulers so that the production potential is sufficient to haul the desired tonnage, but not excessive. If too many trucks are assigned to the shovel, they will wait for service at the shovel and the system will be suboptimal.

The cycle time of the haulage units is determined from measurements made in the field or from reasonable estimates in similar situations. The cycle time is ordinarily divided into its common elements, and each element is estimated separately. One version of the cycle-time equation is:

$$t = t_{te} + t_{we} + t_{se} + t_l + t_{tl} + t_{wd} + t_{sd} + t_d \tag{8.5}$$

where:
t = total haulage unit cycle time, sec
t_{te} = travel empty time from dump to loader, sec
t_{we} = wait time at the loader, sec
t_{se} = spotting time at the loader, sec
t_l = loading time at the loader, sec
t_{tl} = travel loaded time, loader to dump, sec
t_{wd} = wait time at the dump, sec
t_{sd} = spotting time at the dump, sec
t_d = time to dump, sec

To estimate the cycle time, it is best to measure the time elements in the field by means of a time study. However, the spotting, loading, and dumping times can be estimated using manufacturers' data. Some useful nominal values for trucks and tractor trailers are as follows (Terex Corporation, 1981):

Conditions	Spotting Time (min)		Dumping Time (min)	
	Trucks	Trailers	Trucks	Trailers
Favorable	0.15	0.15	1.0	0.3
Average	0.30	0.50	1.3	0.6
Unfavorable	0.50	1.00	1.8	1.5

Travel times are another matter. They must either be measured in the field, estimated using hauler rimpull curves, or estimated using truck simulation software available from the manufacturers or other sources.

If the dump is properly designed and the number of haulage units is not excessive, the wait times can be ignored, and the cycle time becomes

$$t = t_{te} + t_{se} + t_l + t_{tl} + t_{sd} + t_d \tag{8.6}$$

To determine the number of trucks that will adequately handle the tonnage without introducing excess waiting into the system, the number of haulage units is the smallest number n that meets the following condition:

$$t \leqslant n(t_l + t_s) \tag{8.7}$$

We can see in the following example how this applies to the shovel and trucks previously chosen.

Example 8.4. The 25 yd^3 (19 m^3) shovel selected in Example 8.2 is to be used with the 120 ton (109 tonne) trucks chosen in Example 8.3. The following information has been gathered in time studies at a similar operation or with the use of a manufacturer's truck simulation program.

Spotting time: 0.3 min	Dumping time: 1.1 min
Travel empty: 4.8 min	Travel loaded: 7.5 min

Assuming that there is no wait time in the system, calculate the total truck cycle time and the optimal number of trucks in the system.

SOLUTION. The loading time can be calculated, based on the number of swings to load the truck and the swing time of the shovel:

$$t_l = 4 \text{ swings} \times 34 \text{ sec/swing} = 2.27 \text{ min}$$

The total truck cycle time is then

$$t = 4.8 + 0.3 + 2.27 + 7.5 + 0.3 + 1.1 = 16.27 \text{ min}$$

The value of $t_l + t_s$ is 2.57 min, making the optimal value of $n = 7$ using Eq. 8.7. Note that this cycle should be satisfactory. The shovel will occasionally wait for the trucks, but the wait should not be excessive.

The previous example is greatly simplified by the assumption that the time elements in the cycle are constant. This may not be true in practice because the spotting, loading, and dumping times are likely to vary considerably. In addition, travel times on the haul roads will occasionally vary due to interference from other vehicles. When stochastic variation is introduced, wait times will increase, and the problem requires analysis by computer simulation or another more advanced systems analysis tool. For our purposes, we can be confident that the shovel-truck system we analyzed will be reasonably close to optimal.

PROBLEMS

8.1 A limestone mine has the following equipment at its primary production face:

Excavator: a 15 yd^3 (11.5 m^3) front-end loader with 40 sec cycle time
Trucks: five 100 ton (90 tonne) trucks
Material: average blasted rock with medium-hard digging conditions
Working conditions: average
Haulage cycle time: 20.0 min

(a) What is the sustained production rate of the system in tons/hr, assuming 60 min of production per hour?
(b) Which equipment (loader or trucks) governs?
(c) What improvements would you recommend to increase mine production without changing equipment?

8.2 A mining company is considering new electric shovels to load 240 ton (218 tonne) trucks for a large open pit mine supplying ore for a heap leaching method. The shovel sizes that seem appropriate are the 45 yd^3 (34.4 m^3) and the 54 yd^3 (41.3 m^3) models. Other parameters of the operation are as follows:

Digging conditions: medium hard
Ore density (solid): 2.1 tons/yd^3 (2.5 tonnes/m^3)
Ore density (loose): 1.35 tons/yd^3 (1.6 tonnes/m^3)
Swing angle: 90°
Number of passes desired: four to six

Answer the following:

(a) Which of the two shovels is better suited to the trucks under the conditions listed?

(b) What is the maximum shift capacity of the shovel if it can work 50 min/hr for 7 hours?

8.3 For the mine outlined in Problem 8.2, determine the optimal number of trucks to service the shovel if the cycle time of the trucks is 21 min. At the point when the optimal number of trucks are operating, answer the following:

(a) What determines the output of the production face, the shovel or the trucks?

(b) What is the expected output of the production face?

8.4 For Problem 8.1, plot the expected output in tons/shift if the number of trucks varies from 1 to 10. Assume that production is ongoing for 50 min/hr and 7 hr/shift. Will this curve correctly point out the optimal number of trucks to be used with the 15 yd^3 (11.5 m^3) loader?

9

UNDERGROUND MINE DEVELOPMENT

9.1 NATURE AND SCOPE OF THE TASK

9.1.1 Role of Underground Mining

If the appeal of surface mining lies in its mass production and minimal-cost capabilities, then the attraction of underground mining stems from the variety of ore deposits that can be mined and the versatility of its methods to meet conditions that cannot be approached by surface mining. Underground mining cannot compete with surface mining today in its cost or share of U.S. mineral production. Reviewing Table 6.2, we see that in terms of ore tonnage, less than 5% of metals and nonmetals, and less than 39% of all coal is produced in underground mines. Thus, underground mining is relegated to a secondary role for many commodities. However, the United States depends heavily on underground mining for most of its supply of lead, potash, trona, and zinc. In addition, much of its coal, gold, molybdenum, salt, and silver come from underground mines.

Predicting underground production in upcoming years is quite difficult. Many analysts argue that most of the near-surface deposits have already been mined. In addition, they point out the great expense of meeting all environmental requirements on the surface. However, the ever-increasing productivity of surface mining equipment and the greater ease with which surface mining equipment can be upsized and automated makes it more likely that surface mining will continue to increase as a percentage of our total production. Whether it increases or decreases, it is safe to assume that underground mining will continue to play an important role in supplying mineral resources in the future.

9.1.2 Uniqueness of Underground Mining

A number of development steps are required for underground mining that are not necessary in surface mining; these are outlined in Section 4.2. A review of the governing factors indicates that there is less concern for location factors (climate, in particular, is less a factor) and environmental consequences. The most critical factors are ore and rock strength, the presence of groundwater, and the rock temperature gradient in the locality. The social, economic,, political, and environmental factors in underground mining are often quite different from those in surface mining. A more skilled labor force may be required, financing may be more difficult because of increased risk, and subsidence may become the most important environmental concern.

The extent of access development required is also quite different. In surface mining, removal of significant amounts of overburden may complicate development. On the other hand, limited excavation and relatively small openings are necessary for most underground mines. However, the development openings may be considerably more costly on a tonnage basis, and more types of excavations may be required. In addition, underground mines normally must be provided with an artificial atmosphere as a means of ensuring that the quantity and quality of air are always adequate.

On occasion, underground development openings double for exploration purposes, and vice versa. Those openings driven in advance of mining can provide valuable exploration information and afford suitable sites for additional exploration drilling and sampling. Likewise, openings driven for exploration purposes can be utilized to develop the deposit. For example, the exploration shaft and drifts shown in Figure 3.6 would almost certainly serve subsequently to open up the deposit.

9.1.3 Types of Underground Openings

Underground development openings can be ranked in three categories by order of importance in the overall layout of the mine:

1. Primary: *Main openings* (e.g., shaft, slope)
2. Secondary: *Level or zone openings* (e.g., drift, entry)
3. Tertiary: *Lateral or panel openings* (e.g., ramp, crosscut)

Generally, the openings are driven in this order; that is, from main development openings to secondary level or zone openings, to tertiary openings like lateral or panel openings. However, many variations exist with different mining methods. For example, in coal mining, the entries and the associated crosscuts are always driven at the same time, regardless of the category of the entries. It should also be noted that underground mining often employs a distinctive nomenclature. The following lists define a number of the terms commonly used to describe underground workings and other aspects of underground mining.

Other sources that can be consulted for terminology of mining include Gregory (1980), Hamrin (1982), and American Geological Institute (1997).

Deposit and Spatial Terms

Back: Roof, top, or overlying surface of an underground excavation

Bottom: Floor or underlying surface of an underground excavation

Capping: Waste material overlying the mineral deposit

Country rock: Waste material adjacent to a mineral deposit

Crown pillar: Portion of the deposit overlying an excavation and left in place as a pillar

Dip: Angle of inclination of a deposit, measured from the horizontal; also *pitch* or *attitude*

Floor: Bottom or underlying surface of an underground excavation

Footwall: Wall rock under the deposit

Gob: Broken, caved, and mined-out portion of the deposit

Hanging wall: Wall rock above a deposit

Pillar: Unmined portion of the deposit, providing support to the roof or hanging wall

Rib: Side wall of an excavation; also *rib pillar*

Roof: Back, top, or overlying surface of an excavation

Sill pillar: Portion of the deposit underlying an excavation and left in place as a pillar

Strike: Horizontal bearing of a tabular deposit at its surface intersection

Wall rock: Country rock boundary adjacent to a deposit

Directional Terms

Breast: Advancing in a near-horizontal direction; also the working face of an opening*

Inby: Toward the working face, away from the mine entrance

Outby: Away from the working face, toward the entrance

Overhand: Advancing in an upward direction*

Underhand: Advancing in a downward direction*

Excavation Terms

Adit: Main horizontal or near-horizontal underground opening, with single access to the surface

Bell: Funnel-shaped excavation formed at the top of a raise to move bulk material by gravity from a stope to a drawpoint

* Customarily used to modify *stoping* or *mining*.

Bleeder: Exhaust ventilation lateral

Chute: Opening from a drawpoint, utilizing gravity flow to direct bulk material from a bell or orepass to load a conveyance

Crosscut: Tertiary horizontal opening, often connecting drifts, entries, or rooms; oriented perpendicularly to the strike of a pitching deposit; also *breakthrough*

Decline: Secondary inclined opening, driven downward to connect levels, sometimes on the dip of a deposit; also *declined shaft*

Drawpoint: Loading point beneath a stope, utilizing gravity to move bulk material downward and into a conveyance, by a chute or loading machine; also *boxhole*

Drift: Primary or secondary horizontal or near-horizontal opening; oriented parallel to the strike of a pitching deposit

Entry: Secondary horizontal or near-horizontal opening; usually driven in multiples

Finger raise: Vertical or near-vertical opening used to transfer bulk material from a stope to a drawpoint; often an interconnected set of raises

Grizzly: Coarse screening or scalping device that prevents oversized bulk material from entering a material transfer system; constructed of rails, bars, beams, etc.

Haulageway: Horizontal opening used primarily for materials handling

Incline: Secondary inclined opening, driven upward to connect levels, sometimes on the dip of a deposit; also *inclined shaft*

Lateral: Secondary or tertiary horizontal opening, often parallel or at an angle to a haulageway, usually to provide ventilation or some auxiliary service

Level: System of horizontal openings connected to a shaft; constitutes an operating horizon of a mine

Loading pocket: Transfer point at a shaft where bulk material is loaded by bin, hopper, and chute into a skip

Longwall: Horizontal exploitation opening several hundred feet (meters) in length, usually in a tabular deposit

Manway: Compartment of a raise or a vertical or near-vertical opening intended for personnel travel between two levels

Orepass: Vertical or near-vertical opening through which bulk material flows by gravity

Portal: Opening or connection to the surface from an underground excavation

Raise: Secondary or tertiary vertical or near-vertical opening, driven upward from one level to another

Ramp: Secondary or tertiary inclined opening, driven to connect levels, usually in a downward direction, and used for haulage

Room: Horizontal exploitation opening, usually in a bedded deposit

Shaft: Primary vertical or near-vertical opening, connecting the surface with underground workings; also *vertical shaft*

Slope: Primary inclined opening, usually a shaft, connecting the surface with underground workings

Slot: Narrow vertical or inclined opening excavated in a deposit at the end of a stope to provide a bench face

Stope: Large exploitation opening, usually inclined or vertical, but may also be horizontal

Sublevel: Secondary or intermediate level between main levels or horizons, usually close to the exploitation area

Transfer point: Location in the materials-handling system, either haulage or hoisting, where bulk material is transferred between conveyances

Tunnel: Main horizontal or near-horizontal opening, with access to the surface at both ends

Undercut: Low horizontal opening excavated under a portion of a deposit, usually a stope, to induce breakage and caving of the deposit; also a narrow *kerf* cut in the face of a mineral deposit to facilitate breakage

Winze: Secondary or tertiary vertical or near-vertical opening, driven downward from one level to another

Figure 9.1 illustrates a number of development openings that may be utilized in a metal mine. Figure 9.2 shows some of the types of main openings for a coal mine. Note that ramps in metal mines and slopes in coal mines are much more prevalent today than in the past, because they enable the use of the most efficient transport equipment.

9.2 MINE DEVELOPMENT AND DESIGN

Mine development must proceed considering all aspects of the overall mine design. Because of the complexity and expense of underground mining, extreme care must be exercised in making decisions during development that may also affect subsequent production operations. Hamrin (1982, 2001) and Bullock and Hustrulid (2001) discuss many of the important considerations in mine design and development. The most crucial matters of concern are discussed in the following sections.

9.2.1 Mining Method

Once the decision has been made to use underground mining, attention is focused on the selection of an exploitation method. Development should not

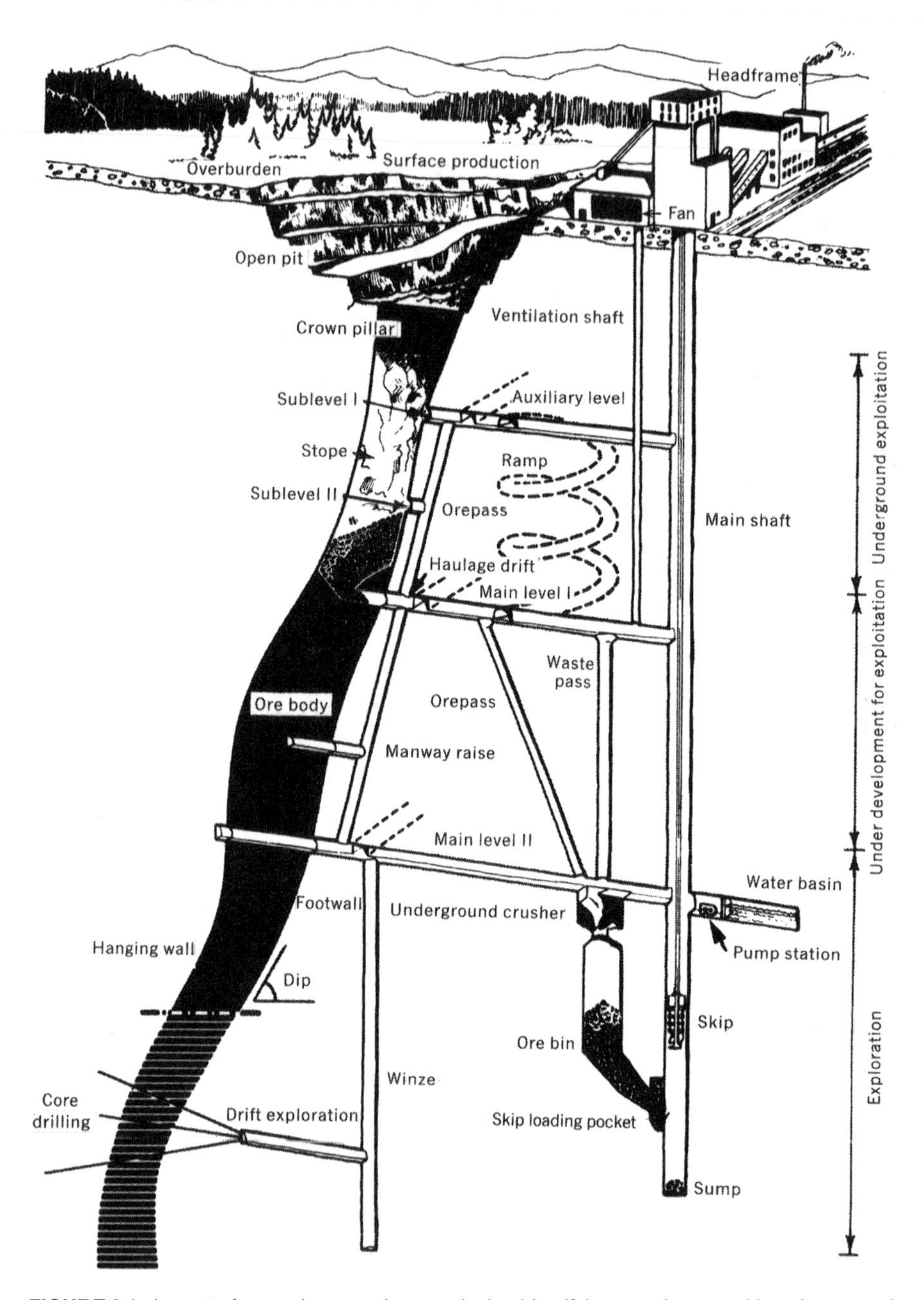

FIGURE 9.1. Layout of an underground noncoal mine identifying openings, working places, and stages in the life of the mine. A surface mine that underwent transition to underground is also shown. (After Hamrin, 1982. By permission from the Society for Mining, Metallurgy, and Exploration, Inc., Littleton, CO.)

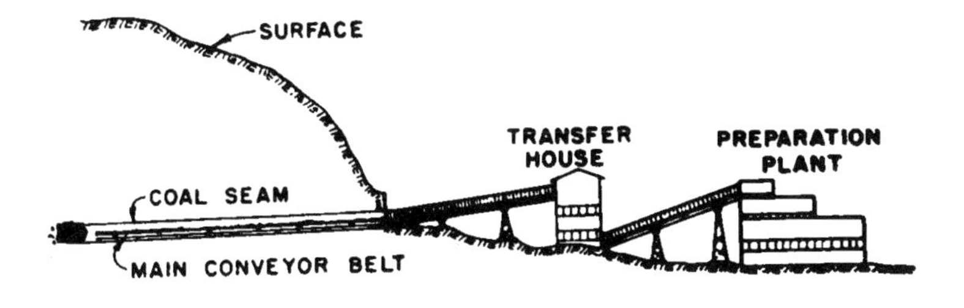

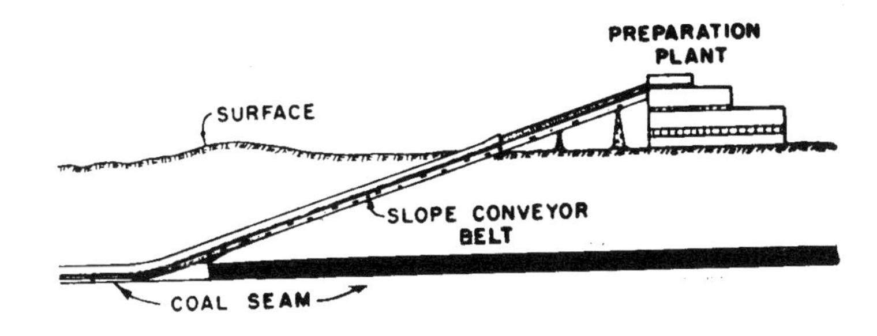

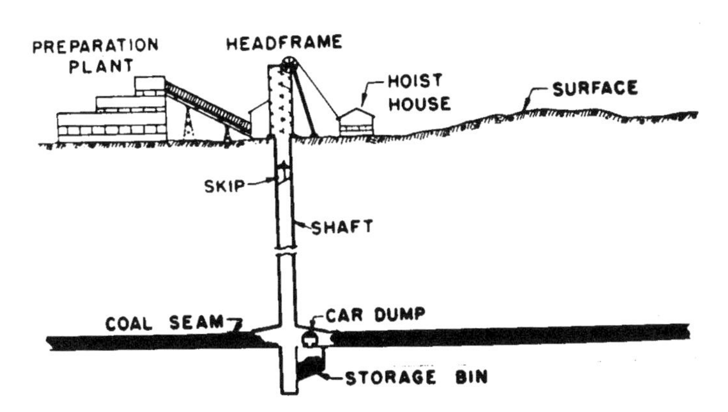

FIGURE 9.2. Layouts of main access openings for an underground coal mine. (*Top*) Drift. (*Center*) Slope. (*Bottom*) Shaft. (After Stefanko and Bise, 1983. By permission from the Society for Mining, Metallurgy, and Exploration, Inc., Littleton, CO.)

proceed until a mine production plan has been adopted, and the first step is to decide which class of underground method is most suitable: unsupported, supported, or caving. Note also that solution mining, although not common, is still a choice for an underground deposit (see Section 8.4). As discussed in Section 4.8, selection of a mining method hinges on natural and geologic conditions related to the mineral deposit, on certain economic and environmental factors, and on other factors that may have a bearing on the specific deposit to be exploited.

The reason that the choice of a method is so crucial is that it largely governs the type and placement of the primary development openings. If disturbance of the surface due to subsidence is anticipated—inevitable with caving methods and possible with other methods—then all the access openings must be located outside the zone of fracture bounded by the *angle of draw*. If the integrity of the ground overlying the active mining area can be ensured for the life of the mine, then the primary openings can be located more centrally above the deposit.

9.2.2 Production Rate and Mine Life

A variety of geologic and economic conditions determine the optimum rate of production from a mineral deposit of known reserves and, hence, the life of the mine. These include market conditions and selling price of the commodity, mineral grade, development time, mining costs, means of financing, government support and taxation policies, and a number of other factors.

The most common goal in modern mine planning is to optimize the *net present value*, that is, to mine the deposit so that the maximum internal rate of return is achieved (Arrouet, 1992). One way of accomplishing this is to select the production rate on the leases of the net present value of the after-tax cash flows.

All other things being equal, the higher the production rate, the shorter the mine life. Formerly, mine life spans were often measured in decades. With the higher cost of borrowed money and greater investment risks, mine lives are now often limited to a few years. Although many low-grade, high-tonnage operations are still financed with a life expectancy of 15 to 30 years, it is not uncommon for a high-grade, low-tonnage operation to be designed with a two- to four-year life span (Glanville, 1984). Because of the relatively high cost of labor and borrowed money, however, the trend in the industry is still toward large mines and high production rates. Nilsson (1982a) explores the reasons and concludes that such trends are likely to continue.

9.2.3 Main Access Openings

A number of initial decisions related to the primary development openings of a mine must be made early in the mine planning stage. They concern the type, number, shape, and size of the main openings. By necessity, these decisions are ordinarily made at the time when the primary materials-handling system is chosen. Factors influencing the decisions include the depth, shape, and size of the deposit; surface topography; natural and geologic conditions of the ore body and surrounding rock; mining method; and production rate. Wise decisions made at the outset avoid later changes in development openings, changes that are always disruptive and expensive. Such changes are common in mines with great longevity but should not occur in short-lived operations.

9.2.3.1 Types of Openings. In regard to types of openings, there are only three common choices that are used with regularity: (1) shafts (vertical and near vertical), (2) declines (slopes or ramps), and (3) adits or drifts. Vertical shafts have always been among the most common types of openings for deep mines. However, nearly all shafts are now vertical because near-vertical shafts are more costly and difficult to develop.

Declines have been among the fastest growing primary development openings because of their association with principal materials-handing systems. In coal mines, these openings are called *slopes* (see Figure 9.2) and are associated with the belt conveyor transport of coal out of the mine. Except in very deep mines, these openings provide for low-cost primary transportation of the coal out of the mine, eliminating the need for a hoisting system. The slopes in a coal mine are generally driven at an angle of about 15°. For this reason, a slope is usually 3.6 times as long as a vertical shaft and costs more than a shaft. However, a slope will pay for the extra cost in a few years with the savings to be obtained by substituting a belt conveyor for a hoist. Similar logic can be applied to a *ramp* in a metal mine. These openings are ordinarily driven to allow free access to any level of the mine with diesel-powered equipment. They can be driven in a spiral form (see Figure 9.1) or in a rectangular form, with linear sections alternating with curved sections to achieve the desired access to each level of the mine. In each case the idea is to provide a means of utilizing mobile equipment throughout the mine without limitation. Ramps are not ordinarily used for primary haulageways, but rather as necessary development openings to facilitate use of diesel equipment in the mine.

Drifts or *adits* are used in coal, nonmetal, and industrial minerals operations in any situation where the deposit can be easily accessed through entries or adits of a horizontal type (see Figure 9.2). The proper situation would normally occur in coal or nonmetal deposits, but can be readily found in metal deposits located in mountainous regions. Clearly, this method of development can apply to many deposits and is often a cost-effective method of developing a mine if the possibility exists.

9.2.3.2 Standard Development Practices. In the development of any given mine, the mine planner ordinarily utilizes the types, sizes, and number of openings that are optimal for the mining conditions of the deposit. These factors vary, depending on the production rate of the mine, the depth of the deposit, the type of equipment to be utilized in the mining system, and the primary method of removing the mineral product from the mine. In drift and adit mines, the standard practice is to evaluate the equipment size and production rate of the mine to determine the size and number of openings needed. Coal mines are required to have a number of openings, and this may result in more entries being driven into the deposit than are necessary with a metal or nonmetal mine. However, drifts or adits into a deposit are normally the least expensive manner of developing a mine, and this may be the primary reason for choosing such a development method.

In shaft development, practices have changed over the last few decades, with bored shafts being much more common and the round shaft becoming all but universal for large shafts and new mines. In coal mining, about half of all shafts are now sunk by blind-boring methods (see Section 9.4). It has become common for shafts up to 16 ft (4.9 m) to be drilled in this fashion (Maloney, 2000). In addition, many coal mining companies now use bleeder shafts to simplify their longwall ventilation systems. Nearly all of these are drilled using blind-boring rigs. Raise-bored shafts are also possible, but access must be available underground to make this feasible. The need for a mucking and transportation system underground also makes this alternative less favorable. The blind-boring method should be considered for any formations that are in the soft to medium-hard category.

For large shafts in coal mines and for most metal and nonmetal mine shafts, conventional shaft-sinking methods are still commonly used. Circular shafts with concrete linings are the most common, with the advantages being favorable ventilation characteristics, a suitable method of support for the opening, and a maximum cross-sectional area for the development dollar. The diameters of the shafts are quite variable, with diameters of 18 ft (5.4 m) to 30 ft (9.1 m) being most common. In most of these shafts, unreinforced concrete liners of at least 1.0 ft (0.3 m) thickness are used for support of the opening (Maloney, 2000; Pond, 2000). Some elliptical shafts have been used in the coal fields, with sizes up to 18 ft (5.4 m) by 32 ft (9.7 m) (Pond, 2000).

The slopes used in today's coal mines are normally driven with two compartments, an upper compartment to accommodate a belt conveyor and a lower compartment containing track that is normally used for personnel and supply transport. The two compartments are separated by a horizontal concrete divider and are supported by bolts, wire mesh, or steel arches. Slopes in coal mines are often horseshoe-shaped, with widths of 17 to 20 ft (5.2 to 6.1 m) and heights of 13 to 14 ft (4.0 to 8.5 m) (Emerick, 2000).

Ramps are declines that are used in many metal and nonmetal mines to access the various levels with the use of rubber-tired equipment (see Figure 9.1). In most cases, the ramps lead from the surface to every major level in the mine. In Figure 9.1 a spiral configuration is shown, but many companies prefer to use a rectangular shape for the ramp to provide greater safety and speed. Ramps are normally sized to accommodate the largest equipment to be used, with added room for ventilation, drainage, and personnel (Pond, 2000).

In determining the types of openings to be used, the choice may be clear, based on the types of service that an opening will provide. However, the choice may also require an economic analysis to better determine the cost of the opening and the resulting cost of transport. For example, a slope costs more than a shaft for any given depth of deposit (Bullock, 1982), but the optimal choice must also consider the resulting transport economics provided by the two types of openings. It is therefore important to consider all these variables.

9.2.4 Interval Between Levels

One of the next major decisions in the development of a mine is the configuration of the sections or faces. In many metal operations, the deposit will be mined from levels that are spaced at an appropriate distance. The costs of developing and exploiting the mineral deposit become very important in this decision. Although this subject has not been discussed at length in recent references, Thomas (1978) and Nilsson (1982b) have outlined the nature of the problem rather well.

Factors that affect the costs of development and exploitation are numerous and revolve around the geologic conditions of the mineral deposit and the country rock, the mining method, the development layout, the production rate, financial considerations, and other variables. Figure 9.3 (Hartman, 1987) provides a general picture of the economic functions associated with an analysis of the optimal interval level. Note that the development cost goes down and the exploitation cost goes up as the interval increases. Somewhere in the middle, the optimum interval can normally be located. The trend over several decades has been to utilize fewer levels, spaced farther apart and supplemented by minimum-cost sublevels as required. Whereas levels were often 100 to 300 ft (30 to 90 m) apart in many mines in the past, many of

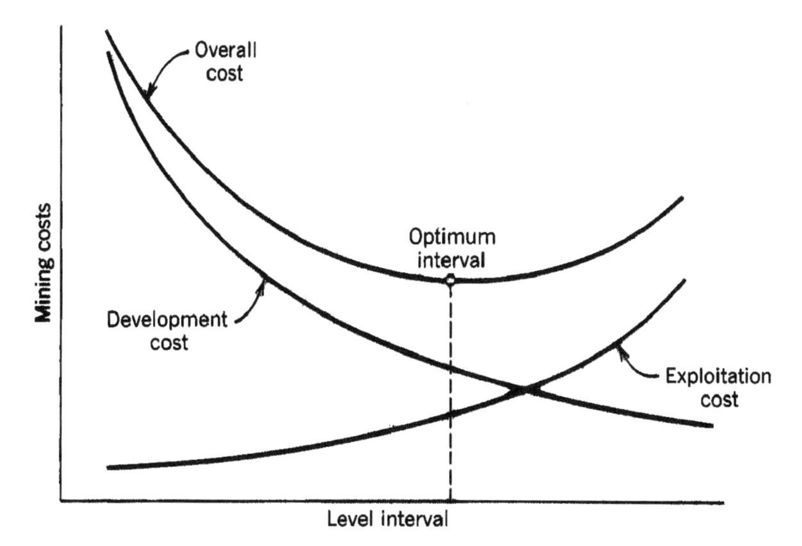

FIGURE 9.3. Determination of the optimum interval between levels for a hypothetical multilevel mine.

today's larger mines employ levels from 300 to 800 ft (90 to 240 m) apart to serve bigger blocks of ore (Nilsson, 1982b).

9.2.5 Secondary Development

Construction of some secondary (level or zone) and many tertiary (lateral or panel) development openings can be postponed until the exploitation stage is reached. In fact, the development of these openings is best postponed until they are needed, as such development normally does not produce an economic return. Planning for these openings as a part of the general mine plan, however, must be completed during the conceptual phase of the development stage.

Some of the openings used for noncoal mines of a vertical nature may be seen in Figure 9.1. The primary secondary openings in vertical mines are drifts, crosscuts, and raises, with occasional use of other openings as part of the secondary and tertiary development. On each level of the mine, a minimum number of drifts and crosscuts are driven to provide access, haulage, and ventilation. These are normally located in the footwall for ground control purposes (Bullock, 1982). Raises, winzes, ramps, and other special-purpose openings (orepasses, ventilation raises, manways, etc.) are similarly located between levels as required. In addition, such stope development (drawpoints, finger raises, manways, etc.) as necessary is commenced. These features are geared to serve the mining method chosen at the onset of the secondary development.

Coal mine development normally presents a different and simpler situation if a single, flat (or nearly flat) seam is being exploited. An idealized development plan showing primary and subsequent openings for a large coal mine appears in Figure 9.4. The main shaft or slope is used only for materials handling and materiel supply and is centrally located; a separate shaft for personnel is placed close by (both are protected from ground movement by a sizeable barrier pillar). Horizontal development for access and haulage is carried out by driving multiple headings. The hierarchy of these headings is normally (1) main entries, (2) submains or butt entries, (3) panel entries, and (4) room or longwall entries. *Butts* are submains driven at right angles to the butt cleat of the coal. Note that the terminology of this hierarchy differs somewhat from one area to another. For example, some mines designate the submains as "rights" and "lefts" off the mains (e.g., first right submain, second right submain, etc.). Additional variations in nomenclature exist. The entry sets, consisting of multiple entries and crosscuts, provide for ventilation, haulage, and other necessary services in the production sections. Multiple openings are required by law, to better address the problems of gas in the seam. (Multiple openings are not generally used in hard-rock mines because of the higher cost of excavation and less gassy conditions.) The production areas off the entry sets are generally called *panels*, both in longwall mining and in room-and-pillar mining.

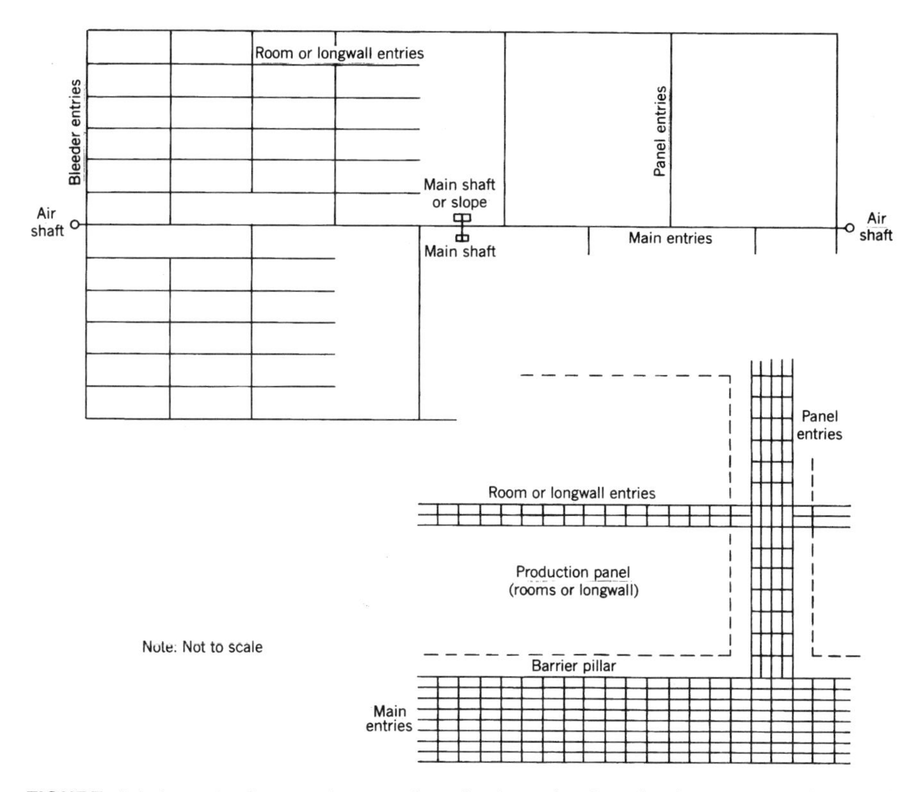

FIGURE 9.4. Layout of an underground coal mine, showing development openings and production panels.

The mine shown in Figure 9.4 is operated on the retreat; entries are driven to the extremities of the property, and exploitation is carried out on the retreat. Most mines employ a combination advance-retreat system to reduce development time. The plan shown can be used for either of the two common U.S. coal mining methods, room-and-pillar or longwall. This layout or a similar one is also suitable for mining horizontal, tabular deposits of metals and nonmetals by the room-and-pillar or the stope-and-pillar method. It should be noted that an effective way to operate a mine in a pitching seam is to use a similar layout with conventional haulage methods, in which the entry sets are driven at an angle to the strike to reduce the grade of the openings. Driving up-pitch is preferred when there are water problems.

9.3 MINE PLANT LAYOUT

An underground mine requires three groups of physical plant installations: the surface plant, the shaft plant, and the underground plant. Construction of the mine plant is essentially completed during the development stage and, because time is of the essence, must be carefully planned, using the critical path method (CPM) or the project evaluation and review technique (PERT) (see Section 4.5). We next examine some critical aspects of the three plant components, focusing on mine access through shafts.

9.3.1 Surface Plant

The *surface plant* consists of a variety of facilities to provide the mine with necessary services. These include access roads and parking, transportation facilities, power supply, water supply, service and maintenance buildings, mineral processing plant, bulk storage, and waste disposal facilities for air, water, and solids. The surface facilities that are unique to understand mining are the shaft collar and enclosure, headframe, bins, and hoist house. In coal mining, the entire assembly associated with a hoisting operation is called the *tipple*. This term is a remnant of the days when mine rail cars were hoisted on cages, and the cars were "tipped" to empty their loads. In many cases, the most prominent structure is the headframe. It houses the sheaves for the hoist and the ropes to the conveyances. Storage bins and the hoist are occasionally mounted in the headframe. Design of the hoisting system is covered later in this chapter.

Facilities for slope or decline access differ little if hoisting is used for materials handling. If, however, as is usually the case, a belt conveyor or trucks provide transport from the mine, then hoisting is eliminated and the surface plant is much simplified. Drift access is essentially horizontal and also dispenses with hoisting.

9.3.2 Shaft Plant

The *shaft plant* consists of the facilities installed for materials handling of ore, coal, or stone and associated waste and the means of transport of miners and materiel. It generally also includes systems for ventilation, drainage, power supply, and communications. Some of this paraphernalia will be located within the shaft or close by and must be planned as part of the shaft layout. Because at least two shafts are required for most mines, certain of the functions identified above can be divided between the two openings. Initially, in the shaft-sinking stage and the period of early horizontal development underground, it is necessary that all of the aforementioned functions, including ventilation, be provided through the main shaft. The ventilation air is supplied by vent tubing installed in the shaft, connected to a fan on the surface. Occasionally, the shaft is compartmented and bulkheaded to provide the airflow.

When connection between the two surface accesses has been completed, permanent arrangements of the shaft plant can be made, and the hoist components can be installed in the shaft. Conveyances called *skips*, generally

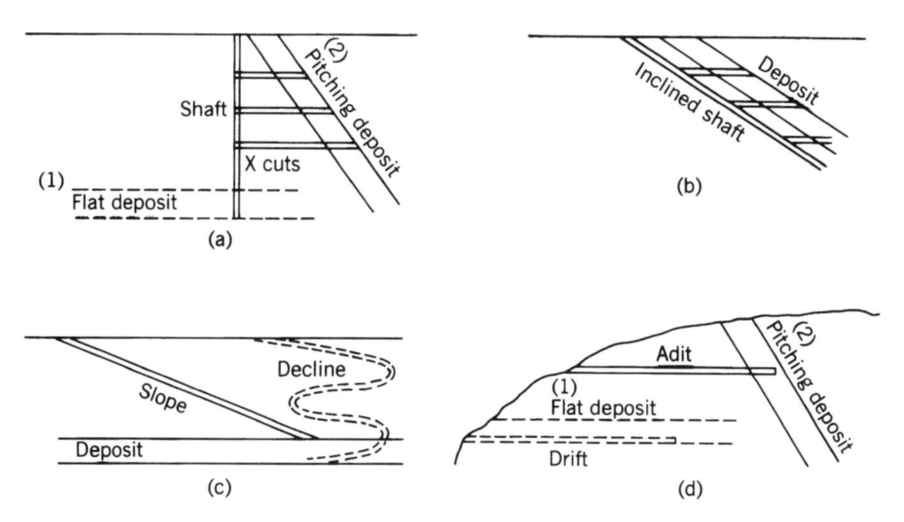

FIGURE 9.5. Alternative placement of main access openings for different occurrences of mineral deposits. (a) Shaft with hoisting. (b) Inclined shaft with hoisting. (c) Slope or decline with haulage. (d) Adit or drift with haulage.

arranged in balanced pairs, are installed on wire ropes and attached to the hoist. The skips will be used to hoist the ore and waste out of the mine. In addition, other compartments in the shaft will contain *cages*, that is, conveyances that operate independently of the hoist to transport workers, supplies, and equipment in and out of the mine.

Note that we have been discussing only shaft mines so far. There are several alternatives to shafts, such as inclined shafts, slopes, and adits as shown in Figure 9.5. Their applications can be summarized as follows:

1. *Vertical shaft with hoisting.* For deep horizontal ($<30^\circ$), vertical, or steeply inclined ($>70^\circ$) deposits; bad natural conditions; high production; long life.

2. *Inclined shaft with hoisting.* For moderately inclined (30°–70°) deposit; moderate conditions; low to moderate production and life; shortens horizontal development and allows exploration during sinking.

3. *Slope or decline with haulage* (conveyor or truck). For shallow or medium-depth horizontal deposits; good to moderate conditions; moderate to high production; long life; rope hoisting can be installed and rail haulage can be used; limited to about 12% to 14% with trucks and about 16° with a conveyor (unless special equipment is used).

4. *Drift or adit with haulage* (conveyor, truck, rail, etc.). For shallow, outcropping, horizontal deposit or steeply inclined deposit in area of high relief; varied conditions; high production; long life.

9.3.3 Underground Plant

The secondary and tertiary underground development openings that are constructed in and adjacent to the deposit have been discussed earlier. These are necessary to complete the development and exploitation of the mine levels or zones. In addition, there is a group of specialized openings and equipment, the *underground plant*, required at the shaft where (1) horizontal openings intersect with vertical ones and (2) transition from haulage and hoisting occurs. This is called a *shaft station.*

In addition, the underground plant may consist of various installations to make the system work efficiently and safely. These may include storage bins, loading pockets, sumps, power distribution equipment, fans, underground maintenance facilities, and numerous other installations that provide auxiliary services to the underground operations.

9.4 CONSTRUCTION OF DEVELOPMENT OPENINGS

The construction of underground openings is specialized and expensive. Consequently, this phase of mine development has become increasingly mechanized and efficient in recent years in order to reduce costs. In coal mining, the development openings produce a return as they are ordinarily driven in the coal. In metal mining, this is seldom the case, and thus the development openings will be more costly. The sinking of shafts and the driving of slopes and ramps are often performed by outside construction firms that have extensive experience, specialized equipment, and personnel knowledgeable of any possible problems. Other development activity may also be contracted out if the mining company thinks that a specialist can do the job at a lower cost. On the other hand, most metal mines drive their own drifts and raises, and most coal companies perform all the development in the coal seam. The following sections describe some of the cycles and equipment used and the costs related to development.

9.4.1 Cycle of Operations

Present-day production cycles of operation and equipment used to drive openings in underground mining—development and exploitation, coal and noncoal, and cyclic and continuous—are summarized in Table 9.1 (Hartman, 1987). But the emphasis has changed in many categories of development toward more mechanized methods and more rapid excavation procedures. The following discussion considers the unique aspects of construction for several openings; note that continuous methods are usually applicable only to soft or medium-hard rock.

9.4.1.1 Shaft. Shaft sinking is one of the most difficult of all development methods; restricted space, gravity, groundwater, and specialized procedures

Table 9.1 Equipment for Production Cycle of Operations in Underground Mining

Opening	Continuity of Operation	Cycle of Operations			
		Drilling	Blasting	Loading	Haulage
Development					
Shaft	Cyclic	Percussion jumbo	Dynamite, slurry	Clamshell, loader, cactus grab, backhoe	Bucket, skip
	Continuous	Shaft-boring machine (full face or pilot hole + ream)			
Slope, decline, ramp, winze	Cyclic	Percussion jumbo	Slurry, ANFO	Shovel loader, slusher, load-haul-dump (LHD) device	Skip, truck, LHD, conveyor
	Continuous	Tunnel-boring machine			
Raise, incline	Cyclic	Stoper, jumbo, mobile platform	ANFO, slurry	Gravity, chute, loader	Truck, mine car, conveyor
	Continuous	Raise-boring machine (full face or pilot hole + ream)			
Drift, crosscut, adit, tunnel	Cyclic	Percussion jumbo, airleg	ANFO, pneumatic loader	Shovel loader, slusher, LHD	Truck, LHD, mine car, conveyor
	Continuous	Tunnel-boring machine, roadheader			
Entry, crosscut (coal)	Cyclic	Cutting machine, rotary auger drill	Permissible	Gathering arm loader	Shuttle car, conveyor
	Continuous	Continuous mining machine			
Exploitation					
Stope	Cyclic	Jumbo, stoper, airleg, downhole	ANFO or slurry	Gravity, slusher, shovel loader, LHD	Truck, LHD, mine car, conveyor
	Continuous	N/A (use continuous miner if stope-and-pillar and soft rock)			
Room, crosscut (coal)	Cyclic	Cutting machine, rotary auger drill	Permissible, airdox, cardox	Gathering arm loader	Shuttle car, conveyor
	Continuous	Continuous mining machine			
Longwall (coal)	Cyclic	N/A (no longer used)			
	Continuous	Shearer, plow			Conveyor, mine car

make the task quite formidable. It is for this reason that most mining companies utilize contractors for this job. Both conventional mining practice and boring procedures are used for shaft sinking. Conventional practice is employed in nearly all shafts greater than 16 ft (4.9 m) in diameter, as well as in most smaller shafts in very hard rock or in difficult geologic conditions. On the other hand, blind boring or raise boring of shafts in many soft-rock conditions is quite common for diameters of less than 16 ft (4.9 m).

In conventional shaft sinking, it is efficient to design the system so that the drilling and mucking operation can be conducted at the same time as the lining of the shaft and the extension of the utilities. Accordingly, a *sinking stage* is often used to advance the shaft (see Figure 9.6). A number of mucking methods

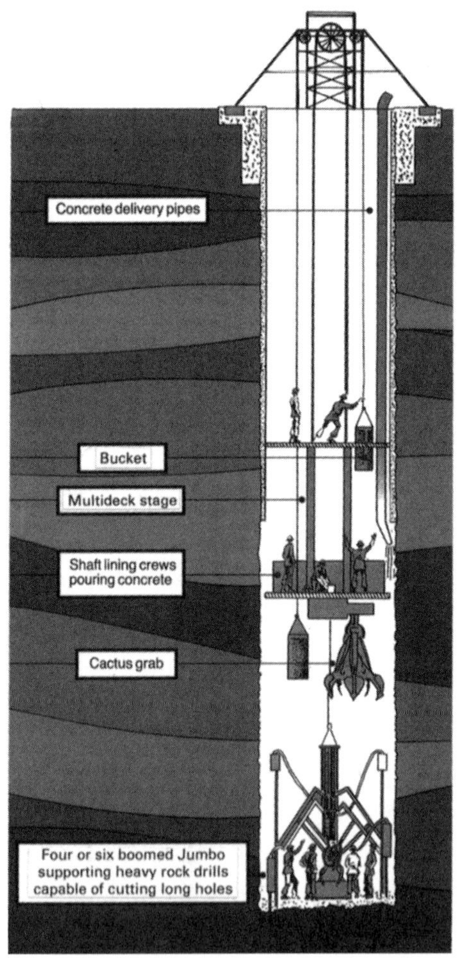

FIGURE 9.6. Shaft sinking using a sinking stage and a drill rig (jumbo). *Source:* Mining Survey, 1983. By permission from Chamber of Mines of South Africa, Johannesburg.

can be used at the bottom of a large-diameter shaft, including a cactus grab (shown in Figure 9.6), a clamshell, an overhead loader, and a backhoe. Russell (1982) provides some capacities for these types of loaders used in the bottom of a shaft. Large ovaloidal shafts of 18 ft by 30 ft (5.5 m by 9.1 m) in coal measures can be produced for $3500 to $4500 per ft ($11,500 to $14,800 per m), and a 20 ft conventional round shaft can cost $2500 to $3500 per ft ($8200 to $11,500 per m). These costs include the shaft lining, normally concrete, and were provided by Crooks (2000). Costs in hard-rock operations may be more expensive and more variable.

In boring shafts up to 16 ft (4.9 m) in coal measures or similar formations, blind boring is gaining in usage (Maloney, 1993; Zeni, 1995). In this method, a drilling rig is set up on the surface and a hydraulic motor drives the drilling head. Drilling mud and an air lift system are used to get the drill cuttings to the surface. Shafts of this size are limited in depth to about 1000 ft (300 m) in good conditions but can go deeper with smaller diameters. Blind-boring techniques are routinely employed to produce bleeder shafts for longwall operations with diameters up to 8 ft (2.4 m) (Maloney, 1993; Zeni, 1995). These are used with high-pressure fans to provide more positive ventilation in the large longwall panels utilized today.

The most important auxiliary operation in a conventionally sunk shaft is ground control (usually a reinforced concrete lining, gunite or shotcrete, steel sets or arches, cast iron tubbing, or bolts and steel mesh are used), with control of groundwater a close second. Grouting or freezing of the earth around the shaft may be required in extremely wet or plastic formations. Bored shafts are often left without support in structurally sound formations. Ventilation is mandatory during the sinking of a shaft if miners must work at the face or the quality of the atmosphere must be maintained.

9.4.1.2 Slopes, Declines, and Ramps. A decline is usually sunk at a low slope angle ($<20°$) so that it can be driven conventionally or bored using a tunnel-boring machine. Most slopes and ramps, however, are driven conventionally to suit the use for which they are intended. In coal mines, the slopes are generally driven conventionally in a horseshoe shape with two compartments, the upper one for the conveyor and the second below for track and personnel. The standard size is approximately 17 to 20 ft (5.2 to 6.1 m) wide and 13 to 14 ft (4.0 to 4.3 m) high, with a concrete divider between the two compartments (Emerick, 2000). These openings cost about $3000 to $4000 per ft ($9800 to $13,100 per m) of length for most coal measures (Crooks, 2000).

Ramps for metal and nonmetal mines are more variable, with spiral and rectangular configurations. The size of these openings and the design of the curves must be carefully matched to the equipment used in the mine and must allow room for the tubing that is used for ventilation. They must also allow for other utilities, as well as drainage. Both loaders and trucks are ordinarily used in these openings, and safety must be the primary concern in their design.

The slopes are normally 12 to 14%, depending on the climbing capability of the equipment (Pond, 2000). Ramps are always driven conventionally, with a horseshoe shape being common.

Support for declines varies, depending on the ground conditions. Concrete linings are not normally used except in the most difficult of conditions. Bolting (often with mesh between bolts), shotcrete, and steel supports are commonly employed, matched to the conditions along the ramp or slope. Firms specializing in development openings are normally engaged, as they have the right equipment and know-how to efficiently produce these openings.

9.4.1.3 Raises. In some metal and nonmetal mining methods, the efficient driving of raises is essential to the economic success of the method. Raises are normally placed near the stopes employing specialized cyclic or continuous operations. The direction of advance is normally upward, using gravity for muck movement. However, the operating conditions are quite adverse, making raising by conventional methods very difficult and definitely to be avoided. To overcome the difficulties, two mechanized methods are normally used to improve the process. The first is the raise climber, a track-mounted machine for conventional advance. The second is the raise borer, an entirely remote method of drilling or boring the raise using mechanized equipment.

Figure 9.7 shows the use of a raise climber in vertical, inclined, and large-area raises. The raise climber is mounted on rails and can be mechanically driven up the raise. The scaling, support, drilling, and explosives-loading operations can all be performed from the platform. The raise climber is then trammed down the raise and swung into the drift below for protection from the blast. The raise climber was introduced in Sweden in 1957 (*Engineering and Mining Journal*, 1992a) and is currently used throughout the world to efficiently drive vertical openings.

The ultimate in raise development methods is the use of the raise borer. This device has been employed for several decades to produce raises where applicable in mining operations. The raise borer mechanism is a compact drilling tool that produces a raise in two distinct steps. First, a small-diameter hole of about 9 in. (0.23 m) is drilled from an opening at the top of the raise to an existing opening below. A reaming head is then attached to the drill string, and the raise is bored to its final diameter by reaming upward to the upper opening. The process of reaming is shown in Figure 9.8. Raises are normally completed in diameters of up to 8 ft (2.4 m), but shafts of up to 18 ft (5.5 m) in diameter have also been developed in mines using this method (Hood and Roxborough, 1992).

In addition to the standard raise borer, a blind-boring tool is now also used for producing raises. The blind boring of raises makes perfect sense if the lower opening does not yet exist. One blind-boring device, known as a BorPak, is actually a mini-TBM (mini-tunnel-boring machine) that uses a cutting action similar to that of a standard TBM (*Engineering and Mining Journal*, 1995b).

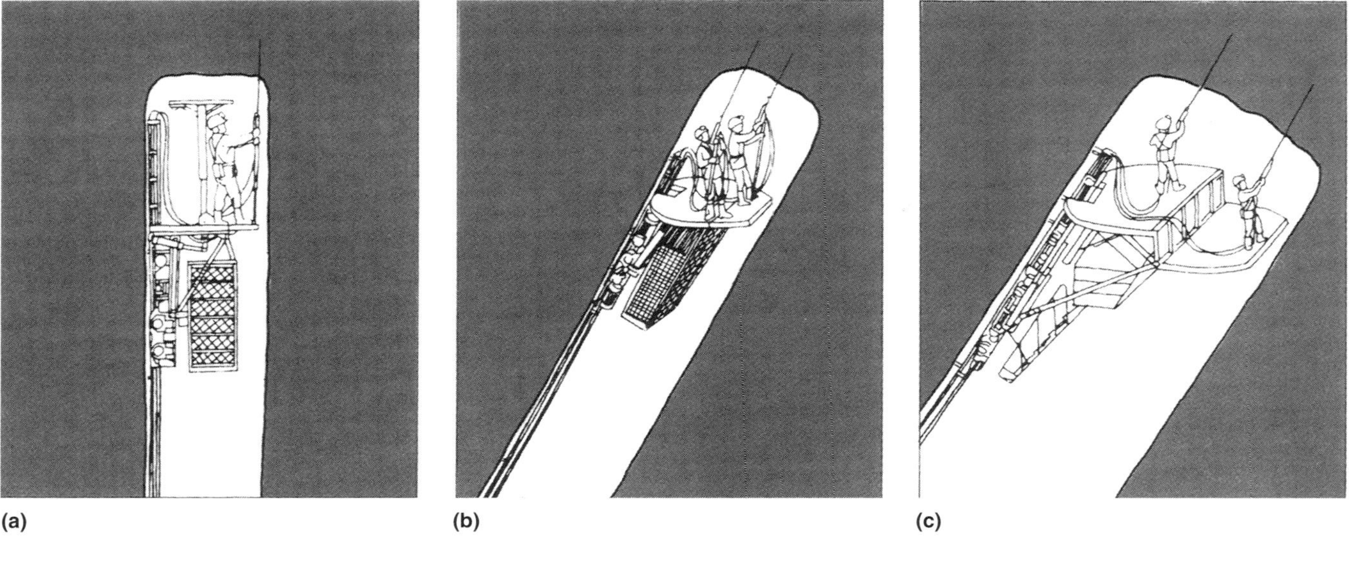

FIGURE 9.7. Raise driving by Alimak raise climber in (a) vertical, (b) inclined, and (c) large-area raises. By permission of Alimak AB, Skelleftea, Sweden.

FIGURE 9.8. A raise-boring machine performing the reaming operation. By permission of Atlas Copco Rock Drills AB, Orebro, Sweden.

9.4.1.4 Drifts, Crosscuts, Adits, and Tunnels. All of the horizontal openings—drifts, crosscuts, adits, and tunnels—are developed by a process called drifting or tunneling. The traditional method of performing this operation is to drill and blast the face, load the muck into a haulage device, and then provide support and ventilation to the newly advanced face. Advancing the opening in this manner is costly in terms of time and labor. As a result, the use of a TBM may be attempted where rock conditions are suitable and the difficult problems of handling a TBM in an underground mine can be overcome. A TBM is shown in Figure 9.9. Note that the device is a huge piece of mechanical equipment that cannot easily be turned or moved around a mining operation. Thus, it has more commonly been utilized in civil engineering tunnels than in mines. Tilley (1991), Bullock (1994), and Alexander (1999) describe TBM applications in mining. A case study on the application of TBMs to mining is presented in Section 9.4.2.

9.4.1.5 Entry and Crosscut Development (Coal). Development of the entries and crosscuts in coal mines is performed primarily by variations of the room-and-pillar mining method, using continuous miners and traditional coal mine haulage equipment. Because this mining method is described in Chapter 10, detail will be left to that discussion. Suffice it to say that the process of developing these openings in a coal mine is more similar to standard production practice than it is in metal or nonmetal mining.

9.4.2 Case Study: TBM Development at the Stillwater Mine

The Stillwater Mining Company of Nye, Montana, has been utilizing tunnel-boring machines in its platinum-group metals operations since the 1980s. The two mines it initiated in this area were both developed in part using TBMs. The deposit is a reef (a ridge-like structure) of primarily platinum and

FIGURE 9.9. A tunnel-boring machine outfitted with disk cutters. By permission of The Robbins Company, Kent, WA.

palladium that is about 28 mi (45 km) long and 1 mi (1.6 km) deep. The rock in this area varies in compressive strength from 9000 to 27,000 lb/in.^{2} (60 to 190 MPa). Mining methods used in these mines included mechanized ramp-and-fill, sublevel stoping, captive cut-and-fill, and slusher cut-and-fill (Dyas and Marcus, 1998).

The first mine at Stillwater employed a Robbins TBM with a diameter of 13 ft (4.1 m). About 160,000 ft (48,000 m) of lateral footwall drifts were to be driven by the TBM. In addition to the drifts, raises were developed to 6 ft (1.8 m) in diameter using a Robbins raise-borer. TBM and raise-borer advance rates were reported by Tilley (1991) and showed the great potential that exists. The second mine, known as the East Boulder Project, used two 15 ft (4.6 m) diameter TBMs manufactured by Construction and Tunneling Services. These machines were designed to have a relatively short turning radius of 200 ft (60 m) to help overcome the primary shortcoming of using a TBM in mining operations. Alexander (1999) illustrates how the TBM can be turned around in the mine using this design feature of the machine. Although raise borers are often used in mining practice, TBMs are seldom called on to perform development work. This case study shows that rapid excavation methods can be both productive and cost-effective for mine development work.

9.4.3 Equipment Selection

The factors involved in the selection of equipment for a particular mining situation were discussed in Section 5.4.3. The general variables of concern there must be coupled with the specific requirements of the mining method to achieve a proper match of equipment and the mining method. The underground equipment used in metal and nonmetal mining is quite different from that used in coal production. Accordingly, these uses are discussed separately in the following sections.

9.4.3.1 Metal and Nonmetal Mine Equipment. The equipment employed in noncoal mines has undergone much change in the last few decades, with an emphasis on hydraulic drills, diesel loaders, bigger equipment sizes, and more automation. Hamrin (1982, 1998) provides a good basic outline of the choices in drilling and loading equipment, as summarized in Figure 9.10. The drill type chosen must be well matched to the type of drilling to be conducted. The basic drill types shown in Figure 9.10 have all been upgraded into more efficient, more automated, and more productive units. Most of the drills used in underground production are now of the hydraulic type. The choice of a drilling setup was more completely discussed in Section 5.2.3.3; review that section for additional insights.

The loaders shown in Figure 9.11 are also an interesting study. The trend in recent years has been toward bigger and more productive loaders, with emphasis on diesel-powered equipment where the mining method permits. Note that additional choices may be available to the engineer selecting

Mining Method	Breast, Large-scale (Stope and Pillar Mining)		Overhand, Small-scale (Shrinkage Stoping)	
Drilling and Blasting Technique	 Drifting and slashing  Frontal benching	 Vertical or downward benching	 Overhand stoping	 Frontal stoping, breasting
Applicable Drilling Equipment	 Machanized drifting jumbo	 Mechanized airtrac drill	 Hand-held stoper drill	 Hand-held airleg drill
Drilling Data				
Hole diameter, in. (mm)	1.5-2.0 (38-48)	2.5-3.0 (64-76)	1.1-1.3 (29 -33)	1.1-1.3 (29-33)
Hole depth, ft (m)	10 18 (3.0-5.5)	as required	6.5-8.2 (2.0-2.5)	6.5-11.5 (2.0-3.5)
Drilling Equipment Performance				
With pneumatic rock drill ft/hr (m/hr)	200-250 (60-75)	50-80 (15-25)	25-40 (8-12)	30-50 (10-15)
With hydraulic rock drill ft/hr (m/hr)	300-360 (90-110)	(25-35)	na	na
Drilling – Blasting Factor				
yd^3 (m^3) rock broken per drilled ft (m)	0.6-0.8 (1.5-2.0)	1.2-1.6 (3.0-4.0)	0.3-0.4 (0.7-0.9)	0.3-0.4 (0.7-0.9)

FIGURE 9.10. Unit operations and equipment for development and exploitation in underground mining: drilling. Performance data included. (After Hamrin, 1982. By permission from the Society for Mining, Metallurgy, and Exploration, Inc., Littleton, CO.)

equipment for underground metal mines, such as a gathering-arm loader and slushers. However, gathering-arm loaders are used primarily in softer rocks and have little applicability in most mining operations. Slushers are unproductive as compared with diesel equipment and are therefore used only where no other equipment is suitable.

9.4.3.2. Coal Mining Equipment Selection. The selection of equipment for coal mining depends heavily on the dip of the coal seams to be mined. Because most of the coal seams in the United States are relatively flat, we will

Mining Method	Overhand, Large-scale (Sublevel Stoping)		Caving (Sublevel Caving)
Drilling and Blasting Technique	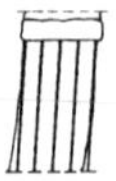 Parallel drilling		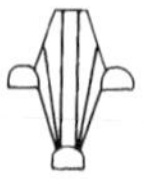 Fan drilling
Applicable Drilling Equipment	 Mechanized airtrac with downhole hammer and high pressure		 Mechanized fandrill
Drilling Data			
Hole diameter, in. (mm)	4.0-4.5 (105-115)	6.0-6.5 (152-165)	1.9-2.0 (48-51)
Hole depth, ft (m)	160-200 (50-60)	160-200 (50-60)	40-50 (12-15)
Drilling Equipment Performance			
With pneumatic rock drill ft/hr (m/hr)	160 (50)	160 (50)	650-800 (200-240)
With hydraulic rock drill ft/hr (m/hr)	na	na	800-1000 (240-300)
Drilling – Blasting Factor			
yd^3 (m^3) rock broken per drilled ft (m)	3-4 (8-10)	6-7 (14-18)	0.7-0.9 (1.8-2.3)

FIGURE 9.10. *Continued.*

concentrate on the choice of equipment for that type of mining. A more detailed discussion of this topic is available in Lineberry and Paolini (1992). For room-and-pillar mining and for development of longwall operations, the pattern is to choose a continuous miner unit suited to its use (sometimes with an on-board roof-bolting system), a section haulage system, a roof bolter (if not associated with the continuous miner), and any auxiliary equipment that is necessary.

The traditional choice for a continuous miner is a drum-type machine with remote control and an integrated dust collector, which will eliminate about 90% of the dust generated by the continuous miner. Figure 9.12 shows one of these units. These machines have been around in basic form for 40 years, with

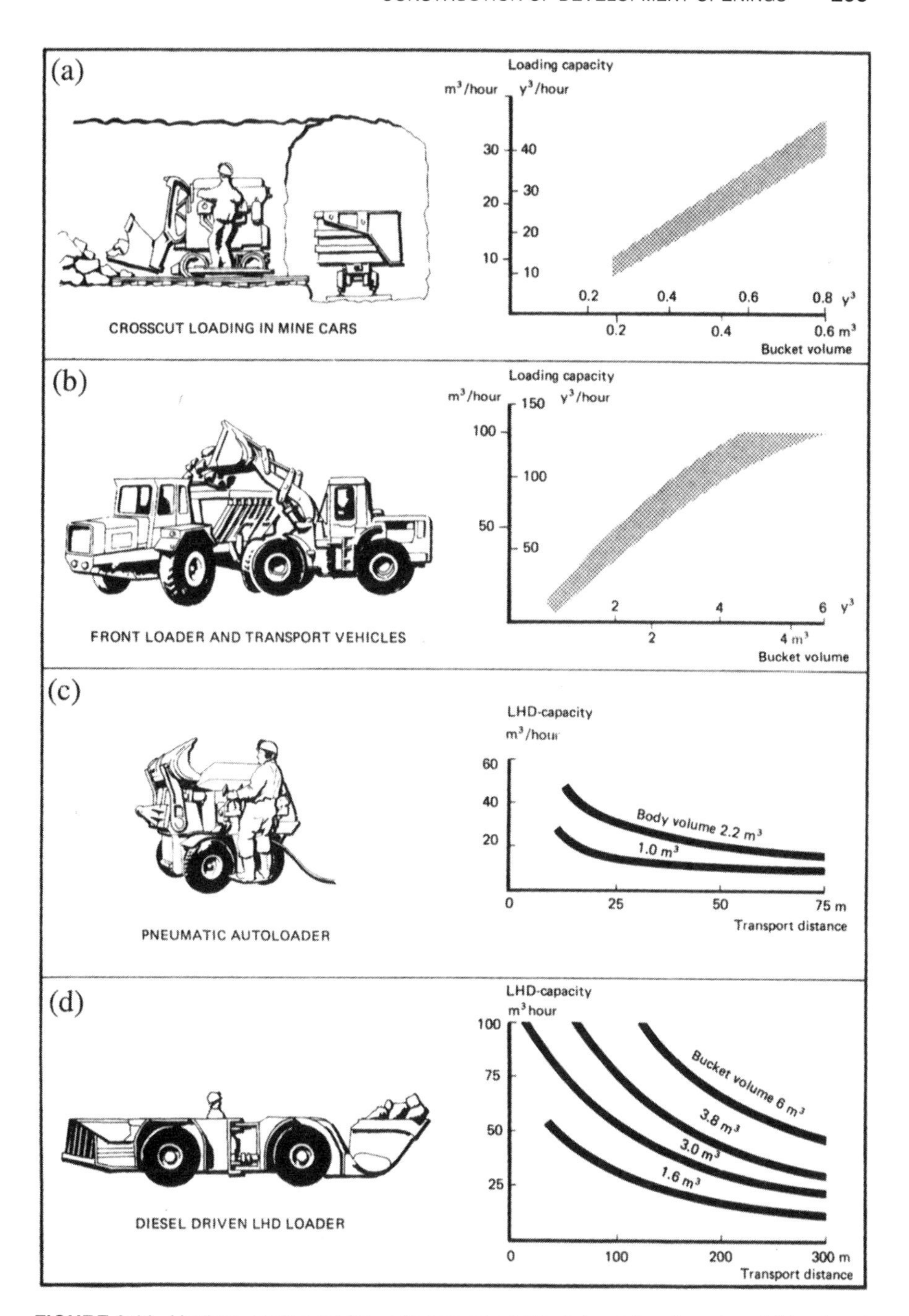

FIGURE 9.11. Haulage equipment for underground mining: (a) overhead mucker, (b) front-end loader, (c) pneumatic autoloader, and (d) LHD unit. (After Hamrin, 1982. By permission of the Society for Mining, Metallurgy, and Exploration, Inc., Littleton, CO.)

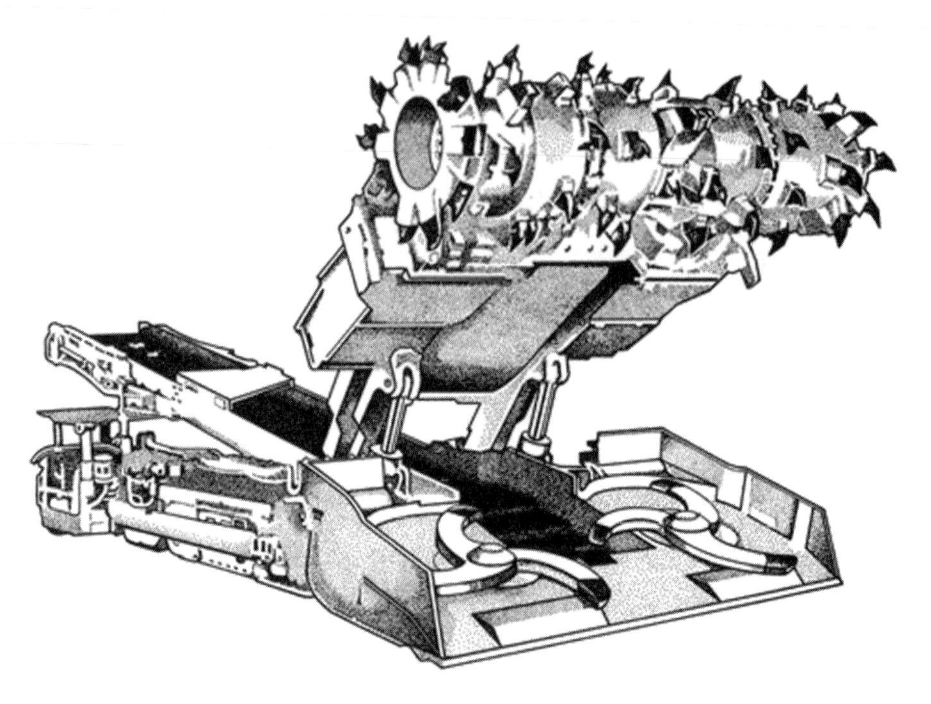

FIGURE 9.12. Drum-type continuous miner. Illustration courtesy of Joy Mining Machinery, Franklin, PA.

many improvements in capability, maintainability, and safety added over their years of use. As a result, they are nearly always chosen for room-and-pillar mining of coal. For haulage, there are a number of choices: electric shuttle cars, diesel shuttle cars, full-dimension mining equipment (mobile bridge carriers and bridge conveyors), and other continuous conveyor systems. Conveyors have advantages if they can be made to work continuously. However, a number of constraints often interfere with this goal. As a result, shuttle cars are the predominant choice, with electric vehicles being more common than diesel. Figure 9.13 illustrates a typical shuttle car. For roof control, the normal choice is a two-boom bolter, with each bolter location protected by a hydraulic roof support system (see Figure 9.14).

For longwall mining, the choice is often not what type of equipment to use, but perhaps which manufacturer or model. Of the approximately 59 longwalls being operated in the year 2000 in the United States, all but 2 use an electric-powered shearer as the coal winner, with the other longwalls using a bidirectional plow. Figure 9.15 shows the basic configuration of a double-drum shearer. For roof support, most longwalls use highly sophisticated hydraulic

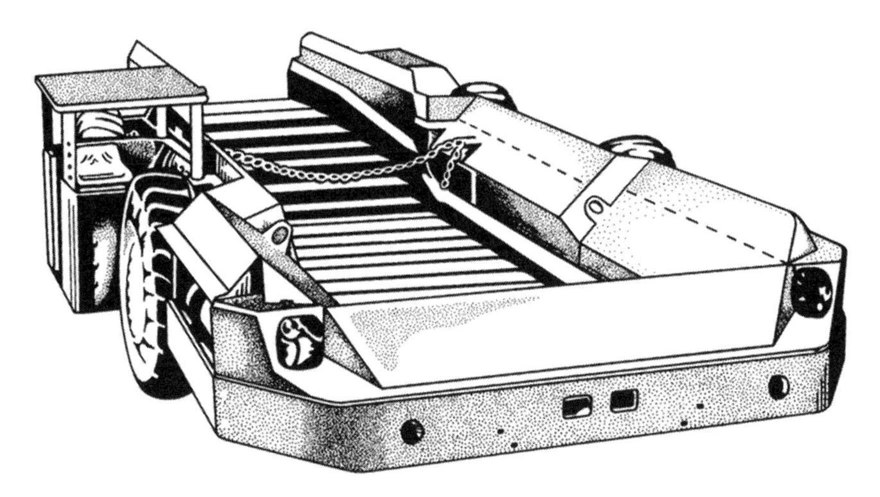

FIGURE 9.13. Electric-powered shuttle car. Illustration courtesy of Joy Mining Machinery, Franklin, PA.

FIGURE 9.14. Two-boom roof bolter. Photo courtesy of DBT America, Inc., Houston, PA.

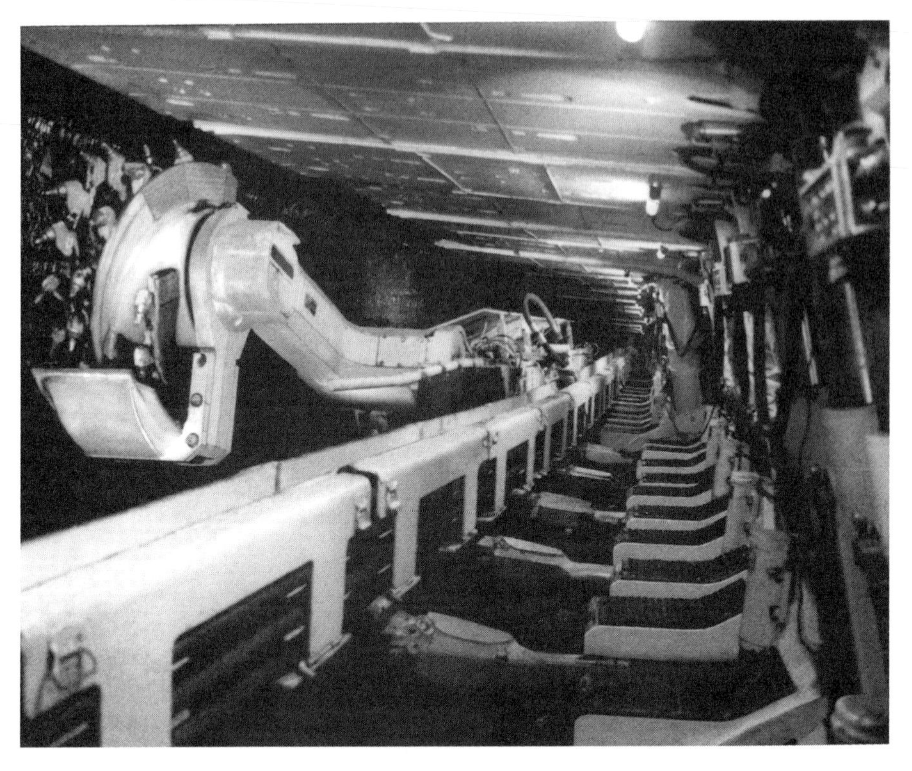

FIGURE 9.15. Longwall shearer. Illustration courtesy of Joy Mining Machinery, Franklin, PA.

shields that are semiautomated. These shields (a typical one is shown in Figure 9.16) can respond to the movement of the shearer, moving forward when needed and providing reliable support of the roof. As the primary coal mover, longwall systems employ a chain conveyor on the face. These conveyors are about 33 to 53 in. (0.84 to 1.35 m) wide and are normally powered by an electric motor at each end. More information on equipment used in longwalls in the United States can be found in Fiscor (2000).

9.4.4 Selection of Ground Control Methods

Ground control is one of the primary tasks that any mining engineer must address; it is fundamental to both safety and productivity in the mine. The choices for ground control (often called roof control in coal mines) are fairly numerous, but several dominate the picture. The following are the most common choices:

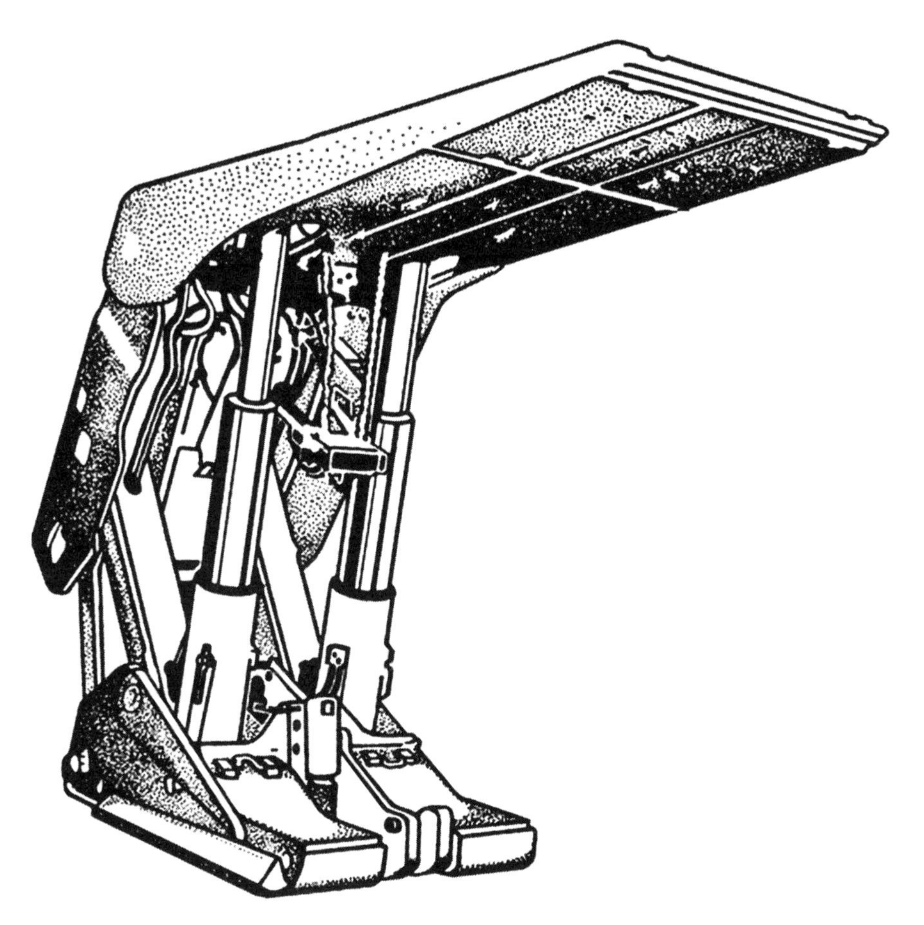

FIGURE 9.16. Longwall shield. Illustration courtesy of Joy Mining Machinery, Franklin, PA.

Rock (Roof) Bolts and Variations

Mechanical point-anchor bolts
Resin-anchored bolts
Split-set bolts
Swellex bolts
Combination-anchor bolts
Bolts with steel mesh
Bolt trusses
Grouted steel cable
Grouted wooden dowels

Mechanical Supports
Timber sets
Steel sets
Yieldable steel arches and ring sets
Resin grouts

Concrete Supports
Gunite (portland cement and sand) sprayed-on linings
Shotcrete (sprayed-on concrete) linings
Formed concrete linings
Concrete grout
Concrete crib rings

Miscellaneous Supports
Wood cribs
Temporary hydraulic supports
Powered mobile hydraulic supports
Freezing techniques
Chemical grouts

Note that all these support systems find use in underground mining methods. However, some are used with more frequency than others. For example, the various methods of rock bolting are used extensively and will find application in nearly every mine. Figure 9.17 shows two variations of the common roof bolt. The mechanical-anchor bolt was the earliest in use; the resin-anchored bolt is probably more commonly used today. The split-set bolt and the Swellex bolt are more effective in ground that is fractured or subject to slippages. Other support methods are shown in Figure 9.18. The roof truss, hydraulic supports, and cribs are more commonly used in coal operations, whereas the others may be more heavily utilized in noncoal mines.

9.5 SHAFT HOISTING SYSTEMS

The hoist, together with its associated plant for an underground mine, is the single most important and expensive element of the mine plant (Butler and Schneyderberg, 1982). Because it is also the most sophisticated part of the entire plant, it is not feasible to build in redundancy (except for the prime mover), yet it must perform with close to 100% reliability. Its cost may be on the order of 5 to 10% of the entire development budget. Obviously, wise engineering judgments are called for in the design of the hoist plant.

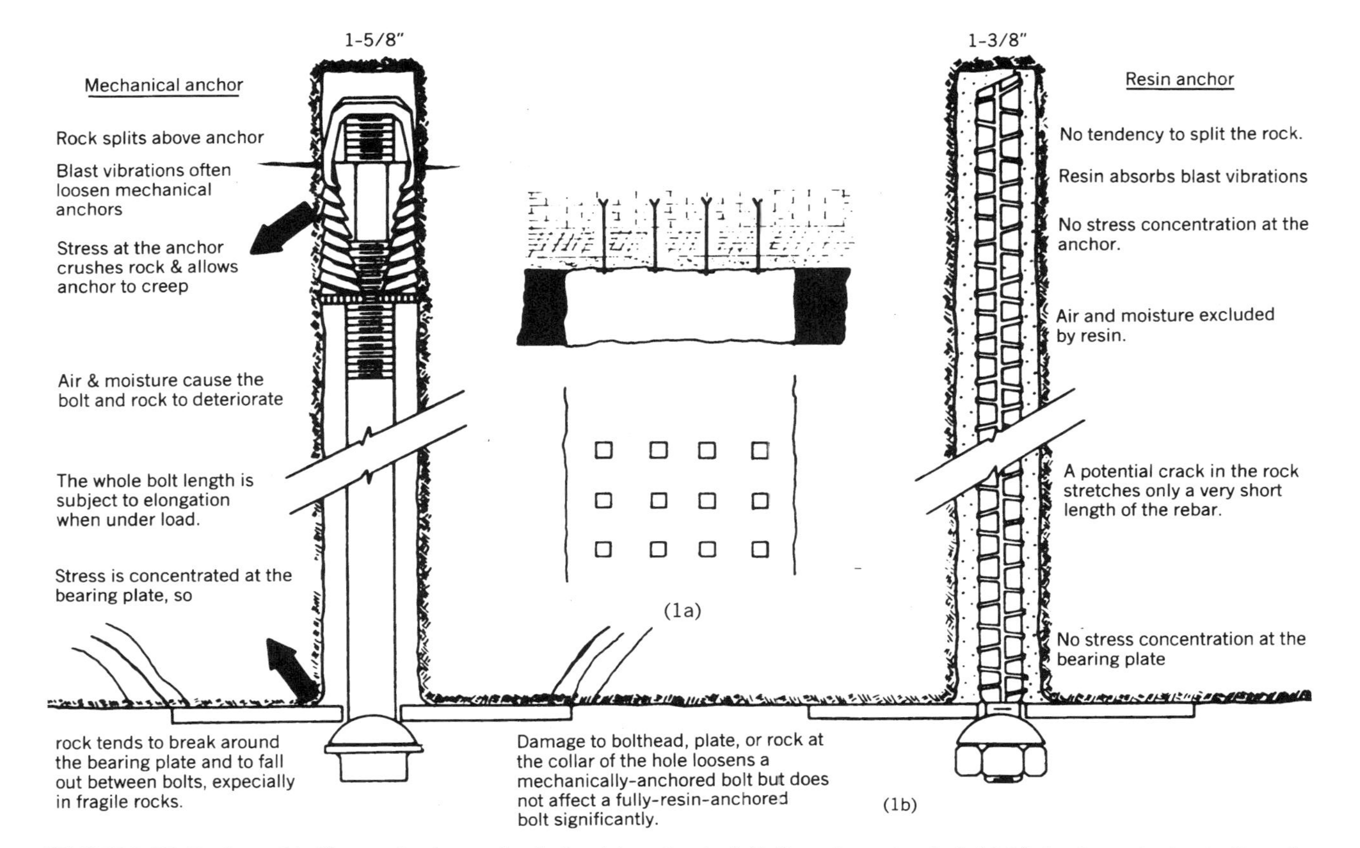

FIGURE 9.17. Basic roof-bolting methods: mechanical point-anchor bolt (*left*), resin-anchor bolt (*right*), basic coal mine bolting plan (*center*). (After Lucas and Adler, 1973. By permission from the Society for Mining, Metallurgy, and Exploration, Inc., Littleton, CO.)

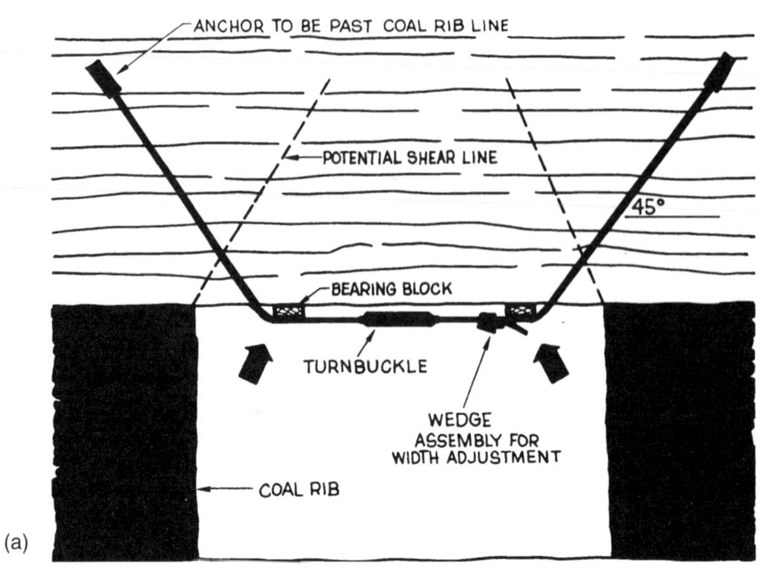

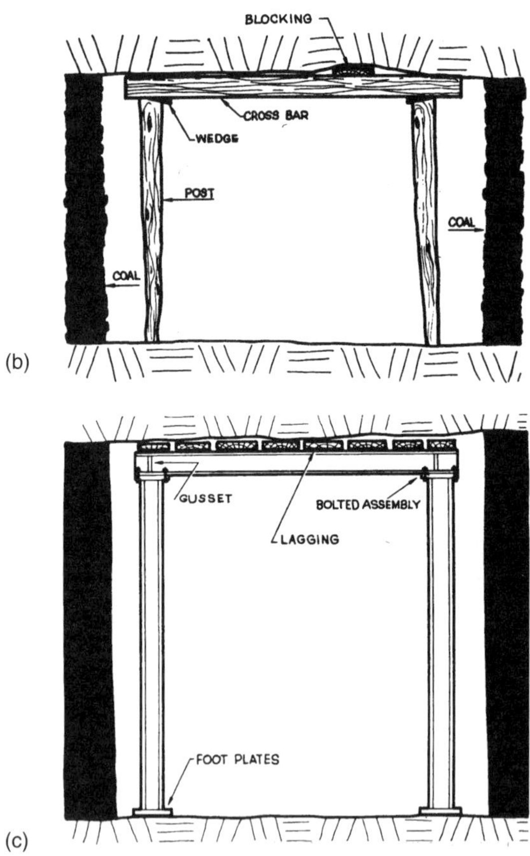

FIGURE 9.18. Underground roof control systems: (a) roof bolt truss, (b) timber set, (c) steel set, (d) yieldable arch, (e) temporary hydraulic jacks, (f) wood crib. Figure 9.18(f) from Barczak and Gearhart (1994). Others from Lucas and Adler, 1973 (by permission from the Society for Mining, Metallurgy, and Exploration, Inc., Littleton, CO.)

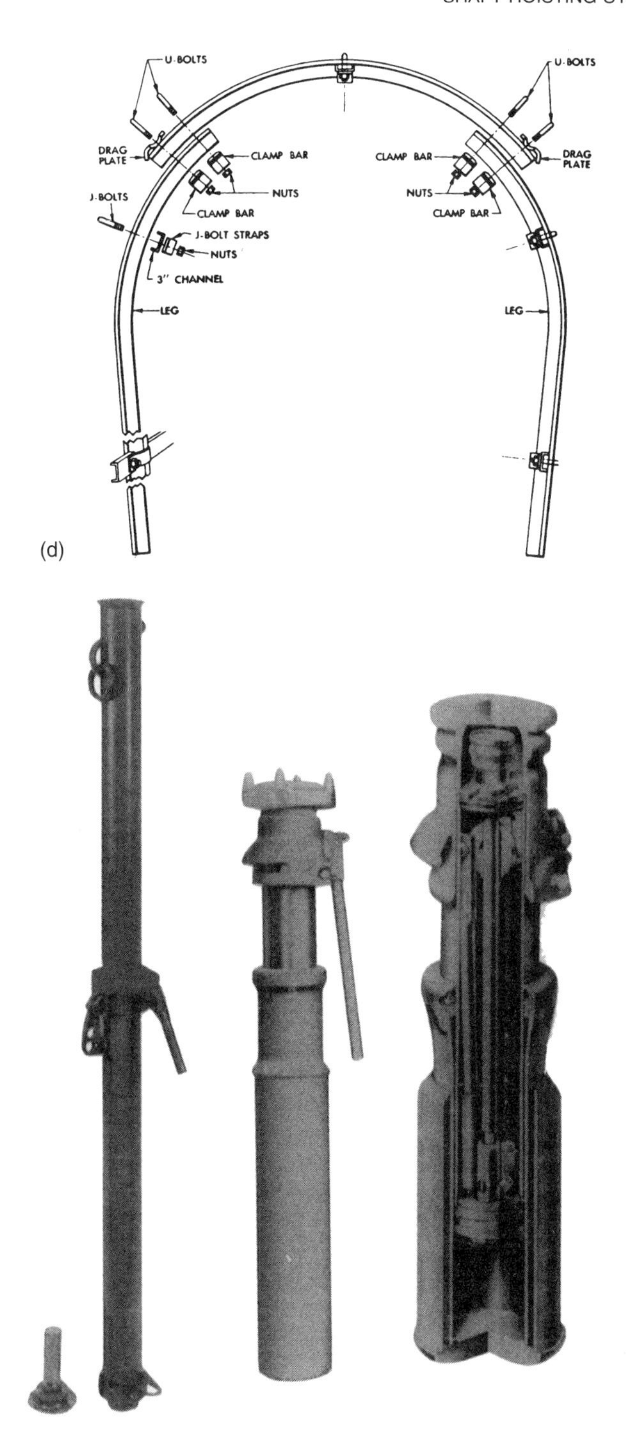

FIGURE 9.18. *Continued.*

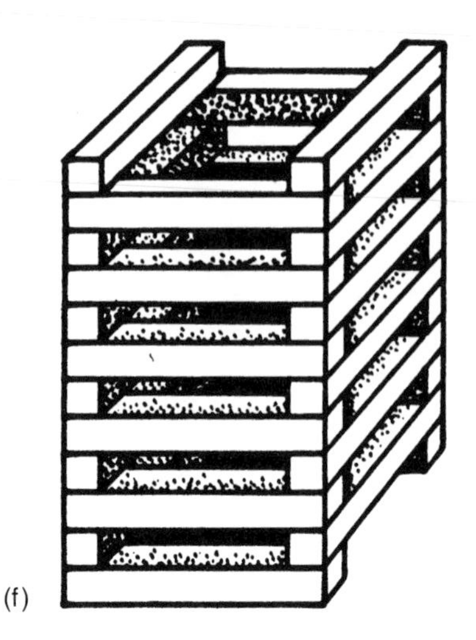

FIGURE 9.18. *Continued.*

9.5.1 Components of Hoist Plant

The *hoist plant* consists of all those components of the mine plant that are necessary to elevate ore, coal, stone, or waste and to raise and lower personnel and materiel in the mine. It is made up of some of the constituents of the mine plant that were discussed in Section 9.3. Classified by location, they consist of the following:

1. Surface plant
   a. Hoist room (headframe- or ground-mounted)
      (1) Hoist drum or sheave (imparts motion to rope)
      (2) Hoise electrical and mechanical equipment (prime mover, brake, clutch, controls)
      (3) Hoist ropes (steel wire strands, woven in a pattern or lay)
   b. Headframe (tower or A-frame, steel or reinforced concrete)
      (1) Idler sheaves
      (2) Storage bins (ore and waste)
      (3) Skip dump mechanism (overturning or bottom dump)

2. Shaft plant
   a. Skips (bulk transport)
   b. Cages, elevators (personnel, materiel)
   c. Shaft guides (tracks for skips and cages)

3. Underground plant
   a. Dump and storage bin
   b. Crusher (if size reduction required for hoisting)
   c. Loading pocket
   d. Personnel and materials-handling facilities

Figure 9.19 assists us in visualizing these components of the hoist plant, assuming a vertical shaft as the main access opening. With little modification, the same layout would also suffice for an inclined shaft using hoisting (a slope, ramp, or drift mine differs, presumably because hoisting would not be used).

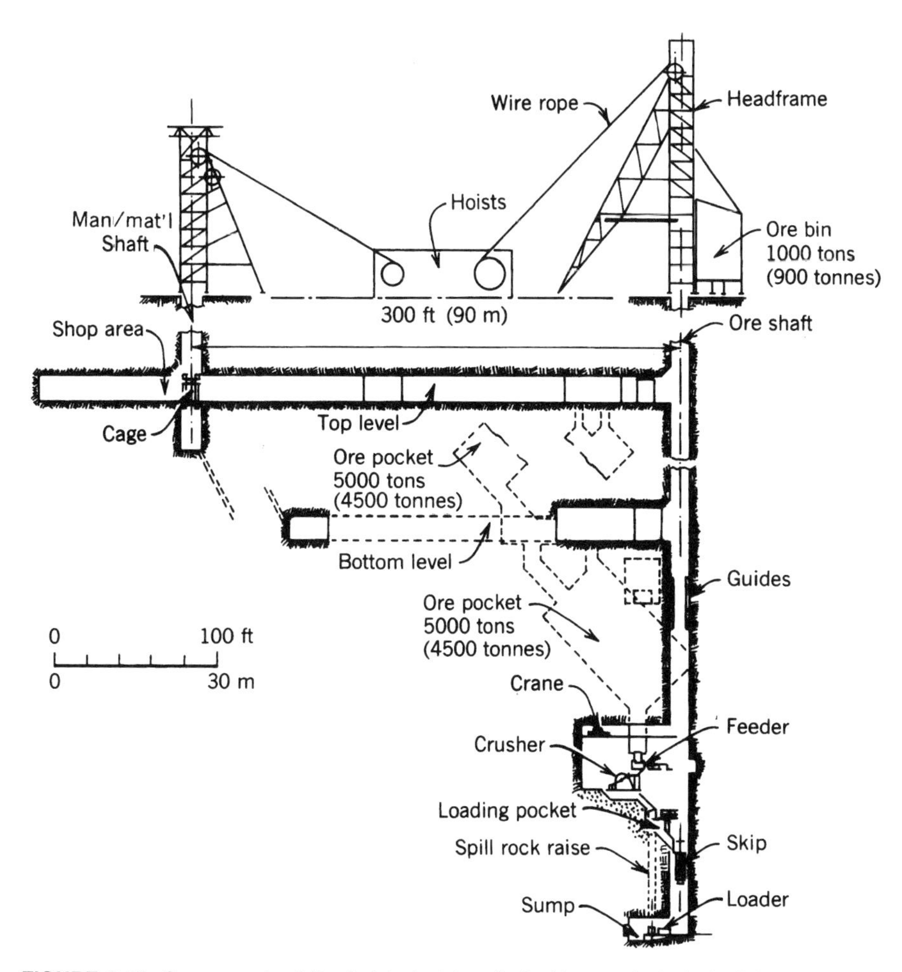

FIGURE 9.19. Components of the hoist plant installed with a vertical shaft. (After Lucas and Haycocks, 1973. By permission from the Society for Mining, Metallurgy, and Exploration, Inc., Littleton, CO.)

9.5.2 Hoisting System

It is with the *hoisting system* itself—those components of the hoist plant located in the hoist room—that engineering design is mainly needed. Three key factors govern hoist selection:

1. Production rate, or tonnage to be hoisted per unit of time
2. Depth of shaft
3. Number of levels to be accessed

There are only two hoist types commonly used today: drum and friction. A *drum hoist* stores the rope not extended in the shaft on the face of the drum. A *friction hoist* passes the rope (or ropes) over the top of the drive wheel but does not store it. Figure 9.20 shows a general diagram of both types. In addition, a special drum hoist design for deeper shafts is occasionally required. The

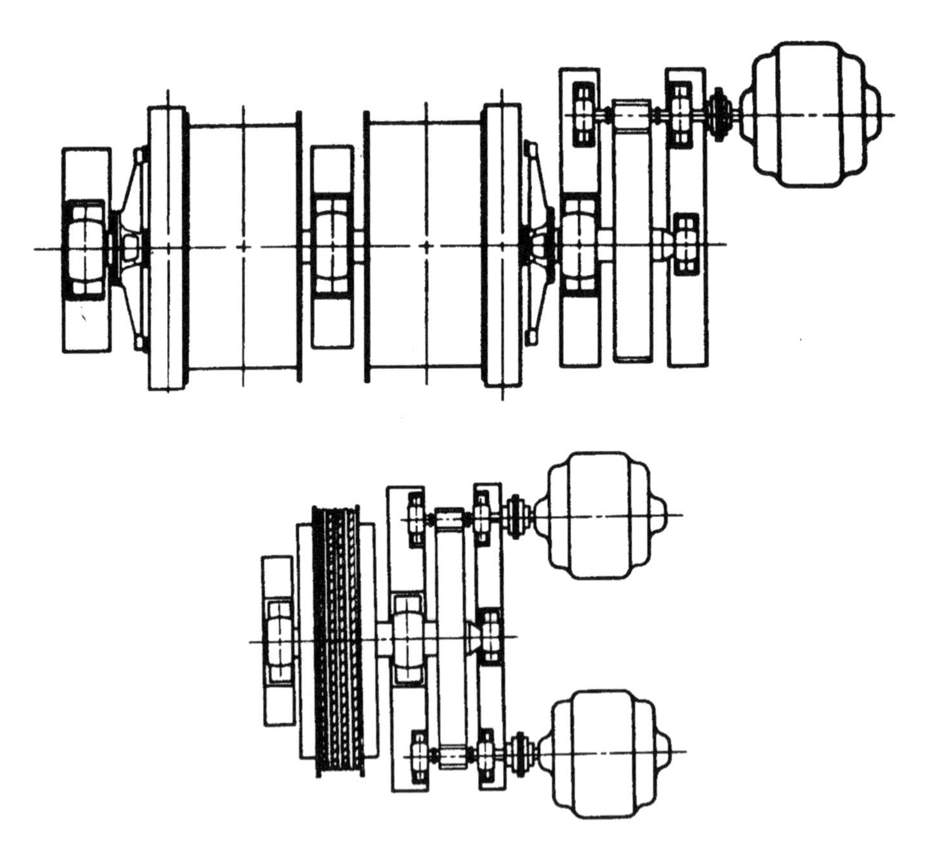

FIGURE 9.20. Diagrams of mine hoists. (*Top*) Double-clutched, double-drum hoist. (*Bottom*) Multirope friction-sheave hoist. (After Russell, 1982. By permission from the Society for Mining, Metallurgy, and Exploration, Inc., Littleton, CO.)

multidrum Blair hoist was devised for deep mines in South Africa. The configurations for two Blair hoists are shown in Figures 9.21d and 9.21e. While there is no hard set of rules for selecting a type of hoist, Harmon (1973) and Russell (1982) have provided general applications (Table 9.2). In addition, Culp (2002) provides general guidelines. He makes note that technical criteria are often overshadowed by regional biases. This leads one country to prefer a friction hoist for the same application where another would choose a drum hoist. Here is Culp's summary of common hoist configurations and their most common applications.

9.5.2.1 Single-Drum, Unbalanced Hoist. A single-drum hoist is the simplest type, with one rope and one conveyance (Fig. 9.21a). Because there is no balancing load, it requires a very high ratio of power to capacity. It is the most flexible configuration because there is no consideration required for the position of a balancing conveyance. The most common applications are shaft-sinking hoists, auxiliary hoists (for escape and other low-utilization applications), and shallow service hoists.

9.5.2.2 Single-Drum, Counterweighted Hoist. When configured with two ropes, winding in opposite directions so that one conveyance ascends while the other descends, the conveyances balance each other and the power requirements decrease significantly (Fig. 9.21b). This configuration is also known as an "over/under" hoist. The balanced loads decrease power requirements because one conveyance offsets the other so that power is required only to lift

Table 9.2 Applications of Hoisting Methods

	Drum	Friction Sheave	Multidrum
Optimum depth, ft (km)	<6000 (1.8)	<3000 (0.9)	>6000 (1.8)
Maximum skip capacity,[a] tons (tonnes)	28 (25)	85 (77)	56 (51)
Maximum output,[a] tph (tonnes/hr)	900 (820)	2800 (2540)	1800 (1630)
Features	Single rope Multilevel Medium depth Wide use in U.S.	Multirope Single level Limited depth High production Best efficiency Wide use outside U.S.	Multirope Great depth

[a] Skip capacity based on MSHA prescribed factors of safety and maximum rope loads.

Sources: Harmon, 1973; Russell, 1982. By permission of the Society for Mining, Metallurgy, and Exploration, Inc., Littleton, CO.

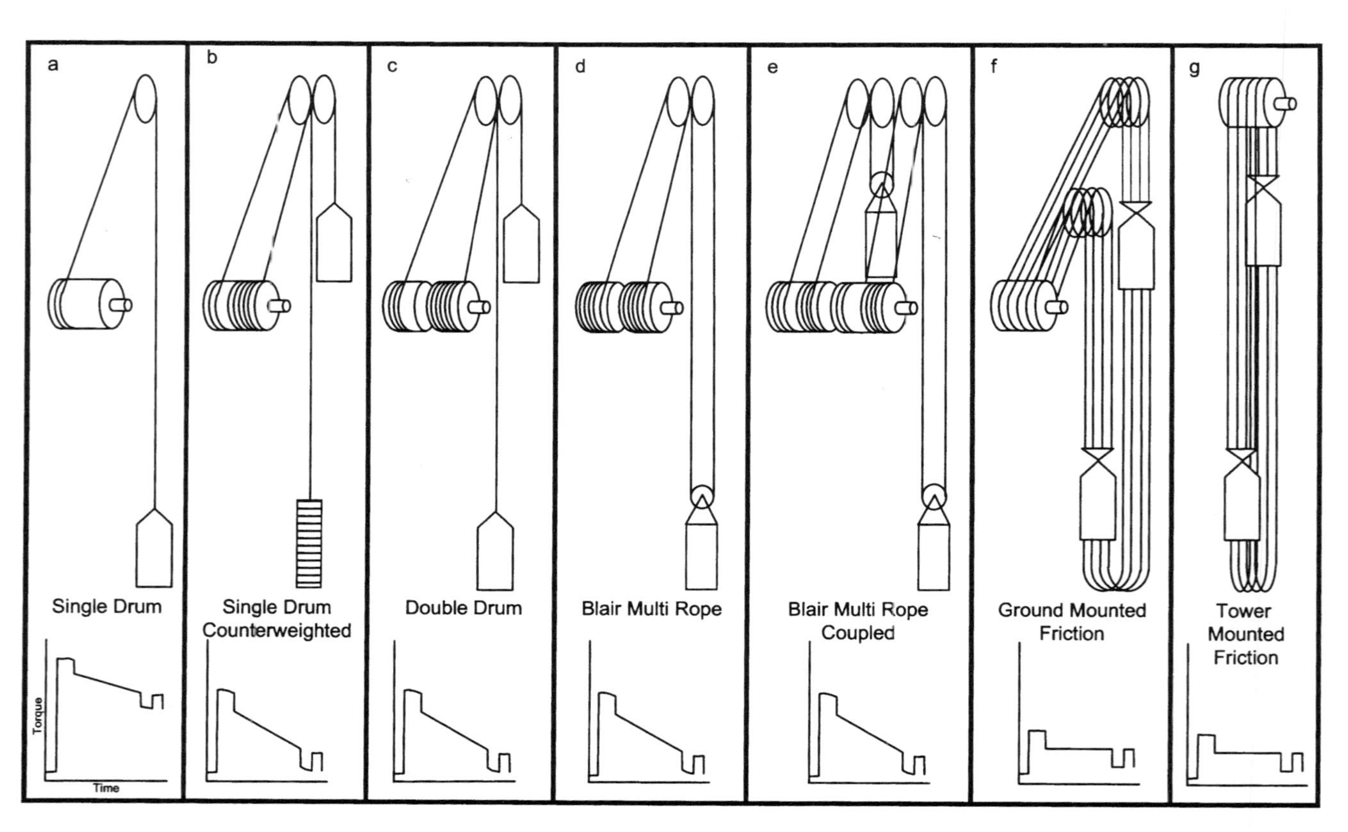

FIGURE 9.21. Hoist arrangements and duty cycles for common hoisting systems. (Culp, 2002. By permission from ASEA, Inc., Montreal, Canada.)

the payload and rope, but not the conveyance. The major disadvantage is that the rope lengths are fixed, requiring coordination between both conveyance positions. For example, if one conveyance is at the top in the dumping position, the other must be in the loading position at the bottom. Also, as the rope stretches, especially when new, it must be adjusted in length to take out the stretch so that the conveyances simultaneously land at their correct position. A common configuration uses one conveyance and a counterweight like a building elevator. This eliminates the need to precisely coordinate two conveyance positions. On a single-drum hoist, two skips can counterbalance one another, but the configuration is practical for only one hoisting level.

9.5.2.3 Double-Drum Hoist. The disadvantages of a single-drum over/under hoist are eliminated with a double-drum configuration (Fig. 9.21c). Here the ropes are wound on separate drums sharing one drive shaft. By setting the brakes on one drum and disengaging a clutch, the drums can be disconnected. A single-clutch hoist will have one fixed drum and one winding drum, while a double-clutch hoist can fix or wind either drum. This simplifies realigning the conveyances after rope stretch. Double-drum hoists are ideal for hoisting from multiple levels.

9.5.2.4 Blair Multirope Hoists. While traditional drum hoists can lift loads from very deep depths ($>$10,000 feet or 3000 m), the limitation is that each conveyance is held by only one rope. As the hoisting distance increases, more of the rope load capacity is used simply to hold or lift the rope. As a result, at 10,000 feet (3000 m) of depth, very little rope strength can be applied to lifting the payload. A Blair hoist solves this problem by looping a single rope around a sheave wheel at the conveyance attachment and bringing it back up the shaft thereby doubling the load capacity (Figs. 9.21d and 9.21e). There are many ways to attach the other end of the rope. It may be "blocked" or fixed in place at the top so that the hoisting speed is half the winding speed. It may also be wound on the second drum in the same direction as the other side so that they both ascend or descend together. This increases the hoisting speed to the full drum speed and results in an unbalanced load. The load can be balanced by adding a second pair of drums linked either electrically or mechanically to the first set (Fig 9.21 e). Blair hoists are not currently used in the United States, but have applications elsewhere, particularly in South Africa.

9.5.2.5 Friction Hoists. Friction, or Koepe hoists make use of the fact that loads in balance often generate sufficient rope pressure on the sheave to be driven by friction. This allows multiple smaller ropes to share the load (Figs. 9.21f and 9.21g). Following the rules for rope diameter-to-drum diameter, smaller ropes decrease the size of the drum required, the required torque and power, and consequently, manufacturing costs. Additionally, since friction hoists are not subject to the squeezing forces of a drum hoist, the drum weight can be smaller. Friction hoist costs are typically 60% of the cost of a drum hoist for the same hoisting capacity. The number of ropes is usually four or six, but there are hoists with as few as one or as many as ten ropes.

As a drum hoist begins to hoist a load from the bottom of the shaft, the rope load is all on one side resulting in large starting torques. Because friction hoist conveyance positions are fixed relative to each other, tail ropes can be used to counterbalance the rope loads throughout the hoisting cycle. Thus the load balance remains constant throughout the shaft and starting torque is much lower. Friction hoists, therefore, require a smaller motor to hoist the same load reducing both capital and operational costs.

The disadvantages of friction hoists are that conveyance positions are fixed relative to each other, similar to a balanced single-drum system. Therefore, they are best used in either a conveyance and counterweight configuration for multilevel mines or with dual conveyances for single-level mines. Secondly, the tension between the multiple ropes needs to be kept nearly equal to reduce wear. As rope tension differences increase with the number of ropes employed, this reduces the practical hoisting depth to about 5000 feet (1500 m). Lastly, the payload must not be more than 0.5 times the weight of the ropes and conveyance (on one side of the system) or the rope may slip on the winder. For example, if the rope, tail rope and empty conveyance on each side weighs 50,000 lb. (22,700 kg) total, the payload is limited to 25,000 lb. (11,350 kg). Another way of expressing this is that the tension in the rope(s) hoisting the load may not be more than 1.5 times the tension in the rope(s) supporting the empty conveyance. This means that in shafts shallower than about 750 feet (230 m), where there is less rope load, payloads are limited. Dead weight can be added to both sides to increase the allowable payload, but at an increased cost.

9.5.2.6 Performance Characteristics. With their different principles, shapes, and numbers and arrangements of ropes, the various hoisting systems exhibit quite different performance characteristics. In comparing operating costs, the parameter that determines energy consumption, and hence cost, is the *duty cycle* of the hoist. This is a plot of the instantaneous power requirements versus hoisting time. In Figure 9.21, rope-velocity diagrams and duty cycles for the principal hoisting systems are shown. The main differences are associated with drum shape and the use or absence of a tail rope.

Usually the most attractive system for high-speed, high-production, low-energy skip hoisting from a single level at moderate depths is the friction-sheave method. Using counterweights, it is also adaptable to multilevel hoisting, and with ground mounting of the sheaves can be used at depths to 5000 ft (1.5 km) or more. Details of the installation of a friction-sheave hoist and its hoist room and headframe appear in Figure 9.22.

9.6 SPECIAL TOPIC: DESIGN OF HOISTING SYSTEM

The design process for a mine hoisting system should be understood by the mining engineer, even though the design and installation are contracted to an

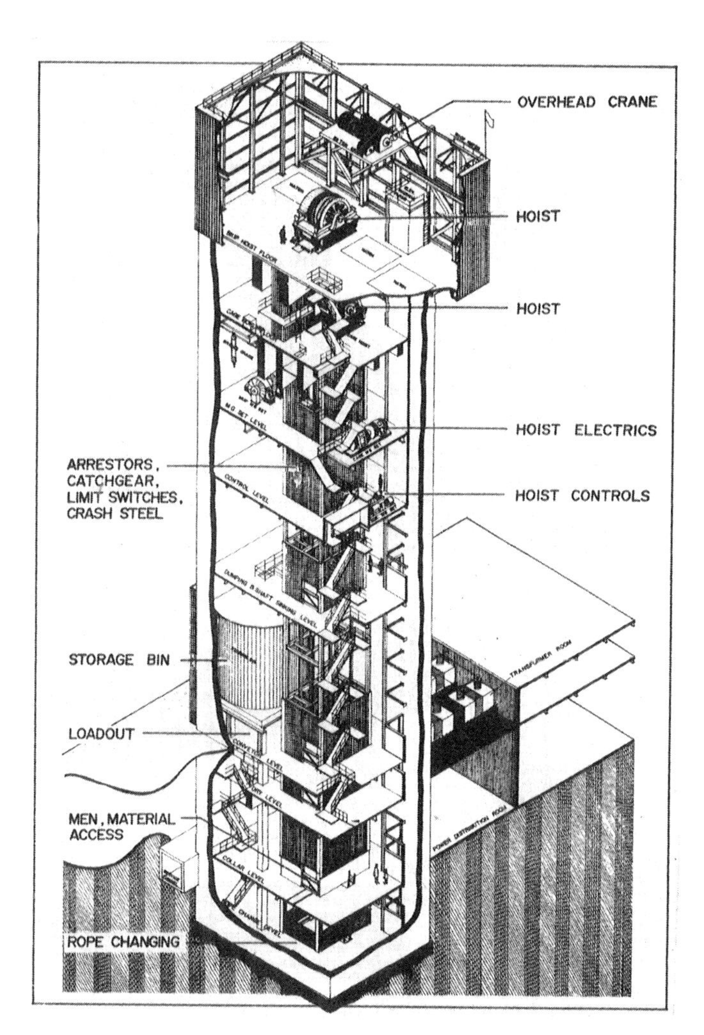

FIGURE 9.22. Multirope, friction-sheave hoisting system, mounted in headframe. (After Butler and Schneyderburg, 1982. By permission from the National Mining Association, Washington, DC.)

engineer-constructor firm and the equipment bid to a hoist manufacturer. Typically, the mining company developing the mine assigns its own engineering department to monitor the entire process, including both the planning and construction of the surface hoist plant. The design process is examined in detail and illustrated by the following example. The design procedure, equations, and discussion are taken in large part from Harmon (1973).

1. *Balanced hoisting.* All mine hoisting systems are operated in balance to reduce moments, torque, and power demand on the hoist. Generally, two conveyances (skips and/or cages) are suspended from one hoist; sometimes,

when more than one level is to be serviced, a counterweight replaces one conveyance. It is designed with a weight equal to the dead load of the skip or cage plus one-half the live load. To further balance the loads, a tail rope can be installed; although rare with a drum hoist, it is normally used with a friction-sheave hoist to reduce slippage as well as moments.

2. *Slippage in friction sheave hoisting.* Slippage occurs in a friction-sheave hoist if the ratio of the rope tensions exceeds a theoretical limit (Figure 9.23). The relationship is

$$\frac{T_1}{T_2} \leqslant e^{\mu\theta} \tag{9.1}$$

where T is rope tension, the subscripts 1 and 2 refer to the loaded and empty skips, respectively, e is the natural logarithmic base, μ is coefficient of friction = 0.45–0.50, and θ is angle of wrap from π radians (180°) for a headframe-mounted hoist to 1.3π radians (240°) for a ground-mounted hoist. In Eq. 9.1, the limiting ratio is 1.5–1.6 for a headframe-mounted hoist and 1.8–1.9 for a ground-mounted hoist. To increase θ, a deflection sheave can be used. To increase μ, the sheave lining or rope lubricant can be varied.

3. *Wire rope size.* Wire rope has a complex structure; several of the more common types of hoist rope are shown in Figure 9.24. Basically, there are three

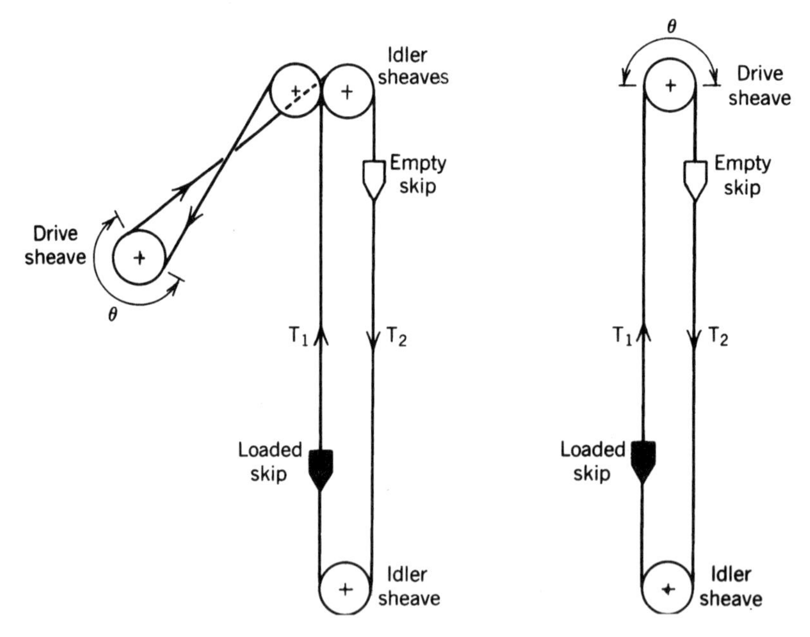

FIGURE 9.23. Analysis of friction-sheave hoist to determine whether rope slippage occurs. (*Left*) Ground-mounted hoist. (*Right*) Headframe-mounted hoist.

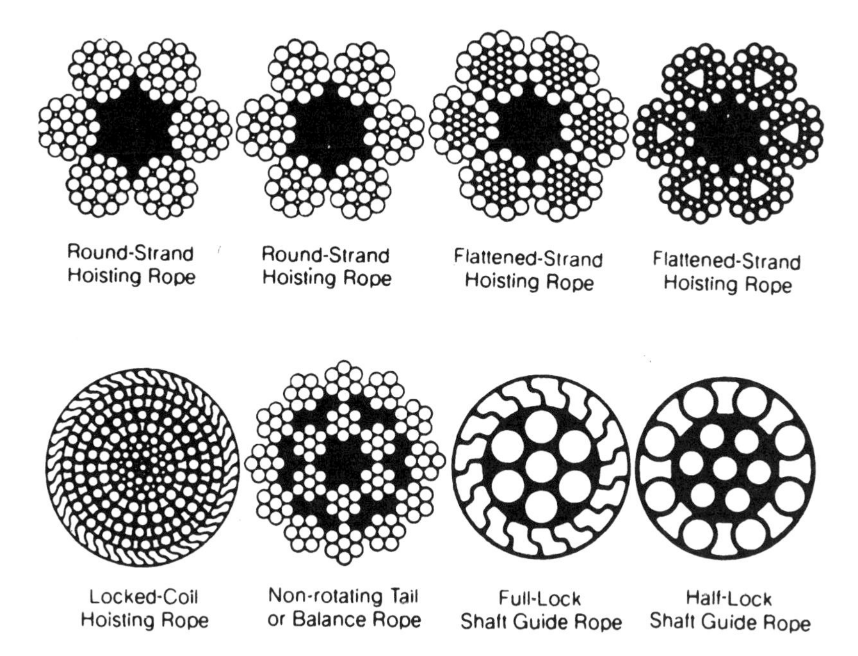

FIGURE 9.24. Construction of wire rope. The three types most commonly used in mine hoisting are round-strand, flattened-strand, and locked coil. (By permission from SIEMAG-Nordberg Hoisting Technology, Milwaukee, WI.)

types used for mine hoisting: round-strand, flattened-strand, and locked-coil. In general, round-strand rope is used with drum hoists, flattened-strand with friction-sheave hoists, and locked-coil for any system at depths over 3200 ft (0.96 km).

In designing a hoisting system, the two properties of wire rope that are most important are weight per unit length and breaking strength. These properties are given for the three popular hoist ropes in Table 9.3. Note that properties for two qualities of steel are included for round-strand and flattened-strand rope.

In selecting the proper size of wire rope for a hoisting application, the most critical consideration is the factor of safety (Figure 9.25). Values in the United States are established by the Mine Safety and Health Administration (MSHA) and vary with the type of hoist, depth, and whether personnel are being hoisted; the stepped curve gives the minimum allowable values to hoist personnel. Standards set by the American National Standards Institute (ANSI) are the maximum load for personnel or materiel. Because the weight of the rope must be taken into account in determining the total load on the hoist, the design process in selecting rope size becomes one of trial and error. Eventually, depth imposes a limit in single-lift hoisting, because rope weight increases with depth, exceeding rope strength at some critical depth.

Table 9.3 Properties of Hoisting Wire Rope

	Round Strand 6 × 19 Class Improved Plow Steel			Flattened Strand 6 × 27 Type H, 6 × 30 Type G			Locked Coil Improved Plow Steel	
		Breaking Strength, tons (tonnes)			Breaking Strength, tons (tonnes)			
Diameter in. (mm)	Weight, lb/ft (kg/m)	Normal	High-Strength	Weight, lb/ft (kg/m)	Normal	High-Strength	Weight, lb/ft (kg/m)	Breaking Strength, tons (tonnes)
$\frac{3}{4}$ (19.0)	0.95 (1.41)	23.8 (21.6)	26.2 (23.8)	1.01 (1.50)	26.2 (23.8)	28.8 (26.1)	1.37 (2.04)	35.0 (31.8)
$\frac{7}{8}$ (22.2)	1.29 (1.92)	32.2 (29.2)	35.4 (32.1)	1.39 (2.07)	35.4 (32.1)	39.0 (35.4)	1.87 (2.78)	46.0 (41.7)
1 (25.4)	1.68 (2.50)	41.8 (37.9)	46.0 (41.7)	1.80 (2.68)	46.0 (41.7)	50.6 (45.9)	2.43 (3.62)	61.6 (55.9)
$1\frac{1}{8}$ (28.6)	2.13 (3.17)	52.6 (47.7)	57.9 (52.5)	2.28 (3.39)	57.9 (53.5)	63.7 (57.8)	3.30 (4.91)	76.1 (69.0)
$1\frac{1}{4}$ (31.8)	2.63 (3.91)	64.6 (58.6)	71.0 (64.4)	2.81 (4.18)	71.0 (64.4)	78.1 (70.9)	3.75 (5.58)	92.0 (83.5)
$1\frac{3}{8}$ (34.9)	3.18 (4.73)	77.7 (70.5)	85.4 (77.5)	3.40 (5.06)	85.5 (77.6)	94.0 (85.3)	4.78 (7.11)	115 (104)
$1\frac{1}{2}$ (38.1)	3.78 (5.63)	92.0 (83.5)	101 (91.6)	4.05 (6.03)	101 (91.6)	111 (101)	5.65 (8.41)	135 (122)
$1\frac{5}{8}$ (41.3)	4.44 (6.61)	107 (97.1)	118 (107)	4.75 (7.07)	118 (107)	130 (118)	6.88 (10.24)	155 (141)
$1\frac{3}{4}$ (44.4)	5.15 (7.66)	124 (112)	136 (123)	5.51 (8.20)	136 (123)	150 (136)	7.56 (11.25)	182 (165)
$1\frac{7}{8}$ (45.3)	5.91 (8.80)	141 (128)	155 (141)	6.33 (9.42)	176 (160)	191 (173)	9.00 (13.39)	212 (192)
2 (50.8)	6.72 (10.00)	160 (145)	176 (160)	7.20 (10.71)		215 (195)	9.77 (14.54)	240 (218)
$2\frac{1}{8}$ (54.0)	7.59 (11.30)	179 (162)	197 (179)	8.13 (12.10)		240 (218)		
$2\frac{1}{4}$ (63.5)	8.51 (12.66)	200 (181)	220 (200)					

Source: Russell, 1982. By permission of the Society for Mining, Metallurgy, and Exploration, Inc., Littleton, CO.

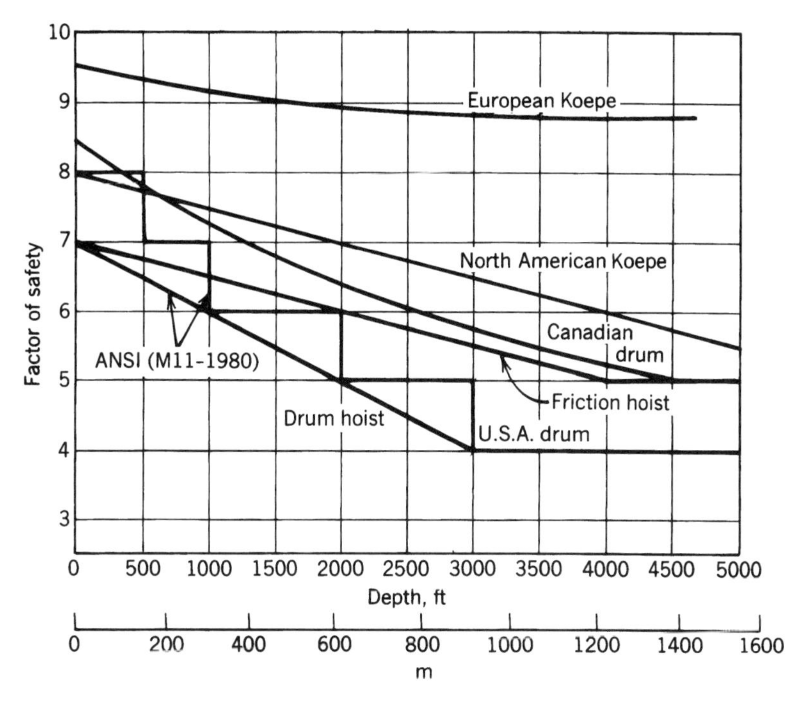

FIGURE 9.25. Required factors of safety for wire rope in mine hoisting. (By permission from SIEMAG-Nordberg Hoisting Technology, Milwaukee, WI.)

4. *Sheave and drum diameter.* To minimize flexing and stressing of the wire rope as it is wound over a sheave or drum, a recommended minimum ratio of drum or sheave diameter to rope diameter should be observed (Figure 9.26). Because the cost of wire rope is modest, there may be occasions, especially in shallow shafts, when it is more cost-effective to select a smaller drum diameter and replace the hoist rope oftener.

5. *Rope fleet angle.* This is the angle subtended by the hoist rope and the centerline from the idler sheave to the drum. To reduce rope abrasion in the sheave groove, the fleet angle is restricted to 1 to $1\frac{1}{2}°$. The principal effect of such a limit is to restrict the width of the drum.

6. *Skip size versus hoisting velocity.* To achieve a desired production rate in a shaft, the design engineer seeks a balance between skip size and hoisting velocity. The ultimate limit on skip size is rope strength, and on hoisting velocity it is energy consumption. As a compromise, it is generally advantageous to hoist the largest skip load possible at the lowest possible rope velocity.

7. *Hoisting cycle.* The relationship of time to distance in hoisting is referred to as the *hoisting cycle.* Calculation of time and distance elements is accomplished with the following formulas:

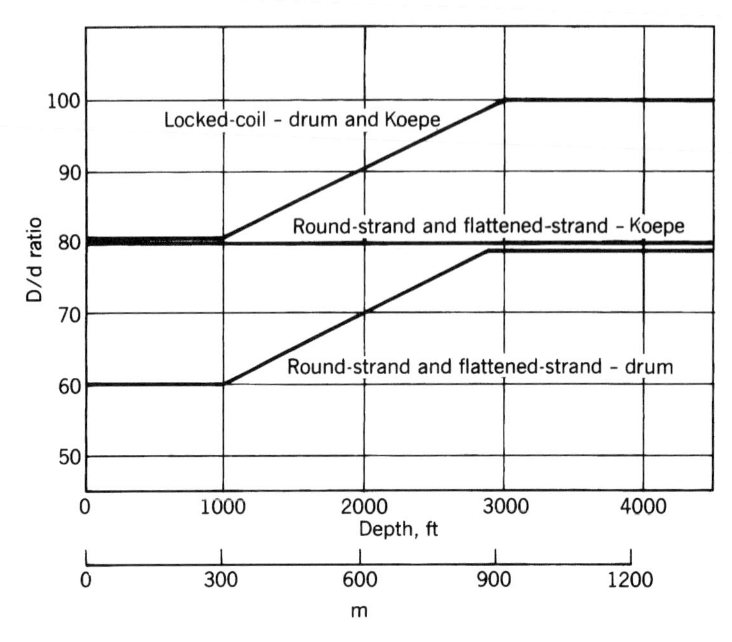

FIGURE 9.26. Suggested minimum ratio of drum-to-rope diameter for varying hoisting depths. (By permission from SIEMAG-Nordberg Hoisting Technology, Milwaukee, WI.)

a. *Acceleration time* $t_a = \dfrac{V}{a}$ (9.2)

where V is hoisting velocity and a is acceleration rate; generally, acceleration rate and time equal retardation rate r and time t_r, respectively.

b. *Acceleration distance* $h_a = \frac{1}{2}at_a^2$ (9.3)

$= \text{retardation distance } h_r$

c. *Constant-velocity distance* $h_v = h_t - h_a - h_r$ (9.4)

where h_t is total hoisting distance (from loading pocket to headframe bin).

d. *Constant-velocity time* $t_v = \dfrac{h_v}{V}$ (9.5)

e. *Cycle time* $t_t = 2(t_a + t_v + t_r + t_d)$ (9.6)

per round trip, where t_d is load or dump time.

8. *Duty cycle.* The relationship between hoist motor power requirements and hoisting cycle times is called the *duty cycle.* Plots of the duty cycle for a drum hoist and a friction-sheave hoist are compared in Figure 9.27. The

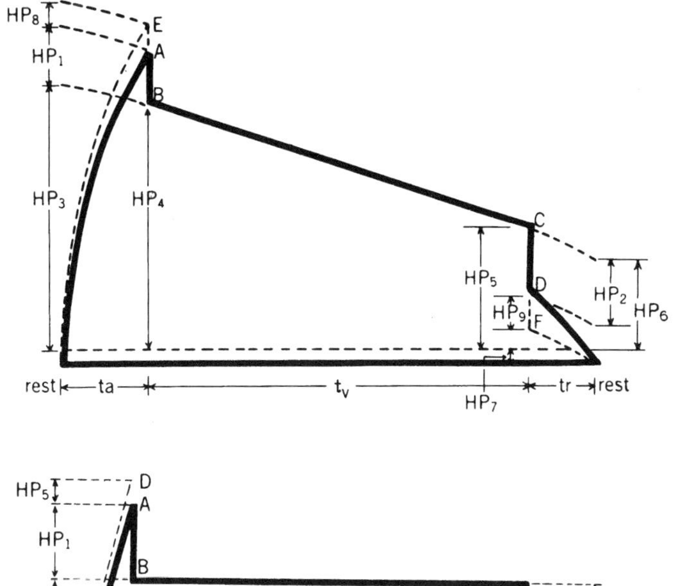

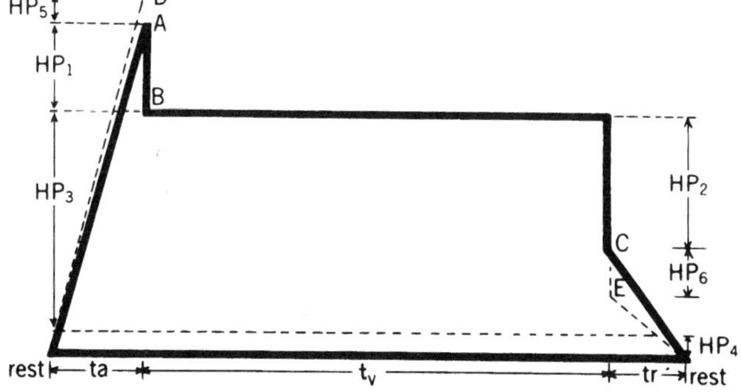

FIGURE 9.27. Plots of duty cycles, power versus time, for (*top*) a drum hoist and (*bottom*) a friction-sheave hoist, (After Harmon, 1973. By permission from the Society for Mining, Metallurgy, and Exploration, Inc., Littleton, CO.)

sloping section of the drum duty hoist reflects the unbalanced load of the hoist rope. Integrating the area under the curve provides the energy consumption for the hoisting-duty cycles. The following calculations enable the key points on the duty cycle to be determined for a friction-sheave hoist (the power equations differ for a drum hoist; see Harmon, 1973):

a. *Rope weight* $W_r = w_r(h_t + h_h)$ (9.7)

where w_r is rope weight per unit length and h_h is distance from bin to idler or drive sheave at the apex of the headframe; if multiple ropes are used, multiply by number of ropes.

b. *Total weight of load* $W_l = W_r + W_s + W_o$ (9.8)

where W_s is skip dead weight and W_o is skip live load.

c. *Design load* $L = \mathrm{FS} \times W_l$ (9.9)

where FS is factor of safety.

d. *Rope strength* $S \geqslant L$ (9.10)

e. *Equivalent effective weight* W_e of the rotating equipment, reduced to the rope center for different drum diameters (read value from Figure 9.28).

f. *Total suspended load* $W = W_e + W_o + 2W_s + 2W_r$ (9.11)

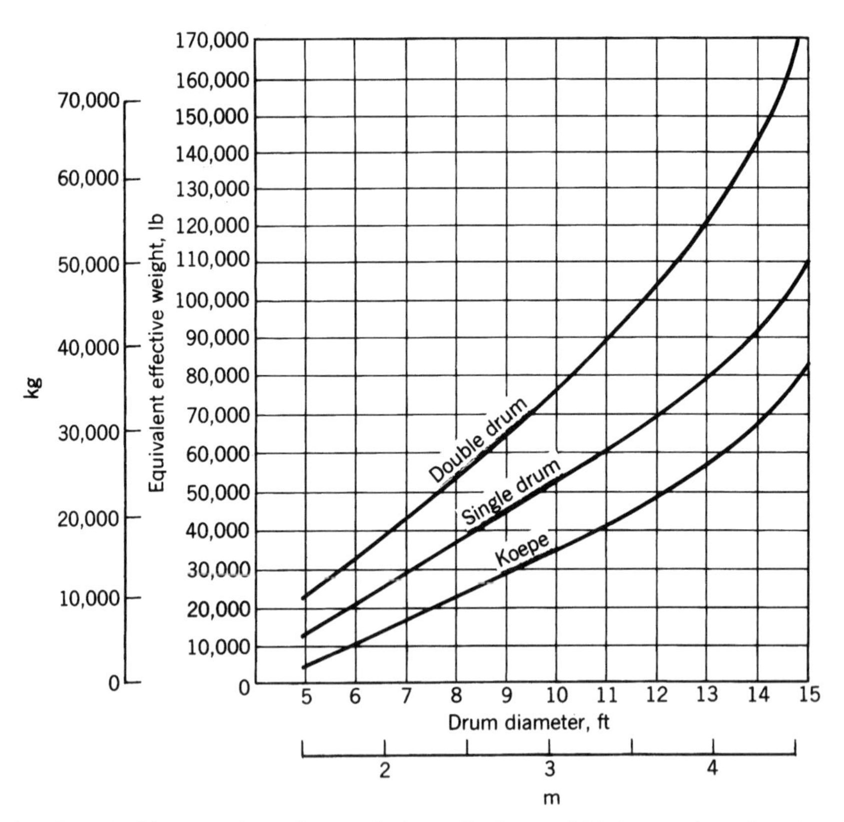

FIGURE 9.28. Chart to determine equivalent effective weight for varying diameter of drum or friction sheave. (By permission from SIEMAG-Nordberg Hoisting Technology, Milwaukee, WI.)

g. *Key points on duty cycle* (refer to Figure 9.27):

$$P_1 = \frac{WV^2}{550gt_a} = \frac{WV^2}{17{,}700t_a} \tag{9.12}$$

where P is power in hp, the subscript refers to a point on the duty cycle, and g is acceleration due to gravity = 32.2 ft/sec² (9.807 m/sec²).

$$P_2 = -\frac{WV^2}{17{,}700t_r} \tag{9.13}$$

$$P_3 = \frac{W_o V}{550} \tag{9.14}$$

$$P_4 = \frac{W_o V}{550}\left(\frac{1-\eta}{\eta}\right) \tag{9.15}$$

where η is hoist efficiency as a decimal; assume 90% for a friction-sheave hoist.

$$P_A = P_1 + P_3 + P_4 \tag{9.16}$$

$$P_B = P_3 + P_4 \tag{9.17}$$

$$P_C = P_2 + P_3 + P_4 \tag{9.18}$$

$$P_5 = \frac{1.2(0.75P_A)}{t_a} = \frac{0.9P_A}{t_a} \tag{9.19}$$

$$P_6 = -\frac{0.9P_A}{t_r} \tag{9.20}$$

$$P_D = P_A + P_5 \tag{9.21}$$

$$P_E = P_C + P_6 \tag{9.22}$$

h. Root-mean-square (rms) power for ac motor:

$$P_{\text{rms}} = \sqrt{\frac{P_D^2 t_a + P_B^2 t_v + P_E^2 t_r}{0.5t_a + t_v + 0.5t_r + 0.25t_d}} \text{ hp} \tag{9.23}$$

For a dc motor, the coefficients in the denominator are 0.75, 1, 0.75, and 0.5, respectively.

i. *Approximate energy consumption for duty cycle:*

$$E = \frac{0.7457P_B(t_a + t_v)}{3600\eta} \text{ kW-hr/trip} \tag{9.24}$$

9. *Auxiliary electrical and mechanical equipment.* For further particulars of hoist drums, their electric motor drives and control devices, and their associated mechanical equipment such as brakes and clutches, refer to Harmon (1973) or Russell (1982).

Example 9.1. A friction-sheave hoist operates two skips in balance under the following conditions:

Shaft depth	1000 ft (305 m)
Skip live load	5 tons (4.5 tonnes)
Skip dead-load/live-load ratio	1.2
Hoist ropes	4 @ 1 in. (25.4 mm), flattened strand, normal strength
Sheave diameter	10.3 ft (3.14 m)
Hoisting velocity	20 ft/sec (6.1 m/sec^2)
Hoisting cycle	$t_a = 10$ sec $t_v = 39.75$ sec $t_r = 8$ sec $t_d = 10$ sec
Hoist efficiency	90%

Check for rope slippage, and determine the rms rating of ac electric motor to drive the hoist and the approximate energy consumption per trip.

SOLUTION. Use Eq. 9.6 to find the cycle time:

$$t_t = 2(10 + 39.75 + 8 + 10) = 135.5 \text{ sec}$$

$$\text{Skip dead load } W_s = 1.2(5) = 6.0 \text{ tons} = 12{,}000 \text{ lb}$$

$$\text{Rope weight (Eq. 9.7) } W_r = 4(1.80)(1000) = 7200 \text{ lb}$$

using Table 9.3 to obtain the rope weight per unit length.

$$\text{Total weight of load (Eq. 9.8) } W_l = (7200 + 10{,}000 + 12{,}000) = 29{,}200 \text{ lb}$$

Check rope slippage (Eq. 9.1):

$$\text{Loaded skip } T_1 = W_l = 29{,}200 \text{ lb}$$

$$\text{Empty skip } T_2 = W_s + W_r = 12{,}000 + 7200 = 19{,}200 \text{ lb}$$

$$\frac{T_1}{T_2} = \frac{29{,}200}{19{,}200} = 1.52 < 1.6$$

which is satisfactory for a headframe-mounted hoist.

For sheave diameter = 10.3 ft, read equivalent effective weight (Figure 9.28) $W_e = 37{,}000\,\text{lb}$.

Total suspended load (Eq. 9.11) $W = 37{,}000 + 10{,}000 + 24{,}000 + 14{,}400$

$$= 85{,}400\,\text{lb}$$

Calculate points on duty cycle (Eqs. 9.12–22):

$$P_1 = \frac{(85{,}400)(20)^2}{17{,}700(10)} = 193 \text{ hp}$$

$$P_2 = -\frac{(85{,}400)(20)^2}{17{,}700(8)} = -241 \text{ hp}$$

$$P_3 = \frac{(10{,}000)(20)}{550} = 364 \text{ hp}$$

$$P_4 = \frac{(10{,}000)(20)}{550}\left(\frac{1 - 0.90}{0.90}\right) = 40 \text{ hp}$$

$$P_A = 193 + 364 + 40 = 597 \text{ hp}$$

$$P_B = 364 + 40 = 404 \text{ hp}$$

$$P_C = -241 + 364 + 40 = 163\,\text{hp}$$

$$P_5 = \frac{0.9(597)}{10} = 54 \text{ hp}$$

$$P_6 = -\frac{0.9(597)}{8} = -67 \text{ hp}$$

$$P_D = 597 + 54 = 651 \text{ hp}$$

$$P_E = 163 - 67 = 96\,\text{hp}$$

Using these values, the duty cycle can be plotted, as in Figure 9.27. Calculate rms power for an ac motor (Eq. 9.23):

$$P_{\text{rms}} = \sqrt{\frac{(651)^2(10) + (404)^2(39.75) + (96)^2(8)}{0.5(10) + 39.75 + 0.5(8) + 0.25(10)}} = 459 \text{ hp (342 kW)}$$

Calculate approximate energy consumption (Eq. 9.24):

$$E = \frac{0.7457(404)(10 + 39.75)}{3600(0.90)} = 4.63 \text{ kW-hr/trip}$$

PROBLEMS

9.1 (a) Calculate the daily production of a shaft equipped with a balanced friction-sheave hoisting system, given the following:

Shift time	7.2 hr
Shifts/day	3
Skip capacity	12 tons (11 tonnes)
Cycle time, 1 trip	85 sec/skip

(b) Also calculate the approximate energy consumption per skip hoisted, if the average power consumed is 975 hp (730 kW), the hoist efficiency is 85%, the acceleration time is 6.5 sec, and the constant velocity time is 63.5 sec.

9.2 (a) A hoisting cycle consists of the following elements:

Constant-speed travel time	80 sec
Acceleration = retardation time	6 sec
Load = dump time	8 sec
Shift time	7.5 hr
Shifts/day	3
Skip capacity	12 tons (11 tonnes)

For a balanced hoisting system, calculate the expected daily production from the shaft.
(b) Also calculate the approximate energy consumption per skip hoisted if the average power consumed is 1200 hp (800 kW) and the hoist efficiency is 87%.

9.3 Select the proper size of friction-sheave hoist, calculate the hoisting and duty cycles, and determine the approximate unit energy consumption and cost for the following hoisting system:

Two balanced skips and cages, sheave in headframe

Production rate	6300 tons/day (5715 tonnes/day)
Working time	3 shifts/day, 7 hr/shift
Shaft depth (1 level)	2250 ft (686 m)
Headframe height	100 ft (30 m)
Bin height	50 ft (15 m)
Hoist rope	4 round strand
Skip + cage dead load	0.67 live load (skip capacity)
Load time = dump time	6.0 sec
Hoisting speed	2250 ft/min (11.43 m/sec)
Acceleration = retardation	7.5 ft/sec^2 (2.29 m/sec^2)
Power cost	5¢/kW-hr

Determine the following design elements:

- Components and total cycle time (1 round trip), sec
- Skip capacity, tons (tonnes) (to nearest 0.1 ton or tonne)
- Rope size, in. (mm) (standard size, normal strength)
- Rope slippage (if unsafe, recommend design change but do not recalculate)
- Sheave diameter, ft (m) (to nearest ft or m)
- Hoist duty cycle (plot of hp or kW vs. time)
- rms power rating of ac hoist motor, hp (kW)
- Unit energy consumption, kW-hr/trip and kW-hr/ton (kW-hr/tonne) (approximate)
- Unit energy cost, ¢/ton (¢/tonne) (approximate)

Compare with a drum hoist requiring a 1650 hp (1230 kW) motor, 3 kW-hr/ton (3.3 kW-hr/tonne), and 15¢/ton (16.5¢/tonne), and explain the difference.

9.4 Select the proper size of friction-sheave hoist, calculate the hoisting and duty cycles, and determine the approximate unit energy consumption and cost for the following hoisting system:

Two balanced skips and cages, sheave in headframe	
Production rate	7500 tons/day (6804 tonnes/day)
Working time	3 shifts/day, 7.5 hr/shift
Shaft depth (1 level)	1850 ft (564 m)
Headframe height	150 ft (46 m)
Bin height	100 ft (30 m)
Hoist rope	4 flattened strand
Skip + cage dead load	0.8 live load (skip capacity)
Load time = dump time	8.0 sec
Hoisting speed	1800 ft/min (9.14 m/sec)
Acceleration = retardation	6.0 ft/sec^2 (1.83 m/sec^2)
Power cost	4¢/kW-hr

Determine the following design elements:

- Components and total cycle time (1 round trip), sec
- Skip capacity, tons (tonnes) (to nearest 0.1 ton or tonne)
- Rope size, in. (mm) (standard size, normal strength)
- Rope slippage (if unsafe, recommend design change, but do not recalculate

Sheave diameter, ft (m) (to nearest ft or m)
Hoist duty cycle (plot of hp or kW vs. time)
rms power rating of ac hoist motor, hp (kW)
Unit energy consumption, kW-hr/trip and kW-hr/ton (kW-hr/tonne) (approximate)
Unit energy cost, ¢/ton (¢/tonne) (approximate)

Compare with a drum hoist requiring a 1650 hp (1230 kW) motor, 3 kW-hr/ton (3.3 kW-hr/tonne) and 15¢/ton (16.5¢/tonne), and explain the difference.

10

UNDERGROUND MINING: UNSUPPORTED METHODS

10.1 CLASSIFICATION OF METHODS

Mineral exploitation in which all extraction is carried out beneath the earth's surface is termed *underground mining*. Underground methods are employed when the depth of the deposit, the stripping ratio of overburden to ore (or coal or stone), or both become excessive for surface exploitation. Once economic analysis points to underground methods, the choice of a proper mining procedure hinges mainly on (1) determining the appropriate form of ground support, if necessary, or its absence; and (2) designing the openings and their sequence of extraction to conform to the spatial characteristics of the mineral deposit.

Choice of an underground mining method is often closely related to the geology of the deposit and the degree of ground support necessary to make the methods productive and safe. Generally, three classes of methods are recognized—unsupported, supported, and caving—based on the extent of support utilized (see Table 4.1). The *unsupported* class, the subject of this chapter, consists of the methods in which the rock is essentially self-supporting and for which no major artificial support (such as artificially placed pillars or fill) is necessary to carry the load of the overlying rock. The weight of the overlying rock plus any tectonic forces is generally called the *superincumbent load*. This load will be too high in many rock masses, but for the unsupported methods, we assume that the geologic materials can sustain the load. The definition of unsupported methods does not preclude the use of roof bolts or light structural supports of timber or steel, provided that such support does not alter the load-carrying capacity of the natural rock.

Theoretically, the unsupported class of methods can be used in any type of mineral deposit (except placers, unless frozen in place) by varying the ratio of span of opening to the width of pillar. However, on a practical basis, such practice would ordinarily reach its limit on an economic basis rather quickly. Hence, the ratio of span to width of pillar is limited to those values that will also result in favorable productivity. The unsupported methods are therefore applicable to deposits with favorable characteristics. These methods, however, are still the most widely used underground, accounting for nearly 80% of the U.S. subsurface mineral production.

There is some disagreement on the best way to classify underground methods, as discussed in Section 4.8.3. For our purpose, we will employ the classification scheme outlined in Table 4.1. In that classification, the following are considered unsupported methods:

1. Room-and-pillar mining
2. Stope-and-pillar mining
3. Shrinkage stoping
4. Sublevel stoping

In underground mining, unlike surface mining, there is little distinction in the cycle of operations for the various underground methods (except in coal mining), the differences occurring in the direction of mining (vertical or horizontal), the ratio of opening-to-pillar dimensions, and the nature of the artificial support used, if any. Of the unsupported methods, room-and-pillar mining and stope-and-pillar mining employ horizontal openings, low opening-to-pillar ratios, and light-to-moderate support in all openings. Shrinkage and sublevel stoping utilize vertical or steeply inclined openings (and gravity for the flow of bulk materials), high opening-to-pillar ratios, and light support mainly in the development openings. We now turn to specific descriptions of the four unsupported methods.

10.2 ROOM-AND-PILLAR MINING

Room-and-pillar mining is a very old method applied to horizontal or nearly horizontal deposits that has been adapted and refined over the years. As indicated in Table 4.1, the method is widely used in both coal and noncoal mining. In room-and-pillar mining, openings are driven orthogonally and at regular intervals in a mineral deposit—forming rectangular or square pillars for natural support. When the deposit and method are both rather uniform, the appearance of the mine in plan view is not unlike a checkerboard or the intersecting streets and avenues of a city. As discussed in Sections 9.3 and 9.4, development openings (generally called entries) and production openings (called rooms) closely resemble each other; both are driven parallel and in multiple, and when connected by crosscuts, pillars are formed. Driving several openings at one time increases production and efficiency by providing multiple working places. In addition, it provides for better ventilation and transportation routes at the working faces.

By its very nature, room-and-pillar mining is ideally suited to the production of coal, potash, sodium chloride, trona, limestone, and any metallic deposits that occur in horizontal seams. At the time of the first edition of this book in 1987, room-and-pillar mining produced about 85% of all the coal mined underground in the United States. That percentage dropped to 52% in 1998 (Energy Information Administration, 2000) because of the growth of longwall at the expense of the room-and-pillar method. For other commodities, room-and-pillar remains strong, as it is an ideal horizontal mining method that can be adapted to productive practices in many situations.

A layout of a room-and-pillar mine used in the mining of bituminous coal is shown in Figure 10.1. In this case, a set of five main entries allows access to the production panel through panel entries. The entries in coal mining are limited to 20 ft (6 m) in width and are generally driven 60 to 100 ft (18.3 to 30.5 m) apart, center to center. The panel itself is normally 400 to 600 ft (60 to 120 m) in width, limited primarily by the cable reach capability of electric shuttle cars. The length of the panel varies, with 2000 to 4000 ft (600 to 1200 m)

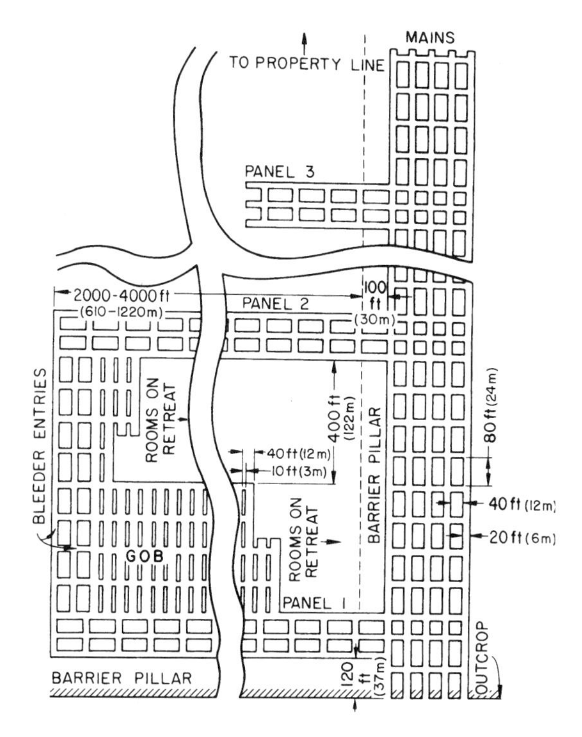

FIGURE 10.1. Room-and-pillar mining, driving rooms on the retreat without pillar recovery. (After Stefanko and Bise, 1983. By permission from the Society for Mining, Metallurgy, and Exploration, Inc., Littleton, CO.)

being common. In the case of Figure 10.1, note that the panel pillars are being mined. This is called *pillaring* or *caving*. The normal practice in pillaring is to drive the rooms and crosscuts on advance (first mining) and to pillar on retreat (second mining). The caved area then becomes known as the *gob*. Pillars are not removed if the surface must be supported; pillaring is often omitted for other reasons as well.

Note that in room-and-pillar mining of coal and other soft minerals, two common variations are practiced. So-called *conventional* mining is cyclical, employing mobile mechanized equipment to carry out the production unit operations. In underground coal mining, only about 5% of the coal is mined in this fashion (Energy Information Administration, 2000). With *continuous mining*, separate unit operations of drilling, cutting, blasting, and loading are replaced by a single high-performance continuous miner. Approximately 48% of all underground coal in the United States is produced by continuous mining (Energy Information Administration, 2000). Where room-and-pillar mining is practiced in hard-rock formations, only conventional mining is utilized, because of the inability of a continuous miner to cut harder mineral deposits.

Productivity in room-and-pillar mining has been interesting to follow. Prior to the enactment of the Coal Mine Health and Safety Act of 1969, productivity had risen steadily since World War II. It then declined rather precipitously (below 200 tons/shift or 180 tonnes/shift), following which it began rising again. Because of better equipment and management, productivity continues its upward ascent. Stefanko and Bise (1983) reported that productivity was about 500 tons/shift (460 tonnes/shift) during the last decade before their publication, but today most room-and-pillar mines produce 1000 to 2000 tons/shift (900 to 1800 tonnes/shift). In stone and metal mining, productivity also continues to increase. In this case, the improvements are due to better and larger equipment, more automation, and improved management practices.

Rock mechanics in room-and-pillar mining is another topic of great concern. Fortunately, much is known, and excellent tools are available to the mining engineer to control the roof and ribs in this method. Of particular interest, Stefanko and Bise (1983) and Farmer (1992) provide much in the way of background on the application of rock mechanics to room-and-pillar mining.

10.2.1 Sequence of Development

Chapter 9 covered general underground development in some detail. The procedure for development of flat-bedded deposits, to which room-and-pillar mining is applicable, can be found in Sections 9.2 to 9.4. A complete mine layout with all the major development openings is shown in Figure 9.4. Additional insights into the process of development can be gained from study of Figure 10.2.

The plan view shown in Figure 10.2 represents the section entries, room entries, and rooms associated with a production panel within the mine. The

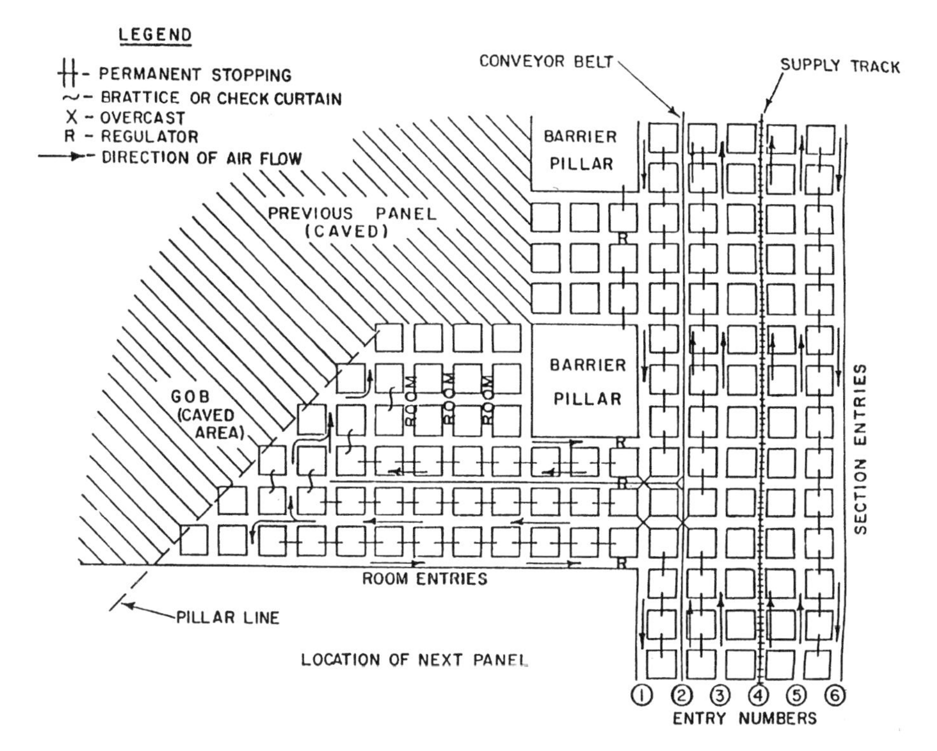

FIGURE 10.2. Room-and-pillar mining system showing all openings and the section pillar line.

room entries, room, and the associated crosscuts are mined on advance and the pillars are mined on the retreat. This mining plan is what was once called the Pittsburgh block section, common when room-and-pillar mining with electric shuttle cars was practiced in the Pittsburgh seam. This or similar systems are still utilized using continuous miners, but only if a longwall operation is not feasible. Note also the numbering on the entries in the section entry set. In coal mining, the entries in any set are numbered from left to right as one looks inby (toward the faces and away from the outside of the mine). Also note that if someone is looking at the face, that person is looking *inby*; but if turned in the opposite direction and facing toward the outside of the mine, that person is looking *outby*.

10.2.2 Cycle of Operations

10.2.2.1 Conventional Mining. The cycle of operations in room-and-pillar mining of coal with conventional equipment is modified from the basic cycle by insertion of the cutting operation to improve coal breakage during light blasting. This is accomplished as shown in Figure 10.3. The cut at the bottom of the face shown in Figure 10.3(a) is the most common. The general

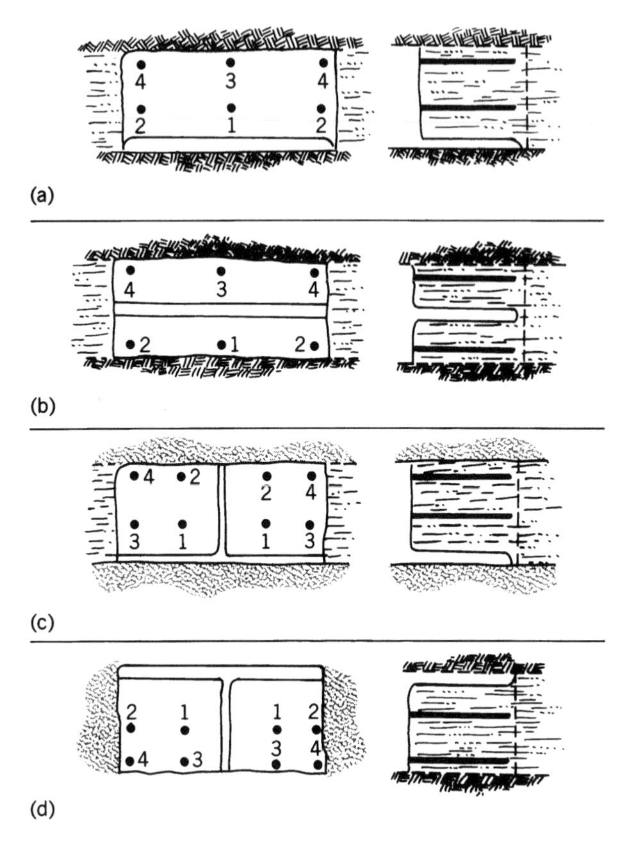

FIGURE 10.3. Drilling and cutting patterns for coal breakage. Holes are numbered in order of firing. (a) Undercutting. (b) Center cutting. (c) Undercutting and center shearing. (d) Top cutting and center shearing. (After Chironis, 1983. Copyright (c) 1983, McGraw-Hill, Inc., New York.)

production cycle then becomes

$$\text{Production cycle} = \text{cut} + \text{drill} + \text{blast} + \text{load} + \text{haul}$$

Several auxiliary operations (roof control, ventilation, and cleanup) must also be accomplished at the same time as these production tasks. To use this method effectively in coal mining, the mining section must have a multitude of faces so that the cutting machine operator, the driller, the blaster, and the loader can all work on separate faces without undue delays. To accomplish this, the mining section may be laid out as shown in Figure 10.4, where a seven-entry development section is illustrated. The numbers represent the sequence of faces that each crew member visits to perform his or her job. To improve efficiency of the operations, computer simulation is often used to optimize the system. Note that this method of coal mining is not used much

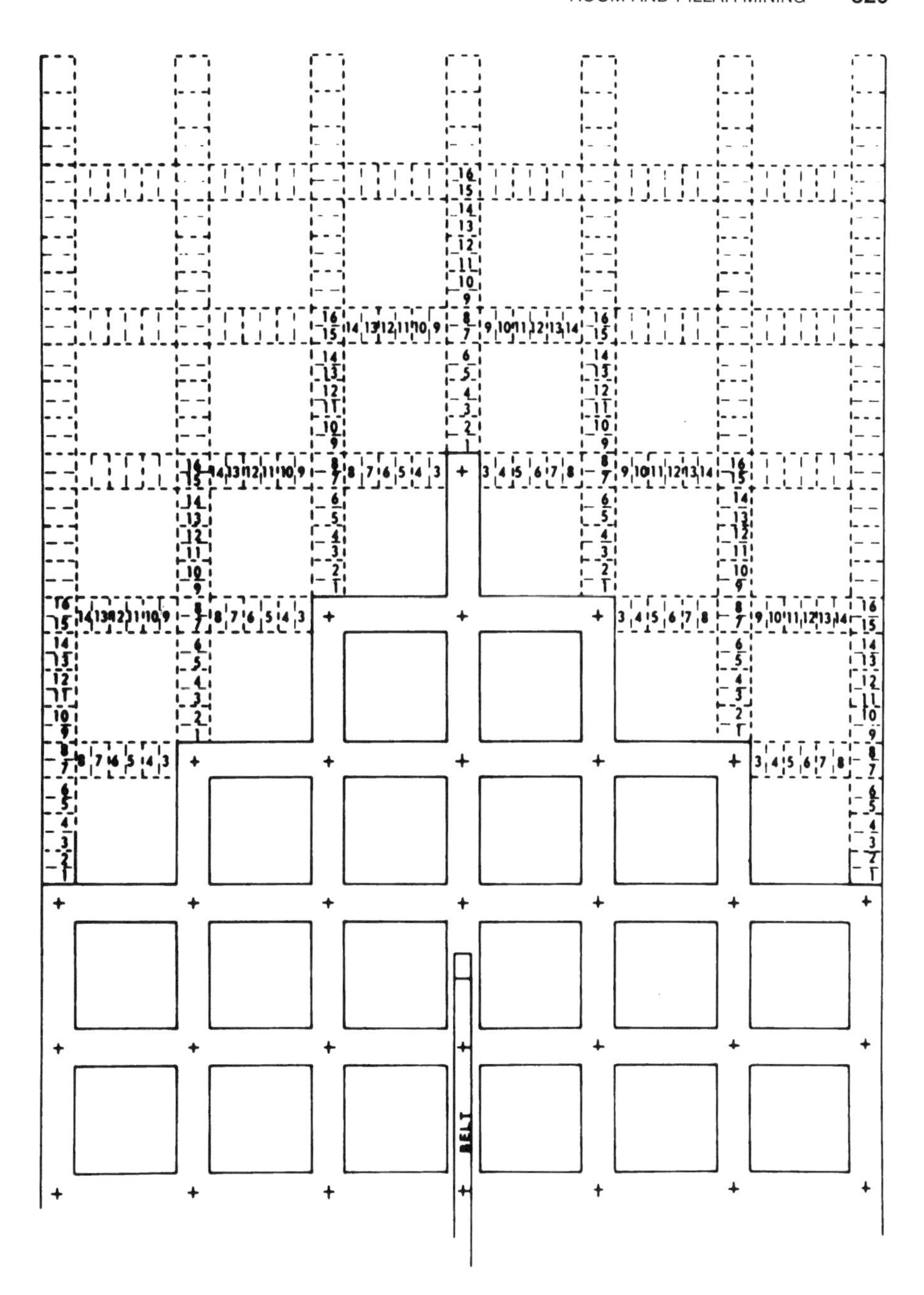

FIGURE 10.4. Cut sequence of continuous miner in seven-entry room-and-pillar mining. The echelon pattern advances the belt conveyor entry first. (After Bullock, 1982. By permission from the Society for Mining, Metallurgy, and Exploration, Inc., Littleton, CO.)

today; only about 4.5% of U.S. underground coal is exploited in this manner.

The use of conventional mining in noncoal operations is quite common, with many mines using the method in limestone, uranium, lead, zinc, sodium chloride, and other mineral commodities. The basic operations are primarily the same, with no cutting performed. The following information compares the typical kinds of equipment used:

Operation	Coal Mining	Noncoal Mining
Cutting	Cutting machine	Not performed
Drilling	Drag-bit rotary drill	Hydraulic percussion drill jumbo
Blasting	Permissible explosive	ANFO
Loading	Gathering-arm loader	Diesel front-end loader
Hauling	Shuttle car	Diesel truck or LHD

Room-and-pillar practice in noncoal mining is more variable than in coal. The mining is more often attempted in dipping seams (called *pitch mining*) and may also be performed in a coordinated manner in multiple seams. An excellent review of variations and the equipment used can be found in Bullock and Mann (1982) and Hustrulid and Bullock (2001).

10.2.2.2 Continuous Mining. Continuous mining is a method that has been practiced, primarily in coal, since about 1950. The method uses a simplified cycle of operations, as follows:

$$\text{Production cycle} = \text{mine} + \text{haul}$$

As in conventional mining, the auxiliary operations of roof control, ventilation, and cleanup must also be performed. The method depends heavily on the ruggedness and reliability of the continuous miner. The continuous miner breaks and loads the coal mechanically and simultaneously eliminates the steps of cutting, blasting, and loading. It is not, however, truly a "continuously" operating machine. Maintenance, moving from face to face, and waiting for other production and auxiliary operations contribute to its inefficiency. Although more efficient than conventional mining, the method is also subjected to simulation studies to improve its productivity.

Continuous mining is practiced in coal, salt, potash, trona, and other soft-rock applications. It is not normally effective in mining hard-rock formations. However, if a road header is employed in place of a continuous miner, the method can be used in harder rocks. The road header is a powerful continuous miner with a smaller cutterhead designed to attack tougher materials. Its capabilities (Copur et al. 1998) allow for additional use of room-and-pillar mining where the geologic conditions are unsuitable for a traditional rotating-drum continuous miner.

10.2.2.3 Auxiliary Operations. Auxiliary operations in room-and-pillar mining are very similar for all mineral commodities and with all production equipment. The following auxiliary operations (listed in Table 5.12) are generally required:

1. *Health and safety:* strata gas control (e.g., methane drainage in coal), dust control (rock dusting, water sprays, dust collector), ventilation (line brattice or vent fan and tubing), noise abatement
2. *Environmental control:* flood protection, water treatment, subsidence control, remote monitor to sense atmospheric contamination
3. *Ground control:* scaling of roof, roof control (roof bolts, timber, arch, crib, hydraulic jack), controlled caving
4. *Power supply and distribution:* electric substation, diesel service station
5. *Water and flood control:* pump station, drainage system (ditches, pipelines, sumps)
6. *Cleanup and waste disposal:* scoop, waste storage, hoisting, dumping
7. *Materiel supply:* storage, delivery
8. *Maintenance and repair:* shop facilities, parts warehouse
9. *Lighting:* stationary lights (where needed), equipment-mounted lights
10. *Communications:* radio, phone
11. *Construction:* haulage, stoppings, overcasts
12. *Personnel transport:* mantrips, shuttle jeeps, cages

As mentioned earlier, those auxiliary operations considered essential to safety, such as ventilation and roof support, are an integral part of the production cycle in coal mining.

10.2.2.4 Conditions. The natural and geologic conditions of a mineral deposit that are well suited for room-and-pillar mining are the following (Boshkov and Wright, 1973; Bullock and Mann, 1982; Hamrin, 1982; and White, 1992):

1. *Ore strength:* weak to hard (generally does not limit the method)
2. *Rock strength:* moderate to strong
3. *Deposit shape:* tabular
4. *Deposit dip:* low ($<15°$), prefer flat
5. *Deposit size:* large areal extent; prefer seams with less than 15 ft (4.5 m) thickness
6. *Ore grade:* moderate
7. *Ore uniformity:* fairly uniform, particularly in thickness
8. *Depth:* shallow to moderate (<1500 ft or 450 m for coal, <2000 ft or 600 m for noncoal, <3000 ft or 900 m for potash)

10.2.2.5 Characteristics. The following advantages and disadvantages are summarized from more detailed information in Stefanko and Bise (1983), Bullock and Mann (1982), Hamrin (1982), White (1992), and Bibb and Hargrove (1992).

Advantages

1. Moderately high productivity; 3.5 tons (3.2 tonnes) per employee hour is average for U.S. mines using room-and-pillar continuous mining; 2.9 tons (2.6 tonnes) per employee hour is typical for conventional room-and-pillar mining (Energy Information Administration, 2000).
2. Moderate mining cost (relative cost about 10 to 25%, depending on commodity; average is 20%).
3. Moderately high production rate.
4. Fair to good recovery with pillar extraction (70 to 90%).
5. Low to moderate dilution (0 to 40%).
6. Suitable to mechanization.
7. Concentrated operations (although multiple faces needed to make it work).
8. Versatile for variety of roof conditions.
9. Ventilation enhanced with multiple openings.

Disadvantages

1. Caving and subsidence occur with pillar recovery.
2. Method inflexible and rigid in layout, not selective without waste disposal.
3. Poor recovery (40 to 60%) without pillar extraction.
4. Ground stress and support loads increase with depth.
5. Fairly high capital investment associated with mechanization.
6. Extensive development required in coal because of multiple openings.
7. Potential health and safety hazards exist, especially in coal mines.

10.2.2.6 Applications and Variations. Room-and-pillar mining is predominantly used in coal mining but finds some use in many other commodities. Descriptions of the method as used in coal can be found in Stefanko and Bise (1983) and Farmer (1992). In noncoal mining, there are a number of variations of the room-and-pillar method. Bullock and Mann (1982) and Hustrulid and Bullock (2001) outline its application to a number of different mineral commodities. Other authors outline the use of room-and-pillar mining in zinc (Suttill, 1991a; Walker, 1992), trona (Brown, 1995), and gold (Pease and Watters, 1996).

10.3 STOPE-AND-PILLAR MINING

Strikingly similar to room-and-pillar mining, the stope-and-pillar method is the most widely used of all underground hard-rock mining procedures. *Stope-and-pillar mining* is the unsupported method in which openings are driven horizontally in regular or random pattern to form pillars for ground support. This is one of the large-scale mining methods, accounting for about 50% of U.S. underground noncoal production.

The stope-and-pillar mining process is similar to room-and-pillar mining in several ways. However, most references (Stewart, 1981; Hustrulid, 1982; and Haycocks, 1992) generally differentiate stope-and-pillar mining if it meets at least two of the following qualifications:

1. The pillars are irregularly shaped and sized and either randomly located or located in low-grade ore.
2. The mineral deposit is <20 ft (6 m) in thickness, the openings are higher than they are wide, or a benching or slabbing technique must be utilized.
3. The commodity being exploited is a mineral other than coal. Although some noncoal deposits are mined by the room-and-pillar method, no coal deposits are mined by the stope-and-pillar method.

It is not surprising that there is confusion between the two methods, in that they are rather similar. Adding to the confusion is the fact that three other terms have been used over the years for stope-and-pillar mining: *open stoping*, *breast stoping*, and *bord-and-pillar mining*. For students of mining engineering, we suggest that the simplest rule is to use the term *room-and-pillar* when referring to coal and *stope-and-pillar* when referring to noncoal mining. The only exception would be noncoal mines with a very regular layout of openings and a single level of mining within the deposit. These would be termed *room-and-pillar mines*.

Stope-and-pillar mining has followed a number of mining plans, depending on the shape and thickness of the deposit, the types of equipment used, and the need for productivity. Figure 10.5 shows a stope-and-pillar mine using two benches. The first level of mining is the upper level, which is mined using a conventional cycle of operations. The first level of mining allows the crew close inspection of the back for roof control purposes. When the second level of mining is completed, the roof has been carefully stabilized, reducing the possibility of a roof fall from far above.

There are a number of variations of stope-and-pillar mining. In Figure 10.6, three variations are shown, each utilizing two benches to mine the seam. In thick seams where the benches are greater in depth, it is common for the blastholes to be drilled from above. Putting in vertical holes in this fashion generally allows for larger-diameter holes and better blasting economics. This method is shown in Figure 10.6(c). The stope-and-pillar mining method

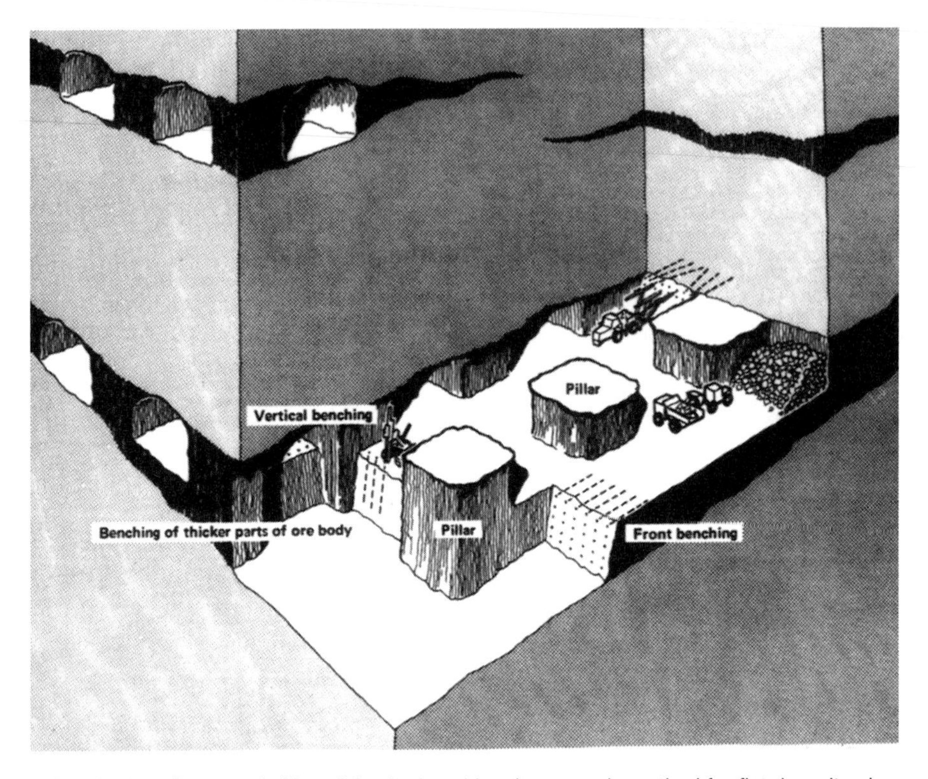

FIGURE 10.5. Stope-and-pillar mining by benching. Large-scale method for flat deposit using drill rigs, front-end loaders, and trucks. (After Hamrin, 1982. By permission from the Society for Mining, Metallurgy, and Exploration, Inc., Littleton, CO.)

provides many faces and method variations. For mining on a pitch, stope-and-pillar mining can be conducted using slushers for stope haulage. However, the method is lacking in productivity and is avoided unless no other alternatives exist.

There are several reasons that the amount of development in stope-and-pillar mining is less than that in room-and-pillar. First, the strict laws requiring multiple openings do not apply to hard-rock mining unless strata gases are present. Second, the development openings must often be driven through barren rock, increasing the cost of development and the desire to avoid unnecessary openings. Consequently, fewer development openings are driven and mining on the advance is both common and cost-effective.

There is less tendency to recover any pillars in stope-and-pillar mining. There are three reasons for this practice: (1) The pillars are relatively small and more difficult to recover safely, (2) the pillars are irregular in size and do not lend themselves to a systematic recovery operation, and (3) caving to the surface would produce much in the way of damaging subsidence (Morrison and Russell, 1973; Hamrin, 1982). Partial extraction of pillars (sometimes

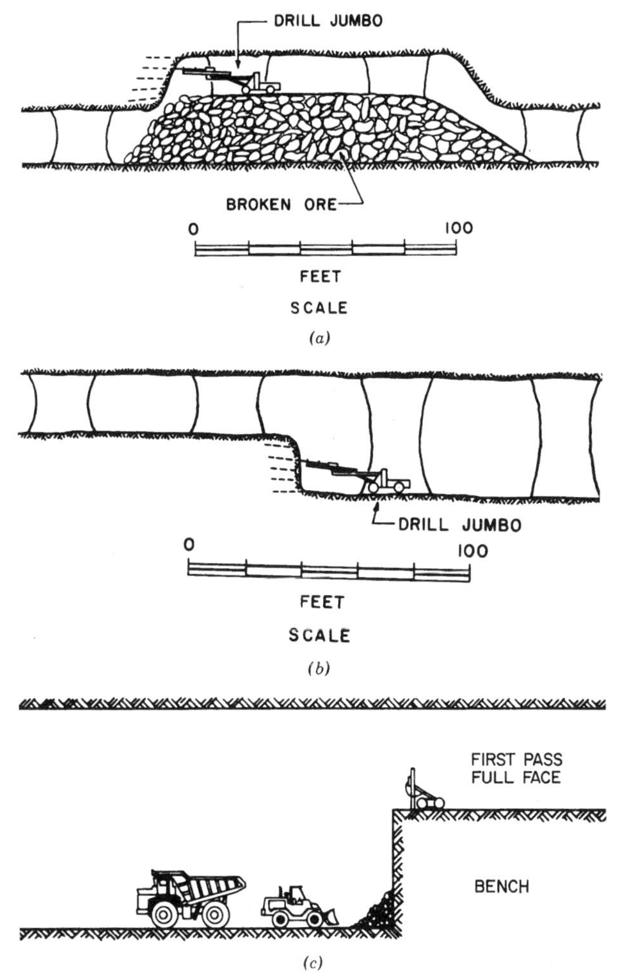

FIGURE 10.6. Methods of exploiting thick deposits by stope-and-pillar mining. (*a*) Breast stoping followed by overhand slabbing. (*b*) Breast stoping followed by benching; horizontal drilling. (After Lucas and Haycocks, 1973. By permission from the Society for Mining, Metallurgy, and Exploration, Inc., Littleton, CO.) (*c*) Breast stoping followed by benching; vertical drilling (Dravo Corporation, 1974).

called *pillar robbing* or *pillar slabbing*) is practiced in some mines, especially if the pillars are larger than necessary to provide good support.

Design of the pillars for proper support of the back is very important in stope-and-pillar mining. The use of standard-sized pillars will make the task easier, and it is recommended that this practice be followed. However, design of the pillars must be conducted in either case. Further discussion of this topic is beyond the scope of this book. Students may refer to complete discussions in Jeramic (1987) and Brady and Brown (1993).

10.3.1 Sequence of Development

The choices of main access openings for stope-and-pillar mining, a method restricted to relatively shallow or moderate depths, are similar for those for room-and-pillar mines. If the depth is relatively shallow, a belt conveyor may be planned and the primary opening may be a slope. For greater depths, a hoist may be used and the primary openings will likely be vertical shafts. The used of diesel-powered equipment may require that a ramp be used instead. Additional openings associated with the belt slope or ramp may be vertical shafts in many cases.

Depending on the geometry and attitude of the ore body, secondary openings are constructed on levels connecting the shaft with the production openings. If required by regulations or good mining practice, parallel drifts and connecting crosscuts may be driven. If the deposits are discontinuous and occur on different horizons, then truck haulage and ramps may be selected to provide maximum flexibility.

10.3.2 Cycle of Operations

Nearly all stope-and-pillar mines use conventional mining practice, with the production cycle involving:

$$\text{Production cycle} = \text{drill} + \text{blast} + \text{load} + \text{haul}$$

The production equipment used in stope-and-pillar operations is much like that used for noncoal room-and-pillar mining. Generally, the trend is toward larger and automated equipment powered with diesel engines. The cycle of operations consists of the following:

Drilling: Hydraulic or pneumatic drill jumbos are heavily favored; rotary drill rigs can be used in softer rocks.

Blasting: Ammonium nitrate and fuel oil (ANFO), gels, or emulsions; charging by hand or by pneumatic loader; firing by electric, nonelectric, or detonating fuse.

Secondary blasting: Drill and blast; impact hammer; drop ball.

Loading: A load-haul-dump (LHD) device and front-end loader are very common; shovel, overhead mucker, and slusher occasionally used.

Haulage: Truck, LHD, belt conveyor, shuttle car.

Continuous mining equipment is not ordinarily used in stope-and-pillar operations. When it is necessary to employ continuous miners, the cycle of operations is essentially identical to that for room-and-pillar mining described in Section 10.2.2.2. The auxiliary operations that are performed are also very similar to those for room-and-pillar mines. The most important functions are health and safety (dust control, ventilation, noise abatement), ground control, power supply and distribution, water handling, and flood control.

10.3.3 Conditions

The conditions listed here are derived from a number of sources, particularly Morrison and Russell (1973), Bullock (1982), Hartman (1987), and Haycocks (1992):

1. *Ore strength:* moderate to strong
2. *Rock strength:* moderate to strong
3. *Deposit shape:* tabular, lens-type deposit
4. *Deposit dip:* preferably flat; dips $<30^\circ$ mineable
5. *Deposit size:* any, preferably large areal extent, moderate thickness or bench if greater (maximum of 300 ft or 90 m)
6. *Ore grade:* low to moderate, most commonly
7. *Ore uniformity:* variable; lean ore or waste left in pillars if possible
8. *Depth:* up to 2000 ft (900 m) in competent rock, up to 3000 ft (1450 m) in very strong rock

10.3.4 Characteristics

The information on stope-and-pillar mining in this section is derived predominantly from Morrison and Russell (1973), Hamrin (1982), Lyman (1982), and Haycocks (1992).

Advantages

1. Moderate to high productivity (30 to 70 tons per employee-shift or 27 to 64 tonnes per employee-shift).
2. Moderate mining cost (relative mining cost about 10%).
3. Moderate to high production rate.
4. High degree of flexibility; method easily modified; operate several levels at one time.
5. Lends itself readily to mechanization; suitable for large equipment.
6. Not labor-intensive; extensive skills not required.
7. Selective method; permits lean ore or waste to be left in the pillars.
8. Multiple working places easy to arrange.
9. Early development not extensive.
10. Fair to good recovery (60 to 80%) without pillaring.
11. Low dilution (10 to 20%).

Disadvantages

1. Ground control requires continuous maintenance of back if rock is not strong; high back difficult to scale and support; ground stress on pillars and openings increases with depth.
2. Large capital expenditure required for extensive mechanization.
3. Difficult to provide good ventilation because of large openings.
4. Some ore lost in pillars.
5. Recovery of pillars difficult or impossible.

10.3.5 Applications and Variations

Stope-and-pillar mining finds many uses in the exploitation of metallic and nonmetallic mineral deposits. As previously stated, it is the most popular of the underground noncoal methods in the United States. Is has been used extensively in limestone in Pennsylvania, marble in Georgia, copper in Michigan (Anderson, 1981), zinc in Tennessee (White, 1979), lead in Missouri (Haycocks, 1992), oil shale in Colorado, uranium in Utah and Canada, and iron in France (Hoppe, 1978). It is used widely throughout the world.

In addition to its wide application to different minerals, stope-and-pillar mining allows much in the way of variation in its exploitation plan. The variations shown in Figures 10.5 and 10.6 are only a few of the many ways the method can be applied. It also adapts very well to moderately dipping seams. *Pitch mining* is carried out if the deposit dip exceeds the gradeability of the mobile equipment in use (Hamrin, 1982). Generally, this variation applies between dips of 15° to 30°. When the thickness of the deposit is greater than 20 ft (6 m), then the mine often applies the concept of *bench mining*, in which the ore is removed in multiple lifts or benches (Bullock, 1982). It is this ability to adapt to a multitude of conditions that makes stope-and-pillar mining cost-effective and popular with mining companies dealing with horizontal or nearly horizontal mineral deposits.

10.4 SHRINKAGE STOPING

We now encounter the first of the so-called vertical stoping methods, those carried on essentially in a vertical or near-vertical plane at an angle greater than the angle of repose of the broken ore. *Shrinkage stoping* is an overhand method in which the ore is mined in horizontal slices from bottom to top and remains in the stope as temporary support to the walls and to provide a working platform for the miners. Because the ore swells when breakage occurs, about 30 to 40% of the broken ore in each stope must be drawn off during mining to provide sufficient working space in the stope. This means that 60 to 70% of the ore must be left in the stope as a working platform for mining activities. As a result, a significant amount of capital is tied up in the remaining broken ore in the stope.

Because of its simplicity and small scale, shrinkage stoping was formerly a very popular method of noncoal mining for veins of modest size. Rising costs of labor and the trend toward diesels and mechanization have largely displaced shrinkage stoping (Lucas and Haycocks, 1973). However, shrinkage stoping is still utilized as a small- to moderate-scale method in lead, trona, potash, limestone, salt, and uranium (White, 1992). The tonnage produced, however, is less than 1% of U.S. mineral production.

Some authors (Boshkov and Wright, 1973; White, 1992) classify shrinkage as a supported method because broken ore is left in the stope to provide ground support. This is, however, a temporary method and does not provide much in the way of support to the hanging wall. Consequently, we designate the method as an unsupported method.

The key design parameters in shrinkage stoping are the dimensions of the stope, largely governed by the size and shape of the deposit. In a relatively narrow ore body, the stopes are placed longitudinally with respect to the vein; in a wide or large ore body, the stopes are placed transversely. Stope widths vary from 3 to 100 ft (1 to 30 m), lengths from 150 to 300 ft (45 to 90 m), and heights from 200 to 300 ft (60 to 90 m) (Lucas and Haycocks, 1973; Lyman 1982; and Haptonstall, 1992). Although rock mechanics will enter the picture in determining size of the stope, the openings used are generally relatively small and are not excessively stressed. Therefore, the major concern is to maintain a manageable-sized stope that ensures a smooth flow of ore by gravity and effective draw control.

10.4.1 Sequence of Development

The nature of vertical stoping methods is that production is often carried out over a considerable variation in elevation. Consequently, several levels are required in most mines, with the levels being spaced 200 to 600 ft (60 to 180 m) apart (see Section 9.2). Normally, each level will have a haulage drift driven parallel to the vein. If the stope height is less than the level interval, then sublevels may be constructed, connected by ore passes. Shrinkage stopes may be driven transverse to the vein if the vein is quite thick. Then haulage laterals or loading crosscuts are driven to the drawpoints below the stopes.

Development of shrinkage stopes is outlined by Lyman (1982), Hamrin (1982), and Haptonstall (1992). The two main tasks in preparing the stope for production are to (1) construct a means of drawing ore in which muck flows by gravity to the bottom of the stope and (2) provide a horizontal undercut at the sill level into which the ore initially breaks and subsequently flows. Normally, either finger raises or boxholes are used at the bottom of the stope to allow flow to the haulage level.

Over the years, three methods were employed to draw ore out of the stopes. The first is the use of chutes, as shown in Figure 10.7. This method was widely

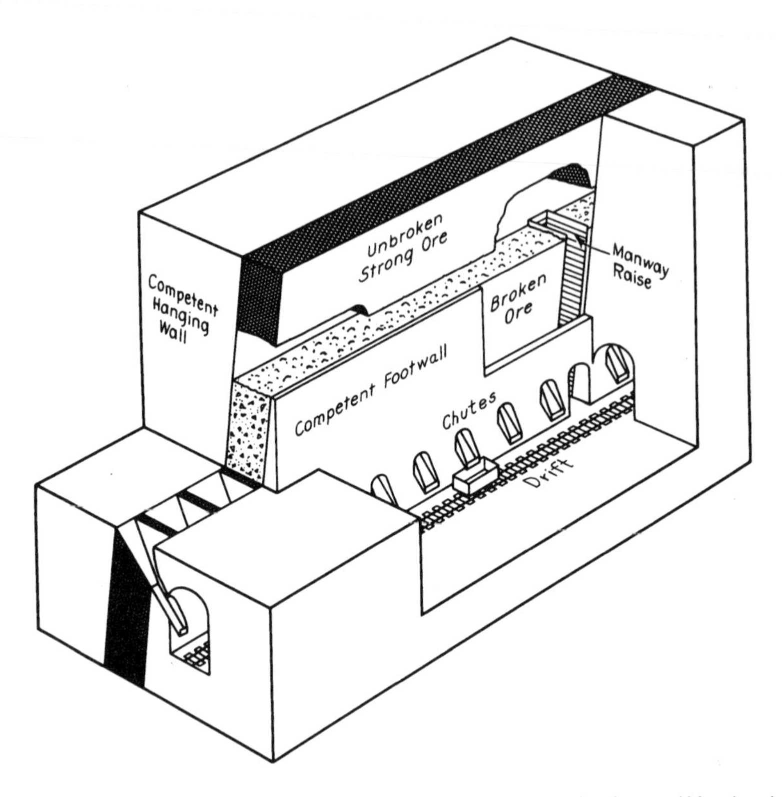

FIGURE 10.7. Shrinkage stoping using gravity draw and chutes to load cars. (After Lewis and Clark, 1964. Copyright © John Wiley & Sons, New York.)

used in the past, but has been all but eliminated because of high costs and low productivity. The second is shown in Figure 10.8, where a slusher is utilized in conjunction with boxholes to move ore from the bottom of the stope to the haulage drift. This is another version of shrinkage that is disappearing because of economic issues. The third method is to use loader crosscuts at the bottom of the stope, as shown in Figure 10.9. This figure illustrates a track system, not widely applied today. Instead, diesel loaders or LHDs are ususally employed to perform the loading operation. Haulage is also likely to be diesel-powered equipment.

10.4.2 Cycle of Operations

10.4.2.1 Production Cycle. Production in a shrinkage stope must revolve around working on an uneven floor that is periodically disturbed and lowered by drawing operations in the haulage drift below. It is important to maintain a safe and adequate platform for work, neither too high nor too low, so that rock breakage operations can be conducted properly.

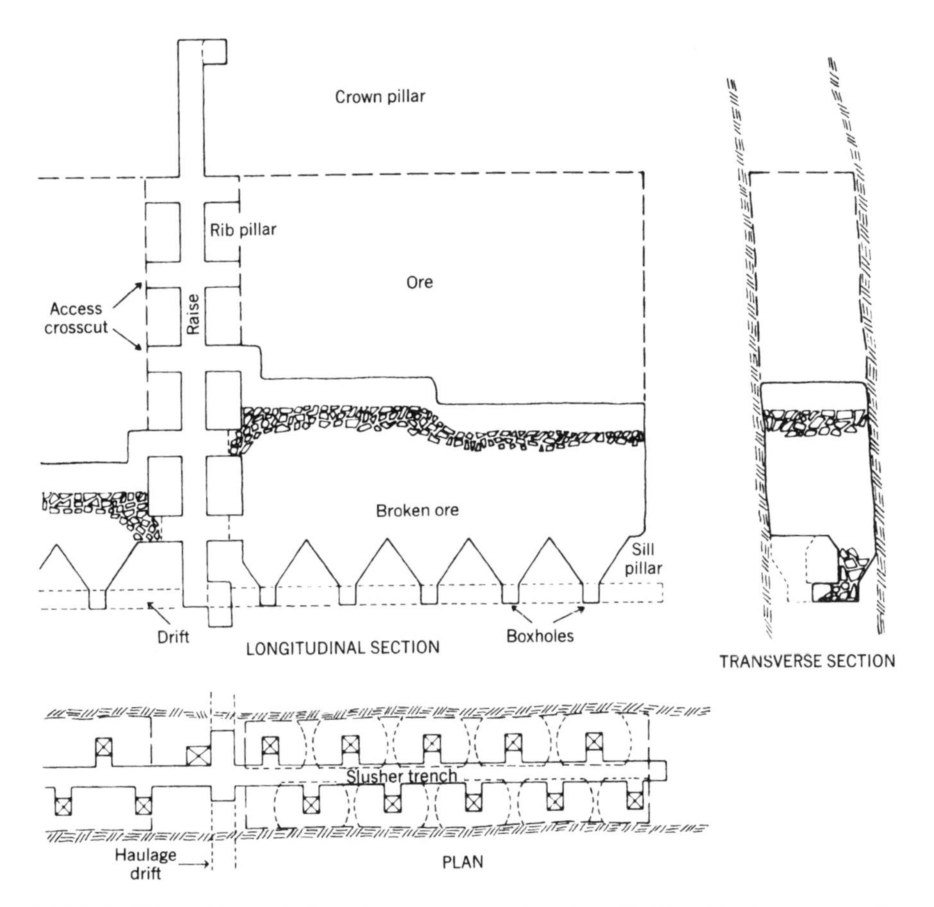

FIGURE 10.8. Shrinkage stoping using a scraper and slusher drift. (After Henderson, 1982. By permission from the Society for Mining, Metallurgy, and Exploration, Inc., Littleton, CO.)

The crew must work in the stope between the back and the broken muck. In narrow stopes, this may require that they perform all drilling using handheld drills, which will increase the costs. In stopes with more space, the drilling will preferably be completed by hydraulic drilling equipment for efficiency reasons. After the holes are charged but prior to blasting, drawing of ore from the stope should occur. Before reentering the stope area, any necessary ground control operations are carried out. This usually consists of bolting or bolting with wire mesh to increase the holding ability of the bolts.

Operations in shrinkage stopes employ the following steps in the production cycle:

Drilling: pneumatic airleg drill or stoper; small hydraulic drill jumbo

Blasting: ANFO, gels, or emulsions; charging by hand, pneumatic loader, or pumping system; firing electrically or by detonating cord

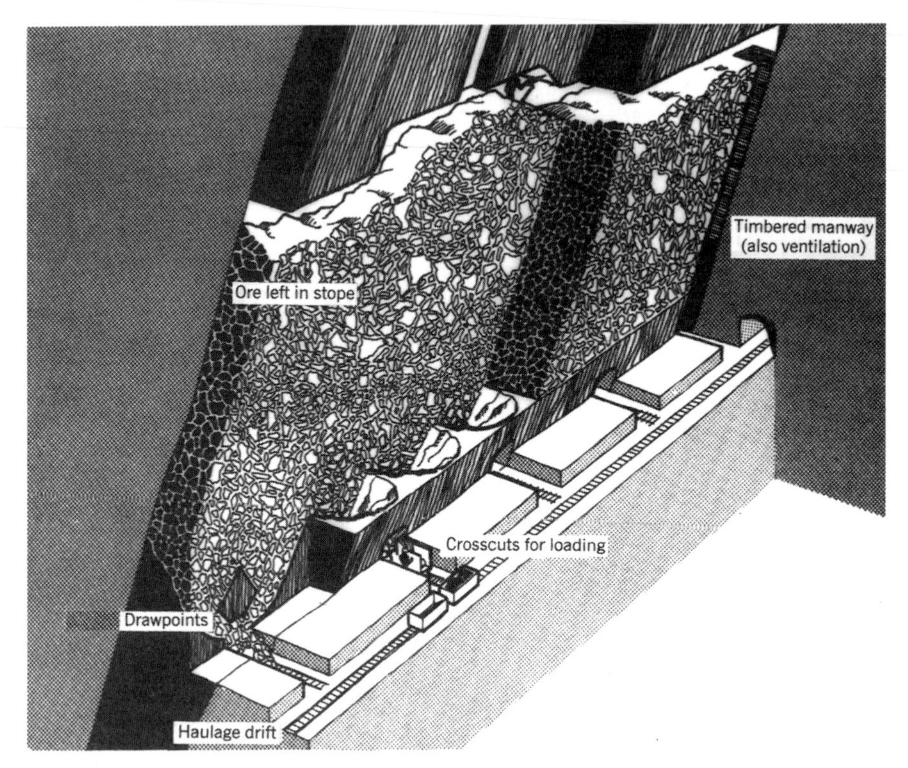

FIGURE 10.9. Shrinkage stoping using drawpoints and loaders. (After Hamrin, 1982. By permission from the Society for Mining, Metallurgy, and Exploration, Inc., Littleton, CO.)

Secondary breakage: drill and blast, packaged boulder charge, impact hammer

Loading: gravity flow, front-end loader, LHD, overhead loader, slusher

Haulage: LHD, truck, rail

Although most of the ore is drawn out at the bottom of the stope, one operation (Kral, 1997) provides for ore removal by slushing the excess off the top of the broken ore mass. Continuous extraction is not normally a part of shrinkage stoping, because of the nature of the operations and the hardness of the rock. The only opportunity for this type of operation is in the development stages of the method.

10.4.2.2 Auxiliary Operations. The usual list of auxiliary operations for underground mining pertains—see the complete list for room-and-pillar mining in Section 10.2 and the abbreviated list for stope-and-pillar in Section 10.3.

10.4.3 Conditions

The following list is based primarily on descriptions of shrinkage stoping operations in Lyman (1982), Haptonstall (1992), and White (1992).

1. *Ore strength:* strong (other characteristics: should not pack, oxidize, or be subject to spontaneous combustion)
2. *Rock strength:* fairly strong to strong
3. *Deposit shape:* tabular to lenticular, regular dip and boundaries
4. *Deposit dip:* fairly steep ($>45^\circ$; prefer 60 to 90° to facilitate ore flow)
5. *Deposit size:* narrow to moderate width (3 to 100 ft or 1 to 30 m), fairly large extent
6. *Ore grade:* fairly high
7. *Ore uniformity:* uniform, blending not easily performed
8. *Depth:* shallow to moderate (<2500 ft or 750 m)

10.4.4 Characteristics

The following summary is based on general knowledge of shrinkage stoping operations and specific information found in Hamrin (1982), Lyman (1982), White (1992), and Haptonstall (1992).

Advantages

1. Small-to-moderate-scale operation.
2. Ore is drawn down by gravity.
3. Method conceptually simple, can be used for a small mine.
4. Low capital investment, little equipment required for the basic method.
5. Little ground support is required in stope.
6. Stope development is moderate.
7. Works very well in veins with widths of <8 ft (2.4 m).
8. Fairly good recovery (75 to 85%).
9. Low dilution (10 to 20%).

Disadvantages

1. Low to moderate productivity range.
2. Moderate to fairly high mining cost (relative cost about 45%).
3. Labor intensive; not easy to mechanize.
4. Rough footing in stope, dangerous working conditions.
5. Most of the ore ($>60\%$) tied up in stope.
6. Ore is subject to oxidation, packing, and spontaneous combustion.
7. Selectivity is only fair.

One of these disadvantages requires more discussion. The oxidation of metallic ores can be of great concern, particularly with sulfides. The oxidation process is exothermic, resulting in heat buildup in packed ore material. Although the ore itself can burn under some conditions, the real danger is any combustible material that may be found in the mine. In addition, sulfide ore dusts have been known to explode (McPherson, 1993, pp. 874–875), which is therefore of concern whenever an operation is dealing with sulfide ores, particularly when the ore is left in a packed mass of broken material. This applies to any mining method in which ore is kept for some time in broken form.

10.4.5 Applications and Variations

Application of shrinkage stoping in the United States is decreasing because of economic forces. However, under the right set of conditions, it may still be the best choice for a particular mineral deposit. Two publications have recently described its application in the Gold Road Mine in Arizona (Silver, 1997) and the Nixen Fork Mine in Alaska (Kral, 1997). At the Nixen Fork Mine, shrinkage is used for veins less than 8 ft (2.4 m) wide. Slushers are used in this mine to remove the excess ore from the top of the stope. At the Gold Road Mine, the method was established to reduce dilution resulting from sublevel stoping and to increase ore grade to the mill. This mine uses diesel equipment below the stopes to load the ore and a 15° ramp to provide main haulage. Marchand et al. (2001) and Norquist (2001) outline shrinkage stoping practices at two additional mines in North America.

Although U.S. production from shrinkage stoping is quite low overall, other countries use the method more frequently, and numerous articles in the mining literature have summaries of these operations. The Las Cuevas Mine in Mexico, which mines a large fluorspar deposit, has been described by Walker (1991a). This mine divides a large orebody into blocks, mines shrinkage stopes in each block, and then systematically caves the pillars using longhole drilling techniques.

In other parts of the world, shrinkage stoping has been used by the Snip Mine in British Columbia to mine gold (Carter, 1992), the Porco Mine in Bolivia to mine zinc (Suttill, 1993), and the Colonial Gold Project in Ireland (Meiklejohn and Meiklejohn, 1997). In addition, Phelps (1994) describes the method being used in Peru at the Casapalca Mine, the Morococha Mine, the San Cristobal mine, and the Yauricocha Mine. This makes economic sense in countries where the veins are small and labor is relatively inexpensive.

10.5 SUBLEVEL STOPING

Sublevel stoping is a vertical mining method in which a large open stope is created within the vein. This open stope is not meant to be occupied by the miners; therefore, all work of drilling and blasting must be performed from

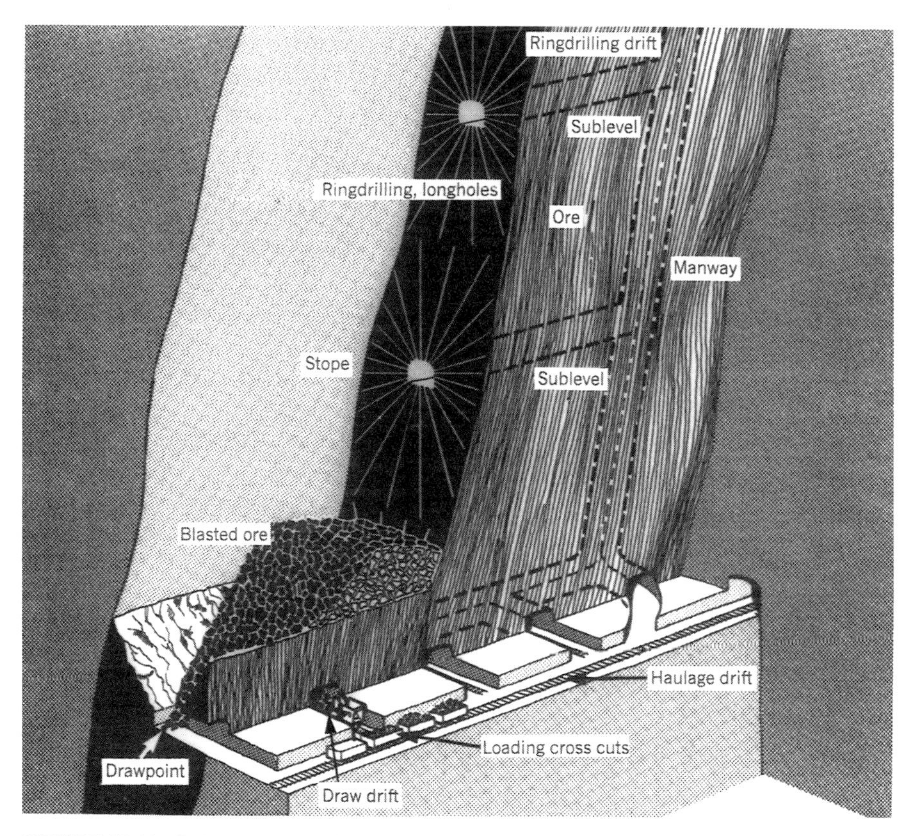

FIGURE 10.10. Sublevel stoping (blasthole method) using ring drilling and blasting into a slot. (After Hamrin, 1982. By permission from the Society for Mining, Metallurgy, and Exploration, Inc., Littleton, CO.)

sublevels within the ore block. Three different variations of sublevel stoping are practiced. The first and most traditional method, often called the *blasthole method*, is shown in Figure 10.10. Using this variation, the miners must create a vertical slot at one end of the stope. They then work in the sublevels to drill a radial pattern of drillholes. After a set of these holes are loaded, blocks of the ore body are blasted into the open stope. The blast should be carefully planned, as it is virtually impossible for miners to go into the stope to correct problems.

The second method of sublevel stoping, illustrated in Figure 10.11, is often referred to as the *open-ending method*. As in the blasthole method, a slot must be developed at one end of the stope. Production is then achieved by drilling parallel holes from top to bottom of the designated stope using a sublevel at the top of the stope that is the width of the stope. As in the blasthole method, vertical slices of the ore are blasted into the open stope. This method generally permits larger drillholes to be used and may be more efficient in terms of explosives consumption.

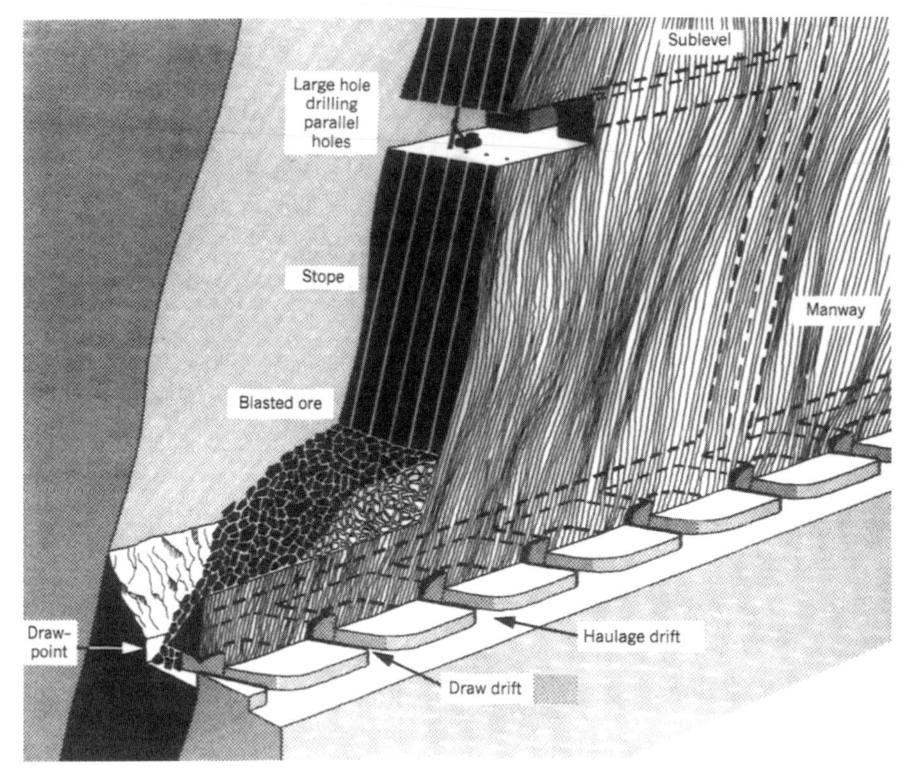

FIGURE 10.11. Sublevel stoping using parallel drilling and blasting into a slot. (After Hamrin, 1982. By permission from the Society for Mining, Metallurgy, and Exploration, Inc., Littleton, CO.)

The third variation of sublevel stoping is a patented method known as the *vertical crater retreat* (VCR) *method*. The procedure, shown in Figure 10.12, uses a drill pattern similar to that of the open-ending method. However, the ore is blasted in horizontal slices using loading and blasting from the sublevel at the top of the stope. This requires that the drillholes first be sealed with plugs that can be put in place from above. The holes are then loaded to a fixed height of charge and blasted. Typically, the horizontal slices blasted are about 15 ft (5 m) in thickness until the blasting horizon approaches the top of the stope. The last blast at the top of the stope is sized to be at least twice the normal blasting thickness so that the working floor of the sublevel is not weakened by blasting until the workers are out of the sublevel for the last time. The VCR method was patented by a Canadian explosives firm that developed the crater testing method that ensures that the blasting procedure is adequate for the rock mass in the stope.

The stopes in sublevel operations vary significantly in size. Some are roughly the size of shrinkage stopes, but stopes up to 100 ft (30 m) by 130 ft (40 m) by 1200 ft (370 m) high have been used at the Mt. Isa Mine in Australia (Suttill,

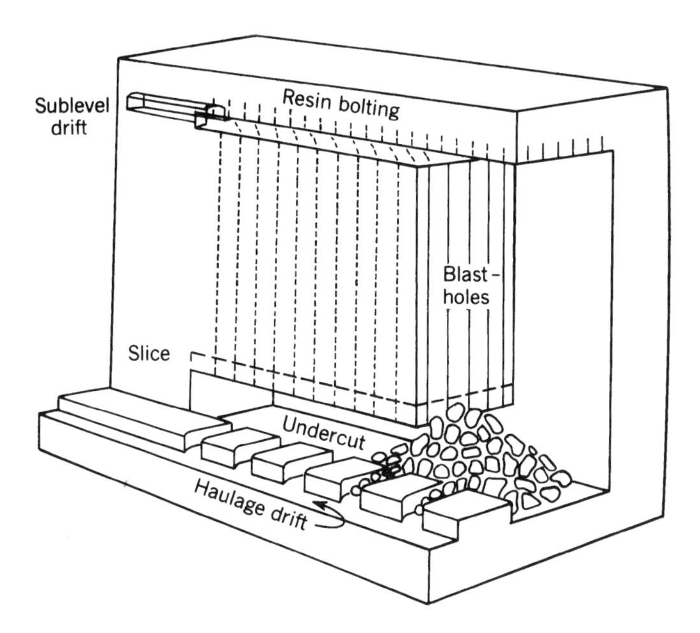

FIGURE 10.12. Vertical crater retreat (VCR) version of sublevel stoping. Large parallel holes are loaded with near-spherical explosive charges, and horizontal slices of ore are blasted into the undercut. (After Green, 1976. By permission from National Mining Association, Washington, DC.)

1991b). The sizes of the stopes that are possible are obviously dependent on the strength of the rock. It should also be noted that many sublevel stoping operations utilize pillar recovery methods, often using cemented fill and cut-and-fill methods to extract the pillars.

10.5.1 Sequence of Development

The general sequence of development in sublevel stoping parallels that in shrinkage stoping and other vertical methods. A haulage drift, crosscuts, and drawpoints are developed below the stope for materials handling, together with interlevel raises for access and ventilation. Either an undercut for VCR stoping or a slot for blasthole or longhole stoping is then developed.

If an undercut is developed (Figure 10.12), the sill development proceeds like that for shrinkage. If a slot is necessary (Figures 10.10 and 10.11), sublevel crosscuts are developed across the stope and a raise is driven at the boundary. This raise is then enlarged to form the slot. In the blasthole method, the sublevel drifts are driven to give the drilling crews access to the entire deposit. In the other two versions of sublevel stoping, a horizontal sublevel the full width of the stope must be driven at periodic intervals to provide room for drill stations.

10.5.2 Cycle of Operations

10.5.2.1 Production Cycle. As in many of the vertical stoping methods, rock breakage and materials handling are carried out in separate sections of sublevel stoping. Drilling and blasting are conducted in the stope sublevel drifts. Loading takes place underneath the stope in drawpoints or crosscuts located off the main haulage drift. Haulage itself is carried out in the main drifts, which are ordinarily parallel to the vein. Coordination of the unit operations in the sublevels and the unit operations in the openings below the stope must be coordinated, but they are conducted largely independently. The cycle of operations follows the basic production cycle (Section 5.1):

Drilling: Normally done with (1) large-hole pneumatic percussion drills, (2) large-hole rotary drills, or (3) small-diameter pneumatic percussion drills. The emphasis today is on the large-hole rigs and long holes.

Blasting: ANFO, emulsions, or gels; charging by pneumatic loader, pumping system, or by cartridges; firing electrically, with detonating cord, or with non-electric initiation systems.

Secondary blasting: Drill and blast, mudcapping, impact hammer.

Loading: Gravity flow to drawpoints; loading with front-end loader, LHD, shovel loader, slusher, belt conveyor (diesel equipment now highly favored).

Haulage: LHD, truck, rail, belt conveyor (rare).

10.5.2.2 Auxiliary Operations. See Sections 10.2 and 10.3 for auxiliary operations.

10.5.3 Conditions

The deposit conditions inherent in the typical sublevel stoping operation have been derived from Mitchell (1981), Hamrin (1982), Mann (1982), and Haycocks and Aelick (1992):

1. *Ore strength:* moderate to strong, may be less competent than for stope-and-pillar mining
2. *Rock strength:* fairly strong to strong
3. *Deposit shape:* tabular or lenticular, regular dip and boundaries helpful
4. *Deposit dip:* fairly steep ($>45^\circ$, preferably 60 to 90°)
5. *Deposit size:* moderate to thick width (20 to 100 ft or 6 to 30 m), fairly large extent
6. *Ore grade:* moderate
7. *Ore uniformity:* fairly uniform
8. *Depth:* varies from fairly shallow to deep (once used at 8000 ft or 2.4 km at Homestake)

10.5.4 Characteristics

The following characteristics have been compiled from material in Mitchell (1981), Hamrin (1982), Hartman (1987), Haycocks and Aelick (1992), and White (1992).

Advantages

1. Moderate to high productivity.
2. Moderate mining cost (relative cost: 20%).
3. Moderate to high production rate.
4. Lends itself to mechanization; not labor-intensive.
5. Low breakage cost; fairly low handling cost.
6. Little exposure to hazardous conditions; easy to ventilate.
7. Unit operations can be carried on simultaneously.
8. Fair recovery (about 75%).
9. Modest dilution (about 20%).

Disadvantages

1. Fairly complicated and expensive development.
2. Inflexible in mining plan.
3. Longhole drilling requires precision ($<2\%$ deviation).
4. Large blasts can cause significant vibration, air blast, and structural damage.

10.5.5 Applications and Variations

Many applications of sublevel stoping can be found worldwide. A significant number of descriptions of sublevel stoping mines can be found in Bullock and Mann (1982) and Hustrulid and Bullock (2001). Additional basic information on ways of using this method is available in Haycocks and Aelick (1992). For specific applications, students may refer to the following publications that describe domestic and foreign mines. In the United States, several articles have delineated the sublevel stoping operations at the Meikle Mine in Nevada (O'Neil, 1996; Werniuk, 1996) and the Carlin East Mine (Driscoll, 1997).

In other parts of the world, sublevel stoping operations have been used in Canada in a number of mines. In particular, the method has been used for nickel in Manitoba (Madsen et al., 1991), for gold in Ontario (Rhealt and Bronkhorst, 1994), and for lead and zinc in the Northwest Territories (Keen, 1992). In Brazil, the Sao Bento Mine has mined gold using the longhole version of sublevel stoping (Shuey, 1998). The blasthole stoping method has been applied to copper mining in Australia at the Mt. Isa complex (Suttill, 1991b) and at the Olympic Dam Mine (Suttill, 1994).

Two of the aforementioned international mines are worthy of special note. First, the Mt. Isa sublevel stopes (Suttill, 1991b) are among the largest anywhere in the world, the largest stopes measuring 100 ft by 130 ft (30 m by 40 m) at the base with a total vertical dimension of 1200 ft (370 m). This height, greater than that of the Eiffel Tower, is only possible because of the strength of the rock. The second mine of great interest is the Polaris Mine, which exploits a lead/zinc ore in the Canadian High Arctic (Keen, 1992). The mine is the most northerly in the world, located far north of any large settlements in the Northwest Territories of Canada. One of the interesting features of the mine is that the backfill in the stopes is allowed to freeze, stabilizing the pillars and allowing for their safe removal.

10.6 SPECIAL TOPIC A: CALCULATION OF PERCENTAGE RECOVERY

During the planning phases of mining, it is often neccssary to estimate the *percentage recovery* or extraction of ore, coal, or stone from a deposit when using a given mining method. Mathematically, it may be calculated as the ratio of mineral extracted to the total minable reserve in the deposit. If a method yields a high recovery, it may be simpler to calculate the unmined material left in pillars rather than that extracted. With a method as systematic as room-and-pillar mining in coal, it is not difficult to estimate the recovery in a panel or for the entire mine. For an approximation, and if the seam height is regular, it is acceptable to base the calculation on areas rather than volumes (or weights).

Example 10.1. A portion of a panel in a room-and-pillar coal mine is shown in Figure 10.13. All openings are 20 ft (6 m) in width, and the mining height is regular. Rooms are driven on 60 ft (18 m) centers and crosscuts on 80 ft (24 m) centers. Calculate the percentage recovery in the panel (1) without pillar recovery and (2) with recovery of chain pillars. Disregard the effect of barrier pillars, and calculate for the smallest repetitive dimensions in the panel.

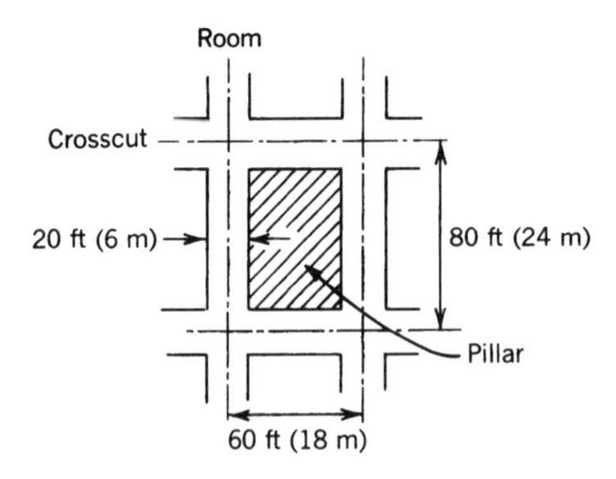

FIGURE 10.13. Portion of a panel in a room-and-pillar mine. See Example 10.3.

SOLUTION. (1) Without pillar recovery:

$$\text{Area of block} = (80)(60) = 4800 \text{ ft}^2$$

$$\text{Area of openings} = 20(80 + 40) = 2400 \text{ ft}^2$$

$$\text{Recovery} = \frac{2400}{4800} = 50.0\%$$

(2) With pillar recovery:

$$\text{Area of openings} = \text{area of block} = 4800 \text{ ft}^2$$

$$\text{Recovery} = \frac{4800}{4800} = 100\%$$

The presence of barrier pillars along entries, usually required in any room-and-pillar mine, will reduce the recovery unless they can be extracted on the retreat. If second mining is not practiced, recovery in room driving can be improved by increasing the width of the openings and decreasing the size of the chain pillars, commensurate with safety in roof control. Note that the actual recovery realized during mining will differ from the calculated recovery because of losses of coal, pillar sloughs, and dilution by rock from roof falls.

10.7 SPECIAL TOPIC B: DESIGN OF MINE OPENINGS

10.7.1 Rock Mechanics

Rock mechanics is the study of the properties and behavior of rock, the nature of the stresses about underground openings, and their relation in the design and support of mine workings and in the induced caving of rock in mine exploitation. All rock at depth is under stress due to the weight of the overlying rock (superincumbent load) and to possible stresses of tectonic origin. In addition, the presence of a mine opening induces or redistributes stresses in the rock surrounding the opening, and this rock (and the opening) will fail if the rock stress exceeds the rock strength (Obert et al., 1960). Thus the problem of designing a stable mine opening reduces to determining (1) the maximum stress in rock surrounding the opening and (2) the strength of the rock in situ.

Rock mechanics is often defined more broadly. The aspect described here—that concerned with time rates of loading that are very long in duration—is referred to as *static rock mechanics*. A different aspect related to rock attack under rates of loading of short time duration and the corresponding behavior of rock is called *dynamic rock mechanics*. The latter includes rock penetration and fragmentation processes of all types, ranging from conventional means of drilling, blasting, and continuous mining to novel methods of applying energy to excavate rock such as fluid, thermal, and electrical attack (Section 5.2).

In this discussion, we shall be concerned with static rock mechanics only, because it is fundamental to the study of all rock mechanics and because the design, stability, and support of underground openings are fundamental to mining itself. We remind ourselves that the ultimate expression of depth as a constraint in mining takes two forms and that one is the inexorable rise in rock stress (the other is the equally unrelenting climb in rock temperature).

Because our treatment of the subject of rock mechanics is abbreviated and restricted to the design of underground openings, a number of simplifying assumptions about rock prove helpful:

1. Rock is perfectly elastic (stress is proportional to strain).
2. Rock is homogeneous (there are no significant imperfections).
3. Rock is isotropic (its elastic properties are the same in all directions).

Although never perfectly true, these assumptions apply reasonably well to many rocks (igneous best, sedimentary least) at moderate depth. Causes of departure are the complex, diversified, and heterogeneous nature of rock itself, the effects of high confining pressure and temperature at great depth, the presence of water or solutions, and the effects of geologic structures (bedding, fractures, folding, joints, alteration, etc.). To a certain extent, uncertainties and departures from theory are compensated for in design by the use of factors of safety.

The following physical properties of rock are employed in many applications of rock mechanics, including the design of mine openings:

Property and Definition	Range
1. Young's modulus of elasticity = stress/strain	
$E = \frac{s}{e}$	$5\text{–}10 \times 10^6$ lb/in.^{2} ($34\text{–}69 \times 10^3$ MPa)
2. Poisson's ratio = lateral strain/longitudinal strain	
$\mu = \frac{e_{lat}}{e_{long}}$	0.1–0.3
3. Unit strengths, based on unconfined uniaxial tests	
a. Compression f_c	5000–50,000 lb/in.^{2} (34–345 MPa)
b. Tension f_t	400–2500 lb/in.^{2} (2.8–17 MPa)
c. Shear f_s	500–4000 lb/in.^{2} (3.4–28 MPa)
4. Specific weight $w = 62.4$ SG (Eq. 3.1) $w = 1000$ SG (Eq. 3.1a)	70–450 lb/ft^3 (1120–7200 kg/m^3)

10.7.2 State of Stress About Mine Openings

We concern ourselves first with the state of stress existing in the rock before a mine opening is driven. The *vertical stress* S_y acting on a horizontal plane is equal to the weight of the overlying rock distributed over a unit area and is determined by the depth and specific weight of the rock:

$$S_y = wL = \left(\frac{62.4\text{SG}}{144}\right)L = 0.433\text{SG} \times L \text{ lb/in.}^2 \tag{10.1}$$

$$S_y = wL = 1000\text{SG} \times L \text{ Pa} \tag{10.1a}$$

For average rock with SG = 2.77, $S_y = 1.2L$ in lb/in.^{2} ($2770L$ in Pa). The *horizontal stress* S_x acting on a vertical plane is a function of the vertical stress:

$$S_x = kS_y \tag{10.2}$$

in which k is a constant varying from 0 to >1.

There are four possible cases for the stress field existing at a point within the earth. They are represented diagrammatically in Figure 10.14 and summarized here:

Case 1

No confining pressure or restraint ($k = 0$); occurs at very shallow depths, close to faults, or adjoining a bench face:

$$S_x = 0 \tag{10.3}$$

Case 2

No lateral deformation or strain, $k = \mu/(1 - \mu)$; occurs at moderate depths in elastic rock; if $\mu = 0.25$, a typical value for rock,

$$S_x = \left(\frac{\mu}{1-\mu}\right)S_y = \frac{1}{3}S_y \tag{10.4}$$

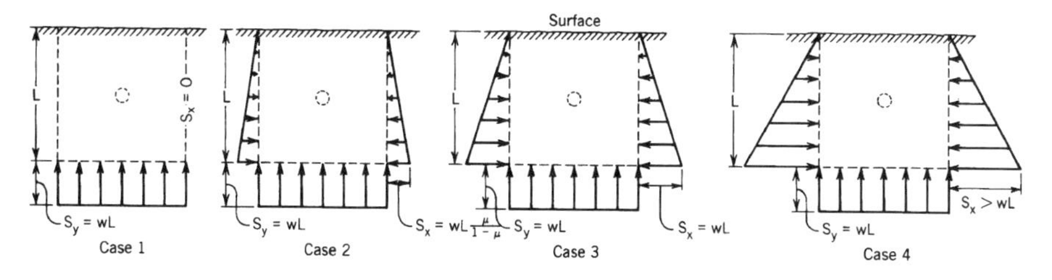

FIGURE 10.14. Stress concentration factor plotted for tangential stress along vertical and horizontal diameters of a circular opening in three different stress fields. (After Caudle iand Clark, 1955).

Case 3

Hydrostatic pressure ($k = 1$); occurs at great depth or in wet, squeezing, and running ground:

$$S_x = S_y \tag{10.5}$$

Case 4

High lateral pressure due to tectonic forces ($k > 1$); occurs in regions of recent orogenic or volcanic activity:

$$S_x = kS_y \tag{10.6}$$

(same as Eq. 10.2) in which k ranges from 2 to 5. Because of its erratic, localized nature, we will disregard case 4.

Assume now that a mine opening penetrates the stress field existing within the earth at a given depth. Knowing the original state of stress enables us to determine the magnitude of the redistributed stresses. This can be done by exact calculation for a circle (the simplest shape of opening), by numerical techniques and the computer for any opening, by model study and experimental measurements in the laboratory (using photoelasticity or holography), or by field measurements in an actual mine (using load cells, stress meters, strain gauges, etc.). The results presented here were obtained in model studies and confirmed by other methods.

Because we are concerned only with the magnitude of the maximum stresses about a mine opening, our task is much simplified. From strength of materials and the theory of elasticity, we draw these conclusions (Obert, 1973b):

1. The maximum stresses occur at the boundary of the opening and act tangentially to it; likewise, the radial stress at the boundary is zero, and the shear stress is equal to one-half the tangential stress.
2. The stresses are independent of the size of opening (but not of the shape or the side ratio).
3. The stresses are independent of the elastic moduli of the material.
4. The peak values of the maximum (tangential) stress acting on a mine opening occur at the midpoints of the top and sides or at the corners of the opening (if any).
5. The stress redistribution about an opening is negligible at a distance of one diameter from the edge of the hole.

We also adopt two conventions. The first is that tension is represented by the (+) sign and compression by the (−) sign. The second is that the maximum tangential stress s_t is expressed as a multiple of the vertical stress acting in the earth S_y:

$$s_t = cS_y \tag{10.7}$$

where c is *stress concentration factor.*

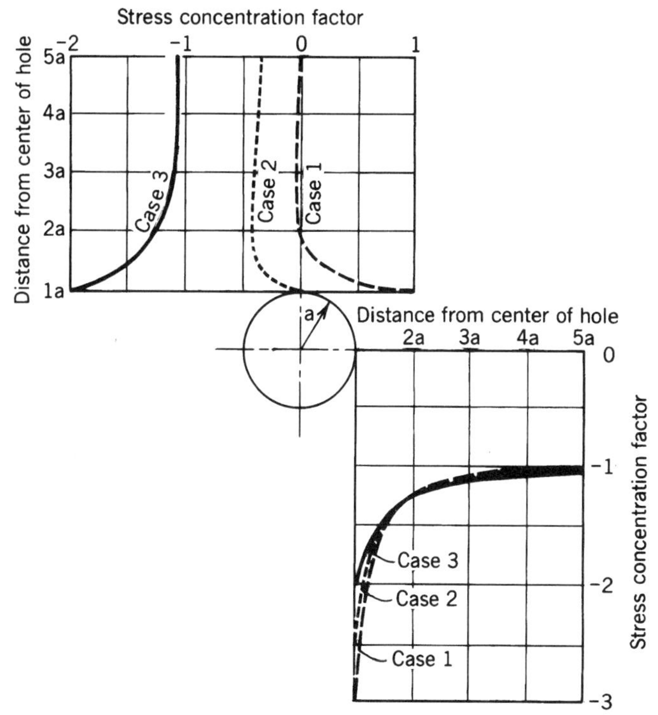

FIGURE 10.15. Stress concentration factor plotted for tangential stress along vertical and horizontal diameters of a circular opening in three different stress fields. (After Caudle and Clark, 1955).

Data obtained from the analysis of a circular opening may be represented as shown in Figure 10.15. Values of the stress concentration factor for the tangential stress are plotted along vertical and horizontal diameters extended into the side walls for cases 1, 2, and 3 (distances are plotted as radii). Alternatively, because the critical stresses occur at the walls of the opening, stress concentration factors may be plotted along the boundaries only (Figure 10.16; *left* for a circular opening and *right* for rectangles of various side ratios R/h and ratio of fillet radius to short dimension r/h of $\frac{1}{6}$). In the symbol convention, R is width, h is height of opening, and r is fillet radius of the rounded corner of a rectangle.

Graphically, we may also depict critical stress diagrams for openings as shown in Figure 10.17 (*left* for circular, *right* for rectangular), in which stress concentration factors at the boundaries of the opening are plotted for the various stress fields. Stress values at the top, corner (if a rectangle), and side of each opening are shown. With a circle, peak values occur at the top and side; with a rectangle, peak values occur at the top and corner. Sketching a stress diagram for existing conditions (shape of opening, stress field) is helpful in mine design in visualizing where high stress concentrations and failure may develop.

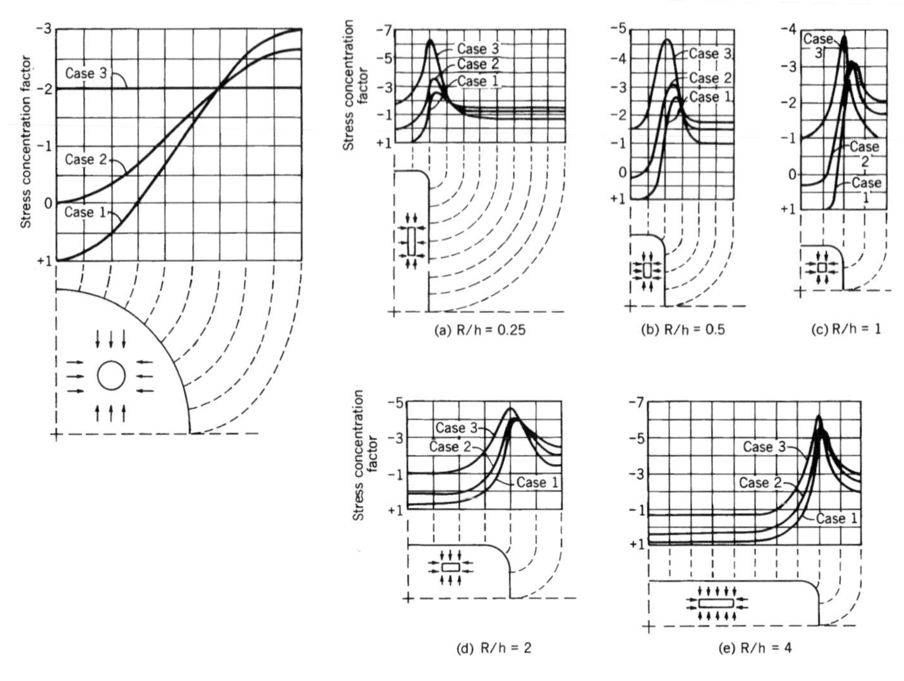

FIGURE 10.16. Boundary stress concentrations for mine openings in different stress fields. (*Left*) Circular opening. (*Right*) Rectangular openings of different side ratio (*R/h*) and ratio of fillet radius to short dimension (*r/h*) of 1/6. (After Obert et al., 1960.)

Theoretical and experimental studies allow us to draw the following conclusions concerning stress about single mine openings (Obert, 1973b):

1. Case 1 stress field is most apt to produce tensile stresses about mine openings, followed by case 2 (case 3 is incapable of producing tensile stresses).
2. Tensile stresses are usually more serious than compressive in causing opening failure; peak tensile values occur at the top of the opening.
3. Rectangular openings produce higher stress concentrations than circular; the more elongated the rectangle and the sharper the corners, the higher the peak stresses.
4. The preferred shapes of opening (starting with the lowest peak stress concentration factors) for different stress fields are the following:
   a. *Case 1:* ellipse (vertical orientation), rectangle (vertical), circle, square, rectangle (horizontal).
   b. *Case 2:* ellipse (vertical), circle, square, rectangle (vertical), rectangle (horizontal).
   c. *Case 3:* circle, square, ellipse (any orientation), rectangle.

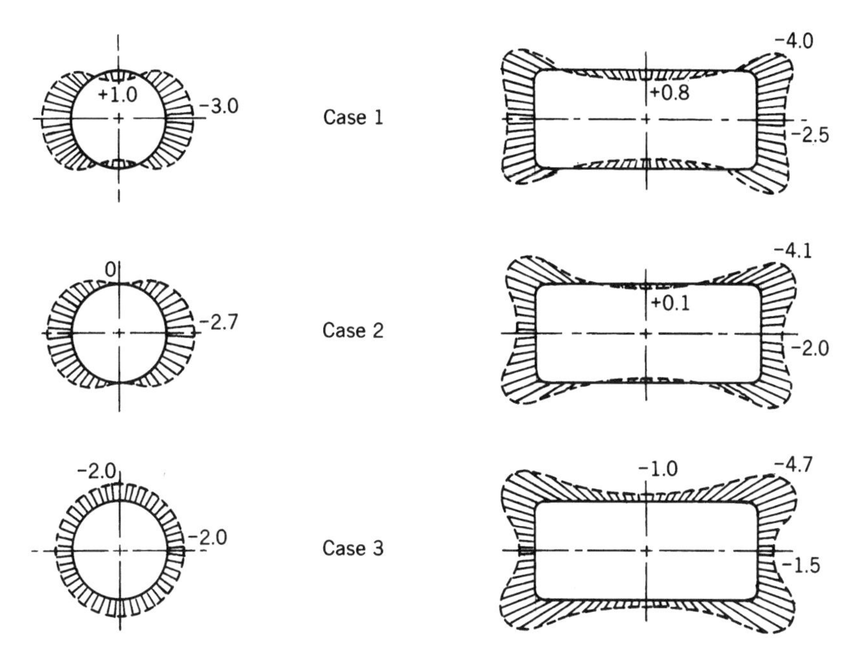

FIGURE 10.17. Critical stress diagrams for mine openings in different stress fields. Stress values at top, corner (if a rectangle), and side of the openings are indicated. (*Left*) Circular opening. (*Right*) Rectangular opening of side ratio $R/h = 2$ and ratio of fillet radius to short dimension $r/h = 1/6$. Plotted for cross section (d, right), in Figure 10.16.

5. Stresses are independent of the size of the opening and properties of the rock.

Although the presence of multiple openings establishes a different redistribution of stresses, the effect is small and may be neglected if the pillar width exceeds the opening width by a factor of 2 (Panek, 1951).

Compilations of critical stresses for the two basic shapes of mine openings, ellipse (including circle) and rectangle (including square), appear in Tables 10.1 and 10.2.

10.7.3 Design Procedure

Our objective in the design of mine openings is now clear: Employing the values of the stress concentration factor given in Tables 10.1 and 10.2 for the basic shapes of mine openings and various stress field cases, calculate the critical (maximum) values of boundary stress to compare with the measured strength properties of the rock to determine whether the opening is safe or will fail. A numerical example demonstrates the procedure.

Table 10.1 Critical Values of Stress Concentration Factor *c* on an Elliptical Boundary in Different Stress Fields

Width-to-Height Ratio = R/h	Case 1, Top	Case 1, Side	Case 2, Top	Case 2, Side	Case 3, Top or Side[a]
0.25	+1.0	−1.5	−2.0	−1.2	−8.0
0.33	+1.0	−1.7	−1.3	−1.3	−6.0
0.5	+1.0	−2.0	−0.7	−1.7	−4.0
1 (circle)	+1.0	−3.0	0	−2.7	−2.0
2	+1.0	−5.0	+0.3	−4.7	−4.0
3	+1.0	−7.0	+0.4	−6.7	−6.0
4	+1.0	−9.0	+0.5	−8.7	−8.0

[a]Top if oriented vertically, side if oriented horizontally.

Source: Modified after Panek, 1951.

Example 10.2. A square mine opening is driven horizontally at a depth of 2500 ft (750 m). The fillet ratio is $\frac{1}{6}$. Rock properties are SG = 2.77, $\mu = \frac{1}{4}$, $f_c =$ 20,000 lb/in.^{2} (138 MPa), and $f_t = 1500$ lb/in.^{2} (10.3 MPa). Determine the vertical and horizontal stresses in the earth, the critical boundary stresses in a uniform stress field for each of the three common cases, and whether the opening will fail. Calculate the factor of safety for one case that is safe.

Table 10.2 Critical Values of Stress Concentration Factor *c* on a Rectangular Boundary in Different Stress Fields[a]

Width-to-Height Ratio = R/h	Case 1, Top	Case 1, Corner	Case 2, Top	Case 2, Corner	Case 3, Corner[b]
0.12	+1.0	−2.3	−0.5	−4.2	−9.3
0.16	+1.0	−2.5	−0.3	−3.7	−7.6
0.25	+1.0	−2.5	0	−3.5	−6.2
0.33	+1.0	−2.6	+0.1	−3.3	−5.2
0.50	+1.0	−2.7	+0.2	−3.1	−4.7
1 (square)	+1.0	−3.1	+0.3	−3.1	−3.8
2	+0.8	−4.0	+0.4	−4.1	−4.7
3	+0.8	−4.6	+0.4	−4.7	−5.2
4	+0.9	−5.4	+0.4	−5.6	−6.2
6	+0.9	−6.8	+0.4	−7.0	−7.6
8	+1.0	−8.6	+0.5	−8.7	−9.3

[a]Ratio of fillet radius to short dimension $R/h = \frac{1}{6}$.

[b]Factors negative and smaller for top and side.

Source: Modified after Panek, 1951.

SOLUTION. Calculate the vertical stress in the earth by Eq. 10.1:

$$S_y = 0.433\text{SG} \times L = 1.2(2500) = 3000 \text{ lb/in.}^2 \text{ (20.7 MPa)}$$

Read stress concentration factors from Table 10.2 for $R/h = 1$ (square opening):

Case 1: $S_x = 0$ (Eq. 10.3)

$$\text{Top } c = +1.0, \quad S_t = +(1.0)(3000) = +3000 \text{ lb/in.}^2 > 1500\text{—fails}$$

$$\text{Corner } c = -3.1, \quad S_t = -(3.1)(3000) = -9300 \text{ lb/in.}^2 < -20{,}000\text{—safe}$$

Case 2: $S_x = \frac{1}{3}S_y = \frac{1}{3}(3000) = 1000$ lb/in.^{2} (Eq. 10.4)

$$\text{Top } c = +0.3, \quad S_t = +(0.3)(3000) = +900 \text{ lb/in.}^2 < 1500\text{—safe}$$

$$\text{Corner } c = -3.1, \quad S_t = -(3.1)(3000) = -9300 \text{ lb/in.}^2 < -20{,}000\text{—safe}$$

Case 3: $S_x = S_y = 3000$ lb/in.^{2} (Eq. 10.5)

$$\text{Corner } c = -3.8, \quad S_t = -(3.8)(3000) = -11{,}400 \text{ lb/in.}^2 < -20{,}000\text{—safe}$$

$$FS = \frac{f_c}{S_t} = \frac{20{,}000}{11{,}400} = 1.75$$

Normally, a factor of safety of 1.25 suffices in the design of mine openings, depending on the confidence limits of the data and the permanence of the openings. If the factor of safety is 1 or less, then provision of artificial support is likely necessary. In designing for induced caving (Chapter 12), the opening shape is chosen to ensure that failure does occur (i.e., $FS < 1$). Usually, the R/h ratio is the only design parameter that can be varied, inasmuch as a rectangular cross section is nearly always employed (unless the opening is bored).

If the rock sequence in which the opening is located departs drastically from elastic, homogeneous, and isotropic conditions, then the approach illustrated here is less applicable. In well-bedded rock, for example, as encountered in the mining of coal and most nonmetallics, it is advisable to compute the factor of safety for the corner stress based on the weaker rock strength, either the bed in which the opening is driven or the bed constituting the roof (back). For more accurate results, a different analytical procedure is recommended (Obert et al., 1960).

Extensive literature on rock mechanics is available, the more noted works being Jaeger and Cook (1976), Goodman (1980), Hoek and Brown (1980), Bieniawski (1984), Jeramic (1987), and Brady and Brown (1993).

PROBLEMS

10.1 A coal mine is using the room-and-pillar method, block version, with rooms and crosscuts 20 ft (6 m) in width, driven on 90 ft (27 m) centers. Entries and entry crosscuts are the same dimensions. If no barrier pillars are left, compute the percentage recovery with and without pillar extraction.

10.2 A panel of 30 rooms in a room-and-pillar coal mine is to be driven from a set of three room entries connecting to panel entries. The rooms are driven on 100 ft (30 m) centers; they are 16 ft (4.8 m) in width and 300 ft (90 m) in length. Room entries are also 16 ft (4.8 m) in width on 100 ft (30 m) centers. Crosscuts between both rooms and entries are placed on 100 ft (30 m) centers. A barrier pillar 100 ft (30 m) in width will be left on three sides of the panel. A sketch of the mining plan appears in Figure 10.18. Compute the percentage recovery for the mining panel in each of the following cases:

a. No pillar recovery

b. Recovery of chain pillars in rooms and entries

c. Room width increased to 20 ft (6 m) and no pillar recovery

10.3 A panel of 30 rooms in a room-and-pillar coal mine is to be driven from a set of three room entries connecting to panel entries. The rooms are driven 20 ft (6 m) wide on 50 ft (15 m) centers and are 300 ft (90 m) in length. Room entries are 15 ft (4.5 m) in width on 50 ft (15 m) centers. Crosscuts between both rooms and entries are placed on 100 ft (30 m) centers. A barrier pillar 50 ft (15 m) in width will be left on three sides on the panel. A sketch of the mining plan appears in Figure 10.19. Compute the percentage recovery for the mining panel in each of the

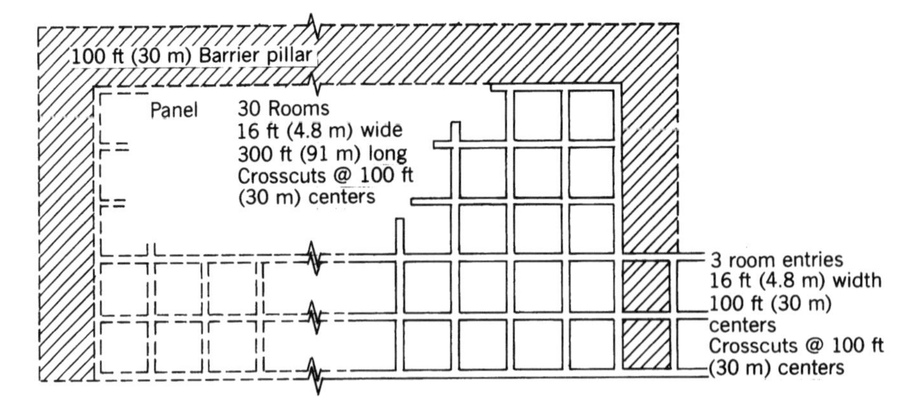

FIGURE 10.18. Panel in room-and-pillar mine. See Problem 10.2.

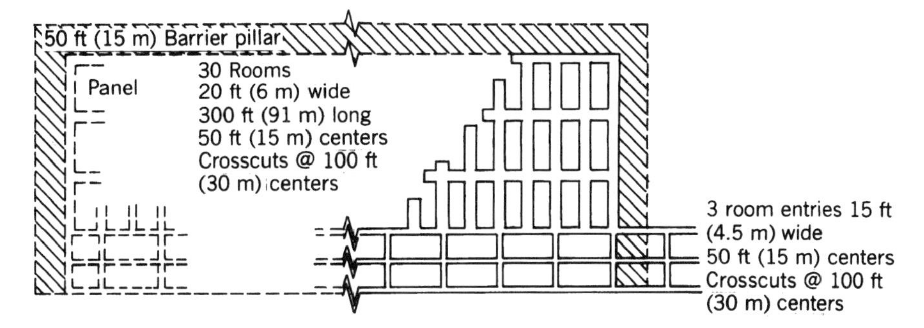

FIGURE 10.19. Panel in room-and-pillar mine. See Problem 10.3.

following cases:

a. No pillar recovery

b. Recovery of room and entry pillars

c. Room width increased to 30 ft (9 m), with timbering, and no pillar recovery

10.4 What should the room and crosscut width be in a room-and-pillar coal mine to achieve 55% recovery, leaving chain pillars but no barriers? Rooms are driven on 110 ft (33 m) centers and crosscuts on 60 ft (18 m) centers. Rooms and crosscuts are the same width. Calculate the width of opening to the nearest foot.

10.5 A mining recovery rate of 60% is required in an underground coal mine using the room-and-pillar method. Exploitation and development are both laid out in block system, and all openings are driven 18 ft (5.4 m) in width. Pillars are not recovered, nor are barriers utilized. Calculate the pillar dimension to provide the required recovery.

10.6 Mine openings of various shapes are to be driven 3000 ft (915 m) underground in a stress field of no lateral strain. The rock overlying and enclosing the openings has the following physical properties:

$$f_c = 22{,}500 \text{ lb/in.}^2 \text{ (155 MPa)}$$

$$f_t = 1200 \text{ lb/in.}^2 \text{ (8.3 MPa)}$$

$$\mu = 0.25$$

$$\text{SG} = 2.77$$

The openings are rectangular in cross section, have a fillet ratio of $\frac{1}{6}$, and are oriented horizontally with side ratios (R/h) of 1, 2, 4, and 8.

a. Compute the rock stresses existing in the earth at that depth.

b. Sketch and label the stress distribution for each opening.

c. Compute the actual critical stresses.

d. Determine which openings, if any, will fail.

e. Compute factors of safety for those that will not fail.

f. Identify the purpose for which each opening might be used.

10.7 A rectangular mine opening is to be driven 5000 ft (1525 m) underground in a hydrostatic stress field. The rock surrounding the opening has the following physical properties:

$$f_c = 30{,}000 \text{ lb/in.}^2 \text{ (207 MPa)}$$

$$f_t = 3000 \text{ lb/in.}^2 \text{ (20.7 MPa)}$$

$$\mu = 0.5$$

$$\text{SG} = 2.40$$

The opening is 8 ft (2.4 m) high and rectangular in cross section, has a fillet ratio of $\frac{1}{6}$, and is oriented with its long axis horizontal.

a. What is the widest possible opening (in even multiples of 8 ft, or 2.4 m) that may be driven?

b. Compute the rock stresses existing in the earth at the given depth.

c. Sketch and label the stress distribution for the opening.

d. Compute the actual critical stresses.

e. Compute the factor of safety for the opening (must exceed 1).

10.8 A single rectangular opening 10 ft (3.0 m) in height is driven in rock having strengths of $f_c = 18{,}000$ lb/in.^{2} (124 MPa) and $f_t = 1500$ lb/in.^{2} (10.3 MPa). Rock specific gravity is 2.3. The opening is located at a depth of 2000 ft (610 m) in a stress field of no lateral pressure and has a fillet ratio of $\frac{1}{6}$.

a. Determine whether the opening will fail when its width is 20 ft (6.1 m).

b. Is there any benefit to reducing the width to 10 ft (3.0 m)? To 5 ft (1.5 m)?

c. What is the maximum safe width of opening?

10.9 Openings 16 ft (4.9 m) wide and 8 ft (2.4 m) high have been driven without failure in a hard-rock mine; the fillet ratio is $\frac{1}{6}$. The rock is massive basalt with a specific gravity of 2.5, compressive strength of 25,000 lb/in.^{2} (172 MPa), and tensile strength of 3500 lb/in.^{2} (24.1 MPa). In a remote extension of the ore body, at a depth of 4500 ft (1370 m), failure of the mine openings is encountered for the first time. Investigate the failure, determine which stress field prevails in that section of the mine, and specify which part(s) of the opening fail(s).

10.10 An attempt was made to drive rooms 28 ft (8.5 m) in width in a 42 in. (1.1 m) seam of coal, but failure occurred. Investigating from a critical stress standpoint, what is the maximum safe width of room (in even multiples of side ratio given in the stress concentration table) that can be driven? The fillet ratio is $\frac{1}{6}$. The following conditions prevail:

Stress field	No lateral strain
Depth	1000 ft (305 m)
Specific gravity of overburden	2.31
Strength of roof strata	
Compressive	18,000 lb/in.^{2} (124 MPa)
Tensile	1200 lb/in.^{2} (8.3 MPa)
Strength of coal	
Compressive	6000 lb/in.^{2} (41.4 MPa)
Tensile	400 lb/in.^{2} (2.8 MPa)

11

UNDERGROUND MINING: SUPPORTED METHODS

11.1 CLASSIFICATION OF METHODS

The second category of underground mining methods is referred to as the supported methods. *Supported methods* are those methods that require some type of backfill to provide substantial amounts of artificial support to maintain stability in the exploitation openings of the mine (Hartman, 1987; Brackebusch, 1992a). Supported methods are used when production openings will not remain standing during their life and when major caving or subsidence cannot be tolerated. In other words, the supported class is employed when the other two categories of methods—unsupported and caving—are not applicable.

Pillars of the original rock mass are the ultimate form of ground control in an underground mine because they are capable of providing near-rigid support (Boshkov and Wright, 1973). In many horizontal methods, natural pillars are relied on for primary support, supplemented by light artificial supports such as bolts and timber. Vertical methods may use natural pillars to surround the stopes, but pillars in the production areas are impractical because they interfere with production operations. Artificial support in the form of pillars of backfill material is therefore used to control the rock mass and make mine output safe and productive.

In the design of artificial support systems for mining methods, an evaluation—preferably quantitative—of the load-carrying capacity of the natural rock structure is a prerequisite. Rock mechanics tests are performed to evaluate the structural properties of the rock, as discussed in Section 10.7. Table 14.2 in the final chapter discusses some of the rock properties and classifies them on the basis of compressive strength. However, to fully understand the capability of the rock to sustain a load, it may be better to determine the rock quality designation (RQD) based on drill core evaluation (discussed in Section

12.4). This variable is important in the decision to choose a supported mining method.

The supported class of methods is intended for application to rock ranging in competency from moderate to incompetent. Three methods (Table 4.1) are in this class:

1. Cut-and-fill stoping
2. Stull stoping
3. Square-set stoping

Cut-and-fill and stull stoping are usually applied to moderately competent rock, and square-setting is normally used only for the least competent rock.

All three of these methods are generally employed for vertical stoping. The methods have declined considerably as a class in the last few decades. However, cut-and-fill still finds a considerable amount of use because it is versatile and allows for a variety of mechanization methods. This class of methods accounts for only a few percent of U.S. underground mineral production.

11.2 CUT-AND-FILL STOPING

The only method of the supported class in common use today, *cut-and-fill stoping*, is normally used in an overhand fashion. The ore is extracted in horizontal slices and replaced with backfill material. However, the method has as many as eight variations (Lucas and Haycocks, 1973), with some of them extracting ore in the underhand direction. The backfilling operation is normally performed after each horizontal slice is removed. An estimated 3% of underground mineral production is derived from cut-and-fill stoping.

The fill material used in this method varies, depending on the support required and the material that may be available to the mine operator. Crandall (1992) lists the major types of fill as follows:

1. Waste fill
2. Pneumatic fill
3. Hydraulic fill with dilute slurry
4. High-density hydraulic fill

Crandall (1992) has identified advantages and disadvantages of each type of fill and methods of delivering them into the stope areas. The primary advantage of hydraulic fill is that it can be supplemented with portland cement to allow the fill to harden into a consistency that approaches hard rock. However, dilute hydraulic slurry creates problems because of the large amount of water used. Thus, mining companies often prefer to use high-density fill when possible.

Like other vertical exploitation openings, cut-and-fill stopes are generally bounded by pillars for major ground support. Because the stopes are filled, however, these pillars often can be recovered, in part or totally. The timing of fill placement is critical to the success of the method, because the fill must be in place to assume most of the superincumbent load on the ore in the stope.

Cut-and-fill stoping is a moderate-scale method. The stope must be designed in much the same way as in other vertical stoping methods. Stope dimensions are ordinarily influenced by mechanization factors, such as ease of access, maneuverability of equipment, and production rate requirements. Stope width ranges widely from 6 to 100 ft (2 to 30 m) and varies with several rock mechanics factors. The smallest equipment available normally sets the minimum width. Stope heights vary from 150 to 300 ft (45 to 90 m), and stope lengths from 200 to 2000 ft (60 to 600 m). However, note that the height in the open part of the stope is seldom more than 30 ft (10 m). Additional information on stope dimensions is available in Pugh and Rasmussen (1982) and Paroni (1992).

Although many variations of cut-and-fill stoping have been identified, only three will be described in detail here. These are overhand cut-and-fill stoping, drift-and-fill stoping, and underhand cut-and-fill stoping. Each of these is fairly widely practiced, both in the United States and internationally.

An *overhand cut-and-fill stoping* operation is shown in Figure 11.1. In the stope area, the miners work under the roof and generally have sufficient head room to move their equipment easily through the stope. The version shown in Figure 11.1 uses ramps to allow the diesel equipment to move from level to level through the stope area. Maximum grade of the ramps is normally 15 to 20% in the stope area. This version of cut-and-fill stoping is applied in rocks that are relatively strong, which will allow the stope to remain open with only bolting of the roof or bolting combined with mesh. Excellent descriptions of this variation of cut-and-fill are available in Pugh and Rasmussen (1982), Paroni (1992), and McLean (2001). Another variation of this method was widely used in the past, with slushers employed to move the ore. However, diesel equipment has now taken over the method as it offers greater productivity. Students should note that this method is sometimes called *breast stoping* or *back stoping*.

Drift-and-fill stoping is a method of mining used for ground conditions that are worse than those for the traditional overhand cut-and-fill stoping. The method can be utilized in either an overhand or an underhand fashion. The mining strategy involves keeping the openings relatively small to reduce the danger of rock failure. Each horizontal slice is removed by drifting forward, using mechanized equipment. After each drift is completed, the drift is backfilled with hydraulic fill to within a few feet (a meter) of the back. This provides excellent support of the hanging wall and footwall rock. A diagram of this method is shown in Figure 11.2 (Paroni, 1992). Note that the vein is four drifts wide, but that fill is placed after each individual drift is completed.

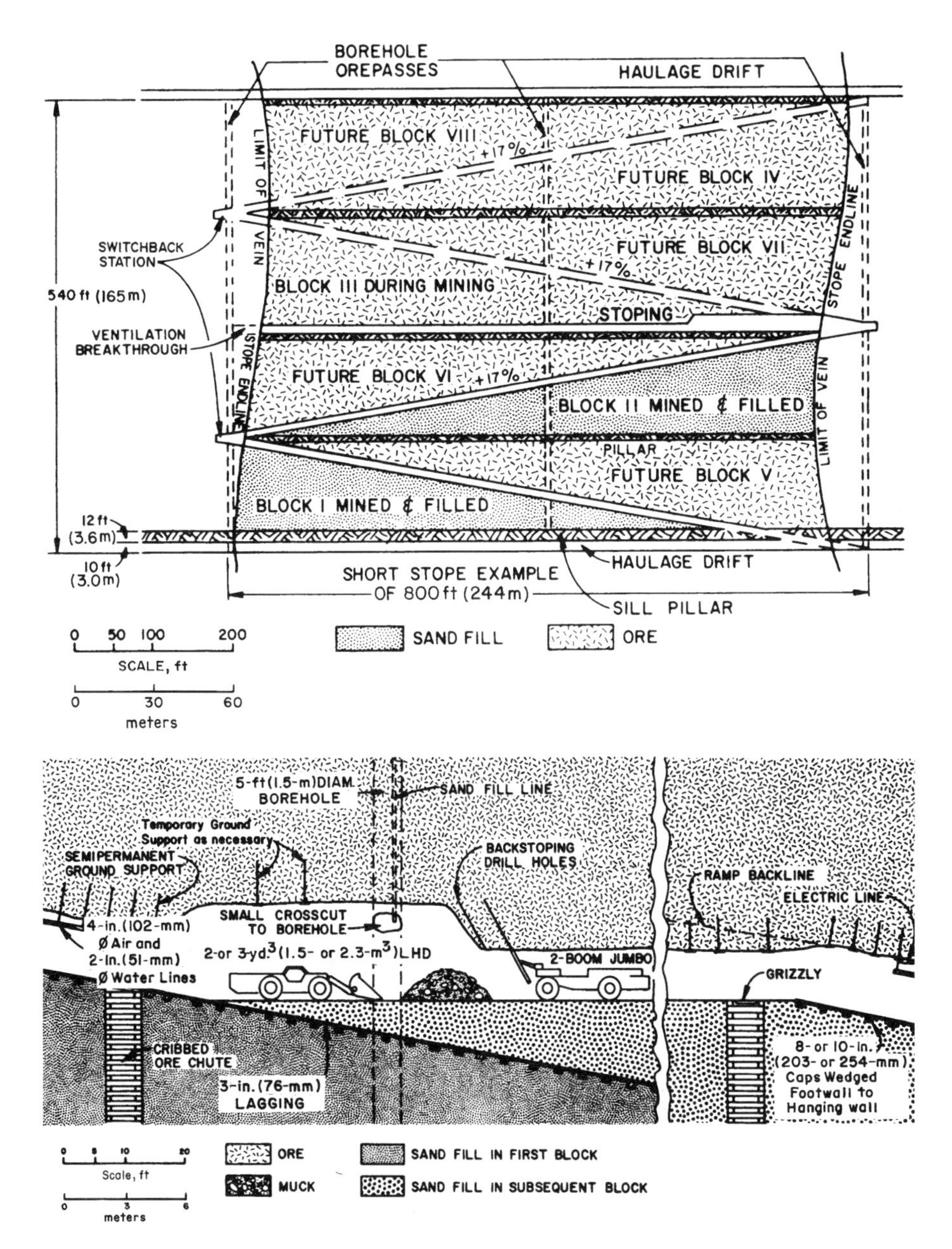

FIGURE 11.1. Highly mechanized version of cut-and-fill mining. (*Top*) Location of ramps to permit mining of alternate blocks of ore; stoping in progress in block III. (*Bottom*) Stoping operations, with drill jumbo and LHD. (After Pugh and Rasmussen, 1982. By permission from the Society for Mining, Metallurgy, and Exploration, Inc., Littleton, CO.)

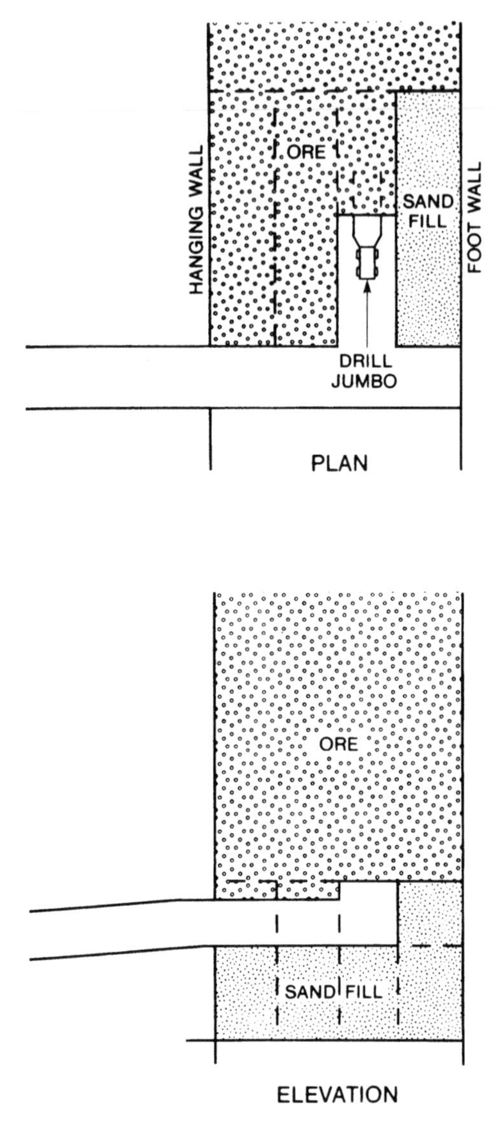

FIGURE 11.2. Drift-and-fill mining using four drifts to span the vein width. (After Paroni, 1992. By permission from the Society for Mining, Metallurgy, and Exploration, Inc., Littleton, CO.)

This provides maximum support and minimizes the probabililty of rock failure. Brackebusch (1992b); Paroni (1992), Bristol and Guenther (1997), Biles et al. (2000), and Sobering (2001) provide good descriptions of the procedures used in drift-and-fill mining.

Underhand cut-and-fill stoping is likewise used in poorer quality rock. It is characterized by the procedure of taking horizontal slices from the top to the

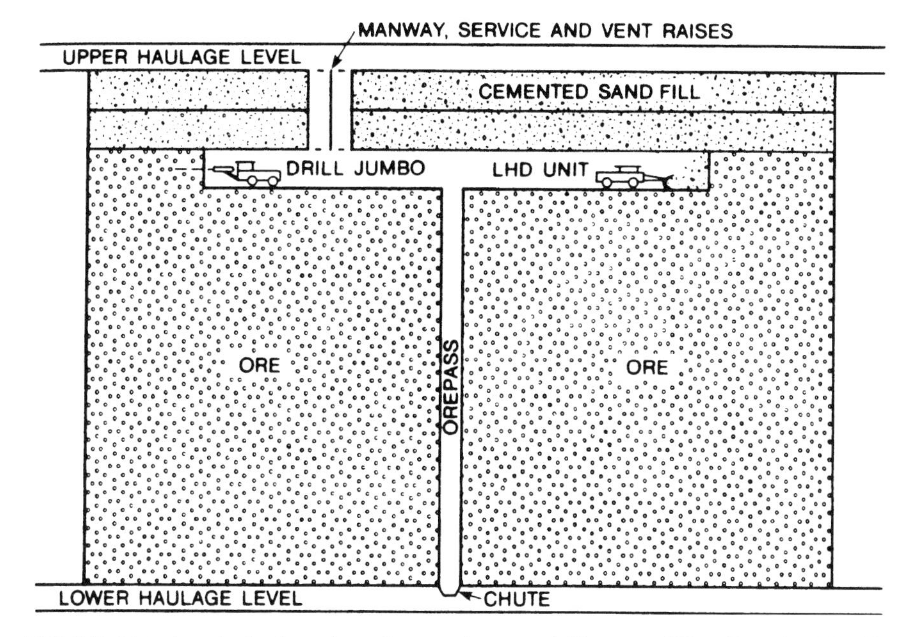

FIGURE 11.3. Longitudinal section of an undercut-and-fill stope. (After Paroni, 1992. By permission from the Society for Mining, Metallurgy, and Exploration, Inc., Littleton, CO.)

bottom of the stope area. To stabilize the surrounding rock after each slice or group of slices, the stope is then filled, usually with cemented hydraulic fill. The fill becomes the roof for subsequent cuts. The method has been practiced with filling placed after every horizontal cut (see Figure 11.3), as well as after a number of cuts have been produced below the fill. The fill is normally a cemented hydraulic mixture to maximize the support capabilities of the material. J. W. Murray (1982), Paroni (1992), Brechtel et al. (2001), Kump and Arnold (2001), Le Roy (2001), and Peppin et al. (2001) outline the methods used in underhand cut-and-fill stoping.

11.2.1 Sequence of Development

Development of cut-and-fill stopes follows the same general plan as other vertical stoping methods, but there are differences. For example, a haulage drift plus drawpoints and crosscuts are driven for access to the ore. However, an undercut and loading bells are not needed, because the ore moves from the stope to the drawpoints by gravity flow through orepasses. Mining begins by taking the first horizontal slice, either at the top or bottom of the stope. If mobile equipment is used in the stope, ramps must be driven or the equipment will be captive in the stope. Some mines use the captive equipment method to reduce the number of development openings that are necessary.

11.2.2 Cycle of Operations

Mechanization using diesel equipment is the normal method of making cut-and-fill stopes both productive and efficient. The production cycle is much like that for stope-and-pillar operations in many ways. The following operations are normally used:

1. *Drilling:* pneumatic or hydraulic percussion or rotary-percussion drills (usually mounted and highly mechanized); handheld drill now avoided except in extremely rare situations; hole sizes normally 2 to 3 in. (51 to 76 mm)
2. *Blasting:* ammonium nitrate and fuel oil (ANFO), slurries, emulsions; charging by cartridge or by bulk methods; firing electrically or by nonelectric methods
3. *Secondary breakage:* drill and blast, mudcapping, impact hammer (all performed in the stope)
4. *Loading:* (in the stope) load-haul-dump (LHD) device, front-end loader, slusher; (on the haulage level) LHD, front-end loader
5. *Haulage:* LHD, truck, rail

The cut-and-fill method also requires that the filling operations be integrated into the production operations. This produces a certain amount of discontinuity because the normal production cycle must be interrupted. Fill barricades are often required, and the filling itself will also interfere with the normal production. Crandall (1992) outlines the various backfilling methods with regard to their requirements and provides information on the advantages and disadvantages of each. Auxiliary operations for cut-and-fill mining are similar to those for other methods. Sections 10.2 and 10.3 outline the general requirements.

11.2.3 Conditions

The following are based on discussions of cut-and-fill stoping by Morrison and Russell (1973), Hamrin (1982), Waterland (1982), Brackebusch (1992b), and Paroni (1992).

1. *Ore strength:* moderately weak to strong (can be quite weak with drift-and-fill; must be stronger for other variations)
2. *Rock strength:* weak to fairly weak
3. *Deposit shape:* tabular; can be irregular, discontinuous
4. *Deposit dip:* moderate to fairly steep (>45°); can accommodate flatter deposit if orepasses are steeper than the angle of repose
5. *Deposit size:* narrow to moderate width (6 to 100 ft or 2 to 30 m), fairly large extent

6. *Ore grade:* fairly high
7. *Ore uniformity:* moderate, variable (can sort waste in stope)
8. *Depth:* moderate to deep (typically <4000 to 8000 ft or 1.2 to 2.4 km)

11.2.4 Characteristics

The following characteristics are compiled from information in Morrison and Russell (1973), Thomas (1978), Hamrin (1982), Waterland (1982), Brackebusch (1992b), and Paroni (1992).

Advantages

1. Moderate productivity (rather low with slushers; much better with diesel equipment).
2. Moderate production rate.
3. Permits good selectivity.
4. Low development cost.
5. Moderate capital investment; adaptable to mechanization.
6. Versatile, flexible, and adaptable (at least eight variations have been identified).
7. Excellent recovery if pillars are recovered (90 to 100%).
8. Low dilution (5 to 10%).
9. Surface waste can be used as fill.
10. Moderately good safety record.

Disadvantages

1. Fairly high mining cost (relative cost about 55%); quite high with slushers, much lower with diesels (Mutmansky et al., 1992).
2. Handling of fill may be up to 50% of mining cost.
3. Filling operations interfere with production.
4. Must provide stope access for mechanized equipment.
5. Tends to be labor-intensive; requires skilled miners and close supervision.
6. Compressibility of fill may cause some ground settlement.

11.2.5 Applications and Variations

Cut-and-fill stoping is one of the most variable of mining methods; Lucas and Haycocks (1973) identify eight varieties. Even more varieties exist today, as mining companies have experimented with better methods of applying mechanized equipment to the method. The classical cut-and-fill method was oriented toward overhand mining with slushers. Underhand cut-and-fill was instituted in the Sudbury Basin to recover pillars in poor ground (J. W. Murray, 1982). This opened the door to other underhand methods that used cemented fill as the roof of the stope.

Today's cut-and-fill operations are mostly oriented toward efficient use of diesel equipment in highly mechanized mines. The three variations of cut-and-fill discussed in Section 11.2 are commonly used, though other variations occasionally come into play. The versatility of cut-and-fill mining is obvious in the wide variety of vein sizes mined. Several articles discuss the ability to mine very narrow stopes of less than 6 ft (2 m) in width (Paraszczak, 1992; Mackenzie and Hare, 1994). Others discuss capability of the method in wide stopes by use of the drift-and-fill or underhand cut-and-fill variations (Barker and Harter, 1997; Bristol and Guenther, 1997).

Many accounts of cut-and-fill operations in the United States are available to the student. Some of the more interesting articles discuss mining gold at the Homestake Mine in South Dakota by mechanized cut-and-fill (Pfarr, 1991), mining silver at the Greens Creek Mine in Alaska by the drift-and-fill method (Walker, 1991b), cut-and-fill variations at Hecla's metal mines in Idaho (Phelps, 1991), and underhand cut-and-fill operations for gold at the Bullfrog Mine in Nevada (Arnold, 1996). More recent publications have discussed cut-and-fill variations practiced at the Jerritt Canyon Mine in Nevada (Bristol and Guenther, 1997; Brechtel et al., 2001), at the Getchell Mine in Nevada (Barker and Harter, 1997), and at the Stillwater complex in Montana (Dyas and Marcus, 1998; Bray et al., 2001). Other applications of the method, both domestic and international, can be found in Bullock and Mann (1982) and Hustrulid and Bullock (2001).

11.3 STULL STOPING

Stull stoping is infrequently used and relatively unimportant today; as a method, it accounts for less than 1% of U.S. underground mine production. However, it possesses features that other methods cannot readily match. For example, the range of deposit dips to which stull stoping is applicable is broad. It varies from the upper limit of grade for a haulage vehicle (about 20% or 11°) to the lower limit for gravity flow (about 45°). In this range, a slusher in a stull stope can mine without difficulty, though not at a low cost. The lower half of this range of deposit dips can be accommodated in some deposits by room-and-pillar mining, using haulages angled somewhat from the strike of the deposit (see Hamrin, 1982, p. 92), but the upper half of the range cannot. Thus, stull stoping has a possible role to play, even though it is not a highly significant one. It is identified as overhand stoping when systematic or random timbering, coupled with pillars of fill, are used to support the hanging wall. The timbers are usually called *posts* (in a near-horizontal deposit) or *stulls* (in a pitching deposit). Stopes that become large are then stabilized by backfilling portions of the open area with fill material. This is shown in Figure 11.4, where

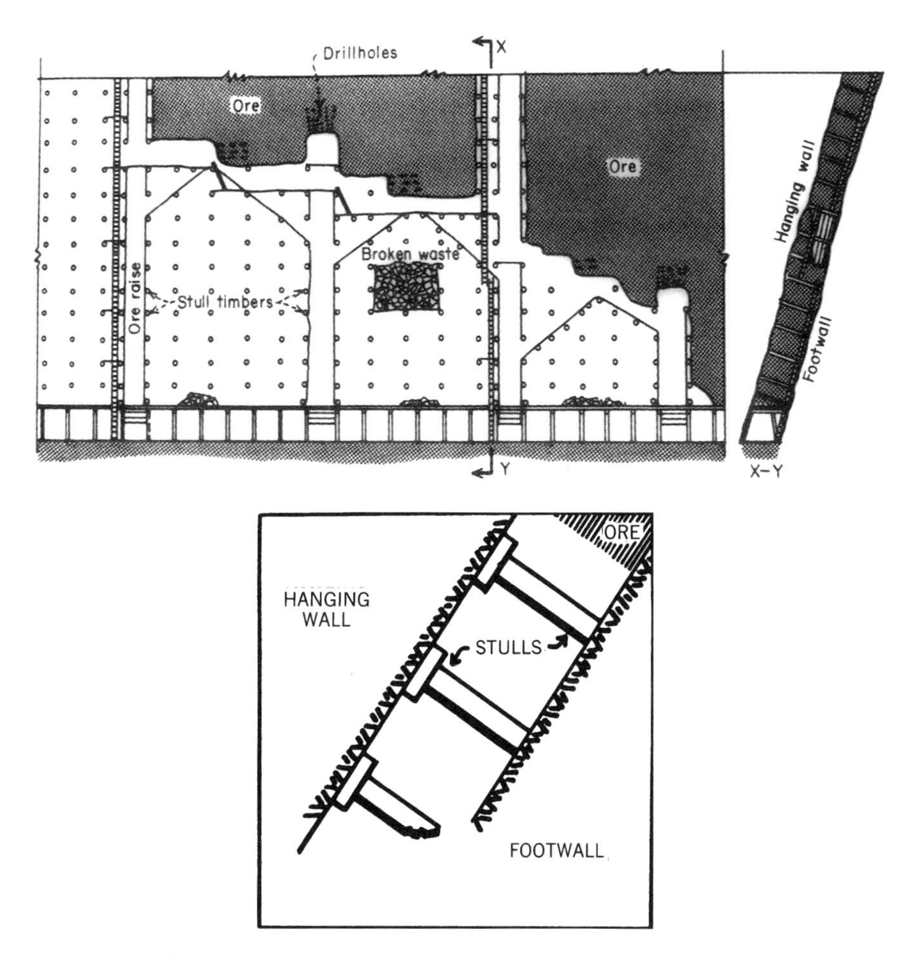

FIGURE 11.4. Stull stoping in an inclined, tabular deposit. (*Top*) Two views of stope. (After Jackson and Hedges, 1939.) (*Bottom*) Details of timbering with stulls. (After Peele, 1941. Copyright © 1941, John Wiley & Sons, New York.)

uncemented waste fill is utilized. Note that the method applies best to thin, pitching, tabular deposits with moderately weak wall rock.

As originally used in underground supported methods, ground control was accomplished by timbering, sometimes supplemented with backfilling for more substantial, and permanent, support. Descriptions of the timber requirements are outlined in Wilhelm (1982) and Taylor (1982). In stull stoping, as in square-set stoping, the timbers also provide a working platform for the miners (Figure 11.4). These are usually wooden platforms supported by the stull timbers. Normally, systematic pillars are not left in a stull stope. Waste pillars or pillars of cemented fill can be used to support the hanging wall. Periodically, the stope may be filled to provide more substantial support and to enhance the recovery of ore (Lucas and Haycocks, 1973; Thomas, 1978). In this modification, stull stoping resembles cut-and-fill stoping or, if the support becomes more elaborate, square-set stoping. It also resembles stope-and-pillar mining when it is conducted in near-horizontal deposits.

11.3.1 Sequence of Development

Development for stull stoping follows the conventional sequence for vertical stoping methods but is kept minimal and simple. Because it is a small-scale method, elaborate development is neither warranted nor necessary. After essential level development is carried out and access to the stopes is provided, raises are driven between levels. Orepasses are formed as the stope advances, so bells and drawpoints are unnecessary (generally, the orepasses terminate in chutes or loader crosscuts to facilitate further handling).

11.3.2 Cycle of Operations

Unlike cut-and-fill stoping, stull stoping cannot easily utilize mechanization. Ore movement in the stopes is primarily by gravity or slushers. The stopes are small and cannot accommodate mobile equipment. Hence, the cycle of operations is labor-intensive and uses the following basic pattern:

Drilling: hand-held pneumatic drills (airleg drill or stoper); hole size 1.5 to 2.5 in. (38 to 68 mm)

Blasting: ANFO, slurries; charging by hand (cartridge) or machine (bulk); firing electrically, by detonating fuse, or by nonelectric methods

Secondary blasting: (in stope) drill and blast, mudcap

Loading: gravity flow to chutes or drawpoints, slusher transport to loadout points

Haulage: LHD, truck, rail

The most critical auxiliary operation is setting stulls for ground support. They must be installed close to the top of the stope as a round is blasted. The flooring is advanced upward as the stope heightens, so as to maintain a convenient work platform. Other auxiliary operations are similar to those identified in Sections 10.2 and 10.3.

11.3.3 Conditions

The following conditions are derived from Morrison and Russell (1973), Thomas (1978), and Taylor (1982):

1. *Ore strength:* fairly strong to strong, more competent than for cut-and-fill stoping

2. *Rock strength:* moderate to fairly weak
3. *Deposit shape:* approximately tabular; can be irregular
4. *Deposit dip:* works best at $<45^\circ$; can accommodate flatter deposits at higher cost
5. *Deposit size:* relatively thin (<12 ft or 3.6 m)
6. *Ore grade:* fairly high to high
7. *Ore uniformity:* moderate; can sort ore and waste in the stope
8. *Depth:* moderate (<3500 ft or 1.1 km)

The following features are based on information in Lewis and Clark (1964), Thomas (1978), and Taylor (1982):

Advantages

1. Simple method; adaptable to small, irregular ore bodies.
2. Requires little mechanization; low capital cost.
3. Low development cost.
4. Selective; versatile for thin tabular deposits at any angle.
5. Good recovery if pillars are mined (>90%).
6. Low dilution (5 to 10%).

Disadvantages

1. Low productivity.
2. Low production rate.
3. High mining cost (relative cost about 70%).
4. Labor-intensive and slow.
5. Heavy timber requirements and cost.
6. Limited applications are available.

11.3.4 Applications and Variations

Very little stull stoping is performed in North America at the present time. The method is applied only where other methods are not readily used. In the past, it has been applied fairly regularly (Stout, 1980; Hartman, 1987), but little mention of the method can be found in the modern literature. One article on stull stoping is provided by Laflamme et al. (1994), in which it is compared with other methods. A method similar to stull stoping, called breast-and-pillar mining, has been practiced in the Pennsylvania anthracite fields for a century. This method is conducted much like traditional stull stoping except that the broken coal is used to support the roof. To accomplish this, a fill enclosure called a *stable* is constructed to hold the broken coal. The coal is not removed from this stable until the entire breast has been mined. Stull stoping may be of greater significance in other parts of the world where labor costs are much lower.

11.4 SQUARE-SET STOPING

The final method in the supported class, *square-set stoping*, is probably the least used of all the mining methods we are discussing. In North America, the method has just about disappeared, as it is the most expensive (along with dimension-stone quarrying) of all the mining methods. We include it here because it is important from a historical standpoint and because it is used in other parts of the world. Like block caving, square-set stoping was developed in the United States. Mining engineers find this method quite interesting because it shows how mining can be conducted under the most difficult of conditions if support of the surrounding earth and the removal of ore are conducted with careful planning and execution.

In square-set stoping, small blocks of ore are systematically extracted and replaced by a prismatic framework of timber sets, framed into an integrated support structure and backfilled level by level to provide substantial support to the surrounding rocks. Each timber set consists of a post, cap, and girt, as shown in Figure 11.5. The method is quite versatile and can be conducted in the overhand, underhand, or horizontal direction. It is capable of application to the weakest ore and rock under the heaviest of ground conditions. Cut-and-fill stoping of the drift-and-fill variety is currently competing with square-set stoping, making inroads into the weaker rock category. Hence, the method is finding less usage with the passage of time.

During stoping, the square sets are cut to size on the surface and assembled in the stope. The arrangement of timbers in the overall stope area is also shown in Figure 11.5. Note the details of the framing of the timber and the assembly of each joint. Gardner and Vandenburg (1982) have provided an excellent analysis of the timber sets. Square sets are generally about 6 or 8 ft (1.8 or 2.4 m) in dimension horizontally and 8 or 10 ft (2.4 or 3.0 m) in height, with either 8 × 8 in. (203 × 203 mm) or 12 × 12 in. (305 × 305 mm) cross sections.

Note that in a square set operation, the filling operation is normally conducted about two levels below the back of the stope. The usual arrangement in an overhand stope (the most common type) is as follows:

Upper level	Drilling and blasting
Second level	Transport of the ore to orepasses
Third level	Filling operations

This arrangement allows as much support as possible without having the filling interfere with the production operations. Nonetheless, each operation is labor-intensive, making the method a very expensive proposition.

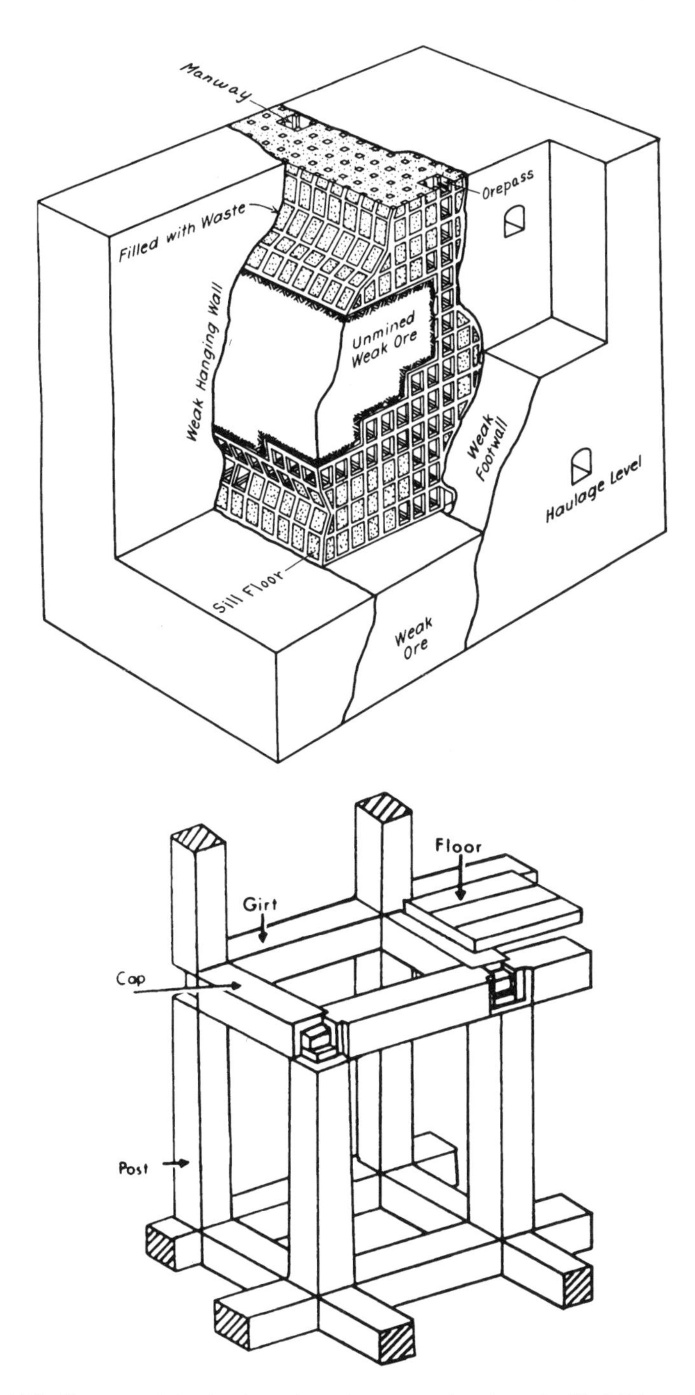

FIGURE 11.5. Square-set stoping in an irregular or massive deposit. (*Top*) Prismatic arrangement of timber in stope, partially backfilled. (After Lewis and Clark, 1964. Copyright © 1964. John Wiley & Sons, New York.) (*Bottom*) Details of square-set timbering. (After Lucas and Haycocks, 1973. By permission from the Society for Mining, Metallurgy, and Exploration, Inc., Littleton, CO.)

11.4.1 Sequence of Development

Because of the characteristically bad ground associated with square-set stoping, only restricted development is carried out. This limits the amount of opening that must be held intact during mining. Fortunately, square-set stoping is also a small-scale method, so few openings are necessary. The minimum requirements are a haulage drift and crosscut to the stoping area; two raises between levels are also mandatory for access and ventilation.

If the stope is mined overhand, development commences at the sill and progresses upward with mining. Undercuts, bells, and drawpoints are omitted; orepasses terminate in chutes to load mine cars directly. Generally, each level is mined out entirely before the next level is started in order to keep the fill as close to the back as possible. In a large stope, several faces can be provided, but a minimum number of openings are normally exposed to the ground pressure. Underhand square-set stoping is avoided because of the added problems of dealing with fill materials above the working level.

11.4.2 Cycle of Operations

Little mechanization is possible within the square-set stope itself. Space is too limited and the load-bearing capacity of the sets is inadequate to accommodate heavy equipment. The only possibilities for stope transport are small slushers or a very small loader. Airleg drills or mounted drifters are used. However, the limitations of the stope space severely restrict productivity. The production cycle (Section 5.1) consists of the following:

Drilling: handheld percussion drills of the airleg type; column-and-bar-mounted drifters

Blasting: ANFO, slurries; charging by hand (cartridges) or machine (bulk); firing by electric, detonating fuse, or nonelectric methods

Secondary blasting: generally not required

Loading: gravity flow to chutes (if possible); small slusher or small loader otherwise

Haulage: rail, LHD

As in stull stoping, the most essential auxiliary operation is ground control, in this case handling and standing timber and filling with waste. Ventilation is also vital. See Sections 10.2 and 10.3 for other auxiliary operations.

11.4.3 Conditions

The following conditions are compiled from Morrison and Russell (1973), Stout (1980), and Gardner and Vandenburg (1982):

1. *Ore strength:* weak to very weak
2. *Rock strength:* weak to very weak

3. *Deposit shape:* any, regular to irregular
4. *Deposit dip:* any, preferably $>45°$ so that gravity flow can be used
5. *Deposit size:* any, generally small
6. *Ore grade:* high
7. *Ore uniformity:* variable (can sort in the stope)
8. *Depth:* deep (up to 8500 ft or 2.6 km)

11.4.4 Characteristics

The following information is summarized from Morrison and Russell (1973), Hamrin (1982), and Gardner and Vandenberg (1982):

Advantages

1. Flexible, versatile, adaptable to a wide variety of conditions.
2. Suitable for the worst ground conditions when caving and subsidence are not permitted.
3. Selective for irregular deposit and variable ore occurrences; waste can be sorted and left as fill.
4. Excellent recovery (nearly 100%).
5. Requires little mechanization; low capital cost.
6. Low development cost.

Disadvantages

1. Very low productivity.
2. Very low production rate.
3. Highest mining cost (relative cost = 100%).
4. Most labor-intensive underground mining method.
5. Very high timber costs.
6. Fire hazards are high.

11.4.5 Applications

No U.S. mines are known to be using the square-set mining method at the current time. In the past, selective mining of high-grade ore areas was conducted via square-setting at the Homestake gold mine in South Dakota (Hartman, 1987), at the Bunker Hill silver mine in Idaho (Songstad, 1982), and at the Burgin lead-zinc operation in Utah (Rausch et al. 1982). These square-set operations appear to have been phased out in the 1980s. The only current operation found in the literature by the authors is the Yauricocha Mine in Peru (Phelps, 1994). This mine produces a copper-lead-zinc-silver ore with a variety of mining methods, each suited to a different part of the ore mass.

11.5 SPECIAL TOPIC A: SELECTION AND COST ESTIMATE OF DRILLING EQUIPMENT

The selection procedures for a production drill were outlined in Section 5.2.3.3. We use them here as general guides, because they are applicable to the selection of equipment for any unit operation. In the present discussion, we deal with underground drilling equipment to demonstrate the broad application of the methodology.

11.5.1 Drill Selection

Although there are a number of performance specifications to be considered in the selection of a production drill, those of primary importance are the following:

1. Type of drill, type of bit, size of bit, power source
2. Drillability (rate of penetration) in rock specified
3. Blasting factor (area of working face broken per drillhole)
4. Drilling factor (length of drillhole required per unit weight of rock broken)
5. Number of drill units per rig or jumbo

Items included in factor 1 are selected with the aid of Chapter 5, especially Tables 5.1 and 5.2. Drillability, the second factor, is usually based on empirical data; Table 11.1 is an example of such data for small-diameter percussion drills. The 1.00 norm is for Barre granite; other rocks are given in relation to the drillability and abrasiveness of Barre granite. The abrasion index is an indicator of bit wear, as drillability is a measure of penetration rate. Use of the blasting factor (3) is an approximate measure expressing the fragmenting ability of a drillhole and is based on field experience. Factor 4 expresses fragmenting ability by another measure. Finally, the number of drills is calculated for the prevailing operating and job conditions as reflected by the preceding factors. A numerical example demonstrates the selection procedure.

Example 11.1. Given the following information regarding a drilling application in a hard-rock mine, select the number of drills required for a mobile, hydraulic, trac-mounted drill rig with a capacity of one to four drill booms:

Mining method	Stope-and-pillar mining
Stope face	Height 20 ft (6.0 m)
	Width 24 ft (7.2 m)
Drill power available	Hydraulic
Drill bits	2.25 in. (57 mm) carbide, cross bits
Rock	Denver granite gneiss

Penetration rate, Barre granite	40 in./min (17 mm/sec)
Blasting factor (BF)	6.8 ft^2/hole (0.632 m^2/hole)
Depth of round or holes	16 ft (4.8 m)
Delay time in drilling/hole	1.40 min
Drill rounds/shift	3
Allowable drilling time/round	2.0 hr
Tonnage factor (TF)	14.0 ft^3/ton (0.437 m^3/tonne)

Also determine the tons (tonnes) of rock broken and the drilling factor (DF).

SOLUTION. From Table 11.1 for Denver granite gneiss, read drillability = 1.52. Therefore, find the penetration rate based on rate in Barre granite:

$$\text{Penetration rate} = (1.52)(40) = 60.8 \text{ in./min}$$

$$\text{Required number of holes} = \frac{(20)(24)}{6.8} = 70.6 \text{ or } 71 \text{ holes}$$

$$\text{Net drilling time} = \frac{(16)(12)}{60.8} = 3.15 \text{ min/hole}$$

$$\text{Gross drilling time} = \text{net time} + \text{delays} = 3.15 + 1.40 = 4.55 \text{ min/hole}$$

$$\text{Capacity/drill} = \frac{(2)(60)}{4.55} = 26.4 \text{ or } 27 \text{ holes}$$

$$\text{Required number of drills} = \frac{71}{27} = 2.63 \text{ or } 3 \text{ drills}$$ (go to the next larger whole number if the fraction is >0.3)

$$\text{Total hole length/round} = 71 \text{ holes} \times 16 \text{ ft/hole} = 1136 \text{ ft/round}$$

$$\text{Total hole length/shift} = \frac{(3)(1136)}{8} = 426 \text{ ft/hr}$$

$$\text{Volume} = 7680 \text{ ft}^3\text{/round}$$

$$\text{Weight} = \frac{\text{volume}}{\text{TF}} = \frac{1136}{14.0} = 549 \text{ tons/round}$$

$$\text{Weight} = (549 \text{ tons/round})\left(\frac{3 \text{ rounds}}{8 \text{ hr}}\right) = 206 \text{ tons/hr (186.9 tonnes/hr)}$$

$$\text{DF} = \frac{\text{hole length}}{\text{weight of rock}} = \frac{1136}{549} = 2.07 \text{ ft/ton (0.68 m/tonne)}$$

In the final selection of rock-breakage equipment for the cycle of operation in a mine, the blasting round is designed along with the drills chosen. In this way, values specified during drill selection for hole size, depth of round, blasting factor, and drilling factor can be confirmed.

TABLE 11.1 Percussion Drillability and Abrasion Index of Various Rocks as Compared with Barre Granite

Rock	Location	Drillability	Abrasion Index
Barre granite	Barre, VT	1.00	1.00
Granite	Dvorshak, ID	1.11	1.14
Granite	California	1.10	0.54
Granite	Newark, NJ	1.05	1.27
Granite	Mt. Blanc, France	0.92	0.86
Granite	Grand Coulee, WA	0.50	2.40
Granite	Bulgaria	0.45	2.29
Granite gneiss	Denver, CO	1.52	1.00
Granite gneiss	Vancouver, BC, Canada	0.89	1.03
Granite gneiss	Hamburg, NJ	0.67	1.46
Quartzite	Capetown, South Africa	1.22	2.70
Quartzite	Corter Dam, GA	1.00	1.40
Quartzite	New Zealand	0.78	1.70
Quartzite	Canada	0.72	3.17
Quartzite	Minnesota	0.56	8.60
Quartzite	Canada	0.33	1.45
Magnetite	Kiruna, Sweden	1.00	1.23
Magnetite	Kirkland Lake, ON, Canada	0.59	1.41
Taconite	Kirkland Lake, ON, Canada	0.84	4.13
Hematite (red)	Sarajevo, Yugoslavia	1.50	0.40
Hematite (dark)	Sarajevo, Yugoslavia	2.20	0.70
Siderite	Sarajevo, Yugoslavia	0.90	0.80
Siderite	Suffern, NY	0.89	0.55
Sandstone	Nova Scotia, Canada	2.70	0.14
Sandstone	Ohio	3.10	0.11
Sandstone	New Zealand	2.30	1.20
Shale	Michel, BC, Canada	0.75	2.80
Shale	Scranton, PA	2.00	0.00
Limestone	Davenport, IA	1.79	0.28
Limestone	Portsmouth, NH	1.77	0.65
Limestone	Saratoga, NY	1.22	0.01
Limestone	Tulsa, OK	1.19	0.10
Limestone	Bellefonte, PA	0.94	0.09
Limestone	Buffalo, NY	0.89	0.09

Source: Russell, 1982. (By permission from Society of Mining Engineers, Inc., Littleton, CO.)

11.5.2 Drilling Cost Estimate

We will develop a cost-estimating procedure in this section. It is based on that developed by the U.S. manufacturers of earth-moving and materials-handling equipment and detailed in their publications (Power Crane and Shovel Association, 1976; Terex Corporation, 1981). For further information on cost estimating for mining equipment, see Church (1981), Hemphill (1981), Russell (1982), Skodack (1982), and Rhoades (1998).

Our cost estimation procedure is not a precise calculation but a close estimate of expected costs. It is intended to provide a valid comparison of candidate machines being considered for a mine task, tentatively chosen by the equipment selection process employed in Section 8.5. Our objective is to determine the unit cost in $/ton ($/tonne) of drilling the faces in a mine. To do that, we develop costs per unit time, generally in $/hr, for both ownership of and operation of the equipment. *Ownership costs* are fixed and consist of depreciation, interest, tax, insurance, and storage. *Operating costs* are variable (although some are calculated as though they are fixed) and include tires (if a rubber-tired vehicle), maintenance, fuel or power, lubrication, and labor. As is customary in dealing with individual operations of the production cycle, only direct costs are considered.

The process is often carried out using the basic cost-estimation form provided in Table 11.2. Note that ownership and operating costs are calculated separately. We should also note that the form is not universally applicable and may have to be altered somewhat for various types of equipment. For example, if the equipment does not have rubber tires, then the tire replacement cost section must be replaced to cover other related costs. Other changes may be necessary to estimate costs of different types of equipment.

The following information is given to provide students with the general practice followed in cost estimation and some rules of thumb to use in estimation of the costs. Note that information can best be obtained by careful study of the actual situation. The rules of thumb should be used only where better information is not available.

A. *Ownership costs*

1. *Depreciation.*
    a. *Purchase price:* Basic cost is for the drill rig or jumbo with one drill boom, cost of additional drills and booms extra.
    b. *Freight:* Basic weight is for drill rig only; weight of each drill boom is additional.
    c. *Unloading:* Estimated at 10% of freight costs.
    d. *Operating period:* Assume 4000 hr/yr unless specified otherwise.
    e. *Economic life:* Assume 24,000 hr unless better information is available.
2. *Interest, taxes, insurance, and storage:* Interest changes with the times, taxes can be estimated at 2% for many operations, insurance and storage can also be estimated at 2%.

B. *Operating costs*

1. *Consumables cost* (replaces tire costs for non-tired equipment).
    a. *Bits:* Base on bit cost, bit life, and hole length drilled per hour.
    b. *Steel (drill rods):* Base on steel unit costs, steel life, and hole length drilled per hour.

TABLE 11.2 Cost Estimation Form

_______ Mining Equipment Unit

A. Ownership Costs

1. DEPRECIATION
   a. Purchase price = $ _________
   b. Salvage value (____ %) = − _________
   c. Freight ____ lb @ $ ____ /cwt = _________
   d. Unloading and moving cost = _________
   e. Delivered price = _________, say $ _________
   f. Operating period ____ hr/yr
   g. Economic life ____ hr = ____ yr(n)
   h. Depreciation = $ _________ (del. price less tire cost) / _________ hr = $ _________ /hr

2. INTEREST, TAXES, INSURANCE, AND STORAGE
   a. Rate = interest 14% + taxes 2% + other 2% = 18%
   b. Average annual investment rate = $\frac{n+1}{2n}$ = _________ = _________ %
   c. Average annual investment = $ _________ × ______ % = $ _________
   d. Annual fixed charge = $ _________ × 18% = $ _________
   Fixed charge = $ _________ / ___________ hr/yr = $ _________ /hr

TOTAL OWNERSHIP COSTS = $ _________ /hr

B Operating Costs

1. TIRE REPLACEMENT COST
   Purchase price 1 set of tires = $ _________
   Tire life _________ hr
   Tire cost = $_________ / _________ hr = $ _________ /hr
2. TIRE REPAIR COST: ____ % × $ _________ tire cost = $ _________ /hr
3. REPAIRS, MAINTENANCE: ____ % × $ _________ deprec. = $ _________ /hr
4. FUEL: _________ gal/hr @ $ ______ /gal
   OR POWER: ____ kW @ $ ______ /kW-hr = $ _________ /hr
5. LUBRICATION: ______ gal/hr @ $ ______/gal = $ _________ /hr
6. AUXILIARY FUEL: $ _________ /hr
7. LABOR: ______ operator @ $ ______ /hr = $ ______ /hr
   ______ oiler @ $ ______ /hr = ________ /hr
   ______ helper @ $ ______ /hr = ________ /hr
   Total $ ________ /hr
   + 35% benefits $ ________ /hr = $ _________ /hr

TOTAL OPERATING COSTS = $ _________ /hr

TOTAL OWNERSHIP AND OPERATING COSTS = $ _________ /hr

UNIT COST $ _________ /hr ÷ _________ ton/hr = $ _________ /ton

2. *Tire and tire repair costs:* Disregard for trac-mounted units.
3. *Repairs, maintenance:* Can use 75% of depreciation cost.
4. *Fuel or power:* Calculate separately for drilling and tramming, if possible.
5. *Lubrication:* Include for hydraulic drill, disregard for pneumatic.
6. *Auxiliary fuel:* Estimate where applicable.
7. *Labor:* Based on number of operators needed.

C. *Overall ownership and operating costs:* See calculation.

D. *Unit costs:* Calculate on unit production and unit hole length bases.

We will now show how this information can be utilized by presenting a rock drill cost estimation problem.

Example 11.2. For the three-boom percussion drill selected in Example 11.1, estimate the overall hourly and unit costs. Assume the following conditions:

Capital costs	
Drill rig, one boom	$300,000
Additional boom	$60,000
Rig weight (with one drill boom)	25,000 lb. (11,340 kg)
Weight, each additional boom	2500 lb. (1134 kg)
Operating costs	
Drilling	
Electric power cost	$0.07/kWh
Power consumption	120 kW
Tramming	
Diesel fuel cost	$1.10/gal
Motor rating	100 hp (75 kW)
Fuel consumption (avg.)	10 gal/hr (37.9 L/hr)
Tram time per shift	30 min
Maintenance	75% of depreciation cost
Consumables	
Drill bit, cost	$100
Bit life	3000 ft (914 m)
Drill steel with coupling, cost	$700
Steel life	10,000 ft (3048 m)
Labor wage rate	$23.00/hr
Maintenance cost	75% of annual depreciation
Lubrication cost	Included in maintenance
Transport cost/cwt	$6 per cwt
Unloading cost	10% of transport cost
Operating time	4000 hr/yr
Economic life	24,000 hr
Interest	14%
Taxes	2%
Insurance, other costs	2%

SOLUTION. Using Table 11.2, the calculations are straightforward. Specific calculations are outlined here and summarized in Table 11.3.

A. *Ownership costs*

1a. *Purchase price* = \$300,000 + (3 − 1)\$60,000 = \$420,000
b. *Salvage* = 0.15 × \$420,000 = \$63,000
c. *Weight* = 25,000 + 2(2500) = 30,000 lb

$$\textit{Freight} = 30{,}000 \text{ lb} \times \frac{\$6}{100 \text{ lb}} = \$1800$$

d. *Unloading and moving cost* = 0.10 × \$1800 = \$180
e. *Delivered price* = a + b + c + d = \$358,980, say \$359,000
f. *Use standard operating period* = 4000 hr/yr
g. *Use standard economic life* = 24,000 hr ÷ 4000 hr/yr = 6 years
2. All calculations self-explanatory

B. *Operating costs*

1. *Tire cost:* omit (trac-mounted machine)
Consumables cost:

a. *Bits:* $\dfrac{\$100}{3000 \text{ ft}} \times 1136 \text{ ft/round} = \$37.87\text{/round}$

b. *Steel:* $\dfrac{\$700}{10{,}000 \text{ ft}} \times 1136 \text{ ft/round} = \$79.52\text{/round}$

c. *Total:* a + b = \$117.39/round

$$\textit{Cost/hr} = \$117.39\text{/round} \times \frac{3 \text{ rounds}}{8 \text{ hr}} = \$44.02\text{/hr}$$

2. *Fuel cost* = 10 gal/hr × \$1.10/gal = \$11.00/hr
3. *Labor:* assume 2 laborers

C. *Overall ownership and operating costs* = A + B = \$152.71/hr

D. *Unit costs: hourly production* = 206 tons/hr

1. $\textit{Cost/ton} = \dfrac{\$152.71\text{/hr}}{206 \text{ tons/hr}} = \$0.74\text{/ton}$

2. $\textit{Cost/ft} = \dfrac{\$152.71\text{/hr}}{426 \text{ ft/hr}} = \$0.358\text{/ton}$

TABLE 11.3 Cost Estimation Form Used in Example 11.2

Hydraulic Drill Mining Equipment Unit

A. Ownership Costs

1. DEPRECIATION
   - a. Purchase price = $ 420,000
   - b. Salvage value (15 %) = − 63,000
   - c. Freight 30,000 lb @ $ 6.00 /cwt = 1,800
   - d. Unloading and moving cost = 180
   - e. Delivered price = 358,980 , say $ 359,000
   - f. Operating period 4000 hr/yr
   - g. Economic life 24,000 hr = 6 yr(n)
   - h. Depreciation = $\dfrac{\$\ 359,000 \text{ (del. price less tire cost)}}{24,000 \text{ hr}}$ = $ 14.95 /hr

2. INTEREST, TAXES, INSURANCE, AND STORAGE
   - a. Rate = interest 14% + taxes 2% + other 2% = 18%
   - b. Average annual investment rate = $\dfrac{n+1}{2n} = \dfrac{7}{12}$ = 58.3 %
   - c. Average annual investment = $ 359,000 × 58.3 % = $ 209,300
   - d. Annual fixed charge = $ 209,300 × 18% = $ 37,700

   Fixed charge = $\dfrac{\$\ 37,700}{4000 \text{ hr/yr}}$ = $ 9.43 /hr

TOTAL OWNERSHIP COSTS = $ 24.38 /hr

B Operating Costs

1. TIRE REPLACEMENT COST
   - Purchase price 1 set of tires = $ ______
   - Tire life ______ hr
   - Tire cost = $\dfrac{\$\ \_\_\_\_\_\_}{\_\_\_\_\_\_ \text{ hr}}$ = $ 44.02 /hr
2. TIRE REPAIR COST: ____ % × $ ______ tire cost = $ ______ /hr
3. REPAIRS, MAINTENANCE: 75% × $ 14.95 deprec. = $ 11.21 /hr
4. FUEL: 10 gal/hr @ $ 1.10 /gal
   OR POWER: ____ kW @ $ ____ /kW-hr = $ 11.00 /hr
5. LUBRICATION: ____ gal/hr @ $ ____ /gal = $ ______ /hr
6. AUXILIARY FUEL: $ ______ /hr
7. LABOR: 2 operator @ $ 23.00 /hr = $ 46.00 /hr
   - ____ oiler @ $ ____ /hr = ______ /hr
   - ____ helper @ $ ____ /hr = ______ /hr

   Total $ 46.00 /hr

   +35% benefits $ 62.10 /hr = $ 62.10 /hr

TOTAL OPERATING COSTS = $ 128.33 /hr

TOTAL OWNERSHIP AND OPERATING COSTS = $ 152.71 /hr

UNIT COST $ 152.71 /hr ÷ 206 ton/hr = $ 0.74 /ton

11.6 SPECIAL TOPIC B: UNDERGROUND BLASTING

We build now on the theory of explosives and explosive detonation developed in Section 5.3. In Section 7.6, the theory was used in the design of surface blasts; we utilize it now for the design of underground blasting rounds, especially in driving development openings in which a single free face exists (production blasts underground very much resemble surface blasting in benches with multiple free faces; see Sections 7.6 and 10.3).

11.6.1 Theory of Rock Failure in Blasting

In Section 5.3 we learned that when explosives detonate, two effects are primarily responsible for breaking rock: the impact of the shock wave and the expansion of the gas bubble. They are effective in varying proportions, depending on the properties of both explosive and rock. In general, powerful explosives (e.g., TNT, nitroglycerin, dynamite) that release more energy fragment by shock wave, and weaker explosives (e.g., ANFO, permissibles) fragment by gas expansion. Correspondingly, shock waves are more effective in hard, brittle rock (e.g., granite, taconite), and gas expansion is more effective in soft, plastic, or jointed rock (e.g., shale, salt, coal).

Shock waves are the initial effect to occur in blasting. As soon as detonation is initiated in the explosive charge, a shock front forms, propagating beyond the charge into the adjoining rock as a stress wave (review Figure 5.2) The shock front has two components: longitudinal and transverse. The *longitudinal wave* travels at higher velocity with amplitude parallel to its motion, whereas the *transverse wave* is slower and weaker with amplitude normal to its motion.

The sequence of events, starting with wave action, that occurs in blasting has been described by, among others, Livingston (1956), Clark (1968), and Hemphill (1981). With detonation (Figure 11.6), the longitudinal stress wave propagates into the rock, traveling to the free face(s) as a compression wave and being reflected back in tension (because of the sharp discontinuity between rock and air, only a small amount of energy escapes into the air as wave energy, vibration, and noise). Because rock is comparatively strong in compression, the initial pulse of the longitudinal wave is effective only in breaking rock close to the charge. Here compressive rock failure is evident by crushing immediately adjacent to the charge and by radial cracks beyond. Major rock fragmentation, however, occurs when the tension wave reflects back from the free face(s); because rock is weaker in tension, it fails in thick tensile slabs as the wave passes, with damage ceasing as the stress level declines. The only effect of the transverse wave, a far weaker wave, is possibly to assist in the formation of release fractures or radial cracks by shear action.

The gas expansion effect sequentially follows the wave action. It propagates at a slower rate, and whereas stress waves produce mainly a shattering action, the gas bubble causes a heaving action. The expanding effect of the high-pressure gas radiating through blasting fractures and preexisting joints in the

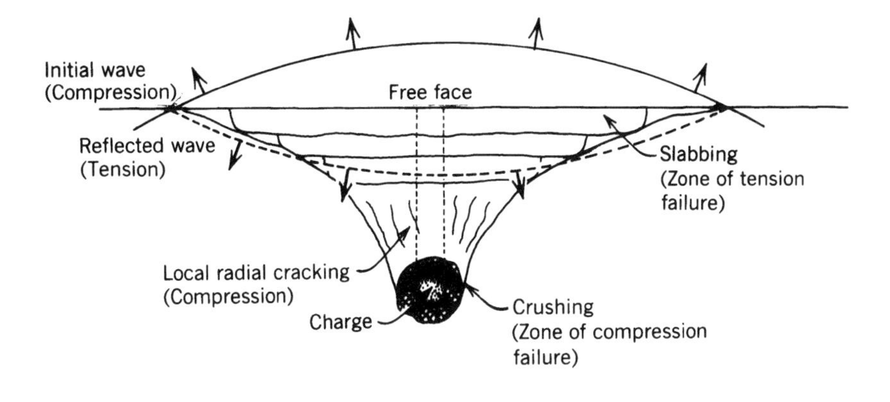

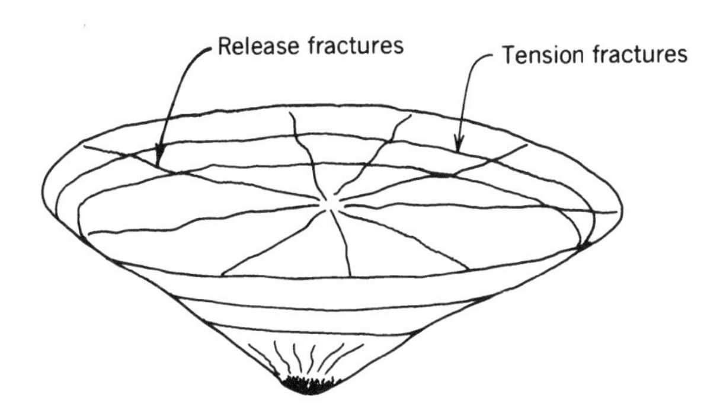

FIGURE 11.6. Sequence of events in crater formation by rock blasting. Shown are the wave action, steps in rock failure, and the resulting crater.

rock is to displace and move the aleady broken rock fragments, thus creating further breakage. It may be concluded that the gas expansion effect in fragmentation is chiefly one of displacement, reflected in throw and the formation of a crater.

Figure 11.6 portrays the sequence of events in rock blasting, the nature of rock failure, and the steps in crater formation. Most blasting authorities agree today that in the majority of cases, it is the combined effects of the stress wave and gas bubble rather than the predominant action of either alone that account for rock fragmentation in the detonation of explosives.

The effect of multiple holes in close proximity and detonated simultaneously is to reinforce the blasting damage produced by a single charge.

11.6.2 Crater Geometry

The dimensional relations of the explosive charge and the resulting crater produced in blasting are unique, affording us the means to analyze or design a blast (Livingston, 1956; Langefors, and Kihlstrom, 1978; R. Gustafsson, 1981).

The so-called cubical law of blasting derives from eighteenth-century military engineering and applies to a single spherical charge detonated in rock with a single free face:

$$W \propto d^3 \tag{11.1}$$

in which W is weight of explosive and d is burden, the depth of charge or the perpendicular distance from the free face to the center of gravity of the charge. Although now discredited, the law assumed that the volume of rock broken (proportional to d^3) varied directly with the weight of explosive, and that the crater was a 45° cone with radius and height equal to d.

Modern theories of blasting recognize that (1) the simple cubical law relationship does not apply and (2) the shape of the crater departs from a cone. Several relationships have been proposed, probably the most satisfactory being the one by Langefors and Kihlstrom (1978):

$$W = ad^2 + bd^3 + cd^4 + \cdots \tag{11.2}$$

where a, b, and c are constants. Because a long series is awkward to solve and decreasing benefits accrue with the addition of each successive term, Eq. 11.2 is often simplified to a empirical, exponential formula, resembling Eq. 11.1 (Livingston, 1956):

$$W \propto d^n \tag{11.3}$$

where n is an exponent that in practice varies from 1.5 for elastic rock to 3.4 for plastic.

To determine the value of n experimentally for a given explosive and a given rock, a series of blasts is detonated for a fixed burden d, increasing the weight of explosive W until a full crater is formed (Figure 11.7). The depth of the crater d_c then becomes equal to the burden d. If the charge weight is insufficient to break any crater (Figure 11.7a), then $d_c = 0$, and the only efffect is to spring the hole around the charge. If the charge is intermediate in size, then $d_c < d$ (Figure 11.7b). When the weight of explosive is sufficient to form a full crater, then $d_c = d$ (Figure 11.7c). The radius of the crater r also bears a fixed relation to the depth d_c.

After a series of tests is conducted, then a family of curves can be plotted for the explosive and the rock. Usually d_c is expressed as a percentage of d, with curves plotted for $W \propto d^n$ from $d_c/d = 0$–100%. (The most efficient blasting, or lowest powder factor, is not obtained at $d_c/d = 100\%$; but in development blasting, efficiency gives way to maximum advance as the governing factor, and $d_c \leqslant d$.) Results of tests for full craters only with different explosives in the same rock are shown in Figure 11.8. Data were obtained for 45% semigelatin dynamite and ANFO in Idaho Springs (Colorado) gneiss. Equation 11.3 becomes

$$W \propto d^{1.85}$$

and $r/d_c = 2.8$ for a full crater in blasting to a single free face.

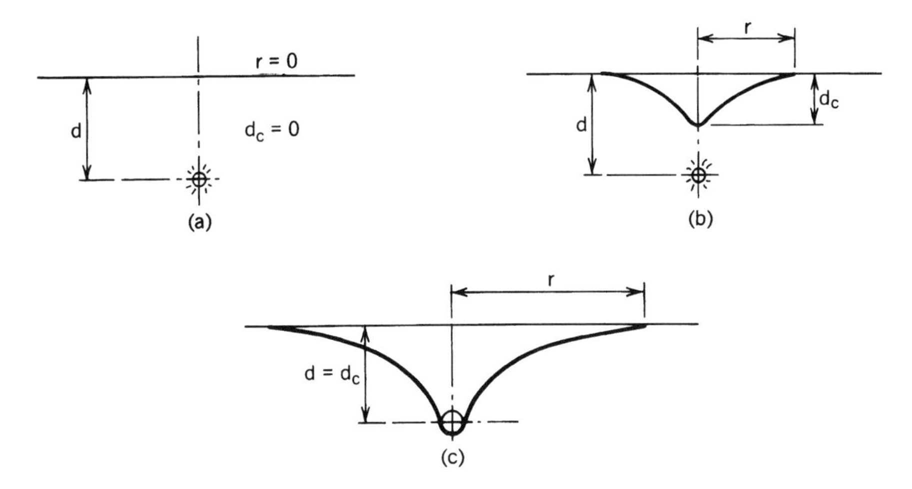

FIGURE 11.7. Crater geometry relations for a given explosive detonated in a given rock. (a) No crater. (b) Normal crater. (c) Full crater. The charge weight increases for a constant burden between (a), (b), and (c). In (a), $d_c = 0$; in (b) $d_c < d$; and in (c), $d_c = d$.

11.6.3 Development Blasting Rounds

Compared with surface bench blasting or underground stope blasting, underground development blasting is difficult, slow, inefficient, and expensive. The reason is, of course, that only a single free face exists. Therefore, the objective of each development round is to open up an additional free face as the first holes, called *cut* holes, are detonated. Successive holes, detonated by the appropriate delays, can then break to the cut, improving the effectiveness and efficiency of the blast. Moreover, it is desirable to break as deep a round as possible in development blasting to achieve the maximum advance, thus further reducing costs.

Major factors in the design of a development blasting round underground are the following:

1. Type of explosive
2. Properties and uniformity of rock
3. Drillhole diameter
4. Hole pattern, number, and depth
5. Amount of charge per drillhole
6. Order of firing and wiring diagram

Hole pattern, or the *blasting round*, is the most important variable to resolve. Rounds are classified by the type of cut employed, as follows (Obert, 1973c;

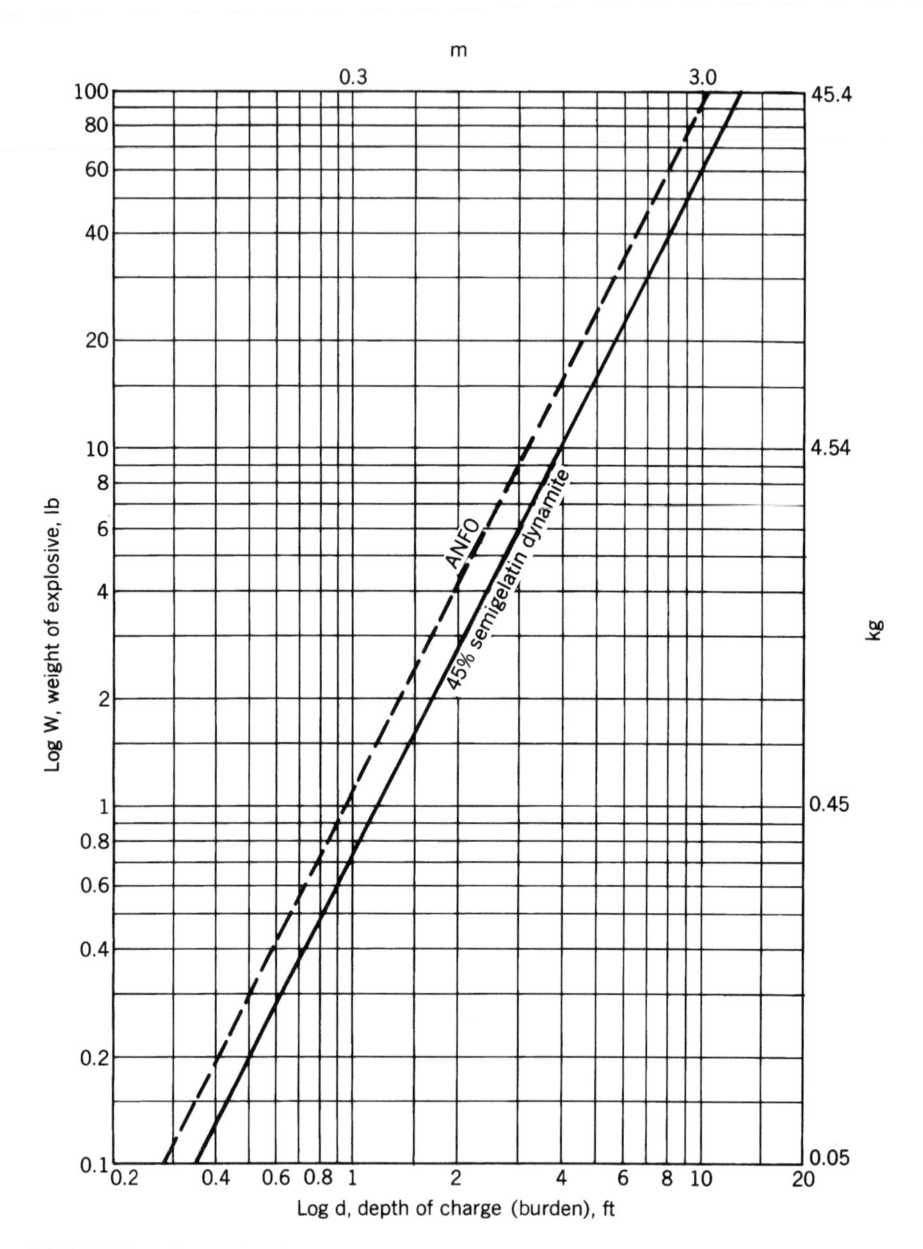

FIGURE 11.8. Crater data for design of blasting round, plotting weight of explosive against burden. Conditions:

Burden for single full crater ($d = d_c$)
$W \propto d^{0.85}$, $r/d = 2.8$
Single free face
Rock: Idaho Springs, CO, gneiss
Explosives: ANFO and 45% semigelatin dynamite
(Modified after Livingston, 1956.)

E. I. du Pont de Nemours and Co., 1977; Gregory, 1979):

1. Machine-cut kerf
2. Angled-cut round
3. Burn-cut rounds
4. No-cut rounds

Kerfs are used only in soft to moderately hard rock (coal, salt, potash, etc.) to provide an additional free face for blasting. A cutting machine of the universal or shortwall type is employed to undercut, top cut, center cut, or shear the face. The use of a kerf substantially reduces the amount of explosive required and lowers the risk of an explosion in a flammable atmosphere, such as in a coal mine. Unfortunately, it is not feasible to employ a machine-cut kerf in hard rock.

The other blasting rounds involve only drillholes, unassisted by cutting. Angled-cut rounds were the first developed and are still the most widely applied. The principal ones used are the V cut, pyramid, and draw, in varying arrangements and combinations of holes (Figure 11.9). Angled cuts are effective in hard, tough rock but cause flyrock and consume large amounts of explosives. The advance with an angled cut is limited to the least dimension of the face, and much skill and care are required to place drillholes properly.

Burn-cut rounds have become popular in recent years because they are easier to locate properly, consume less explosive, and have no limitation on depth of advance. The principle of burn cuts is that certain (usually alternate)

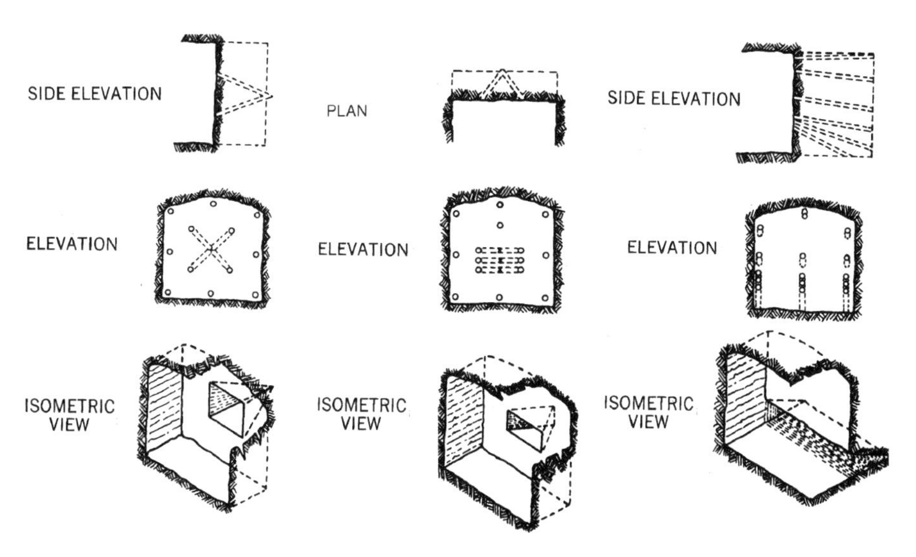

FIGURE 11.9. Angled-cut development rounds. (*Left*) Pyramid. (*Center*) Triple V. (*Right*) Draw. (After Gregory, 1979. By permission from Trans Tech Publications, Ueticon-Zurich, Switzerland.)

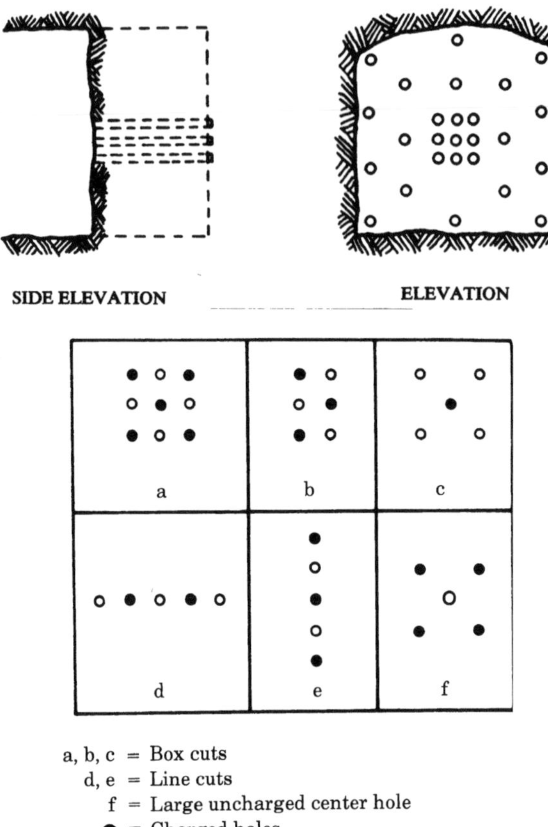

FIGURE 11.10. Burn-cut development rounds. (After Gregory, 1979. By permission from Trans Tech Publications, Ueticon-Zurich, Switzerland.)

holes are left uncharged to provide free faces—thus, drilling rather than blasting is utilized to create free faces. A wide variety of hole patterns and even hole sizes are employed in burn cuts (large center holes are the most satisfactory in providing space for shattered rock to move) (Figure 11.10). Burn cuts are applicable to hard, brittle, homogeneous rock.

No-cut rounds are of experimental interest. Their principle is that holes drilled in a deepening and expanding helical spiral can produce an initial cut and successively larger and deeper cuts by the proper, careful placement of drillholes and sequence of firing (Livingston, 1956).

As with surface blasting, the efficiency of blasting in underground development is measured by several parameters, the two most common being drilling factor and powder factor. Values for current practice are the following (compare with those for surface mining in Section 7.6):

	Drilling Factor, ft/ton (m/tonne)	Powder Factor, lb/ton (kg/tonne)
Underground development, metallic and nonmetallic ores	3–4 (1.0–1.3)	2–4.5 (1–2)
Underground production, metallic and nonmetallic ores	0.5–2 (0.2–0.7)	0.2–0.4 (0.1–0.2)
Underground mining, coal	1–2 (0.3–0.7)	0.8 (0.4)

11.6.4 Design of Blasting Rounds

Information developed in the preceding sections on the theory of blasting and crater geometry can now be employed in the design of underground development rounds. In this section, we shall deal with the design of only one type of blasting round, the angled cut. Crater data presented in Figure 11.8 will be the basis for our design procedure.

Figure 11.11 (*left*) illustrates the steps in crater formation and burden relationships for a V-cut round in a development heading (only a portion of the drillholes and the detonation sequence in the face are shown). The first holes detonated are the V cut (1), next the relievers (2), and then the enlargers (3). In designing each sequence of holes, the burden d is measured from the nearest free face to the center of gravity of the charge. The amount of explosive required is then read from Figure 11.7. Thus, for the cut holes, the burden is d_1, measured to the heading face; the weight of explosive, W_1, corresponds to d_1. When the cut is blasted, the first crater forms. For the relievers, burden d_2 is measured to the first crater, and charge weight W_2 is specified accordingly. Subsequent holes are handled similarly. A comparison with bench blasting in a production round is provided by Figure 11.11 (*right*).

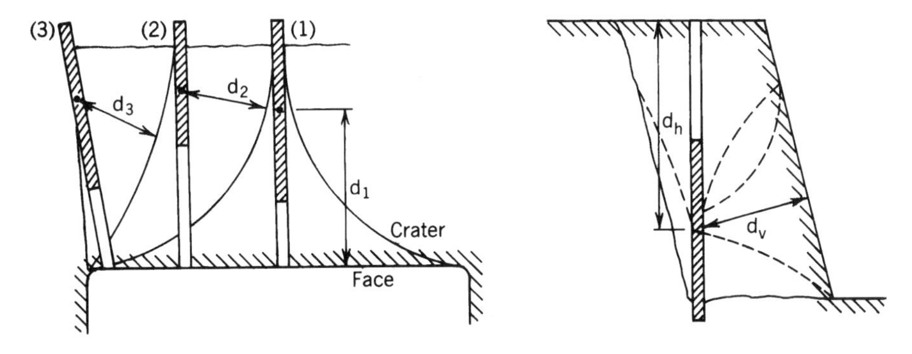

FIGURE 11.11. Relationship of burden to free face and center of gravity of charge in blasting rounds. (*Left*) Underground development. (1) Cut hole. (2) Reliever hole. (3) Enlarger hole. (*Right*) Surface or underground production.

Although the resulting design is approximate and must be adjusted for variations in actual practice, it can be used successfully to obtain a trial blasting round. Because the configuration is only approximate, a further simplification will be made in the following numerical example and problems by locating the center of gravity of all holes at the depth of round or anticipated advance.

Example 11.3. (a) Design the eight-hole V cut of a blasting round intended to pull 6 ft (1.8 m) in an 8 × 10 ft (2.4 × 3.0 m) drift in Idaho Springs gneiss (TF = 12.1 ft^3/ton, or 0.38 m^3/tonne). The explosive is ANFO of specific gravity 0.8, fired electrically with millisecond delays. Drillhole diameter is 3 in. (76 mm).

(b) The following data apply to the entire blasting round:

Total explosives consumption	150 lb (68 kg)
Total length of blasthole drilled	165 ft (50 m)

Calculate the powder factor and drilling factor for the round.

SOLUTION. (a) From Figure 11.8, the ANFO curve, read $W_1 = 34.5$ lb for $d_1 = 6$ ft. Find loading density:

$$\text{Explosive specific weight } w_1 = (62.4)(0.8) = 49.9 \text{ lb/ft}^3$$

$$\text{Hole area } A = \frac{\pi}{4}\left(\frac{3}{12}\right)^2 = 0.0491 \text{ ft}^2$$

For unit length l,

$$\text{Loading density} = (0.0491)(49.9) = 2.45 \text{ lb/ft (3.65 kg/m)}$$

Determine explosives charge per hole in cut:

$$\text{Charge/hole, } W_l = \frac{34.5}{8} = 4.31 \text{ lb}$$

$$\text{Length of charge/hole, } L = \frac{4.31}{2.45} = 1.76 \text{ ft (0.53 m)}$$

Because $L = 1.76$ ft and the center of gravity of the charge is located at the depth of the round $d = 6$ ft, the length of each cut drillhole L_h must be

$$L_h = 6 + \frac{1.76}{2} = 6.88 \text{ ft (2.10 m)}$$

(b) Calculate powder factor:

$$\text{Weight of rock broken, } W = \frac{(6)(8)(10)}{12.1} = 39.7 \text{ tons/round}$$

$$\text{Powder factor PF} = \frac{150}{39.7} = 3.78 \text{ lb/ton (1.89 kg/tonne)}$$

Calculate drilling factor:

$$\text{Drilling factor DF} = \frac{165}{39.7} = 4.16 \text{ ft/ton (1.40 m/tonne)}$$

In practice, drillholes are generally overdrilled (about 6 in., or 152 mm) to ensure that the full depth of round is pulled. This requires that the center of gravity of the charge be shifted closer to the collar of the hole. Moreover, when cartridges rather than bulk explosives are used, loading densities are calculated to the nearest one-half cartridge.

After the cut is designed, the burden on succeeding holes is read from either a plan or elevation view of the round, selected so that a true distance, without foreshortening, is read. Further, it is advisable to check the desired radius of crater against the design r/d_f ratio to ensure that the crater formed by the cut holes breaks to the full dimensions of the face.

PROBLEMS

11.1 The cycle in driving a raise conventionally by drilling and blasting must be completed in a 7 hr shift. Allowing 0.8 hr for scaling and bolting, 1.6 hr for blasting, 0.2 hr for chute pulling, and 20% of shift time for delays, calculate the rating of two stoper percussion drills (in in./min, or mm/sec, penetration rate) required with two operators to achieve a 6 ft (1.8 m) advance per shift. The raise is 6 × 8 ft (1.8 × 2.4 m) in cross section, and 30 $6\frac{1}{2}$ ft (2.0 m) holes are required. Setup time for drilling is 20 min, collaring time is $\frac{1}{2}$ min per hole, and steel change time is 1 min per hole.

11.2 **a.** A percussion drill jumbo is operating in a drift in a hardrock mine under the following conditions:

Rock	Minnesota quartzite
Drilling rate (in Barre granite)	52 in./min (22 mm/sec)
Position, collar, retract time	3.2 min/hole
Drillhole depth	9 ft (2.7 m)
Shift drilling time	7.2 hr

What is the output of one drill per shift, measured in total feet (meters) of hole?

b. If three rounds are required per shift, and each round consists of 68 holes, how many drills must be mounted on the jumbo?

c. If the working face is 18 × 26 ft (5.5 × 7.9 m) and the tonnage factor is 11 ft^3/ton (0.34 m^3/tonne), calculate the tons broken per shift and the drilling factor.

11.3 A percussion drill rig demonstrates the following performance in a mine:

Rock	Bellefonte, Pennsylvania, limestone
Drilling rate (in Barre granite)	28 in./min (11.9 mm/sec)
Position, collar, retract time	2.52 min/hole
Drillhole depth	12 ft (3.6 m)
Number of holes/round	58
Drilling time/round	2 hr
Number of rounds/shift	3
Shift time	8 hr

Pertinent cost data are as follows:

Delivered price	$140,000
Operating period	6000 hr/yr
Life	33,000 hr
Operating cost	$32.00 hr

Determine the following:

a. Number of drills required for the conditions

b. Ownership cost, $/hr

c. Unit drilling cost, $/ft ($/m)

11.4 Given the following information, calculate and compare the unit costs ($/hr or $/ft or m of drillhole, and per ton or tonne of rock broken) of drilling with pneumatic and hydraulic equipment in underground development work, employing multiple headings in hard-rock mining. Select the more economical drill rig for this application. Assume one rig with one to four drills each.

	Pneumatic	Hydraulic
Design and performance data		
Drill rounds/shift (8 hr)	3	3
Drill face	20 × 16 ft (6.1 × 4.9 m)	20 × 16 ft (6.1 × 4.9 m)
Drillhole depth	12 ft (3.7 m)	12 ft (3.7 m)

	Pneumatic	Hydraulic
Bit diameter	2 in. (51 mm)	2 in. (51 mm)
Drilling time/round	2 hr	2 hr
Blasting factor	6.4 ft^2 (0.59 m^2)/hole	6.4 ft^2 (0.59 m^2)/hole
Rock drillability (based on Barre granite)	0.8	0.8
Drilling rate (in Barre granite)	36 in./min (15 mm/sec)	66 in./min (28 mm/sec)
Positioning, collaring, retract time	2.0 min/hole	2.0 min/hole
Rock specific weight	175 lb/ft^3 (2800 kg/m^3)	175 lb/ft^3 (2800 kg/m^3)
Capital costs		
Drill rig, 1 boom	\$84,000	\$125,000
each additional boom	\$18,000	\$24,000
Rig weight	20,000 lb (9070 kg)	22,000 lb (9070 kg)
Weight, each boom	2000 lb (907 kg)	2000 lb (907 kg)
Operating period	4000 hr/yr	4000 hr/yr
Life	20,000 hr	20,000 hr
Operating costs		
Electric power (drilling)	6¢/kW-hr	6¢/kW-hr
Power consumed	110 kW	90 kW
Diesel fuel (tramming)	\$1.05/gal (\$0.26/L)	\$1.05/gal (\$0.26/L)
Tram time/shift	30 min	30 min
Motor rating	100 hp (75 kW)	100 hp (75 kW)
Fuel consumed	0.75 gal/hp-hr (3.81 L/kW-hr)	0.75 gal/hp-hr (3.81 L/kW-hr)
Hydraulic fuel @ \$1.50/gal (\$0.40/L)	0	1.5 gal/round (5.7 L/round)
Maintenance	60%	75%
Consumables		
Drill bit, cost	\$80	\$55
Bit life	500 ft (152 m)	500 ft (152 m)
Drill steel, couplings, hose, etc., cost	\$840	\$560
Steel life	4000 ft (1220 m)	4000 ft (1220 m)
Labor wage rate (use 1 operator/1–4 booms)	\$16.00/hr	\$16.00/hr

11.5 Given the following information, calculate and compare the unit costs (\$/hr or \$/ft or m of drillhole, and per ton or tonne of rock broken) of drilling with pneumatic and hydraulic equipment in underground development work, employing multiple headings in hard-rock mining. Select the more economical drill rig for this application. Assume each rig can have up to four drills.

	Pneumatic	Hydraulic
Design and performance data		
Drill rounds/shift (8 hr)	3	3
Drill face	18 × 24 ft (5.5 × 7.3 m)	18 × 24 ft (5.5 × 7.3 m)
Drillhole depth	14 ft (4.3 m)	14 ft (4.3 m)
Bit diameter	$2\frac{1}{2}$ in. (64 mm)	$2\frac{1}{2}$ in. (64 mm)
Drilling time/round	2.5 hr	2.5 hr
Blasting factor	7.3 ft^2 (0.68 m^2)/hole	7.3 ft^2 (0.68 m^2)/hole
Rock	New Jersey granite gneiss	
Drilling rate (in Barre granite)	32 in./min (14 mm/sec)	56 in./min (24 mm/sec)
Positioning, collaring, retract time	2.25 min/hole	2.0 min/hole
Rock tonnage factor	13.5 ft^3/ton (0.42 m^3/tonne)	13.5 ft^3/ton (0.42 m^3/tonne)
Capital costs		
Drill rig, 1 boom	$92,000	$135,000
each additional boom	$22,000	$28,000
Rig weight	20,000 lb (9070 kg)	22,000 lb (9070 kg)
Weight, each boom	2000 lb (907 kg)	2000 lb (907 kg)
Operating period	4000 hr/yr	4000 hr/yr
Life	28,000 hr	28,000 hr
Operating costs		
Electric power (drilling)	4¢/kW-hr	4¢/kW-hr
Power consumed	110 kW	90 kW
Diesel fuel (tramming)	$1.10/gal ($0.29/L)	$1.10/gal ($0.29/L)
Tram time/shift	20 min	20 min
Motor rating	100 hp (75 kW)	100 hp (75 kW)
Fuel consumed	0.75 gal/hp-hr (3.81 L/kW-hr)	0.75 gal/hp-hr (3.81 L/kW-hr)
Hydraulic fuel @ $1.50/gal ($0.40/L)	0	1.5 gal/round (5.7 L/round)
Maintenance	75%	80%
Consumables		
Drill bit, cost	$90	$60
Bit life	450 ft (137 m)	450 ft (137 m)
Drill steel, couplings, hose, etc., cost	$860	$540
Steel life	3600 ft (1097 m)	3600 ft (1097 m)
Labor wage rate (use 1 operator/1–4 booms)	$16.40/hr	$16.40/hr

11.6 Design the cut holes only of a blasting round for a 4 × 6 ft (1.2 × 1.8 m) crosscut employing an angled cut of your choice (arrangement and number of holes) and based on the charge–burden chart provided for Idaho Springs gneiss and dynamite or ANFO (Figure 11.8). Pull the *maximum* advance possible (what is it?). Name the cut you have selected, and sketch it and the crater in two views. Loading density in the holes is 1.5 lb/ft (2.23 kg/m) for either explosive. Calculate the weight of explosive and length of drillhole required.

11.7 A V-cut blasting round is to be employed in driving a 5 × 7 ft (1.5 × 2.1 m) raise. (a) What is the maximum depth of round that can be pulled? (b) Using two holes in the V, design the cut portion of the round *only* to permit an advance of 4.5 ft (1.4 m). Make use of the charge-burden chart for Idaho Springs gneiss and dynamite, assuming a loading density of 1.4 lb/ft (2.08 kg/m) of drillhole. Sketch the holes and craters in two views, and calculate the footage of hole, charge, and weight of explosive.

11.8 **a.** A 7 × 10 ft (2.1 × 3.0 m) drift is being driven in Idaho Springs gneiss using 45% semigelatin dynamite. Using a V-cut blasting round, determine the *minimum* number of holes in the cut only required to produce a *maximum* advance per round. Specify the advance, hole depth, and number of holes. Holes are overdrilled 0.5 ft (0.15 m) and sealed with 0.5 ft (0.15 m) of stemming. Use the blast design (W vs. d) chart, assuming that the crater breaks to within 0.5 ft (0.15 m) of the end of the hole rather than to the center of gravity of the charge. (In other words, this is a theoretical design, modified for practical application.) Take loading density as 0.9 lb powder/ft drillhole (1.35 kg/m). Sketch the cut design in two views.

b. Design the round to the same specifications using ANFO.

11.9 Using an angled cut of the four-hole, pyramid type, design a complete blasting round for development work in a 6 × 6 ft (1.8 × 1.8 m) drift, employing millisecond delay firing (Figure 11.12):

Rock	Hard gneiss (Idaho Springs formation); tonnage factor 12 ft^3/ton (0.375 m^3/tonne)
Explosive	Semigelatin dynamite, 45% strength; density 122 sticks/box (50 lb, or 22.7 kg); standard $1\frac{1}{4}$ × 8 in. (32 × 203 mm) cartridge
Depth of round	5 ft (1.5 m)
Hole diameter	$1\frac{3}{8}$ in. (35 mm)

Determine the following:

a. Sketch of blasting round (to scale), showing placement of drillholes, order of firing, and shape of craters formed

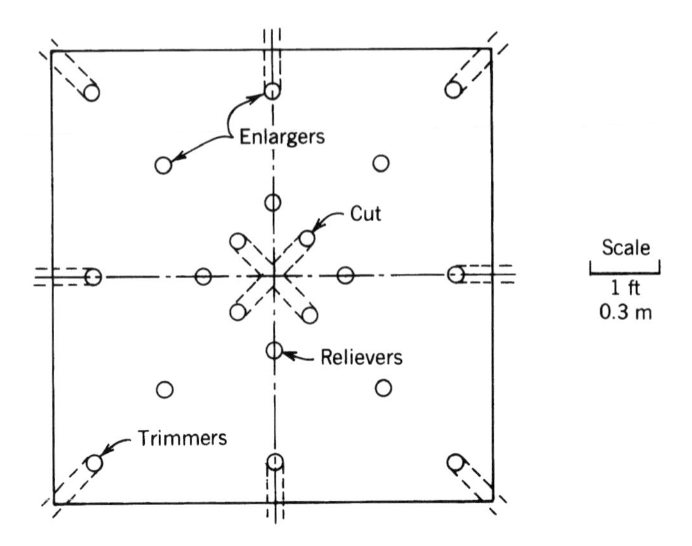

FIGURE 11.12. Pyramid-cut blasting round. See Problem 11.9.

b. Explosives requirement for each hole, calculated to nearest half stick
c. Total number of holes required
d. Total length of drillhole
e. Drilling factor, in ft/ton (m/tonne)
f. Total explosives consumption, in lb (kg)
g. Powder factor, in lb/ton (kg/tonne)

11.10 Using an angled cut of the six-hole V type, design a complete blasting round for development work in a 6.5 × 6.5 ft (2.0 × 2.0 m) drift employing millisecond delay firing (Figure 11.13):

Rock	Hard gneiss (Idaho Springs formation); tonnage factor 12 ft^3/ton (0.375 m^3/tonne)
Explosive	Semigelatin dynamite, 45% strength; density 112 sticks/box (50 lb, or 22.7 kg); standard $1\frac{1}{4} \times 8$ in. (32 × 203 mm) cartridge
Depth of round	6 ft (1.8 m)
Hole diameter	$1\frac{1}{2}$ in. (38 mm)

Determine the following:

a. Sketch of blasting round (to scale), showing placement of drillholes, order of firing, and shape of craters formed
b. Explosives requirement for each hole, calculated to nearest half stick
c. Total number of holes required
d. Total length of drillhole

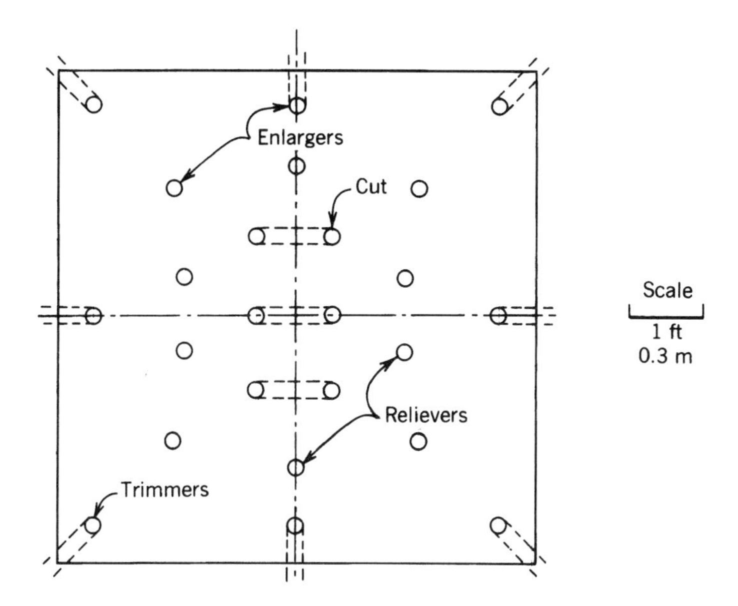

FIGURE 11.13. V-cut blasting round. See Problem 11.10.

e. Drilling factor, in ft/ton (m/tonne)
f. Total explosives consumption, in lb (kg)
g. Powder factor, in lb/ton (kg/tonne)

12

UNDERGROUND MINING: CAVING METHODS

12.1 CLASSIFICATION OF METHODS

We have concentrated our efforts so far on classes of mining methods that require exploitation workings to be held open, essentially intact, for the duration of mining. If the ore and rock are sufficiently competent, unsupported methods are adequate; if ore and rock are incompetent to moderately competent, then supported methods must be used. We now encounter a class of methods in which the exploitation openings are designed to collapse; that is, caving of the ore or rock or both is intentional and the very essence of the method.

We define *caving methods* as those associated with induced, controlled, massive caving of the ore body, the overlying rock, or both, concurrent with and essential to the conduct of mining. There are three current methods that are considered to be caving methods (Table 4.1):

1. Longwall mining
2. Sublevel caving
3. Block caving

Longwall mining is used in horizontal, tabular, deposits, mainly coal; the other methods have application to inclined or vertical, massive deposits, almost exclusively metallic or nonmetallic. The caving class accounts for about 15% of U.S. mineral production, which will likely increase. In regard to cost, this class includes a moderately priced method as well as two of the cheapest of the underground methods.

Because the exploitation openings are deliberately destroyed in the process of mining, the caving class is truly unique. Rock mechanics principles are applied to ensure that caving, in fact, does occur—rather than to prevent the occurrence of caving. In effect, the cross-sectional shape of the undercut area (i.e., the width-to-height, or R/h, ratio) is sufficiently elongated to cause failure of the roof or back (see Section 10.7). Further, development openings must be designed and located to withstand shifting and caving ground, as well as subsidence that usually extends to the surface. Production must be maintained at a steady, continuous pace to avoid disruptions or hangups in the caving action. Good mine engineering and supervision are indispensable to a successful caving operation.

12.2 LONGWALL MINING

Longwall mining is an exploitation method used in flat-lying, relatively thin, tabular deposits in which a long face is established to extract the mineral. The layout of five longwall panels is illustrated in Figure 12.1. Note that the face in Panel 2 is established between the headgate entries and the tailgate entries and that the face is being advanced toward the main entry set (i.e., in the retreat direction). The longwall is kept open by a system of heavy-duty, powered,

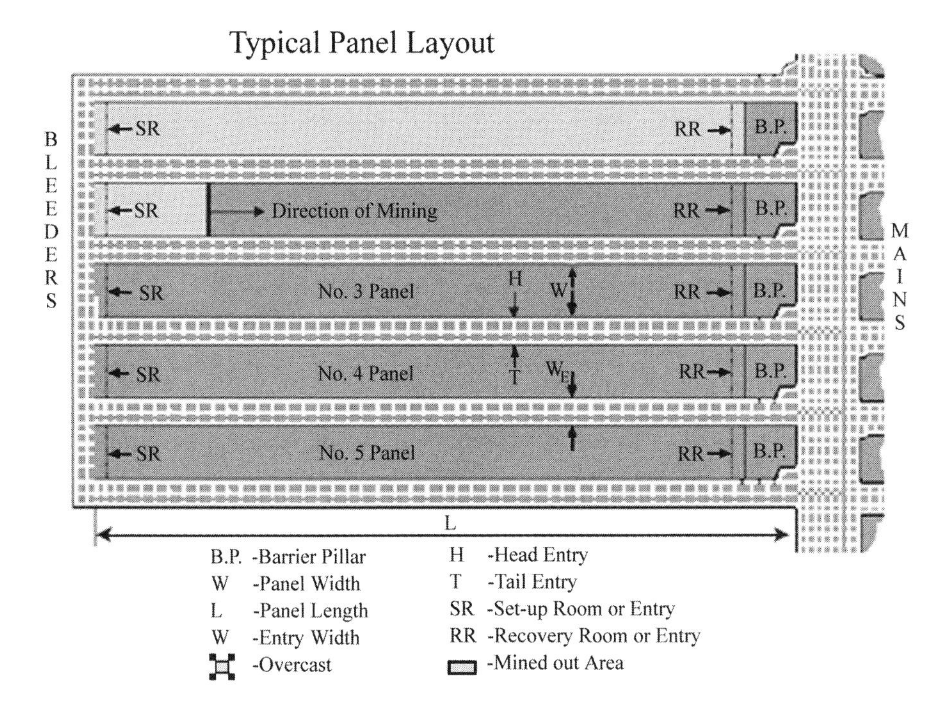

FIGURE 12.1. General layout of a longwall operation. (After Peng and Chiang (1984). *Longwall Mining*. Copyright © 1984, John Wiley & Sons, Inc. Photo provided by Syd Peng.)

yielding supports that form a cantilever or umbrella of protection over the face (see Figures 9.15 and 9.16). As a cut or slice is taken along the length of the wall, the supports are collapsed, advanced closer to the face, and reengaged, allowing the roof to cave behind. The caved area is called the *gob*. In Figure 12.1, all of Panel 1 and a small portion of Panel 2 have been mined and converted to gob.

A very old method, longwall mining originated in the coal mines of Europe in the seventeenth century and is used almost exclusively in many coal-producing countries outside the United States (Lucas and Haycocks, 1973). It has enjoyed success in the United States only since the 1960s, when self-advancing hydraulic support systems were perfected (Trent and Harrison, 1982). Other innovations that have led to its rapid adoption here are the development of mobile, flexible armored conveyors, high-speed continuous miners for development, and roof control and caving practices grounded in sound rock mechanics principles.

In the decades since the 1960s, longwall has grown steadily and now produces 47.8% of the coal mined underground in the United States (Energy Information Administration, 2000). More than 100 longwalls were in operation in the mid-1980s; fewer are now in use because of the tremendous increase in productivity and production of longwall operations. Fiscor (2000) reports that 59 longwalls are currently operating in the United States at 52 mines. Of the longwalls in operation, 57 use a shearer for coal winning; the other two faces use a plow to extract the coal. Longwall is considered to be a large-scale method of exploitation and is one of the cheapest underground methods.

Classified by some as a supported mining method because of the elaborate roof support system employed, longwall mining more properly belongs to the caving class of methods. The reason has to do with the role of caving in the mining process, which is to aid in the breakage of the in-situ material while permitting the immediate roof to cave safely, thus preventing excessive superincumbent loads from damaging the supports (Trent and Harrison, 1982). Ideally, the immediate roof should be thin-bedded and moderately weak so that it will break and collapse; stronger, thick-bedded formations do not preclude the use of a caving method, so long as they occur higher in the geologic column. The floor should be competent to provide a firm foundation for the roof supports. Because it is difficult theoretically or experimentally to determine the suitability of ore, rock, or coal for caving, an empirical scale of *cavability* is often employed (Peng and Chiang, 1984). Its use is not limited to longwall mining but applies to the entire caving class.

Subsidence is an environmental hazard of longwall mining that must be addressed. Depending on the depth of cover and area mined, caving and collapse of the overlying strata will eventually extend from the mining horizon to the surface. Although some damage to the surface must be expected, it can be controlled by maintaining a uniform rate of advance of the longwall and the subsequent caving. If the surface subsides evenly, then destruction of structures and water courses need not occur (Peng and Chiang, 1984).

Rock bursts are another potential hazard in longwall mining, especially at great depths (>2500 ft or 750 m). Bursts are accompanied by the very rapid release of large amounts of strain energy and may have violent, often explosive-like repercussions (Trent and Harrison, 1982). Damage to personnel and property is frequent. Fortunately, most of the longwalls operated in the United States are not subject to great depths and high strain energy conditions. Nonetheless, it is important for longwall operations to be monitored for stress and strain to ensure safe conditions.

Most U.S. longwall operations have three entries on the headgate and three on the tailgate and require at least two continuous miners to drive development openings for each longwall. The panels are likely to be increased in size each year as a means of improving productivity. Fiscor (2000) reports the following data relating to the size of U.S. longwalls:

Panel width	600–1200 ft (883 ft avg.)	182–365 m (269 m avg.)
Panel length	3000–14,600 ft (8580 ft avg.)	911–4450 m (2615 m avg.)
Seam height	4–23 ft	1.2–7 m
Depth of cut (plow)	0.75 ft	0.2 m
Depth of cut (shearer)	2.5–3.5 ft	0.76–1.1 m
Overburden depth	200–2700 ft	61–823 m

Larger panel sizes, automated supports, better face conveyors, higher face voltage, and higher-powered shearers have all been responsible for improved productivity in longwall operations.

Equipment choice in longwall mining has been simplified by the reduction in the number of manufacturers who serve the industry. Only two equipment types are available for extraction: shearers and plows. Shearers are used on 97% of U.S. longwalls today, with the other two longwalls using plows (Fiscor, 2000). Most of the shearers are of double-drum design with face voltages of 2300 or 4160 volts. Nearly all U.S. longwalls use shield-type roof supports, with most being semiautomated to enhance productivity. The armored face conveyors utilized are normally of much higher power than a decade ago, and many have soft-start technology to ease the strain on the system under starting conditions. Graham (1998) has reported on other advances in technology. Annual production values of 3,000,000 to 5,000,000 tons/yr (2,700,000 to 4,500,000 tonnes/yr) have been common in many longwalls.

Longwalls have also been used in trona, uranium, copper, and potash. In most metallic and nonmetallic deposits, the methods are substantially different from that used in coal. Hamrin (1982) shows an application of longwall in hard-rock mining. This variation of longwall is depicted in Figure 12.2. The method is quite different from longwalls in coal in that no mobile supports are used and concrete pillars support the roof. Although the long face is still apparent, the equipment and mining procedures are distinct. Other references to longwall operation in hard-rock mining can be found in Hustrulid and Bullock (2001).

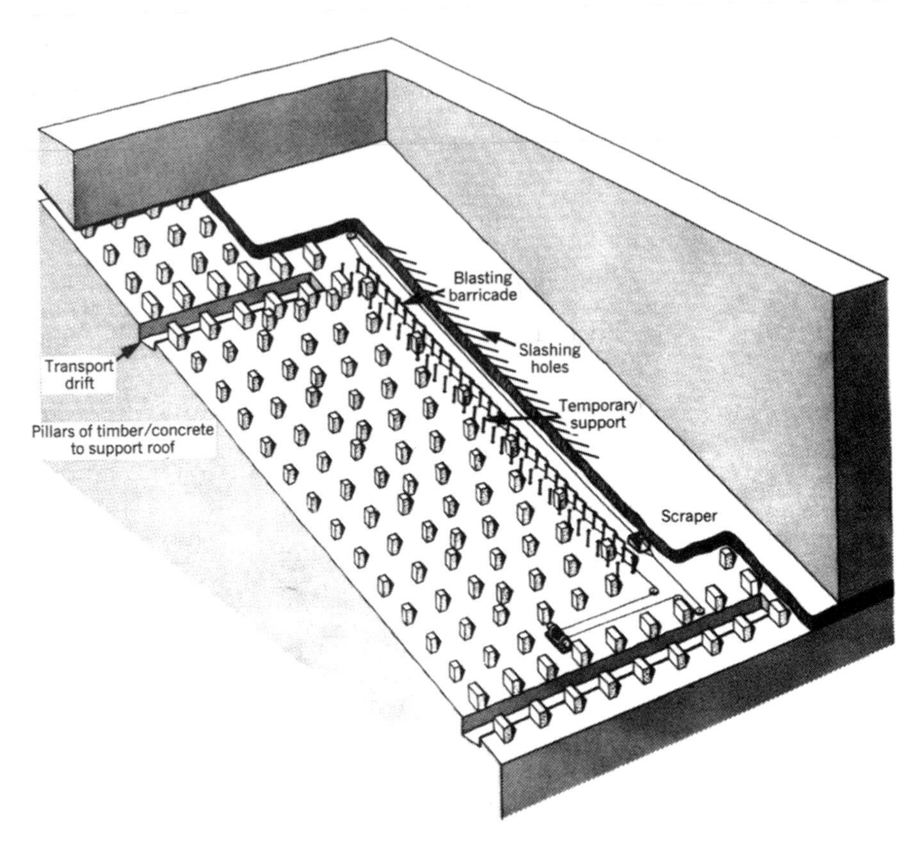

FIGURE 12.2. Longwall method applied to hard-rock mining. (After Hamrin, 1982. By permission from the Society for Mining, Metallurgy, and Exploration, Inc., Littleton, CO.)

12.2.1 Sequence of Development

Longwall development is strikingly similar to development in room-and-pillar mining. By referring to Figure 9.4, we can see how a coal mine is ideally laid out for either method of mining. Main entries are driven across the property, from which orthogonal panel entries divide the coal into large blocks, mineable by either the room-and-pillar or the longwall method. Continuous miners are used to develop the headgate and tailgate entry systems. Generally, continuous miner sections perform this task, though the continuous miners are sometimes equipped with integral roof bolters to enhance the advance rate. After the gate entries are developed, bleeders are normally driven at the inby extreme of the panels to allow for better ventilation of the longwall faces.

When the panel entries are completed and the bleeders established, the longwall face is started from the inner bleeder entry (Trent and Harrison, 1982). As a cut is taken across the face, the hydraulic supports advance, allowing the roof behind to cave. The armored conveyor used to transport coal along the face is snaked forward by the supports, while the mining machine advances by means of a chain or rack and pinion. As the coal arrives at the headgate, it is transferred from the chain-and-flight conveyor to a belt conveyor (the belt is usually installed when the panel entries are driven). Additional support is often required in the entries, where cribs may supplement roof bolts.

12.2.2 Cycle of Operations

The development cycle of operations is essentially the same as in the room-and-pillar mining method using continuous miners. The cycle of operations is outlined in detail in Section 10.2.2. In the actual longwall operation, the cycle of operations is quite simple. The unit operations that must be conducted are the following:

Mining (breaking and loading): shearer (normally a double-drum version), plow

Haulage: armored chain-and-flight face conveyor, normal belt conveyor utilized in the headgate entry set

Auxiliary operations for longwall mining are headed by ground control, in which hydraulically actuated, self-advancing shield supports provide a continuous canopy of protection over the face. Other operations of importance are health and safety (gas control, dust control, ventilation), power supply, maintenance, and materiel supply. Of these, dust control and ventilation are most essential because of the excessive dust concentration inherent in longwall mining. For other auxiliary operations, refer to Table 5.12.

12.2.3 Conditions

The following conditions are derived in part from information in Trent and Harrison (1982), Stefanko and Bise (1983), and Peng and Chiang (1992):

1. *Ore strength:* any, but should crush rather than yield under roof pressure; preferably material that is weak and can be cut by continuous miner
2. *Rock strength:* weak to moderate, must break and cave; ideally, thin-bedded in intermediate roof; floor must be firm, nonplastic
3. *Deposit shape:* tabular
4. *Deposit dip:* low ($<12^\circ$), prefer flat, uniform
5. *Deposit size:* large in areal extent (>1 mi^2 or 260 ha); thin-bedded (3 to 15 ft or 1 to 5 m); uniform thickness
6. *Ore grade:* moderate
7. *Ore uniformity:* uniform (particularly in thickness)

8. *Depth:* moderate (500 to 3000 ft or 150 to 900 m) for coal to very deep (<12,000 ft or 3.5 km) for noncoal

12.2.4 Characteristics

The following lists are based on general knowledge and descriptions in Trent and Harrison (1982), Stefanko and Bise (1983), and Peng and Chiang (1984, 1992):

Advantages

1. Highest productivity for underground coal mines (4.89 tons/employee-hr or 4.4 tonnes/employee-hr)(Energy Information Administration, 2000); outstanding continuity of operations and low labor intensity resulting in high output.
2. Fairly low mining cost (relative cost about 15%).
3. High production rate; large-scale method.
4. Approaches continuity of production, permitting nearly simultaneous cycle of operations to be conducted.
5. Suitable for total mechanization, remote control, and automation.
6. Low labor requirement.
7. Fairly high recovery (about 59% on average) (Energy Information Administration, 2000).
8. Concentrated operations, facilitating transport, supply, and ventilation.
9. Applicable to deep seams under bad roof conditions.
10. Good health and safety factors, especially with regard to roof-fall accidents.

Disadvantages

1. Caving and subsidence (10 to 80% of mined height) occur over wide areas; controllable to some extent.
2. Method very inflexible and rigid in execution; no selectivity except in varying the height of opening somewhat.
3. Mining rate should be uniform to avoid roof support and subsidence problems.
4. High capital cost, totaling about \$30,000 to \$40,000/ft (\$98,000 to \$131,000/m) of face length.
5. Reliance on a single production face can result in costly delays, interruptions in supply.
6. High longwall move costs.
7. Heating in the gob can create temperature-humidity problems and spontaneous combustion.

12.2.5 Applications and Variations

Longwall mining is prospering in the United States, with increased productivity being the major reason for success. There are a number of outstanding longwall mines throughout the country: Bailly and Enlow Fork Mines in Pennsylvania, Upper Big Branch Mine in West Virginia, and the West Elk and Twentymile Mines in Colorado are some that are worthy of note. In addition, longwall mining is doing well internationally, with major activities occurring in China, Poland, South Africa, Russia, Australia, and a number of other countries (Reid, 1997a, 1997b).

Certain variations in longwall operation are evident as the seam dips, although these variations are not practiced in the United States. An important variation of the longwall mining method is *shortwall mining*. This method can be considered an evolutionary variation of longwall mining or a separate method. However, we consider it here to be a variation of the longwall method. It is quite similar in layout to a longwall, as shown in Figure 12.3. However, three variations are evident: (1) the face is typically only 150 to 200 ft (46 to 61 m) in length, (2) coal extraction is performed by a continuous miner, and (3) face haulage is normally performed by shuttle cars.

Specially designed chocks (roof supports) are ordinarily used on a shortwall face to allow a continuous miner to work under the support canopy. In application, the shortwall is limited to seam heights of 3.5 to 12 ft (1.1 to 3.6 m), a depth of 200 to 1700 ft (60 to 500 m), and panel lengths of 2000 to 4000

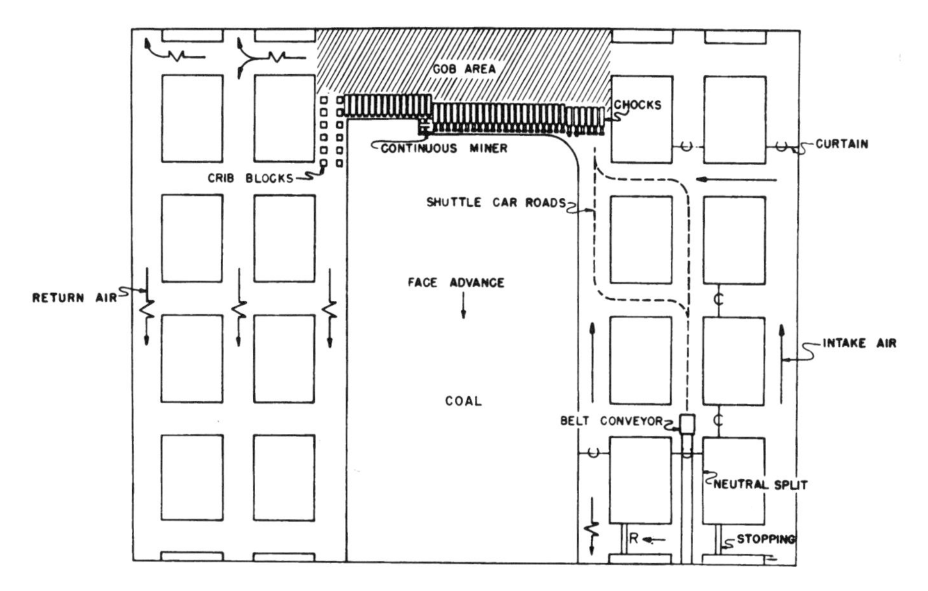

FIGURE 12.3. Shortwall mining of coal. (After Schroeder, 1973. By permission from the Society for Mining, Metallurgy, and Exploration, Inc., Littleton, CO.)

ft (600 to 1200 m). The shortwall system of mining has suffered disuse in recent years because of the better performance of both the longwall and room-and-pillar mining methods.

12.2.6 Case Study: Twentymile Coal Company

The Twentymile Mine of RAG American Coal Co. is located in Routt County in the northwestern part of Colorado. The mine produces about 9,000,000 tons/yr (8,200,000 tonnes/yr) of steam coal from the 96 to 114 in. (2.4 to 2.9 m) Wadge Seam using the longwall mining method. Twentymile has been mining by the longwall method since 1989 and has always been prone to improving productivity through better planning, use of the most efficient equipment, and management of a highly motivated workforce. A diagram of some Twentymile longwall panels is shown in Figure 12.4. Note that the plan includes a three-entry headgate and tailgate with face lengths of about 1000 ft (307 m). The mines started out using panels about 640 ft (195 m) in width, but the width was increased to about 840 ft (256 m) and then to 1000 ft (307 m). In recent years, panels up to 18,000 ft (5490 m) in length have been mined in order to reduce downtime and improve productivity (Ludlow, 2000). With this level of intensity, mine personnel have set world records for longwall production more than a half dozen times.

The current longwall system at Twentymile Mine is designed to produce 5000 tons/hr (4500 tonnes/hr)(Johnson and Buchan, 1999). Face voltage is 4160 volts; the double-drum shearer is powered by two 800 hp (600 kW) motors with two 135 hp (100 kW) traction motors and a 50 hp (35 kW) pump motor. The chain conveyor on the face travels at 360 ft/min (110 m/min) and is powered by three 1000 hp (746 kW) motors (Ludlow, 2000). The section conveyor is 72 in. (1.8 m) in width. Continuous miner development is carefully

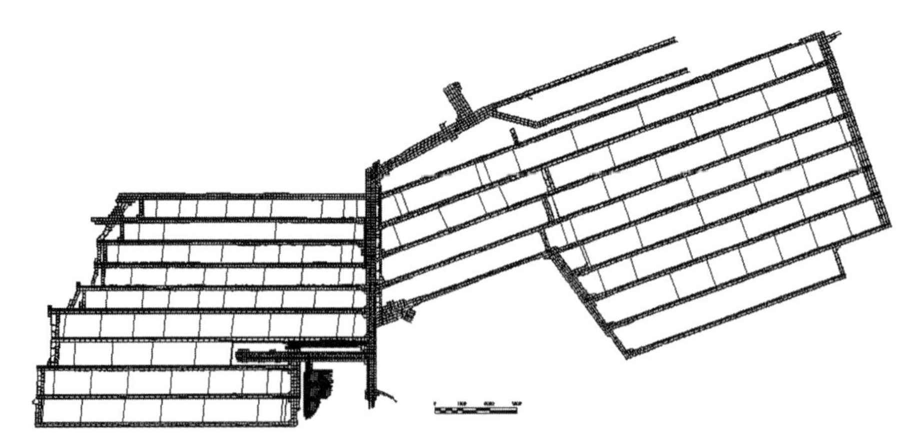

FIGURE 12.4. Layout of the longwall panels at the Twentymile Mine, Routt County, CO. By permission from Twentymile Coal Company, Oak Creek, CO.

planned to keep ahead of the longwall. The development section employs 40 ft (12.2 m) cuts to limit the number of moves and two roof bolters to improve the efficiency of moves from face to face (Johnson and Buchan, 1999). The mine operates according to a management policy of continuing improvement, a process that is sustained by carefully planned meetings to actively involve every employee in the production planning and execution (Ludlow, 2000).

The record-setting performances of the longwall at Twentymile have been attributed to the following (Buchan, 1998):

1. An excellent reserve with low methane content and an excellent top
2. A high-quality suite of longwall equipment
3. A highly motivated workforce created in part by incentive pay

The increase in production and productivity at the Twentymile Mine is a result of continually improving the mining process. Plans for the future include going to 1200 ft (366 m) faces to improve the mining production process even further. This case study parallels many other success stories in longwall mining. The longwall mining method is truly one of the success stories in U.S. mining practice.

12.3 SUBLEVEL CAVING

The two remaining caving methods (sublevel caving and block caving) are applicable to near-vertical deposits of metals or nonmetals. In *sublevel caving*, overall mining progresses downward while the ore between sublevels is broken overhand; the overlying waste rock (hanging wall or capping) caves into the void created as the ore is drawn off. Mining is conducted on sublevels from development drifts and crosscuts, connected to the main haulage below by ramps, orepasses, and raises. Because only the waste is caved, the ore must be drilled and blasted in the customary way; generally fanhole rounds are utilized.

Figure 12.5 illustrates sublevel caving in a steeply dipping ore body. Because the hanging wall eventually caves to the surface, all main and secondary development is located in the footwall. In vertical cross section, the sublevel drifts and crosscuts are staggered so that those on adjacent sublevels are not directly above one another (see Figure 12.6). Thus, fanholes driven from one sublevel penetrate vertically to the second sublevel above. Development and exploitation operations are of necessity carefully planned so that adjacent sublevels are engaged in sequential unit operations, as shown in Figure 12.5.

Modern sublevel caving bears little resemblance to sublevel caving of yesterday (Lewis and Clark, 1964). Formerly a small-scale, labor-intensive method requiring heavy timbering, sublevel caving evolved through research and development into a highly mechanized, large-scale method of mining using only nominal support. Much of the progress took place in Swedish iron mines;

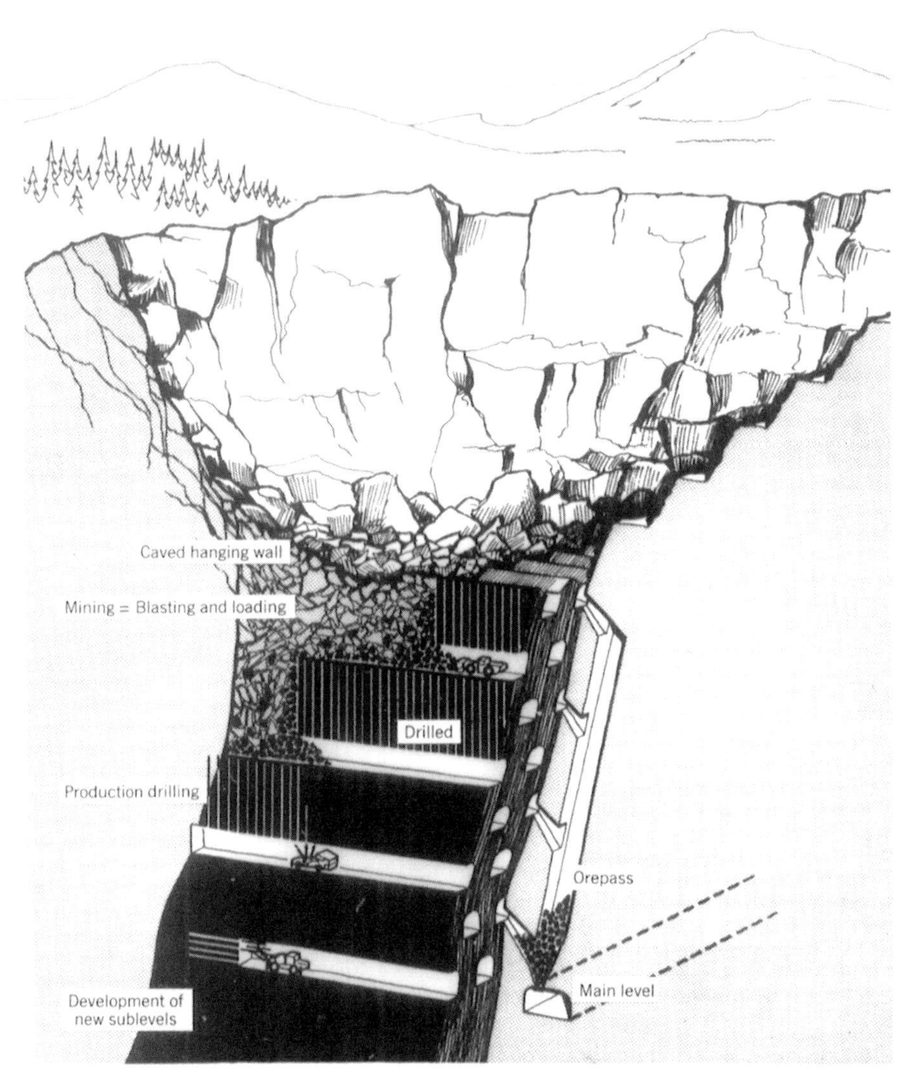

FIGURE 12.5. Sublevel caving in a large, steeply dipping ore body. Different unit operations are conducted on adjacent sublevels. (After Hamrin, 1982. By permission from the Society for Mining, Metallurgy, and Exploration, Inc., Littleton, CO.)

the technology was then exported around the world. In the United States, sublevel caving is not widely used. However, in Sweden, Canada, and Australia, the method finds more extensive use because ore deposits in those countries are more amenable to sublevel caving. Several international mining operations using the sublevel caving mining method are described by Hustrulid and Bullock (2001).

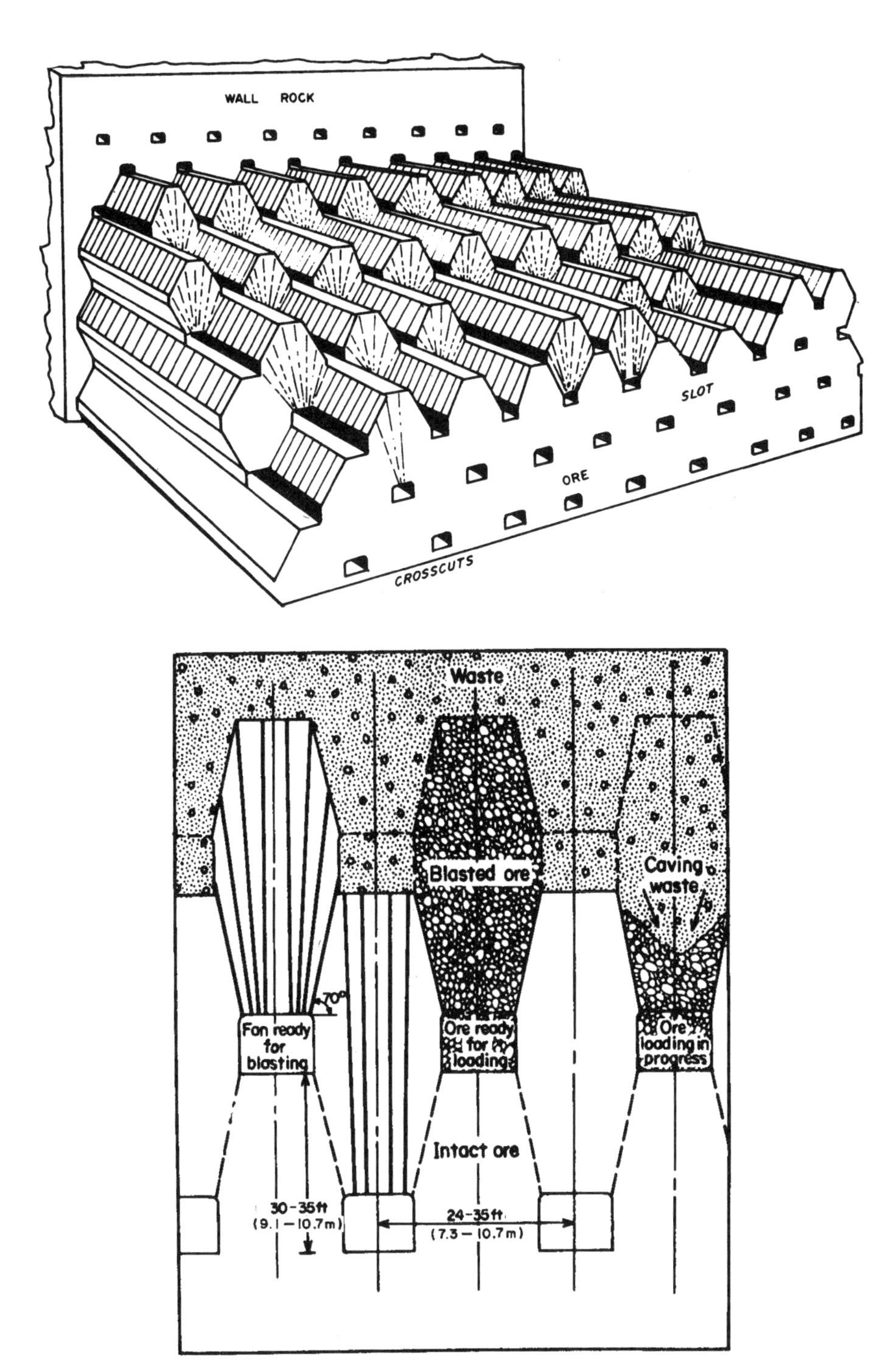

FIGURE 12.6. Layout for sublevel caving. (*Top*) View showing caving slot, sublevel crosscuts, and fanhole drilling round. (After Baase et al., 1982. By permission from the Society for Mining, Metallurgy, and Exploration, Inc., Littleton, CO.) (*Bottom*) Details of sublevel layout and unit operations. (After Cokayne, 1982. By permission from the Society for Mining, Metallurgy, and Exploration, Inc., Littleton, CO.)

Design parameters in sublevel caving are largely a function of *caving mechanics*, the branch of rock mechanics related to the breakage and collapse of consolidated materials in place and their flow downward by gravity. Although the ore has to be drilled and blasted in sublevel caving, the overlying rock forming the capping or hanging wall is undercut and caves. Extremely careful controls must be exercised in drawing the ore to avoid excessive dilution. *Draw control* is the practice of regulating the withdrawal of ore in the sublevel crosscuts so as to optimize the economics of the draw (Cokayne, 1982). Premature cutoff results in poor recovery, and delayed cutoff produces excessive dilution of the ore. Generally, a cutoff grade based on economics is employed to determine when the mucking should cease and the next fan pattern of holes should be blasted (Nilsson, 1982a).

Gravity flow of bulk materials has been studied and analyzed in bins, silos, and chutes (Kvapil, 1982, 1992). The results are now being applied to sublevel caving and to specifying its geometry, dimensions, and layout. Models are very useful to demonstrate flow principles and have been successful in simulating gravity flow in various caving methods. The simplest are two-dimensional, consisting of two vertical, parallel glass plates filled with horizontal layers of colored sand; as the sand flows by gravity through an opening at the bottom, the layers distend and reveal the flow pattern (see Figure 12.7, *left*). As more sand is withdrawn, the ellipse forms (in three dimensions, it becomes an ellipsoid); we refer to it as a *gravity-flow ellipse* (Figure 12.7, *right*). In actual caving, it is this funneling action into the overlying waste that dilutes the ore during drawing.

The dimensions used in the extraction of ore by sublevel caving have evolved over the years into larger, more efficient values. In the early days of mechanized sublevel caving, the sublevel spacings were as close as 30 ft (9 m) and the crosscut spacings were about 20 ft (6 m) (*Engineering and Mining Journal*, 1969; Pillar et al., 1971). However, improved understanding of the flow patterns associated with sublevel caving has resulted in current spacings of 94 ft (28.5 m) between sublevels and 82 ft (25 m) between crosscuts (Wyllie, 1996). Thus, the changes in the spacings have greatly improved the ratio of production tonnage to development tonnage. Other improvements have also improved the situation in sublevel caving. Better utilization of larger diesel equipment, increased use of remote control in the crosscuts, and automation of some of the haulage operations have all been influential in the quest for productivity improvement.

12.3.1 Sequence of Development

All caving methods require extensive development prior to and during mining. In sublevel caving, as much as 15 to 20% of the production takes place during development (Cokayne, 1982; Hamrin, 1982). The major portion of the development is horizontal, both on the haulage level and on the sublevels

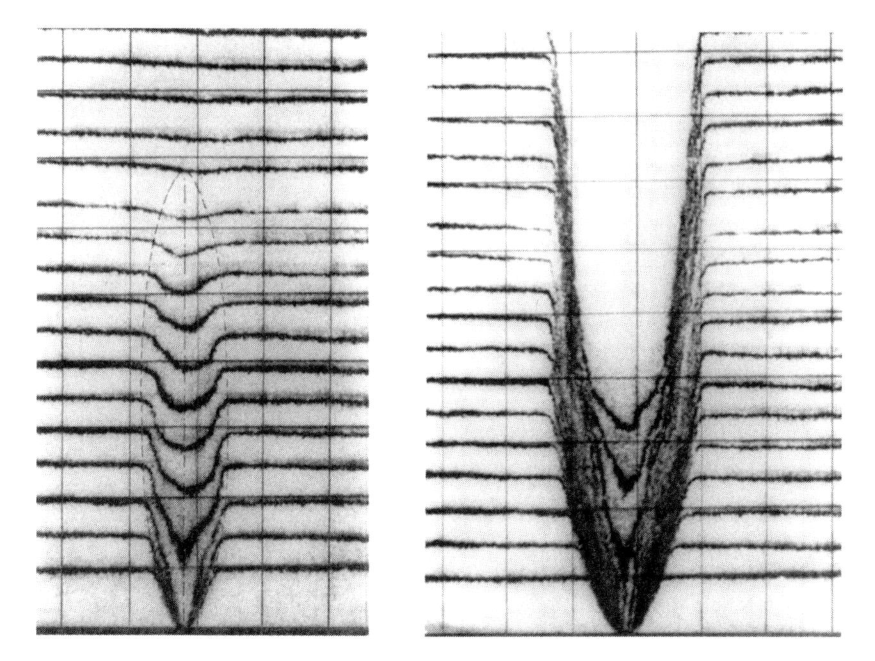

FIGURE 12.7. Model studies representing successive phases of bulk-material flow in sublevel caving. (*Left*) Vertical ellipse formed at the boundary of gravity motion, delineating the active zone above each crosscut. (*Right*) Advanced phase of gravity flow in drawing ore. (After Kvapil, 1982. By permission from the Society for Mining, Metallurgy, and Exploration, Inc., Littleton, CO.)

(drifts, crosscuts), although some is inclined (access ramps) or nearly-vertical (orepasses). The crosscuts can be laid out parallel to the vein in relatively narrow deposits; otherwise, they are laid out transverse to the vein. If the deposit has a flat dip but is sufficiently thick, sublevel caving may still be applicable (Cokayne, 1982).

Haulage levels and sublevels are usually laid out in a grid of drifts and crosscuts similar to those in a room-and-pillar mine. Multiple interconnected openings facilitate transportation and, more important, substantially improve ventilation. In the traditional development plan, crosscuts are driven across the deposit to the hanging wall or cave boundary (see Figure 12.5; note the development on the deepest sublevel). At the end of the crosscuts, slot raises are driven to the cave above, then slabbed off to form slots the shape of the fanhole round. The first blast breaks into the slot; subsequent rounds pull against the broken muck. Several rounds may be blasted simultaneously to initiate the first cave on the uppermost level. Subsequently, multiple blasts in adjacent crosscuts or even an entire sublevel are detonated as needed to meet production requirements.

12.3.2 Cycle of Operations

Sublevel caving employs a conventional cycle of operations in nearly every application. The advent of mechanization in this method was very important because it allowed extensive development to be conducted without greatly enburdening the overall mining costs. Drillholes are often carefully aligned (by surveying techniques), and this has become more important with increased spacings. Materials handling has become the province of diesel load-haul-dump (LHD) equipment, with ever more remote control being utilized. The main haulage in the mine is also often automated, with driverless locomotives used to move the ore to the shaft bottom (Wyllie, 1996). The following operations and equipment choices apply to sublevel caving production operations. Development operations closely parallel those in other hard-rock mining.

Drilling: fandrill jumbos used, two- or three-boomed with pneumatic, hydraulic (oil), or water-powered drills; air or diesel power; hole sizes 2 to 3 in. (50 to 76 mm); drilling factor 0.7 ft/ton (0.2 m/tonne) (Hamrin, 1982)

Blasting: ammonium nitrate and fuel oil (ANFO), slurries; bulk charging by pneumatic loader or pump; firing electrically or by detonating fuse

Secondary blasting (on sublevel): drill and blast, mudcap, impact hammer

Loading: LHD, front-end loader

Haulage: LHD or front-end loader on the sublevel; gravity flow through ore pass; rail or conveyor on haulage level

The important auxiliary operations in sublevel caving are ground control and ventilation. Neither is complex in this method of mining. Support of development openings is the main requirement, and normal light bolting, timbering, or shotcreting is sufficient (excessive support is not appropriate for this method). Ventilation is easily carried out if the multiple sublevel crosscuts are interconnected, as in room-and-pillar mining; control devices and auxiliary ventilation afford good air quality and distribution at the face. Otherwise, vent tubings may be necessary. Other auxiliary operations of importance include health and safety measures, maintenance, power supply, drainage, and materiel supply. See Sections 10.2 and 10.3 for other auxiliary operations.

12.3.3 Conditions

The following deposit conditions are summarized from information in Cokayne (1982), Hamrin (1982), and Kvapil (1982, 1992):

1. *Ore strength:* moderate to fairly strong, requiring blasting; requires sufficient competence to stand without excessive support; less strength than for supported methods, but more than for block caving

2. *Rock strength:* weak to fairly strong; may be blocky, but should be fractured or jointed and cavable; prefer moderate to large fragments, no fines to dilute ore
3. *Deposit shape:* tabular or massive (if elongated along one axis, preferably vertical); may be moderately irregular
4. *Deposit dip:* fairly steep (>60°) or vertical; can be fairly flat if the deposit is thick
5. *Deposit size:* large, extensive vertical or areal extent; thickness >20 ft or 6 m
6. *Ore grade:* moderate
7. *Ore uniformity:* moderate, no sorting possible (some dilution acceptable)
8. *Depth:* moderate (<4000 ft or 1.2 km)

12.3.4 Characteristics

The following characteristics are based on information provided by Cokayne (1982), Hamrin (1982), and Kvapil (1982, 1992):

Advantages

1. Fairly high productivity.
2. High production rate; large-scale method.
3. Fairly high recovery (80 to 90%).
4. Suitable for full mechanization.
5. Somewhat adaptable, flexible, and selective; no pillars are required.
6. Good health and safety factors.
7. Moderate mining cost (relative cost about 15%).

Disadvantages

1. Moderate to high dilution (10 to 35%).
2. Caving and subsidence occur, destroying the surface.
3. Draw control is critical to success of method.
4. High development cost.
5. Must provide stope access for mechanized equipment.

12.3.5 Applications and Variations

The method has been used in Sweden, Canada, Australia, and a number of other countries. However, it has not been widely used in the United States. Applications in the literature include bituminous coal in Hungary (Ravasz, 1984), anthracite coal in Pennsylvania (Green, 1985), copper in British Columbia (Baase et al., 1982), copper and lead in Australia (Hornsby et al., 1982), iron ore in Sweden (Wyllie, 1994, 1996), and nickel in Canada (Espley and Tan, 1994).

In general, variations of the sublevel caving method are not numerous. However, one variation of the method has been practiced at the Fazenda Brasileiro Mine in Brasil to produce gold ore (Souza et al., 1998). This variation is a combination of the sublevel caving development plan with the ore withdrawal concepts of shrinkage stoping. The authors call this method sublevel shrinkage caving. The sublevels are 50 ft (15 m) apart with a crosscut plan similar to that of ordinary sublevel caving.

12.3.6 Case Study: Kiirunavaara Iron Mine, LKAB, Kiruna, Sweden

Kiirunavaara Iron Mine is located above the Arctic Circle in the iron ore mining district of Sweden. The mine is the largest underground operation in the world and has been in production since 1904. The ore body is a disc-shaped sedimentary deposit with a dip of 50° to 70° and a grade of 60 to 67% iron.

This mine has been the proving ground for the modern version of sublevel caving. Many experiments have been conducted to improve the geometry of the development openings, the productivity of the equipment used, and the automated features of the method. The most recent development plan calls for 94 ft (28.5 m) between levels and 82 ft (25 m) between crosscuts. The burden on the blastholes is 9.8 to 11.5 ft (3 to 3.5 m). Automated driverless locomotives are used in the main haulages, and remote control of LHDs improves the efficiency of the loading operations. Robotized drilling machines further improve the efficiency of the underground operations. Additional information on this interesting mine and its development of sublevel caving can be found in H. R. Gustafsson (1981), Heden et al. (1982), Wyllie (1994, 1996), and Quinteiro et al. (2001).

12.4 BLOCK CAVING

The one underground mining method that has the potential to rival surface mining in output and cost is block caving. *Block caving* is the mining method in which masses, panels, or blocks of ore are undercut to induce caving, permitting the broken ore to be drawn off below. If the deposit is overlain by capping or bounded by a hanging wall, it caves too, breaking into the void created by drawing the ore (see Figure 12.8). This method is unlike sublevel caving, in that both the ore and the rock are normally involved in the caving. As in sublevel caving, the caving proceeds in a columnar fashion to the surface. The result is massive subsidence, accompanied by the exceptionally high production rate and great areal extent characteristic of block caving.

A truly American mining method, block caving was invented in the years following World War I to cope with the exploitation of the massive low-grade copper porphyry deposits of the southwestern United States. The technology

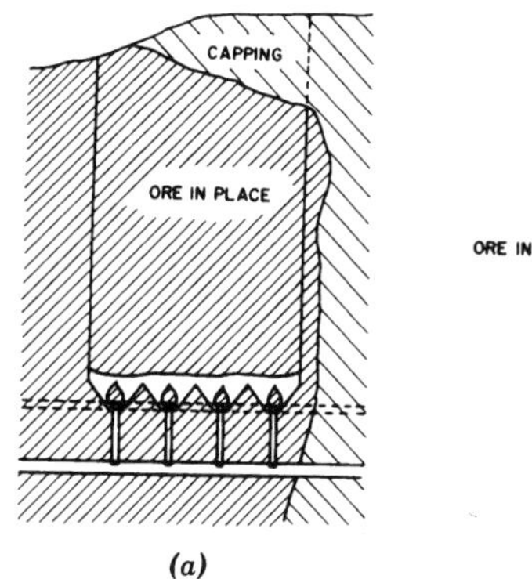

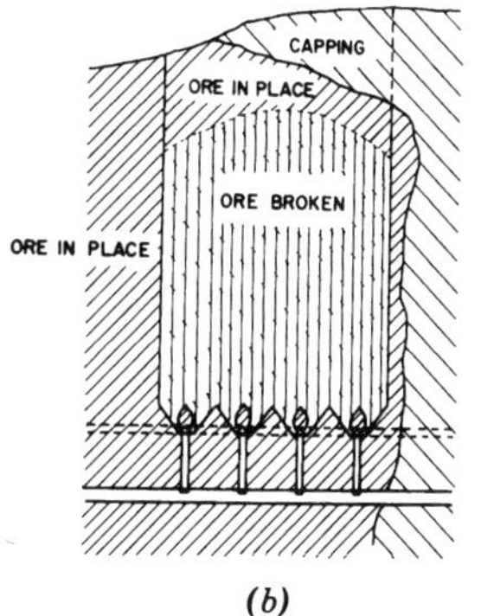

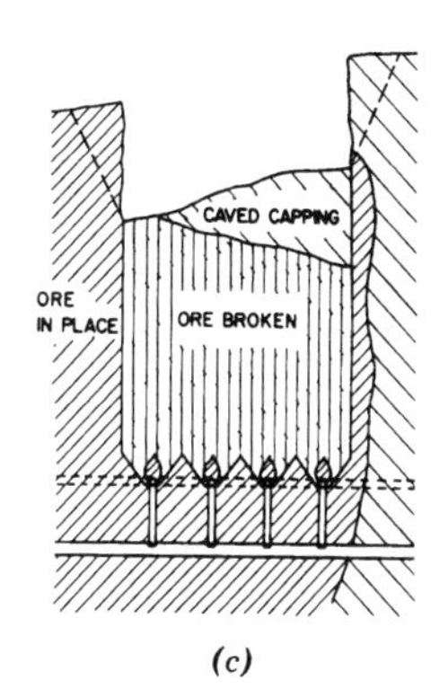

FIGURE 12.8. Progressive stages (a, b, c) of block caving, showing caving of ore and waste. (After Trepanier and Underwood, 1981. By permission from the Society for Mining, Metallurgy, and Exploration, Inc., Littleton, CO.)

rapidly spread to the rest of the world and is now applied in many countries. A large-scale method, block caving is utilized to produce about 10% of U.S. underground metals and nonmetals and 3% of all underground minerals.

As shown in Figure 12.8, the area and volume of ore removed at the bottom of the block during undercutting (and the R/h ratio of the resulting cross section) must be sufficiently large to induce caving in the mass above, which then continues progressively on its own (Tobie and Julin, 1982). Steady drawing of the caved ore from the underside of the block provides space for more broken ore to accumulate and causes the caving action to continue upward until all the ore in the original block has been caved.

Caving mechanics provides the basis for understanding and controlling the operating factors in block caving, as it does in sublevel caving. Determining the *cavability* of an ore body is the first task to be undertaken. Good caving action generally requires that the ore body have fractures in three orientations (Julin, 1992). To investigate the cavability of the ore body, drill cores are obtained throughout the ore body using exploration openings. These cores are then often subjected to *rock quality designation* (RQD) analysis, which measures the percentage of intact (>4 in. or 102 mm) core recovered from a drillhole. The RQD value will help to identify the caving characteristics of the rock mass. The RQD values and other methods of determining the suitability of an ore body to caving have been discussed by McMahon and Kendrick (1969).

Cavability is not just a matter of achieving acceptable fragmentation and optimum operating costs. From a safety standpoint, the ore or capping must not arch over long distances for long periods of time. The formation of stable arches not only disrupts the caving operation but very likely will cause air blast and concussion in the mine when they suddenly collapse.

After cavability is determined for the ore body, the second application of caving mechanics is in draw control and drawpoint spacing (Richardson,

1981). Involved in this problem is the gravity-flow ellipse or ellipsoid referred to in Section 12.3, modified by the inflow of waste as the caving funnel progresses upward into the capping. Figure 12.9 depicts caving action as a function of drawpoint spacing. With theoretically ideal drawpoint spacing (Figure 12.9a), the ellipses are contiguous. Excessive spacing (Figure 12.9b) or deficient spacing (Figure 12.9c) produces zones of draw that may yield unsatisfactory grade control and create weight problems on sill pillars. In plan view, drawpoints may be arranged in a hexagonal (Figure 12.9d) or square pattern (Figure 12.9e). To ensure that the zones of draw completely blanket the ore, drawpoint spacing is reduced somewhat, permitting minor overlap of the zones, as shown in Figures 12.9d and 12.9e.

As the caving action is initiated — whether in a block, panel, or mass — the caving commences as the undercutting of a critical-sized area is completed and then progresses upward; it also progresses in a planned pattern across the ore body. Once caving begins, the only means of regulating it, as well as the production of ore, is through draw control (DeWolfe, 1981). Because of irregularities in the contact between ore and waste and, more important, because of funneling in the ore, some dilution is inevitable if a high recovery is achieved. Effective draw control optimizes the combination of grade control, recovery, and dilution. This is illustrated in Figure 12.10, where caving action across an ore body is shown. Caving is just being initiated by undercutting on the right while full production (100%) is being realized on the left (DeWolfe, 1981). By practicing proper draw control, the interface between the ore and the rock is kept as intact as possible, minimizing dilution and maximizing recovery and grade. Proper drawpoint spacing will make this possible.

The most critical spacings to be designed into the block caving operation are the drawpoint spacings and the spacings between levels. The spacing between levels dictates how high a column of ore is to be extracted through the drawpoints. In past decades, the drawpoint spacings seldom exceeded 40 ft (12 m) (Richardson, 1981) and the columns were generally less than 400 ft (122 m) high. However, the tendency is to design the mining system with bigger blocks of ore to be drawn from each drawpoint. The most recent plan at the Henderson Mine, Colorado, is to use 60 ft (18 m) by 50 ft (15 m) spacings on the drawpoints and a column height of 800 ft (244 m). This reduces the development costs and allows a reduction in the overall mining cost if the plan is successful.

12.4.1 Variations

In block caving one of the critical parameters that determine success is the size and shape of the area that is caved (Ward, 1981; Tobie and Julin, 1982). Three methods are used: block, panel, and mass. In *block caving*, regular rectangular or square areas are undercut in a checkerboard pattern. Usually these blocks are mined in an alternating or diagonal order (see Figure 12.11a) to effect better caving action and draw control. In *panel caving*, ore in continuous strips is mined across the ore body (see Figure 12.11b). Manageable areas are caved

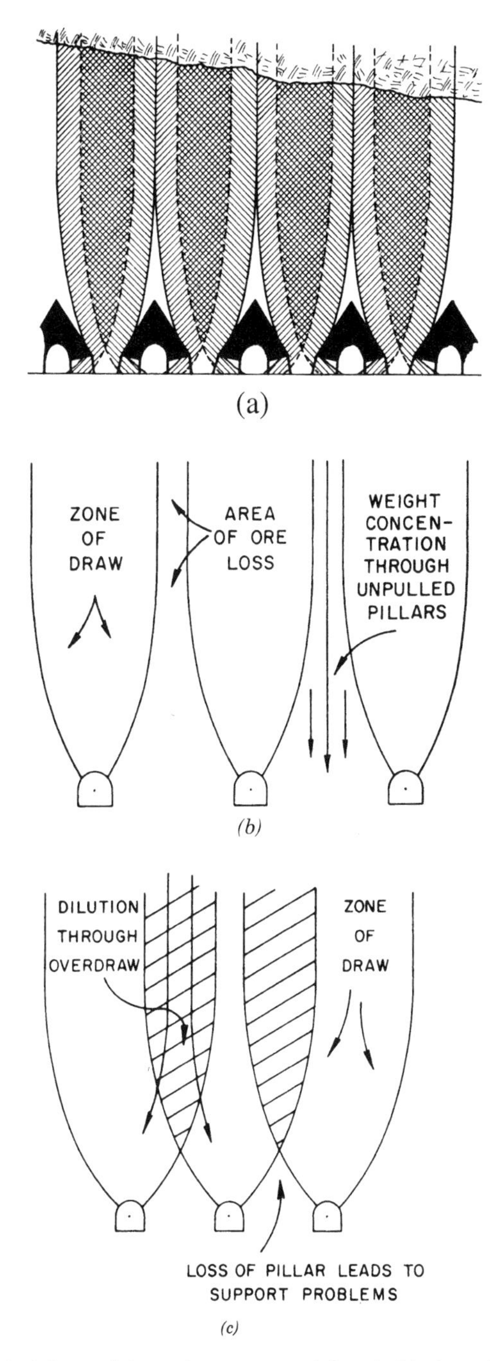

FIGURE 12.9. Effect of drawpoint spacing on zone of draw in block caving. Vertical sections. (a) Theoretical ideal spacing. (After Weiss et al., 1981. By permission from the Society for Mining, Metallurgy, and Exploration, Inc., Littleton, CO.) (b) Excessive spacing. (c) Insufficient spacing. Plan views. (d) Suitable hexagonal spacing. (e) Suitable square spacing. (After Richardson, 1981. By permission from the Society for Mining, Metallurgy, and Exploration, Inc., Littleton, CO.)

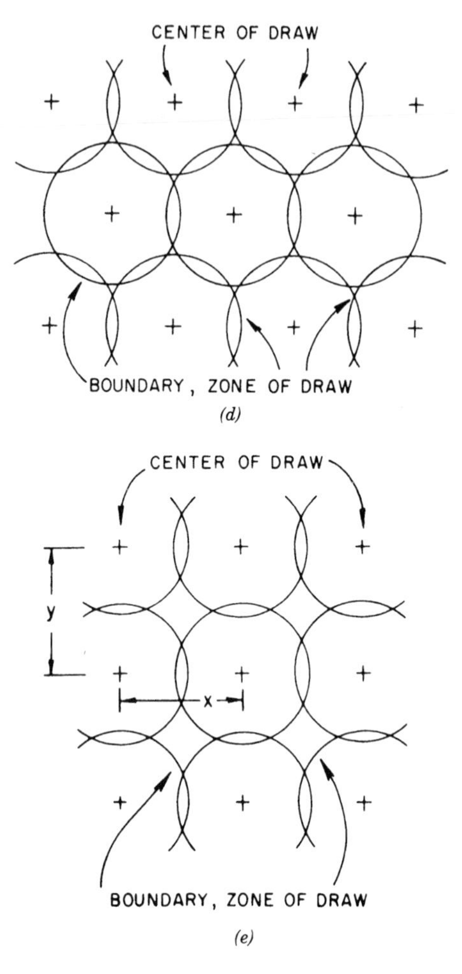

FIGURE 12.9. *Continued.*

simultaneously and retreated in panels. In *mass caving*, there is no area division into blocks or panels; irregularly sized prisms of ore are mined as large as consistent with the caving properties of the ore and the stresses on the openings below (see Figure 12.11c). The plane of contact between the ore and waste is generally inclined as undercutting proceeds on the retreat across the ore body.

Whether a cavable deposit should be mined in blocks, panels, or masses depends mainly on rock competency (Tobie and Julin, 1982). Ore that is weak or highly fractured and breaks fine lends itself to the rapid draw rate attained in block caving. If the fractures are more dispersed or the rock more competent, then panel caving is more suitable. But when the ore body is massive, the rock strong, or the fractures widely spaced, mass caving is usually the choice.

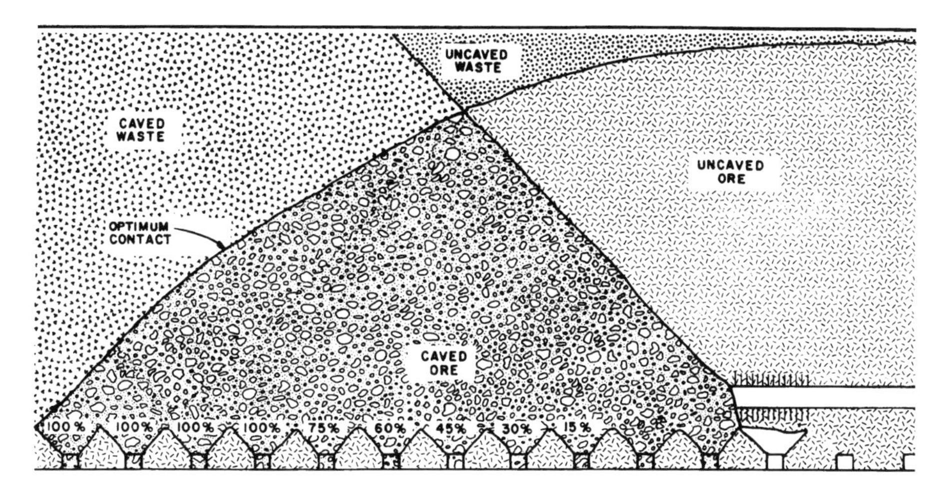

FIGURE 12.10. Progression of cave across an ore body (from left to right), with differential draw-control rates indicated. The plane of contact between caved ore and waste is inclined, typical of panel and mass block caving. (After DeWolfe, 1981. By permission from the Society for Mining, Metallurgy, and Exploration, Inc., Littleton, CO.)

The source of the second variation in block caving is the system of draw or the method of materials handling of the caved ore (Pillar, 1981). Three versions of materials handling are normally used: *gravity*, *slusher*, and *loader*. The original method was the gravity-draw method, which utilizes the forces of gravity and finger and transfer raises to transport the broken ore from the undercut and caving levels to chutes in the haulage drift. This is usually done via a grizzly sublevel where secondary breakage is carried out (see Figure 12.12). The *slusher method*, next in line in terms of the frequency of adoption, utilizes scrapers and slusher drifts to regulate and facilitate the downward flow of ore to drawpoints in the haulage drift (see Figure 12.13). The *loader method* employs mechanical loading equipment (LHDs or front-end loaders) to elevate and deposit material flowing through the drawpoints into rail conveyances or trucks (see Figure 12.14).

The primary determinants of the means of materials handling are the geologic and physical characteristics of the ore body and the financial condition of the mining operation. The most important variable is the fragment size of the caved ore. Gravity draw requires a finely fragmented, easy flowing ore; slushers are suitable for a moderate degree of fragmentation; and mechanical loaders accommodate very coarse ore. Gravity draw is best applied to large, regular deposits and high production rates. If ground control of the development openings is a serious problem, gravity draw is indicated as the development openings are smaller and more easily kept intact. However, the development cost is much higher with the gravity method. The slusher method is often applied to small, irregular deposits. Of course, the fragment size will

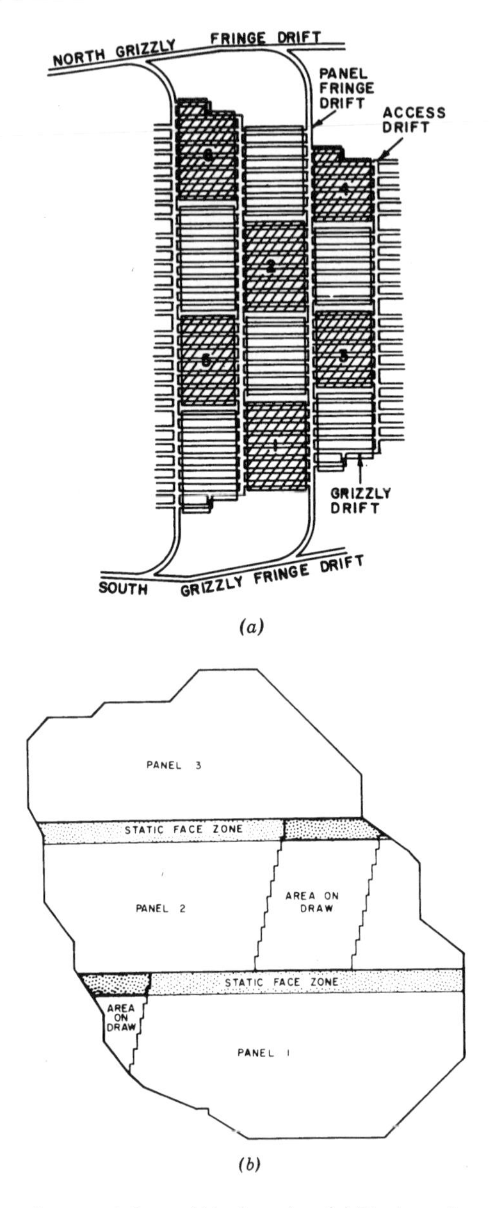

FIGURE 12.11. Three shape variations of block caving. (a) Block version of block caving. (After Tobie and Julin, 1982. By permission from the Society for Mining, Metallurgy, and Exploration, Inc., Littleton, CO.) (b) Panel caving. (After DeWolfe, 1981. By permission from the Society for Mining, Metallurgy, and Exploration, Inc., Littleton, CO.) (c) Mass caving. (After Pearson, 1981. By permission from the Society for Mining, Metallurgy, and Exploration, Inc., Littleton, CO.)

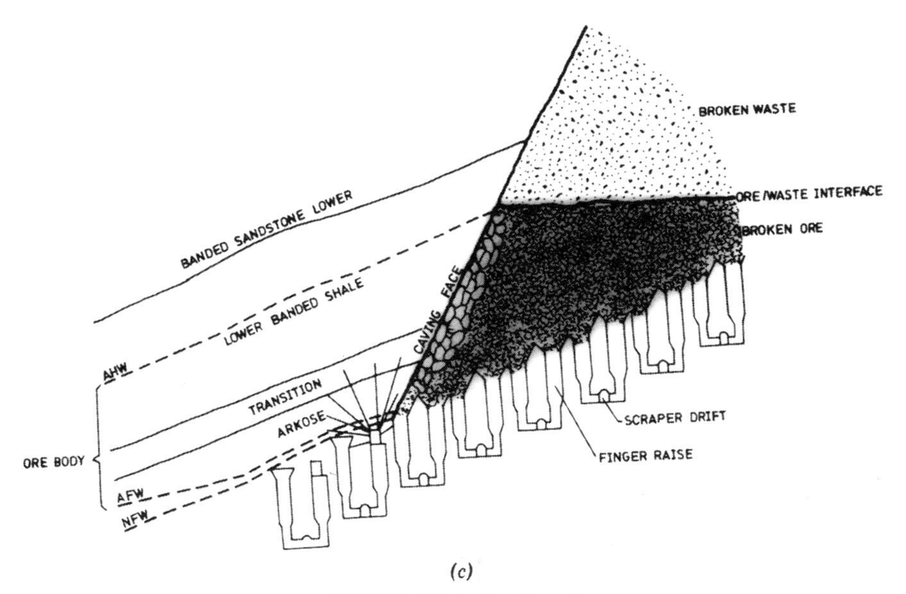

(c)

FIGURE 12.11. *Continued.*

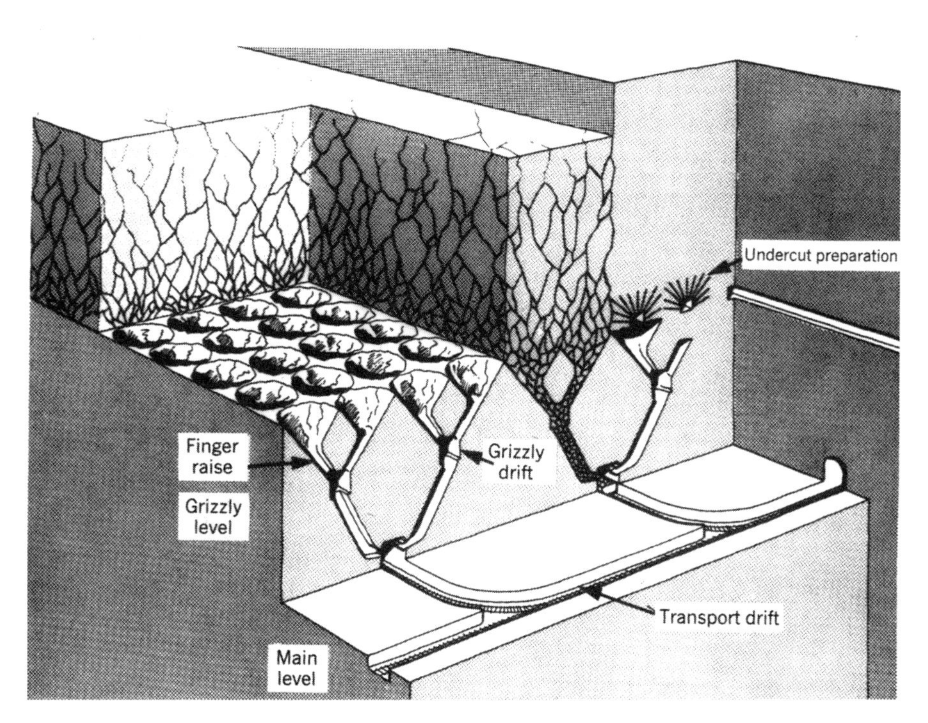

FIGURE 12.12. Block caving, with gravity draw and grizzly drifts. (After Hamrin, 1982. By permission from the Society for Mining, Metallurgy, and Exploration, Inc., Littleton, CO.)

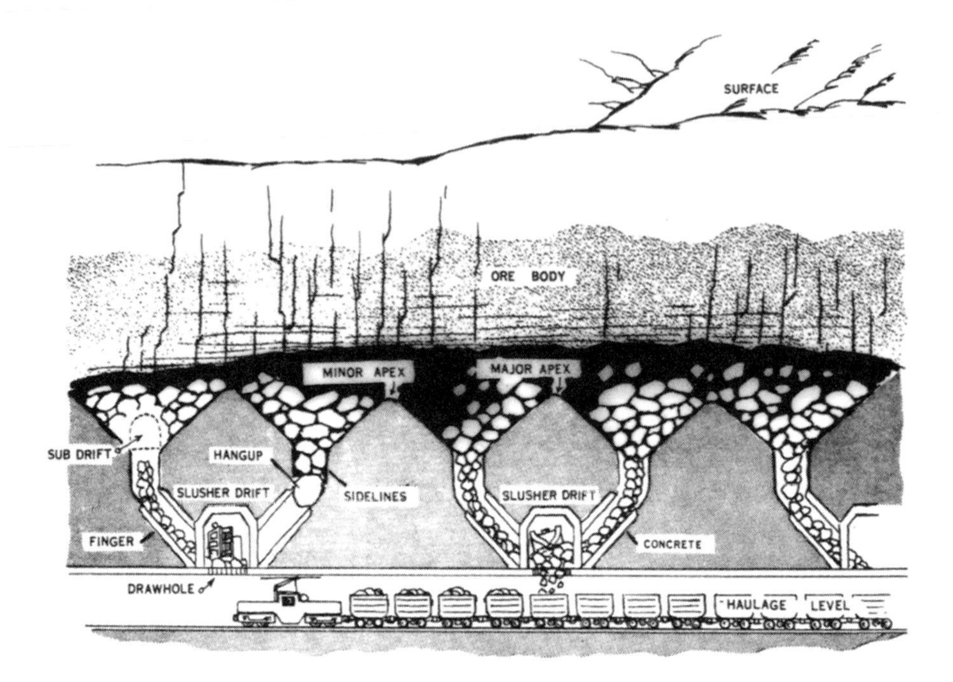

FIGURE 12.13. Block caving, with scraper loading in slusher drifts. (After Johns, 1981. By permission from the Society for Mining, Metallurgy, and Exploration, Inc., Littleton, CO.)

determine whether the method will work in the given geologic environment. For the coarser ore materials, loader caving is the best choice. It also has the advantage of being the most versatile and least expensive in terms of development costs. In the final analysis, the choice of a materials handling method is not determined by costs alone but must take into account compatibility with the natural conditions, ore recovery, and dilution results (Ward, 1981). Of the three methods, the loader method is often the choice for today's block caving operations (Gallagher and Teuscher, 1998).

12.4.2 Sequence of Development

Development for block caving is typically extensive and expensive, but generally less so than that for sublevel caving on a unit-cost basis. Mine level development commences from the shaft station in the usual way, providing for high-speed, high-capacity haulage and ample ventilation airflow capacity. Main haulageways are often paralleled by laterals, interconnected by crosscuts, to ensure good ventilation and to provide adequate lanes or stub crosscuts for loading. One or more additional sublevels are required for grizzly or slusher operation, increasing the development costs for these two methods.

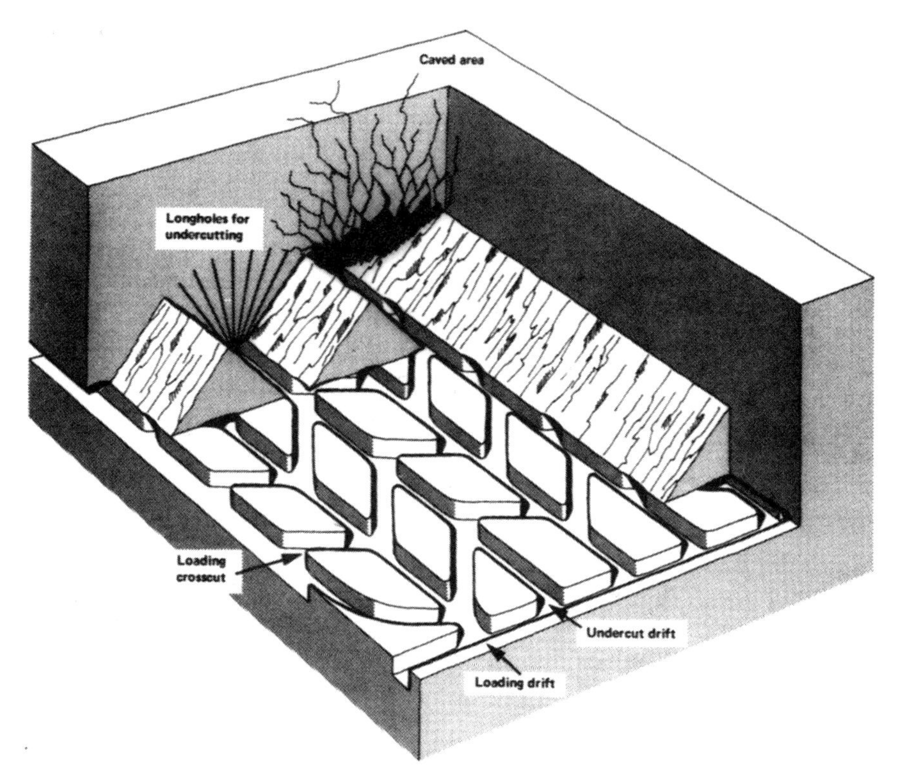

FIGURE 12.14. Block caving, with machine loading in draw drifts. (After Hamrin, 1982. By permission from the Society for Mining, Metallurgy, and Exploration, Inc., Littleton, CO.)

To provide ore-drawing facilities, chutes, drawpoints, or trenches are prepared in the ore body under the block to be mined (Hamrin, 1982). They lie adjacent to the haulage drifts and crosscuts on the main level, as may be seen in Figures 12.8 to 12.14. Finger raises to serve as orepasses are then driven to the grizzly sublevels above. With slusher handling of the ore, the raises may not be necessary as the slusher level is often immediately above the haulage. The most critical development step is undercutting the ore zone, which is carried out in a manner similar to that of a room-and-pillar mining operation, with the pillars being withdrawn in a systematic manner across the cave zone.

Because of the large expanse of opening and tonnage of ore involved, the dangers of premature collapse, hangup, and air blast are ever present. In the past, the procedure of weakening the boundary of the cave zone was often used to overcome these dangers. However, this is seldom employed today (Ward, 1981). Instead, the mining company generally relies on the knowledge of rock mechanics to ensure that the caving proceeds in a predictable manner.

12.4.3 Cycle of Operations

As in most other vertical stoping and caving operations for hard-rock mining, the development and exploitation cycles of operations are separate and distinct. The tendency today is for each to be highly mechanized. For exploitation, a truncated cycle of operations is followed because only the loading and hauling operations are utilized (rock breakage is performed by the caving action). The cycle of operations consists of the following:

Drilling (*undercut*): pneumatic- or hydraulic-powered percussion jumbos; hole size 2 to 3 in. (51 to 76 mm)

Blasting (*undercut*): ANFO, slurries; bulk charging by pneumatic loader or pump, firing electrically or by detonating fuse

Secondary blasting (*on sublevel or in haulage drift*): impact hammer, dynamite bomb, drill and blast, mudcap

Loading (*through bells and ore passes*): gravity flow to chutes; LHD, front-end loader, slusher at drawpoints

Haulage (*on main level*): LHD, rail, truck, belt conveyor

Like the other caving methods, block caving requires elaborate ground-control measures (Hamrin, 1982). Reinforcement and concrete linings, as well as conventional support, are often required in the development openings (e.g., raises, orepasses, slusher drifts, and haulage drifts) that perform a production function. The usual forms of support are bolts, shotcrete, steel sets or arches, cable, or mesh. Reinforced concrete or steel liner is often installed to prevent excessive wear of the openings, especially in slusher and haulage drifts (Watson and Keskimaki, 1992). The next most important auxiliary operation is ventilation; large airflow quantities are essential to manage dust, fumes, and heat control (Tobie and Julin, 1982). Other key auxiliary functions are drainage, materiel supply, maintenance, and power supply (see Table 5.12, Section 10.2, and Section 10.3).

12.4.4 Conditions

The following conditions are derived from discussions in Weiss et al. (1981), Hamrin (1982), Tobie and Julin (1982), and Julin (1992):

1. *Ore strength:* weak to moderate or fairly strong, prefer friable, fractured, or jointed rock, not blocky; should cave freely under own weight when undercut; free running, not sticky if wet, not readily oxidized
2. *Rock strength:* weak to moderate, similar to ore in characteristics
3. *Deposit shape:* massive or thick tabular deposit, fairly regular
4. *Deposit dip:* fairly steep ($>60^\circ$) or vertical; can be fairly flat if sufficiently thick

5. *Deposit size:* very large areal extent; thickness >100 ft (30 m)
6. *Ore grade:* low, ideal for disseminated ore masses; most suitable of underground methods for low-grade deposits
7. *Ore uniformity:* fairly uniform and homogeneous; sorting not possible
8. *Depth:* moderate; >2000 ft (600 m) and <4000 ft (1200 m)

12.4.5 Characteristics

The following characteristics are based on descriptions by Ward (1981), Hamrin (1982), Tobie and Julin (1982), and Julin (1992):

Advantages

1. Relatively high productivity.
2. Fairly low mining cost, least of the underground mining methods (relative cost about 10%).
3. Highest production rate of the underground stoping methods; large-scale method.
4. High recovery (90 to 125%).
5. Rock breakage in production occurs entirely by caving induced by undercutting; no drilling and blasting cost in production.
6. Suitable for gravity draw or fully mechanized materials handling; repetitive, standardized operations.
7. Ventilation is generally very satisfactory; good health and safety factors.

Disadvantages

1. Caving and subsidence occur on a large scale.
2. Draw control is critical to success of the method.
3. Slow, extensive, costly development.
4. Dilution may be high (10 to 25%).
5. Maintenance of openings in production areas is substantial and costly if pillars load excessively.
6. Rigid, inflexible method.
7. Hazardous work because of hangups in the grizzly and slusher sublevels; some risk of air blast throughout the mine.
8. Possible spontaneous combustion in ore or rock during caving if drawing is slow or delayed (risk is high if sulfide content is >45%).

12.4.6 Applications

Block caving is applied primarily to large, disseminated mineral deposits that are too deep for open pit mining. If an underground method must be applied to this type of deposit, then block caving is the logical first choice. Examples of block caving operations are numerous in the literature. Of particular note are the Magma Mine (copper) in Arizona (Zappia, 1981; Snyder, 1994), the Henderson Mine (molybdenum) in Colorado (Gallagher and Teuscher, 1998; Rech, 2001), the El Teniente Mine (copper) in Chile (Julin, 1992; Copier and Meza, 1998; Rojas et al., 2001), the Andina Mine (copper) in Chile (Torres et al., 1981), the King-Beaver Mine (asbestos) in Quebec (Trepanier and Underwood, 1981), and the Mather Mine (iron) in Michigan (Bluekamp, 1981).

12.4.7 Case Study: Henderson Mine, Colorado

The Henderson Mine near Empire, Colorado, is a molybdenum operation located atop the Continental Divide at an elevation of 10,400 ft (3.2 km). The mine was put into operation in 1976 at a cost of $500 million. The deposit consists of two overlapping quartz-molybdenite ore bodies, elliptically shaped in the plan view, with horizontal dimensions of 2200 ft (670 m) by 3600 ft (910 m) and an average thickness of 600 ft (185 m). The grade averages 0.42% MoS_2, with about 242 million tons (220 million tonnes) of reserve initially (Gallagher and Teuscher, 1998).

The original mining plan called for panel caving of 400 ft (122 m) columns of ore with a drawpoint spacing of 53 ft (16 m) by 40 ft (12 m). This resulted in about 143,000 tons (130,000 tonnes) of ore per drawpoint. The loading and haulage operations were to load the trains by diesel loaders and haul the ore to the mill using a 15 mile (24 km) track haulage system. The main problem with the system was that the rail system constrained the production capacity of the mine.

In order to improve productivity, increase production, and overcome the haulage constraint, management at the mine decided to change the system significantly (Gallagher and Teuscher, 1998; Rech, 2001). By means of benchmarking studies, the caving system was overhauled for the planned new deeper level and the main haulage system was changed to a belt conveyor. The drawpoint spacings were increased to 60 ft (18 m) by 50 ft (15 m), and the column height was increased to 800 ft (244 m). This allowed about 377,000 tons (342,000 tonnes) of ore to be taken from each drawpoint, significantly decreasing the development costs. Figure 12.15 shows the relative sizes of the old and new ore columns used at Henderson. The ore in the new plan is gathered using four 80 ton (72 tonne) side-dump trucks and two 40 ton (36 tonne) end-dump trucks. The trucks are to be loaded using pneumatic-power chutes that are operated from the trucks. The trucks will then transport the ore to the loading point on the belt conveyor. The quest for improved productivity in this mine was achieved by increased drawpoint spacing, a larger column height, and the switch to a belt conveyor system.

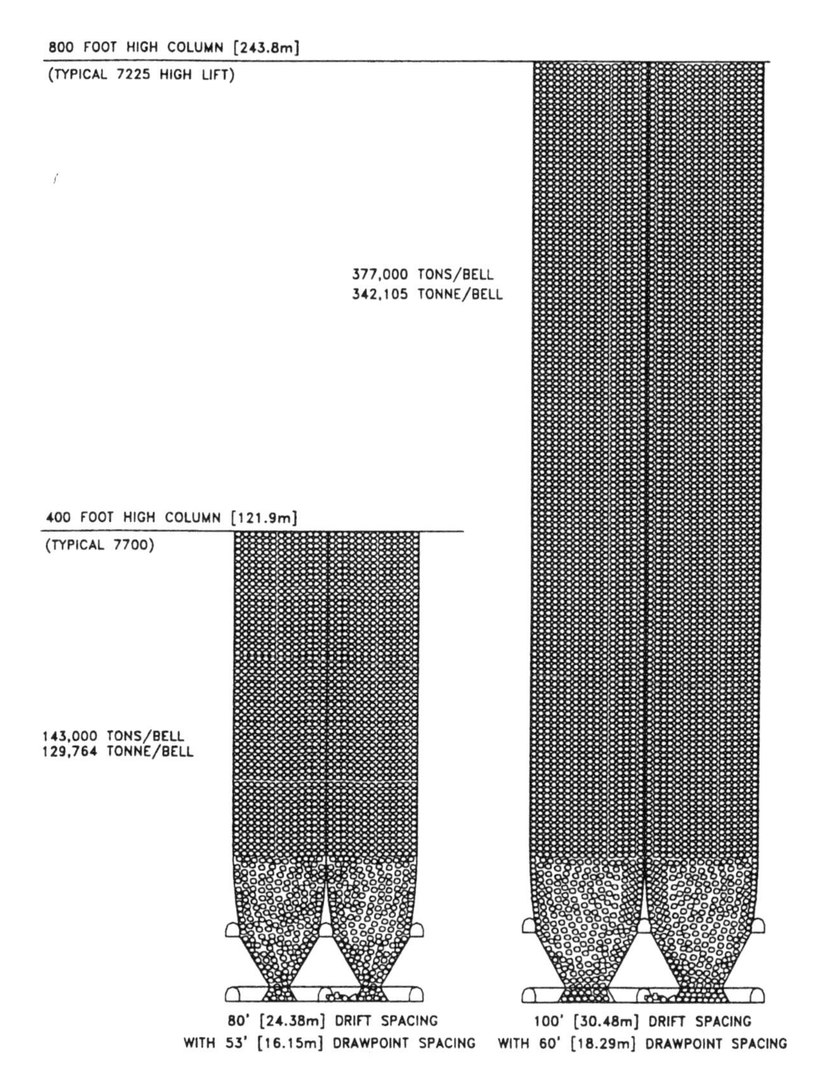

FIGURE 12.15. Old (*left*) and new (*right*) block caving columns at the Henderson Mine. (After Gallagher and Teuscher, 1998. By permission from the Society for Mining, Metallurgy, and Exploration, Inc., Littleton, CO, and Henderson Mining Co., Empire, CO.)

12.5 SPECIAL TOPIC: MINE VENTILATION

The most vital of the auxiliary operations in underground mining is ventilation. It largely maintains the quality and quantity of the atmospheric environment and is the mainstay of the miner's life-support system and the mine's health and safety program.

Mine ventilation is the process of total air-conditioning responsible for the quantity control of air, its movement, and its distribution. Other processes specifically help to accomplish quality control (e.g., gas and dust control) or temperature–humidity control (e.g., air cooling and dehumidification, heating). Because of its versatility, only ventilation carries out aspects of all three control functions. When the control of the atmospheric environment is complete—that is, when there is simultaneous control of the quality, quantity, and temperature–humidity of the air in a designated space—then we are employing *total air-conditioning*.

In recent years, environmental standards in mines have been raised substantially (Hartman et al., 1997). Although threshold limits are based on human endurance and safety, we are increasingly concerned with standards of human comfort as well. The reasons have to do with cost-effectiveness as well as humanitarianism. Worker productivity, job satisfaction, and accident prevention correlate closely with environmental quality.

Conditioning functions and processes commonly used in mines consist of the following:

1. Quality control
   a. Gas control
   b. Dust control
2. Quantity control
   a. Ventilation
   b. Auxiliary or face ventilation
   c. Local exhaust
3. Temperature–humidity control
   a. Cooling and dehumidification
   b. Heating

Processes may be applied individually or jointly. We concentrate here on mine ventilation as the most important and universally used process.

12.5.1 Quality Control

Chemical contaminants—principally gases and dusts—constitute a variety of hazards in the mine atmosphere. Gases may be suffocating, toxic, radioactive, or explosive. Dusts may be nuisance, pulmonary, toxic, carcinogenic, or explosive. Contaminants occur naturally (e.g., strata gases such as methane in coalbeds or radon gas from radioactive ores) or are introduced by mining activity (e.g., diesel or blasting fumes, smoke from fires, all dusts). Even human breathing liberates a contaminant (carbon dioxide) while consuming oxygen, admittedly in small amounts.

The engineering principles of mine air *quality control*, for both gases and dusts, are as follows:

1. Prevention or avoidance
2. Removal or elimination
3. Suppression or absorption
4. Containment or isolation
5. Dilution or reduction

Different practices are employed to implement these control principles. For example, in the room-and-pillar mining of coal with continuous equipment, the two major contaminants are methane gas and coal dust. In this case, corresponding to the five control principles, the following control practices would be considered for application:

Gas	Dust	Both
1. Nonsparking bits	Sharp bits	Good mining practices
Airway sealent		Explosion-proof atmosphere
2. Methane drainage	Dust collector	Bleeders
	Good housekeeping	
3. Sprayfans	Water sprays	Water infusion
	Rock dusting	
4. Sealing old workings	Hood enclosure	Separate split of air
5. Main ventilation	Main ventilation	Main ventilation
Auxiliary ventilation	Auxiliary ventilation	Auxiliary ventilation

Although not all these principles and practices would be utilized in every mine, enough of them would be employed to cope with existing hazards and contain them within threshold limit values promulgated by the Mine Safety and Health Administration (MSHA) and other health and safety authorities (American Conference of Governmental Industrial Hygienists, 2000; Code of Federal Regulations, 2000). Ideally, they should be practiced in the order in which they are given; this results in the best quality control and cost-effectiveness. Note that ventilation is listed last, not because it is least effective but because it is universally applicable and most effective when coupled with other measures.

The *quantity* of ventilation, Q, required to dilute an airborne hazard is determined by the following relation:

$$Q = \frac{Q_g(1 - \mathrm{TLV})}{\mathrm{TLV} - B_g} \tag{12.1}$$

where Q_g is contaminant inflow rate, B_g is concentration of contaminant in normal intake air, and TLV is threshold limit value of the contaminant.

Example 12.1. Calculate the air quantity necessary to dilute methane gas in a coal mine to its TLV when the inflow rate is 250 ft^3/min (0.118 m^3/sec) and its concentration in the intake air is 0.1%.

SOLUTION. Consider the limit for methane to be based on its explosibility (it is also suffocating in high concentrations but nontoxic); MSHA prescribes the TLV = 1%, at which work must cease during the exploitation stage (Code of Federal Regulations, 2000). Employing Eq. 12.1,

$$Q = \frac{(250)(1 - 0.01)}{(0.01 - 0.001)} = 25{,}000 \text{ ft}^3/\text{min } (11.8 \text{ m}^3/\text{sec})$$

Example 12.2. Calculate the air quantity necessary to dilute bituminous coal dust during continuous mining to its TLV when the dust generation rate is 2.5 g/min (0.0055 lb/min) and the concentration in the intake air is 0.5 mg/m^3 (38.9×10^{-6} lb/ft^3).

SOLUTION. Consider the TLV for coal dust, set by MSHA, to be its respirable limit (2 mg/m^3) (Code of Federal Regulations, 2000). Expressed on a percentage basis, the TLV becomes so small that it may be omitted from the numerator in Eq. 12.1. For consistency of units, Q_g is defined as a weight flow rate, and TLV and B_g are weight concentrations (the calculation is carried out more easily in SI units):

$$Q = \frac{2.5 \times 10^3}{2 - 0.5} = 1667 \text{ m}^3/\text{min} = 27.8 \text{ m}^3/\text{sec } (58{,}900 \text{ ft}^3/\text{min})$$

12.5.2 Quantity Control

Quantity control in mine ventilation is concerned with supplying air of the desired quality and in the desired amount to all working places throughout the mine. Air is necessary not only for breathing—a remarkably low 20 ft^3/min (0.01 m^3/sec) per person usually suffices—but to disperse chemical and physical contaminants (gases, dusts, heat, and humidity) as well. Because breathing requirements are easily met, federal and/or state laws provide for a higher minimum quantity of air per person (100 to 200 ft^3/min, or 0.05 to 0.09 m^3/sec), a minimum quantity at the face (3000 ft^3/min, or 1.4 m^3/sec) or in the last crosscut of a coal mine (9000 ft^3/min, or 4.3 m^3/sec), or a minimum velocity at the coal face (60 ft/min, or 0.3 m/sec) (Code of Federal Regulations, 2000). Mine ventilaton practice is heavily regulated in the United States as well as in the rest of the world, especially in coal and gassy (noncoal) mines, and other statutes relate to air quantities required to dilute diesel emissions, blasting fumes, radiation, dusts, battery emissions, and many other contaminants.

Rather surprisingly, legal requirements for minimum airflow seldom govern in determining the design quantity for the mine. Most mines operate with ventilation quantities far in excess of those legislated; the best-ventilated mines circulate millions of ft^3/min (thousands of m^3/sec) of air, attaining ratios of the weights of airflow to mineral produced of 10 to 20 tons/ton (tonnes/tonne).

The design basis for specifying the amount of airflow in working places is usually the critical velocity or quantity required to disperse or dilute contaminants (airflow must be well into the turbulent range). If the quantity needed for dilution is inadequate for effective dispersion or cooling, then the velocity and hence the quantity is increased to an adequate level. (Economics plays a role also because excessive airflows squander fan horsepower.) Critical velocities at the working face range from 100 to 400 ft/min (0.5 to 2.0 m/sec), unless cooling is a consideration, in which case velocities may attain 400 to 600 ft/min (2.0 to 3.0 m/sec). Knowing the area A and velocity V, the quantity can then be determined:

$$Q = VA \tag{12.2}$$

Example 12.3. Calculate the quantity of air required to dilute 80 ft^3/min (0.038 m^3/sec) of carbon dioxide, a strata gas, in the stopes of a metal mine and to maintain a critical face velocity of 150 ft/min (0.76 m/sec). The cross-sectional area of each stope is 15 × 20 ft (4.5 × 6.0 m) and the CO_2 content of the intake air is 0.03%. The TLV established by MSHA for CO_2 is 0.5%.

SOLUTION

1. Calculate the required dilution by Eq. 12.1:

$$Q = \frac{(80)(1 - 0.005)}{0.005 - 0.0003} = 16{,}900 \text{ ft}^3\text{/min } (7.98 \text{ m}^3\text{/sec})$$

2. Calculate the required quantity for effective dispersion by Eq. 12.2:

$$Q = (150)(15)(20) = 45{,}000 \text{ ft}^3\text{/min } (21.24 \text{ m}^3\text{/sec})$$

3. Specify the critical quantity as the larger value (item 2 governs):

$$Q = 45{,}000 \text{ ft}^3\text{/min } (21.24 \text{ m}^3\text{/sec})$$

Theoretically, contamination by the CO_2 in human exhalation ought to be considered too, but because it is so small ($Q_g = 0.1$ ft^3/min, or 47×10^{-6} m^3/sec, per person), it can be neglected in this example.

FIGURE 12.16. Axial-flow fans installed in the exhaust position at the surface of a coal mine. (By permission from Howden Buffalo, New Philadelphia, OH.)

Once quantity requirements in all the working places have been specified, it is then necessary to create a *pressure difference* in the mine to provide the desired flows. Either *natural ventilation*, in which the pressure difference results from thermal energy (like the chimney effect), or *mechanical ventilation*, in which the rotational energy of a fan is converted to fluid-flow energy, may serve as an energy source. In modern mines, only fans are relied upon to provide a pressure difference for ventilation (natural ventilation is too variable, unreliable, and insufficient in magnitude to be utilized). Both centrifugal and axial-flow fans are in common use, and they may be installed in either the blower, booster, or exhaust position (except that underground booster fans are prohibited in coal mines). Axial-flow fans installed as exhausters appear in Figure 12.16.

Calculating the pressure difference required over all working places and airways is prerequisite to the selection of a mine fan. Resembling the well-known Darcy equation from fluid mechanics, the Atkinson equation is generally utilized in mine ventilation:

$$H = \frac{KOLQ^2}{5.2A^3} \tag{12.3}$$

where H is pressure difference, K is a friction factor, O is perimeter, and L is length (actual length plus equivalent length due to shock loss). Friction factors are selected from a table, experience, or actual measurement. Sometimes Eq. 12.3 is written in the form

$$H = RQ^2 \tag{12.4}$$

where R is airway resistance $= KOL/5.2A^3$.

Example 12.4. Given: a rectangular mine opening of cross-sectional dimensions 6×20 ft (1.8×6 m), length 7500 ft (2250 m), and friction factor 45×10^{-10} lb-min^2/ft^4 (8.35×10^{-3} kg/m^3). Calculate the pressure difference for an airflow of 140,000 ft^3/min (66.1 m^3/sec).

SOLUTION

$$Q = 2(6 + 20) = 52 \text{ ft}, \qquad A = (6)(20) = 120 \text{ ft}^2$$

Using Eq. 12.3,

$$H = \frac{(45)(10^{-10})(52)(7500)(140{,}000)^2}{5.2(120)^3} = 3.83 \text{ in. water (953 Pa)}$$

(*Note:* The inch of water is a unit of pressure measurement; 1 in. water = 5.2 lb/ft^2, or 249 Pa.)

Mine ventilation circuits, like electrical circuits, are arranged with airways in series or parallel or as combination series–parallel circuits called *networks.* Simple circuits can be solved mathematically, using principles derived from electrical theory; but because of their complexity, networks are best solved by computer and are beyond the scope of this discussion. In series circuits, H and R values are cumulative and Q is constant. In parallel circuits, Q values are cumulative and inversely proportional to $\sqrt{R}$ for a given airway and H is constant. Referring to Eq. 12.4, for series circuits,

$$H = (R_1 + R_2 + \cdots + R_n)Q^2 = H_1 + H_2 + \cdots + H_n \tag{12.5}$$

and for parallel circuits,

$$R_{eq} = \left(\frac{1}{\sqrt{1/R_1} + \sqrt{1/R_2} + \cdots + \sqrt{1/R_n}}\right)^2 \tag{12.6}$$

and

$$Q_1 = Q\sqrt{\frac{R_{eq}}{R_1}}, \text{ etc.} \tag{12.7}$$

where R_{eq} is the equivalent circuit resistance.

Example 12.5. Four airways are arranged as a series circuit. Their resistances, in units of 10^{-10} in.-min^2/ft^6 (N-sec^2/m^8), are

23.50 (2.627)
1.35 (0.151)
3.12 (0.349)
3.55 (0.397)

If the quantity is 100,000 ft^3/min (47.2 m^3/sec), find the pressure difference across the circuit.

SOLUTION. Use Eq. 12.5:

$$H = (23.50 + 1.35 + 3.12 + 3.55)(10^{-10})(100{,}000)^2$$
$$= 31.52 \text{ in. water } (7.84 \text{ kPa})$$

Example 12.6. If the airways of Example 12.5 are rearranged as a parallel circuit, calculate the pressure difference across the circuit and the quantity of airflow in airway 1.

SOLUTION. Use Eqs. 12.6, 12.4, and 12.7:

$$R_{eq} = \left(\frac{1}{\sqrt{1/23.5} + \sqrt{1/1.35} + \sqrt{1/3.12} + \sqrt{1/3.55}}\right)^2 (10^{-10})$$
$$= 0.214 \times 10^{-10} \text{ in.-min}^2/\text{ft}^6$$
$$H = (0.214)(10^{-10})(100{,}000)^2 = 0.214 \text{ in. water } (53.25 \text{ Pa})$$
$$Q_1 = 100{,}000 \sqrt{\frac{0.214 \times 10^{-10}}{23.5 \times 10^{-10}}} = 9540 \text{ ft}^3/\text{min } (4.50 \text{ m}^3/\text{sec})$$

The H–Q relation (Eq. 12.4) for an airway is termed its *characteristic*; for the entire mine, it is called the *mine characteristic*. A graphical plot of pressure difference versus quantity results in a parabolic-shaped curve; we refer to it as the *mine characteristic curve*.

Example 12.7. Calculate and plot the characteristic curve of the mine of Example 12.4.

SOLUTION. Given $H = 3.83$ in. water for $Q = 140{,}000$ ft^3/min, calculate several points by Eq. 12.4:

Q, ft^3/min (m^3/sec)	H, in. water (Pa)
0 (0)	0 (0)
50,000 (23.6)	0.49 (122)
100,000 (47.2)	1.95 (485)
140,000 (66.1)	3.83 (953)
175,000 (82.6)	5.98 (1488)

Sample calculation for 50,000 ft^3/min:

$$H_2 = H_1 \left(\frac{Q_2}{Q_1}\right)^2 = 3.83 \left(\frac{50{,}000}{140{,}000}\right)^2 = 0.49 \text{ in. water (122 Pa)}$$

The mine characteristic curve is plotted in Figure 12.17a.

Fan performance may also be expressed graphically, useful for display by computer graphics. A *fan characteristic curve* plots its H–Q relationship. The mathematical expression for a fan curve is a quadratic polynomial; it cannot be derived theoretically but must be determined experimentally by a fan test.

Example 12.8. Points on the characteristic curve of a centrifugal fan are as follows:

Q, ft^3/min (m^3/sec)	H, in. water (Pa)
0 (0)	6.0 (1493)
50,000 (23.6)	6.6 (1642)
100,000 (47.2)	6.3 (1568)
150,000 (70.8)	5.5 (1369)
200,000 (94.4)	4.0 (995)
250,000 (118.0)	2.1 (523)
300,000 (141.6)	0 (0)

Plot the fan characteristic curve.

SOLUTION. See the fan curve in Figure 12.17b.

Characteristic curves are used mainly to determine the system operating point when a fan is connected to a mine. The *operating point* is the combination of pressure difference and quantity at which the fan and the mine are in equilibrium. If the fan and mine characteristic curves are plotted to scale on the same H–Q graph, the operating point lies at their intersection. It is a useful procedure to employ for the selection of a fan.

Example 12.9. If the mine and fan whose characteristic curves are depicted in Figure 12.18a and b constitute a ventilation system, determine the operating point graphically.

SOLUTION. See the curves in Figure 12.17c. Read at operating point 1: $Q = 160{,}000$ ft^3/min (75.5 m^3/sec) and $H = 5.0$ in. water (1244 Pa).

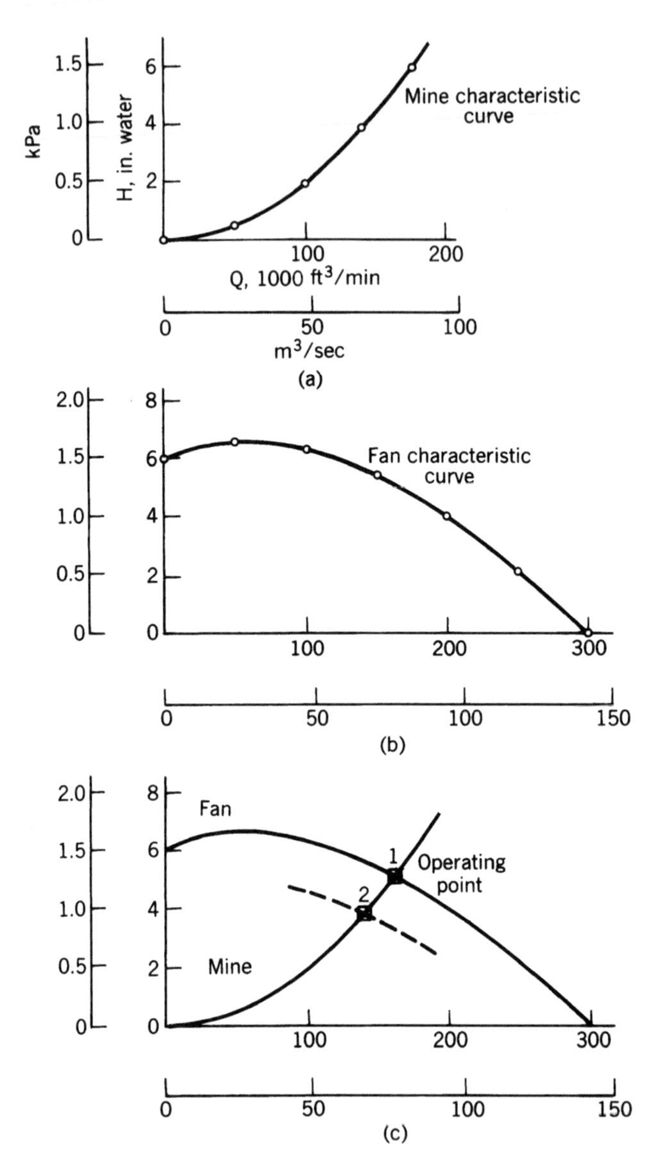

FIGURE 12.17. Mine and fan characteristic curves. See Examples 12.7, 12.8, 12.9, and 12.10.

In selecting a fan for a given mine, it may be the case that no fan curve intersects the mine curve at the required pressure difference and quantity. The fan's performance may be modified, within limits, by changing its rotational speed, blade pitch (if axial-flow), or its inlet vanes (if centrifugal)—or additional fans may be installed in the system. Performance curves or data are necessary to predict the effects on fan performance of changing the latter two parameters, but fan laws enable us to determine the effect of a change in speed n:

$$\frac{Q_2}{Q_1} = \frac{n_2}{n_1} \quad \text{and} \quad \frac{H_2}{H_1} = \left(\frac{n_2}{n_1}\right)^2 \tag{12.8}$$

Thus, if the performance (operating point or characteristic curve) of a fan is desired at another speed, the H–Q relation is readily determined by Eq. 12.8. Likewise, the rotational speed of a fan required to deliver a specified pressure difference and quantity of airflow can also be found by the same equation.

Example 12.10. If the speed of the fan whose characteristic curve is plotted in Figure 12.17c is 1750 rpm, determine the speed at which it must operate to deliver $Q = 140{,}000$ ft^3/min (66.1 m^3/sec) and $H = 3.83$ in. water (953 Pa) when connected to the mine whose characteristic curve is also plotted in Figure 12.17c.

SOLUTION. Plot operating point 2 for the H–Q relationship specified (Figure 12.17c). Calculate the required speed n_2 by Eq. 12.8:

$$n_2 = n_1\left(\frac{Q_2}{Q_1}\right) = 1750\left(\frac{140{,}000}{160{,}000}\right) = 1531 \text{ rpm}$$

The entire fan curve can be transposed to the new speed if desired by applying Eq. 12.8 (a portion is shown).

If many fan and mine curves are involved, as in selecting the most economical fan over the life of a mine, then the graphical simulation or algebraic solution of intersecting curves is best carried out by computer.

12.5.3 Temperature–Humidity Control

In *temperature–humidity control*, we are concerned with the physical quality of air and its heat content. Both sensible and latent heat are involved because normal air is a mixture of dry air and water vapor. Two air-conditioning processes are employed in mines: (1) heating and (2) cooling and dehumidification. The latter is the more complicated and costly process. Ventilation alone is adequate for cooling until the mine air wet-bulb temperature substantially exceeds 80°F (27°C); above 90°F (32°C), human exertion is seriously impaired and work output and safety suffer. Artificial cooling supplements ventilation in very hot environments. Chilled water is prepared on the surface in cooling towers or underground by refrigeration, and then in heat exchangers (coils or sprays) it cools and dehumidifies the air going to the working faces. Heat is the ultimate constraint in mining; it increases inexorably with depth. Major sources of heat in mines are autocompression (more than 5°F/1000 ft of depth, or 10°C/1 km) and wall rock or geothermal gradient (0.5°–3°F/100 ft of depth, or 0.9–5.5°C/100 m). Groundwater, machinery, human metabolism, blasting, and oxidation may also contribute heat. Although temperature–humidity control applications in mines are still relatively uncommon, they can be expected to increase as underground mines deepen.

12.5.4 Summary

Recent sources on mine ventilation, gas and dust control, and air conditioning include Burrows (1982), Bossard (1983), McPherson (1993), Hartman et al. (1997), and Tien (1999).

The broad objective of mine ventilation and air-conditioning—to provide a comfortable, safe atmospheric environment for workers—should never be neglected. However, its role must be placed in the proper perspective: Atmospheric control is only one phase of a broader, more general, environmental-control mission. Mine environmental control can best be administered by a single department that includes health and safety engineering as well, organized in the role of staff to the chief operating officer.

PROBLEMS

12.1 Calculate the minimum width of undercut required in a block caving mine to ensure a successful cave, given the following:

Stress field	No lateral strain
Rock strength, compressive	9700 lb/in.^{2} (66.9 MPa)
tensile	810 lb/in.^{2} (5.6 MPa)
Depth of cover	1500 ft (460 m)
Undercut height	12 ft (3.7 m)
Fillet ratio	$\frac{1}{6}$

12.2 Nitrogen gas occurs as a normal constituent (78.09%) of the atmosphere we breathe and may be found as a strata gas in mines. If nitrogen is liberated at the rate of 8.5 ft^3/min (0.00401 m^3/sec) in a mine and the TLV is 80%, calculate the quantity of dilution required to maintain air quality standards.

12.3 **a.** A circular airshaft 8 ft (2.4 m) in diameter transmits an airflow of 250,000 ft^3/min (118 m^3/sec) into a mine when a fan is connected to it. If a similar new shaft 10 ft (3 m) in diameter having the same length, friction factor, and pressure difference is sunk to replace the original shaft, what airflow will result? Disregard the rest of the mine.

b. The pressure difference for a mine is 4.0 in. water (995 Pa) with an air quantity of 200,000 ft^3/min (94.4 m^3/sec). What is the pressure difference if the flow is halved?

12.4 **a.** A circular ventilation shaft is to be sunk with the following characteristics:

Diameter	16 ft (4.9 m)
Length	1400 ft (427 m)
Friction factor	25×10^{-10} lb-min^2/ft^4 (4.6×10^{-3} kg/m^3)

Calculate the pressure differential with an airflow of 215,000 ft^3/min (101 m^3/sec).

b. If the rest of the mine consists of a main shaft and haulageway having a pressure difference of 0.41 in. water (102 Pa), determine the overall pressure difference if the ventilation shaft is connected in series to the rest of the mine.

12.5 **a.** Develop an expression to calculate the pressure difference across multiple, identical, parallel airways (number = N) in a coal mine, neglecting the effect of crosscuts, seam irregularities, and so on. Start with Eq. 12.3 for the pressure difference in a single airway.

b. Using your expression from part a, calculate the pressure difference in in. water in the following multiple airways:

$N = 5$ airways

Cross section = 5×18 ft (1.5×5.5 m)

$K = 30 \times 10^{-10}$ lb-min^2/ft^4 (5.5×10^{-3} kg/m^3)

$L = 3900$ ft (1190 m)

$Q = 175{,}000$ ft^3/min (82.6 m^3/sec), total flow

12.6 A circular, smooth-lined ventilation shaft 6 ft (1.8 m) in diameter is to be bored vertically, connecting to mine openings 2400 ft (730 m) below. The shaft friction factor is 15×10^{-10} lb-min^2/ft^4 (2.8×10^{-3} kg/m^3), and the desired quantity of airflow is 125,000 ft^3/min (59.0 m^3/sec).

a. Calculate the pressure difference in the ventilation shaft.

b. Plot the mine characteristic curve for the shaft, using five points, on the graph in Figure 12.18.

c. A fan whose characteristic curve is also plotted in Figure 12.18 is available to supply air to the shaft and overcome its pressure difference. Determine the operating point of the system (shaft + fan) graphically, and read the actual quantity of air and pressure difference in the system.

d. If the desired H–Q does not result, what can be done to provide it with the given fan? Calculate the required fan speed if the original speed is 1000 rpm.

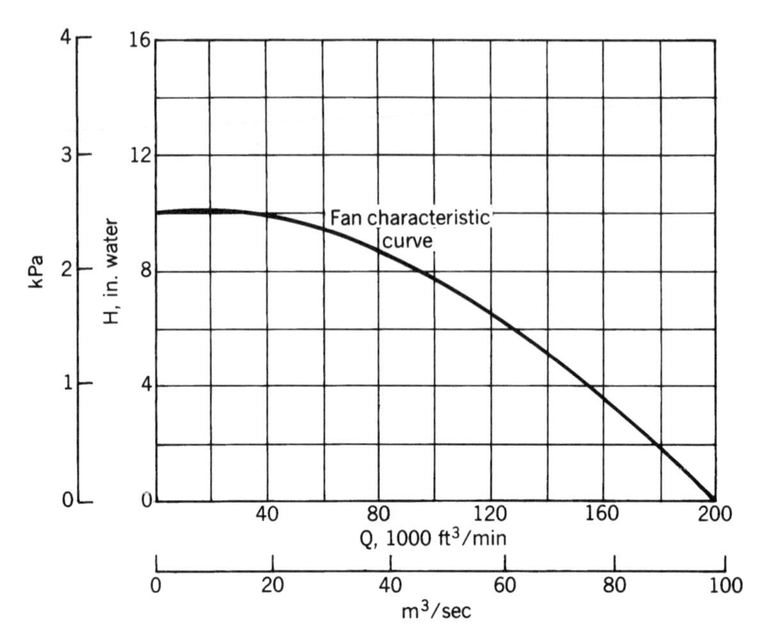

FIGURE 12.18. Fan characteristic curve. See Problems 12.6 and 12.7.

12.7 A circular, bored drift 1600 ft (490 m) in length is connected at its extremities to the surface by circular shafts 1200 ft (360 m) deep; all mine openings are 8 ft (2.4 m) in diameter. Their friction factor is 25×10^{-10} lb-min^2/ft^4 (4.6×10^{-3} kg/m^3), and the desired quantity of airflow is 160,000 ft^3/min (75.5 m^3/sec).

a. Calculate the pressure difference in the entire mine, assuming a series circuit.

b. Plot the mine characteristic curve, using five points, on the graph in Figure 12.18.

c. A fan whose characteristic curve is also plotted in Figure 12.18 is available to supply air in the mine and overcome its pressure difference. Determine the operating point of the system (mine + fan) graphically, and read the actual quantity of air and pressure difference in the system.

d. If the desired H–Q does not result, what speed change will be necessary for the fan if the original speed is 2800 rpm?

13

NOVEL METHODS AND TECHNOLOGY

13.1 CLASSIFICATION OF METHODS

This chapter deals with both mining methods and technology. Several of the mining methods introduced here will be defined as *novel methods*, those methods that employ nontraditional principles or technologies, or exploit uncommon resources, and that are not yet widely accepted in practice. The distinction between traditional and nontraditional methods is not as sharp as we might believe. Just as classical methods evolve, are modified or combined with other methods, or become obsolete and fall into disuse, so novel methods may in time receive the acceptance that warrants their reclassification into one of the traditional categories. Other novel methods may fall into disuse for economic or political reasons or because other methods compete with them successfully. A study of Table 13.1 shows some indications of progress. This table appeared in the first edition of this book (Hartman, 1987). Since the table was first published, the methods that have significantly gained in acceptance have been automation and robotics, rapid excavation, and methane drainage. Both nuclear mining and underground retorting have essentially lost their attractiveness and have become less favorable as methods of the future.

In Table 13.1, we also note that the first two methods (rapid excavation and automation) are not mining methods but technology. They can be applied to a variety of mining methods, and that may be their greatest advantage. Both of these technologies have grown in use since the first edition went into print. The other entries in the table are either mining methods or a group of methods that are considered together. One observation made in assessing the table is that no major new technologies or methods have emerged since 1987. New automation procedures have clearly made headway, and several of the novel

TABLE 13.1 Classification of Novel Mining Methods

Method	Probable Commodities	Locale		Likelihood of Use		
		Surface	Underground	Existing	Promising	Questionable
Rapid excavation	Noncoal (hard rock)		×	×		
Automation, robotics	All	×	×	×		
Hydraulic mining	Coal, soft rock		×	×		
Methane drainage	Coalbed methane	×	×	×		
Underground gasification	Coal		×		×	
Underground retorting	Hydrocarbons		×		×	
Ocean mining	Metal, nonmetal	×			×	
Nuclear mining	All	×	×			×
Extraterrestrial mining	Metal, nonmetal	×	×			×

methods have become almost standard practice (e.g., rapid excavation and methane drainage). However, there have been almost no new or innovative concepts that show considerable promise for the future.

It may be of interest to study how novel mining methods originated. In past times, most new methods evolved almost entirely from operating experience within the industry. However, that is less true today than in the past. Technology is being transferred from other industries and endeavors. Military and space technology often find application in other industries, including mining. Today, automation and robotics are likely to find their inspiration from areas of technology outside mining. Automation can be applied in all areas of industry, and ideas can come from any one of these areas.

Because of their limited application or because of the unknown potential for use, none of the novel methods listed in Table 13.1 warrants as extensive coverage as the traditional methods of surface and underground mining. Accordingly, we will confine our discussion to a brief explanation of each method, identification of its likely uses, and some comments on its current status and future promise.

13.2 RAPID EXCAVATION

Rapid excavation is the concept of replacing the intermittent unit operations of rock breakage and materials handling in tunneling, raising, and shaft sinking with a system of continuous extraction. Notice that we say *concept* rather than *practice* because truly continuous excavation is difficult to achieve. Nonetheless, progress in rapid excavation in the 40 years it has existed has been significant, making this area one of the more successful technological advances in mining practice. Still, there are many improvements yet to be made. One of the deterrents to the elimination of cyclic components is the implementation of auxiliary operations—ground control, ventilation, gas and dust control, drainage, and materiel supply—in the production operations without interrupting the advance of the working face. Figure 13.1 illustrates the principle and components of rapid excavation with a tunnel-boring machine (TBM), often referred to as a *mole*, in which breakage, handling, support, and ventilation are attempted simultaneously. This machine has been used successfully for drifting in mines, though it is more commonly used in producing highway, dam, and sewer tunnels.

Historically, continuous excavating machines have been in existence for a good many years, preceding the invention of dynamite in 1867. The first such machine appears to have originated in Italy in 1846; it utilized percussive energy to produce a slot around the face (Robbins, 1984). TBMs were then used in the earliest attempts to tunnel under the English Channel, starting in 1865; most of them used drag-bit or disk cutters. Modern rock-boring machines were developed in the 1950s, although successful continuous miners for coal were in operation a decade earlier.

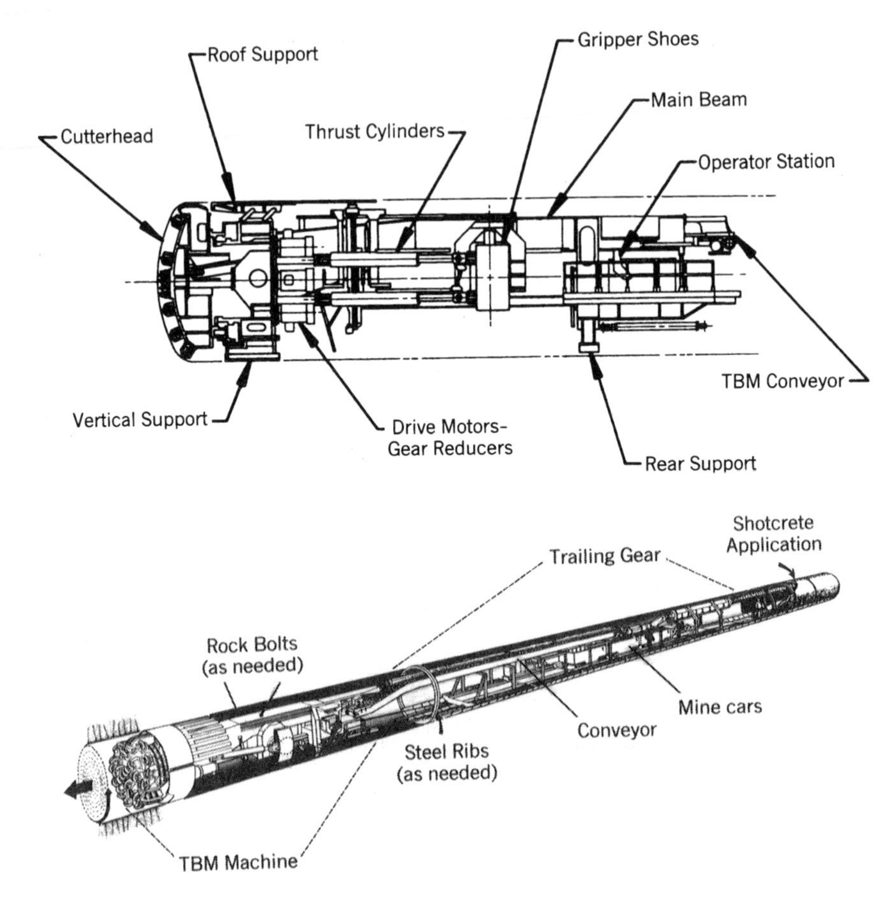

FIGURE 13.1. Diagrams of tunnel-boring machines (TBMs). (Top) Boring machine unit and constituent parts. (After Brockway, 1983. By permission from the Society for Mining, Metallurgy, and Exploration, Inc., Littleton, CO.) (Bottom) TBM with trailing gear. (After Sager et al., 1984. By permission from National Academy of Engineering, Washington, DC.)

The modern history of continuous hard-rock excavation began in the 1960s when the term *rapid excavation* was coined and the mining and civil engineering fields began an intensive interest in the concept and its applications. In 1971, the Executive Board for Rapid Excavation and Tunneling Conferences (RETC) was established to hold meetings and disseminate technical information in this exciting field of technology. Those conferences have continued to this day. A more recent organization, the Institute of Shaft Drilling Technology (ISDT), has been established for information exchange on shaft drilling (Breeds and Conway, 1992). Interest in this field remains high, and the technology continues to advance.

Progress in rapid excavation requires that excavation methods be considered systems, with each component contributing to the overall success of the system. Although the entire operating cycle is involved, the element of rock breakage lies at the heart of rapid excavation. This is nearly always accomplished by mechanical means (see Section 5.2), but other forms of rock attack noted in Table 5.1 may also be candidates (to date, hydraulic fragmentation is the only other mechanism to be used commercially, mainly to supplement mechanical attack) (Souder and Evans, 1983). The continuous miner has become routine in coal mining, though this machine is not normally considered a rapid excavation device. The most extensively used types of rapid excavation devices in mining are illustrated in Figures 9.8 and 9.9.

Progress in rapid excavation can be measured by several performance parameters: (1) hardness of the rock, (2) time percentage of machine availability, (3) diversity of application, (4) rate of advance, (5) specific excavation rate, and (6) cost of advance. Several of these parameters deserve to be discussed further here. First, hardness of rock is an area in which some progress is being made. Tunneling through rocks up to 30,000 psi (207 MPa) has become fairly common (Breeds and Conway, 1992); however, some tunnels have been driven through rock approaching 50,000 psi (350 MPa) in compressive strength (Stevenson, 1999). TBM specific excavation rates have also gone up over the period of rapid excavation usage. Stevenson (1999) has published specific excavation rates for tunnels produced since 1980. Some of his data comes from a publication by Nelson et al. (1994). The specific data points are plotted in Figure 13.2 and 13.3. Note that the metamorphic, granitic, and volcanic rocks have TBM specific excavation rates of 8 to 360×10^{-6} ft^3 of rock extracted per lb of force per cutter per revolution (0.00005 to 0.0023 m^3 of rock extracted per kN per cutter per revolution) and the rates in sedimentary rocks are 8 to 720×10^{-6} ft^3 of rock per lb of force per cutter per revolution (0.00005 to 0.0046 m^3 of rock per kN per cutter per revolution).

Although improvements have been made in these parameters, machine availability is still a problem area. Cox (1973) listed TBM availability at 35 to 50%, and Breeds and Conway (1992) gave a similar range (35 to 45%). The values reported in these publications may not be totally comparable, but they lead us to believe that availability remains a problem in TBM systems. There are many good reasons that the TBM is not engaged in the excavation process a greater percentage of the time. Maintenance, backup equipment, ground control, cutter replacement, and other delays all contribute to the problem. Delays in terms of percentage of total time for each of these subcategories have been reported by Breeds and Conway (1992). This is one of the systems problems that still need much in the way of research and development.

One of the major deterrents to applying rapid excavation in underground mining is the rather massive dimensions of the typical TBM with its trailing gear. This is evident in the application of a TBM to development in an underground mine (Alexander, 1999). In this project the mining company had to work with a TBM manufacturer to reduce the turning radius of its TBM to about 200 ft (60 m) in order to effectively use the device in mine development.

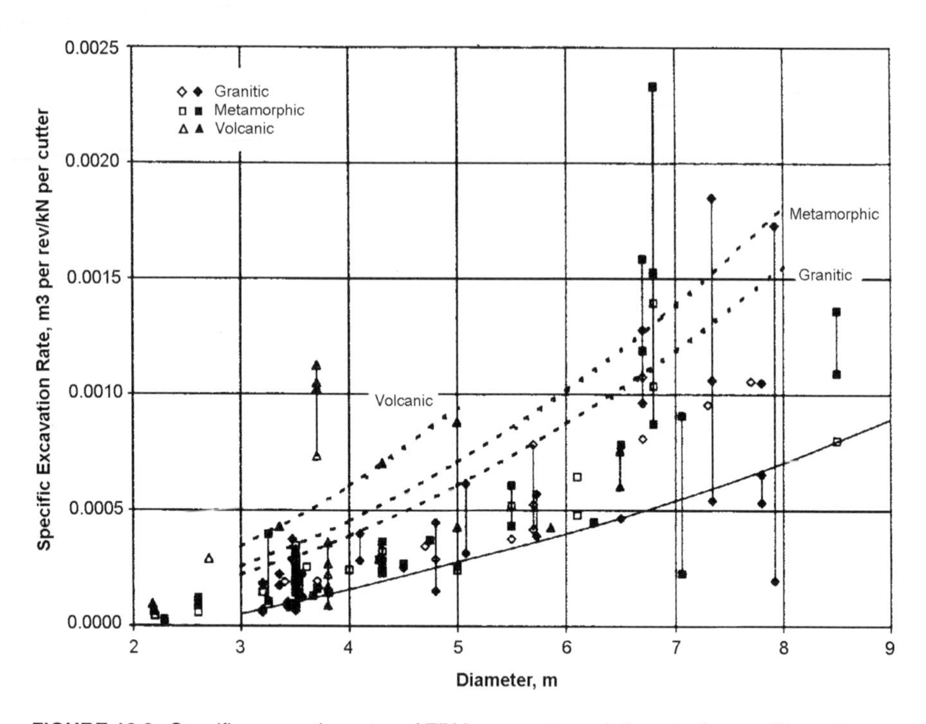

FIGURE 13.2. Specific excavation rates of TBMs versus tunnel diameter for granitic, metamorphic, and volcanic rocks. Data from Nelson et al. (1994) and Stevenson (1999). By permission from the Society for Mining, Metallurgy, and Exploration, Inc., Littleton, CO.

The original turning radius of the machine (350 ft or 106 m) was simply too great to be able to effectively maneuver the machine underground.

Another important limitation of the TBM in mining projects is the economics of conventional versus rapid excavation development. Sharp et al. (1983) indicate the difficulty by plotting the tunnel cost as a function of tunnel or drift length (see Figure 13.4). For tunnels of less than about 22,000 ft (6 km), the TBM normally cannot provide a cost that is lower than the costs of conventional development practices. This eliminates its use for many mine development projects. Although TBMs have been used for development at the Stillwater Complex in Montana (Tilley, 1991; Alexander, 1999) and at the Magma Mine in Arizona (VanDerPas and Allum, 1995), the rapid excavation revolution has still not established this process as commonplace in providing horizontal development openings.

In the advance of vertical openings, rapid excavation has greatly improved mine development, particularly in the area of raise boring. In many mines, raise borers for development of stoping operations are routinely used and have replaced conventional development in all but unusual circumstances (Gertsch, 1994). These machines are typically used for raises up to 1000 ft (305 m) in length. Atlas Copco Robbins has more than 150 in operation around the

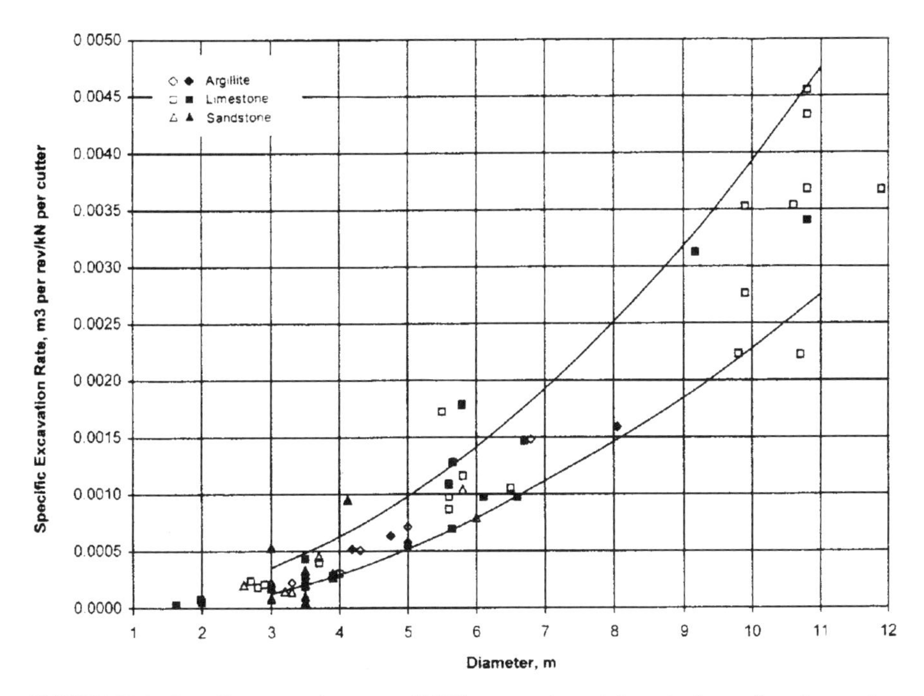

FIGURE 13.3. Specific excavation rates of TBMs versus tunnel diameter for sedimentary rocks. Data from Nelson et al. (1994) and Stevenson (1999). By permission from the Society for Mining, Metallurgy, and Exploration, Inc., Littleton, CO.

world, most of them used for raise development in metal mines (Walker, 1997). In addition, the boring of small-diameter shafts and many larger shafts is much more common today than in the past. Blind-boring applications have greatly improved in coal mining (Maloney, 1993; Zeni, 1995), making the application of rapid excavation principles much more prevalent in this area of mine development. The primary use for rapid excavation in U.S. coal fields is in the boring of shafts up to 16 ft (4.9 m) in diameter.

Perhaps the most opportune area for research into rapid excavation procedures for mining is in specialty machines that can be applied to production mining as well as development. The roadheader is one such machine that finds application in mining from time to time. Breeds and Conway (1992) and Copur et al. (1998) discuss their performance with emphasis on predicting their advance rates under various rock conditions. Perhaps more interesting to mining engineers is the class of machines that may be called hard-rock continuous miners, a machine type that has been around since the 1980s. The most noted of these machines is the Robbins Mobile Miner, which has a vertical rotating cutter head that swings horizontally to mine a face that is roughly rectangular. This shape is useful in production operations in mines (see Figure 13.5). The usefulness of a continuous miner of this type can be

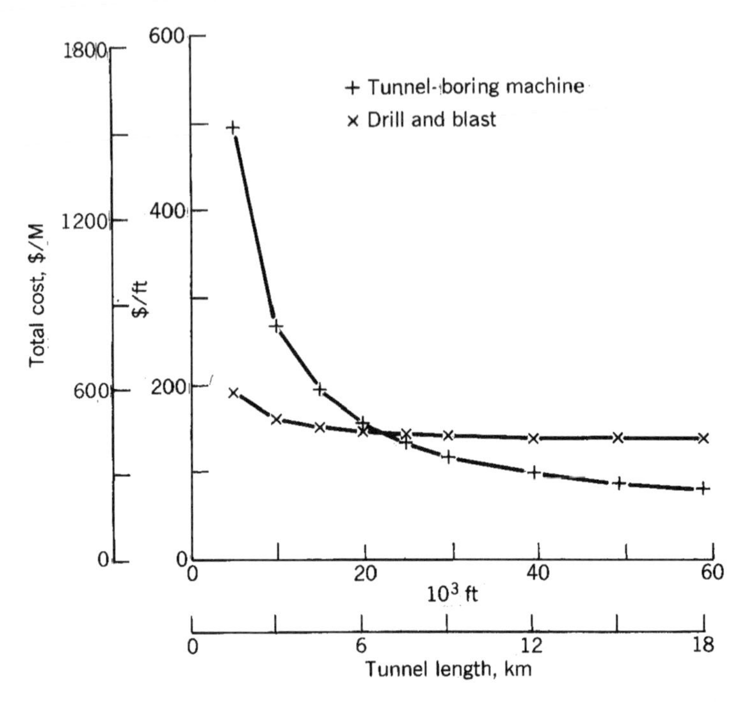

FIGURE 13.4. Cost comparison of conventional cyclic mining and a TBM as a function of tunnel length. (After Sharp et al., 1983. By permission from the Society for Mining, Metallurgy, and Exploration, Inc., Littleton, CO.)

illustrated by demonstrating its application to two mining methods, shrinkage and cut-and-fill stoping. As illustrated in Figure 13.5, the Mobile Miner is used to repeatedly drive rectangular horizontal openings through stopes to produce ore without blasting. Although the use of such miners is not yet common practice, there is a potential that they can be one of the important developments in mining in the future. The attempt by Robbins to automate the Mobile Miner seems to enhance its usefulness in hard-rock mining applications (Turner and Carey, 1993).

In summary, the rapid excavation technology available to mining engineers today provides significant new tools that can be applied to the development of mines and is one area that has become somewhat standard practice in mining. The methods have been under continuous advancement over the last 40 years. The primary application in mining has been in the production of raises. Shaft sinking is also an area in which rapid excavation practices have had an impact. The area in which only modest impact has been made is the production of horizontal openings. This is the result of the many limitations on the use of current equipment. However, this area has potential for the most significant advances in rapid excavation in mining in the future.

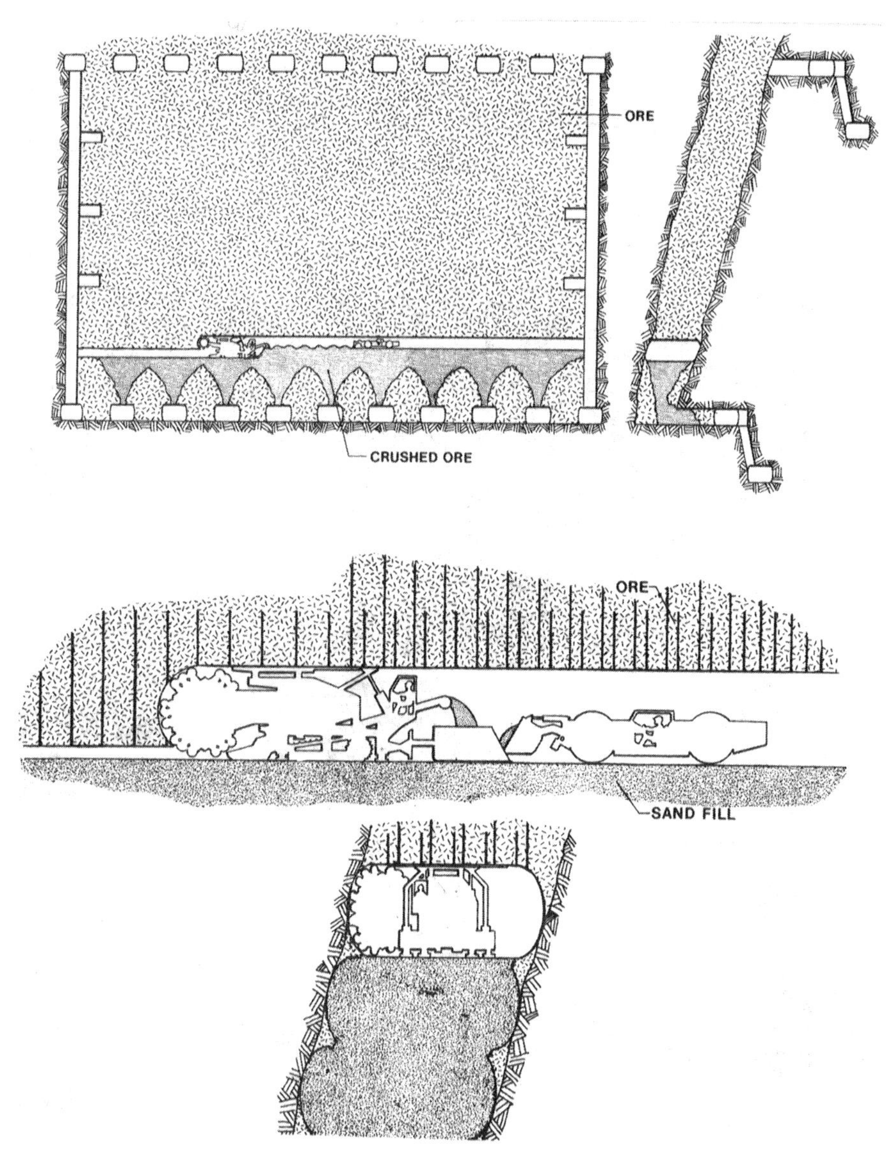

FIGURE 13.5. Conceptual applications of TBMs to hard-rock exploitation. (*Top*) Shrinkage stoping. (*Bottom*) Cut-and-fill stoping. (After Robbins, 1984. By permission from the Society for Mining, Metallurgy, and Exploration, Inc., Littleton, CO.)

13.3 AUTOMATION AND ROBOTICS

Elements of automation and robotics have been utilized in mining for several decades, but their implementation on a routine basis has been long in coming. However, progress in the last decade has been significant. It is probably correct to say that more advances have been made in this area in the last decade than in any other realm of technology. The reasons for such gains are competitive economic forces and the growth in the use of computers and computer-guided systems. The normal growth pattern may be measured along the following scale:

$$\text{Mechanization} \rightarrow \text{remote control} \rightarrow \text{automation} \rightarrow \text{robotics}$$

Mechanization in mining began around 1950 with continuous miners and other high-powered equipment that reduced the labor associated with mining. About two decades later, the industry began to experiment with remote control application in continuous miners to remove the operator from the most dangerous area of the face and allow operation from a safer area some distance from the face. It took another two decades until mining companies began any serious attempts to automate the mining process, although a number of experiments were conducted before that time. During the last decade, however, the pace of adoption of automation concepts has greatly increased, with many mining operations now utilizing this technology to benefit.

In the next sections, we examine the nature of remote control, automation, and robotics and their application in mining. Like rapid excavation, they do not constitute a mining method but a technology, which applies to any mining method, that can improve—even revolutionize—unit operations.

13.3.1 Remote Control

Remote control permits machine operations to be controlled from a distance. This can be performed using a hard-wired control unit or a radio-controlled system. The basic logic of the system primarily involves health and safety considerations. The operator is out of the worst environmental conditions and is ordinarily placed in a safer location while controlling the machine. There are two subcategories of remote control: line-of-sight remote control and teleoperation of the equipment. In both cases, the operator is manually controlling the mining process. Location of the operator in the mining operation within sight of the machinery characterizes the line-of-sight method. The operator uses sight and hearing to control the machine and keep it operating efficiently and within the ore zone. The limitations are that the operator loses some control capabilities as he or she moves farther from the machine, and is still subject to certain health and safety hazards.

Teleoperation is the process of controlling the mining operation from a remote location beyond line-of-sight to guide the equipment through the mining steps. Ordinarily, the operator is located on the surface and never ventures underground; the only personnel in the mine perform routine and breakdown maintenance. This type of remote control is ordinarily associated with a broadband communications system that utilizes a distributed antenna translator (DAT) and a radiating (leaky) cable antenna in the mining section (Fortney, 1996, 2001; Poole et al. 1996). Although the method is still a manual control procedure, the benefits are that the operator has no travel time into and out of the mining face and can operate more than one piece of equipment at times, particularly if some elements of semiautonomous control are employed.

The use of remote controls on continuous miners in coal mining has become standard practice, primarily because the operator can operate remotely while still having good visual contact with the miner's cutting head. The use of remote control for loaders in metal and nonmetal mining is also fairly common, although the advantages are not always as evident because the operator may not have ideal visual contact. In addition, the teleoperation of load-haul-dump devices (LHDs) and loaders seems to be a better strategy in most cases. Overall, remote controls have been widely used with applications to the following (Murphy, 1985; Scott, 1985; King, 1992):

Operation	Commodity	Locale
Rock breakage		
Drill	Metal, nonmetal	Surface, underground
Raise borer	Metal	Underground
Hard-rock miner	Metal	Underground
Materials handling		
Loader	Coal, metal	Underground
Auger	Coal	Surface
Continuous miner	Coal, nonmetal	Underground
Longwall shearer	Coal	Underground
LHD	Metal	Underground
Train	Metal, coal	Surface, underground
Belt conveyor	Coal	Underground
Highwall miner	Coal	Underground
Ground control		
Mobile roof support	Coal	Underground
Roof bolter	Coal	Underground

Remote control, both line-of-sight and teleoperation, has greatly improved the health and safety aspects of operating certain types of equipment. However, in most cases it does not do much to improve efficiency because the method still requires manual control. Efficiency enhancements normally come when more elements of automation are introduced.

13.3.2 Automation

After introducing remote control, the next logical step in productivity and safety improvement is to automate certain jobs or functions so that manual labor is reduced. Automation has been widely pursued in the manufacturing arena for quite some time, with many successful applications in electronics, auto, and other general manufacturing plants. However, the development of automation in mining operations had been held back by the difficulty of the mining environment, the hazards associated with mining, and the lack of direct application of concepts developed in other areas of industry. During the last decade, mining companies have overcome many of the obstacles and have begun vigorous development of automation applications. This has been accomplished primarily in Canada and Sweden, though the industry in other countries has also been involved in the process. It seems as though the automation age has finally come to the mining industry.

Automation is ordinarily introduced into machinery operations in several forms. The first is semiautonomous control, in which a computer controls some of the machine's functions but the overall operation is monitored and controlled by a human observer. This type of automation is commonly used in the metal industry for the control of underground LHDs from the surface (Brophey and Euler, 1994; Baiden and Henderson, 1994; Baiden, 1994; and Piche and Gaultier, 1996). In this system, the operator controls the LHD via teleoperation during the loading process and turns the transport over to a system that advances the vehicle along the haul road using a guidance system consisting of two lasers that allow the machine to follow a reflective tape over the haulage. The basic concept of the system is illustrated in Figure 13.6. In this arrangement, the operator may be able to control two or three loaders from the surface while an additional person, located underground, maintains the equipment and solves problems as they occur.

A second major development in underground mining is the control of drilling rigs from the surface (Baiden, 1996; Wyllie, 1996; Poole et al., 1998). Used mostly for the drilling of fan patterns of holes for sublevel caving, this application is again semiautonomous, the operator controlling the drills at times while some of the operations are conducted automatically. The primary

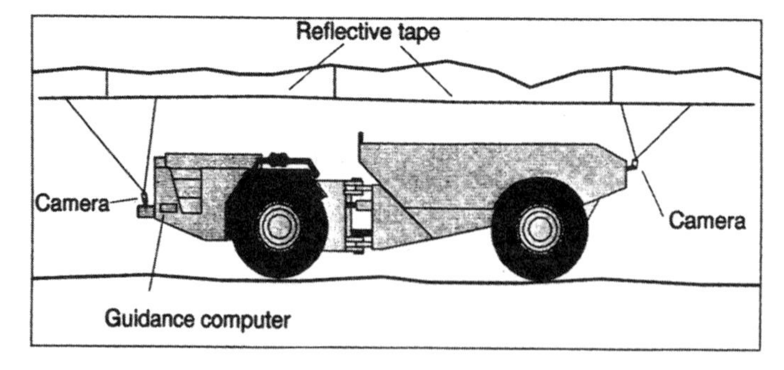

FIGURE 13.6. Vehicle guidance system using reflective tape. *Source:* Piotte et al., 1997, *CIM Bulletin*, V. 90, n. 1006, pp. 78–81. Reprinted by permission of the Canadian Institute of Mining, Metallurgy, and Petroleum.

advantages of this system are that the operator can operate several drills and the performance of the drills is generally significantly better than that of manually operated drills.

Other advances in automation in Canada have been achieved in the potash field of Saskatchewan (Fortney, 1996, 2001). In this operation, five continuous miners, the belt conveyors, and the hoisting system are all part of the automated mining process controlled from a central operator's station. Although efficiency was the primary goal, it has been reported that ground control was also greatly improved because of more uniform openings that reduced the likelihood of roof falls.

All of these applications of automation were carried out through use of a broadband communications system that allowed two-way communication between the equipment and the operator's station. The system requires a distributed antenna translator in the mining section, with a leaky cable antenna distributed throughout the equipment travelways in the section. This allows the operator to send commands to each piece of equipment and receive data back from the equipment.

The emergence of automated equipment has also been evident elsewhere in the world. In Sweden, the train haulage system, the hoists, the drills, and the LHDs are all part of the industry's automation efforts. Again, the mining method is sublevel caving, showing that the application to drilling regular patterns of holes is one of the most appropriate uses of automation. The successful use of these automation procedures throughout the international mining industry indicates that they are clearly applicable to underground mining and that productivity can be increased dramatically in the right application. A number of publications have discussed the increases in production and productivity that have occurred. Some projected productivity figures are available in Baiden (1994). Perhaps more indicative of the progress made are the productivity increases shown by Shuey (1999), such as the increase in productivity at Inco's mines from about 2200 tons/person/yr (2000 tonnes/person/yr) to about 3800 tons/person/yr (3500 tonnes/person/yr) in less than a decade. The increase is due both to increases in production and decreases in the labor requirement. It indicates the power of automation in the mining workplace.

In surface mining, automation is also being utilized to improve productivity. However, the systems associated with automation are different from those for underground mining. The central technology used in surface mining is that of the global positioning system (GPS). This system allows the location and control of equipment in an open pit to an accuracy of 1 cm (0.4 in.) and enables automation of any open pit equipment to proceed where the GPS is applicable. To date, the technology has been applied primarily to surveying functions in a mine; positioning and boom orientation of blasthole drills; positioning, elevation maintenance, and ore grade control at shovels; haulage truck location and dispatching; and autonomous vehicle navigation (Peck and Gray, 1995). The following paragraphs summarize some of these applications.

The positioning of drills and shovels to provide better blasthole patterns and ore grade control has been the objective of the technology developed by Aquila Mining Systems (Peck and Gray, 1995; Greene, 1999). The GPS allows the drills to be properly located and the drill towers to be adjusted to permit accurate drillhole positioning. Many of the same abilities are applied to the location of the shovel and the elevation control to produce ore grades that are more nearly optimal. The GPS also greatly reduces the surveying work that would normally be conducted to achieve this control of the operations. The real-time kinematic GPS technology (also called high-precision GPS) is now used in more than 150 mines worldwide to replace the many surveying and marking functions previously performed by hand (Greene, 1999).

Low-precision GPS technology is now also being applied to haul-truck location and dispatching systems (White and Zoschke, 1994; Peck and Gray, 1995). Previously, radio beacons had to be located at many points along the haul roads to keep track of the advance of each truck. Today this equipment can be replaced by GPS technology to simplify and enhance the dispatching system.

Moreover, GPS can be applied to the operation of autonomous mining trucks (AMTs) in surface mines (Sprouls, 1997). Caterpillar has demonstrated this technology in operating two driverless trucks along a repeatable course in both simulated and actual mining operations. This technology, although not routinely used, is testimony to how far the automation of mining equipment has come during the last decade.

Most of the applications of automation to the aforementioned situations have been used in a semiautonomous fashion with a human operator overseeing the activity to ensure that it is functioning properly. However, some automated systems in mines are truly autonomous and can be considered to be the application of *robotics*, the total control of machinery without human interaction. Of particular interest are hoisting systems, rail transport systems, belt conveyor networks, and processing facilities. Where no safety reductions are involved, a mining company can use truly autonomous control. However, under these conditions, it is normally required that additional safeguards be built into the system to protect the equipment. Such precautions ordinarily consist of some or all of the following systems (Poole et al., 1998): obstacle detection, fire suppression, lighting modifications, overload protection, and machine condition indicators (e.g., vibration, temperature, and fluid level sensors). Most of these are necessary because an operator is no longer present in the vicinity of the equipment to detect problems.

Both evolutionary and revolutionary changes have been evident in the application of automation and robotics to mining. This area of technology is now set to continue enhancing the process of mining. It appears that such technology has captured the hearts and minds of mining people and that its application is limited only by humankind's ability to create new uses for its many features. This technology should be as productive in gains over the next decade as it has been over the last decade. Thus, automation and robotics are areas of technology with great promise for the future.

13.4 HYDRAULIC MINING AND TRANSPORT

13.4.1 Hydraulic Mining

Previous discussion has been directed to hydraulic penetration (Chapter 5) and hydraulicking (Chapter 8). In this section we consider the additional methods of hydraulic extraction and transport.

The range of applications of hydraulic energy to mining and transport is broad, as demonstrated by the following:

Function/Application	Deposit or Material
Hydraulic penetration (drilling, etc.)	Rock
Hydraulicking (extraction)	Placers (sand, gravel)
Hydraulic mining (extraction)	Coal, soft rock
Hydraulic transport (haulage, hoisting)	Sized bulk material

In *hydraulic mining*, a very high-pressure jet of water, steady or pulsed, fragments consolidated mineral or rock in place. Thus, it has application as a primary extraction or mining mechanism, although it is limited at present mainly to softer materials. When combined with mechanical action (for mining, cutting, drilling, or boring), hydraulic attack is an effective technique in mining a variety of materials.

Hydraulic mining utilizes the kinetic energy of a fluid jet to break and excavate material from the solid. Its effectiveness or cutting rate is primarily a function of nozzle (jet) size, flow rate, pressure, force, and power (Fowkes and Wallace, 1968). Other important operating factors are the standoff distance (range), the attack angle, and the jet traverse rate (Jeramic, 1979). In addition, the mechanism of cutting must also be understood. Summers (1992) points out that under some conditions, the cutting rate is improved when the pressure is decreased because of the way in which the water jet exploits the weaknesses of the rock mass. Lower pressure allows the jet to better utilize preexisting cracks to break the rock, thus cutting more rock than with higher pressures.

The range of design parameters (nozzle diameter, pressure, and flow rate) employed in hydraulic mining of a variety of minerals is given in Table 13.2. These values are derived from either commercial operations or extensive experimental work. For the applications shown in the table, the general rules that apply to a successful hydraulic mining operation are as follows:

1. Successful hydraulic mining requires that the threshold nozzle pressure associated with a given substance be exceeded. That pressure is a function of various rock properties, of which the compressive strength is most important. As an approximation, the threshold pressure of a steady jet must be equal to or greater than the rock compressive strength.

TABLE 13.2 Design Parameters for Operational Hydraulic Mining

		Nozzle		
Material	Country	Diameter, in. (mm)	Pressure, lb/in.^{2} (MPa)	Flow Rate, gal/min (m^3/sec)
Soil	U.S.[a]	1.5–6 (38–152)	100 (0.7)	2500 (0.16)
Bituminous coal	U.S.[b]	0.25–0.56 (6–14)	4000 (27.6)	300 (0.02)
	Canada	0.60–1.2 (15–30)	1700 (11.7)	1300 (0.08)
	Germany	0.67 (17)	1300 (9.0)	480 (0.03)
	USSR	0.75–0.87 (19–22)	1500 (10.3)	1980 (0.13)
Anthracite coal	U.S.[b]	0.40–0.46 (10–12)	5000 (34.5)	300 (0.02)
Sandstone, soft	U.S.	0.62 (16)	1000 (6.9)	400 (0.03)
Gilsonite	U.S.	—	2000 (13.8)	—
Hard rock	U.S.[b]	0.02–0.10 (0.5–2.5)	>25,000 (>170)	150 (0.01)

[a]Hydraulicking.
[b]Experimental only.
Sources: Coal Age, 1962; Malenka, 1968; Lucas and Haycocks, 1973; Jackson, 1980; Wood, 1980; U.S. Bureau of Mines, 1985.

2. Cutting rate increases with flow rate, which because of pump characteristics is limited by nozzle pressure. In practice, the two variables are optimized, pressure at a threshold value and flow rate sufficiently high to yield the desired output at a reasonable power level.
3. Nozzle diameter is adjusted primarily to regulate the flow rate.
4. Levels of other operating parameters (force, energy, and power) are determined from the preceding guidelines and from the mining conditions.

Commercial hydraulic mining operations in North America have been conducted in Utah for gilsonite and in British Columbia for coal. The general nature of these operations is illustrated in Figure 13.7. Although the operations were once quite successful, they are no longer active. The only extensive hydraulic mining operations today are found in Russia and China. The parameters of some previous operations are shown in Table 13.3. Significant production was achieved in each of these operations using the hydraulic mining method.

In the future, hydraulic power for cutting rock may be more effectively utilized if it is used as an assist to a conventional mechanical cutting device. For example, a drag bit or a continuous miner's pick outfitted with a properly designed hydraulic jet, aimed very close to the point of attack ahead of the cutter, will greatly improve performance in some rocks (Ropchan et al., 1980). Performance improved by up to 40% but was not achieved in all rock masses. This technology has been applied in a number of other mining machines with varying results (Summers, 1992). Of particular note are applications to road

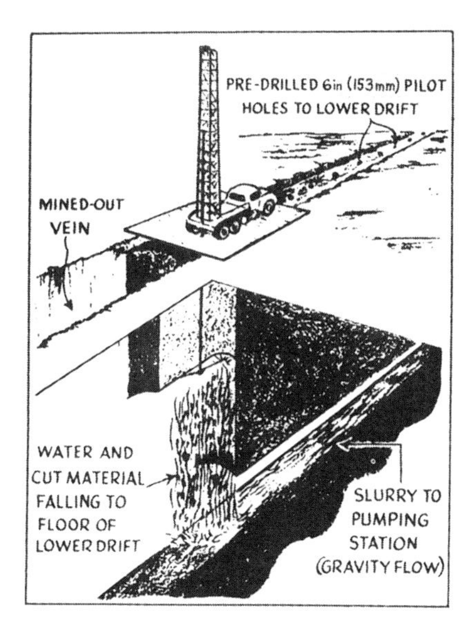

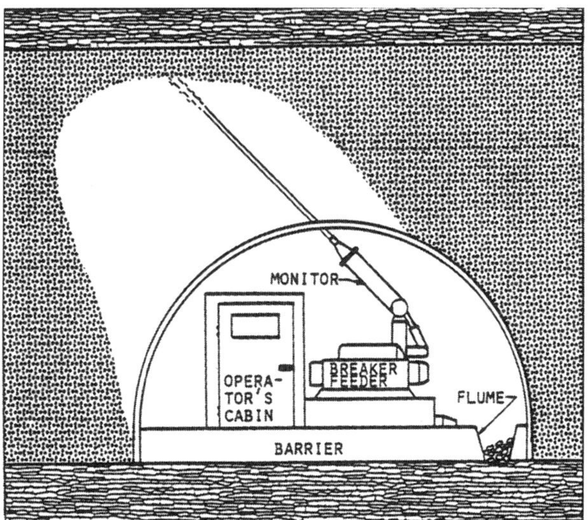

FIGURE 13.7. Hydraulic mining applications. (*Top*) Old method of gilsonite mining with surface rig for underground cutting and collection. (After *Coal Age*, 1962. Copyright © 1962, McGraw-Hill, Inc., New York.) (*Bottom*) Underground coal mining with monitor. (After Evers, 1983. By permission from the Society for Mining, Metallurgy, and Exploration, Inc., Littleton, CO.)

headers, tunnel-boring machines, and longwall shearers. However, the application of hydraulic assist to mechanical cutters does not always result in overall savings.

Another technique that has proven to be a viable technology is borehole slurrying. Savanick (1992) provides an excellent summary of the applications

TABLE 13.3 Production Output and Productivity in Hydraulic Coal Mines

	Canada	USSR		
	Michel	Novokutznetz	Jubilee	Red Army
Mining method	Sublevel	Sublevel	Room and pillar	Room and pillar
Seam thickness, ft	33–49	2.6–66	20–13	3.3
(m)	(10–15)	(0.8–20)	(3–4)	(1)
Seam pitch degrees	25–55	66	10	8–10
Monitor output, tons/min	6.1	3.3	2.2	1.1–2.2
(tonnes/min)	(5.5)	(3.0)	(2.0)	(1.0–2.0)
Mine production, tons/day	8800	13,200	4400	1300
(tonnes/day)	(8000)	(12,000)	(4000)	(1200)
Productivity, tons/employee-shift	28	17	44	22
(tonnes/employee-shift)	(25)	(15)	(40)	(20)

Source: Jeremic, 1979. (By permission from *Mining Magazine*, London.)

of this method. The method has been applied to coal, uranium, coal sands, and phosphate with good results but without favorable economics for commercial applications. The environmental characteristics of the method are generally favorable, which may give it some advantage over other mining methods. The technology can also be useful under certain conditions where other mining methods are not suitable, but the economics will still be a problem in most cases.

13.4.2 Hydraulic Transport

Hydraulic transport is the process of moving solid particles suspended in water in a channel or pipe. The method has been used extensively in processing plants, but less often in mining operations. However, there are sufficient applications to mining to make hydraulic transport a serious contender for use under a number of conditions. Pipelines for coal and ores of copper, gold, iron, limestone, and phosphate have been used worldwide (Link, 1982). Although the technology has been in existence since 1891, the first commercial pipelines were not built in the United States until 1957. At that time, to combat high rail charges, Consol built a 108 mi (174 km) pipeline in Ohio to move coal from its mines to a power plant in Cleveland. The company successfully operated the system until rail charges came down. In addition, a 72 mi (116 km) pipeline was built to connect a gilsonite mine near Bonanza, Utah, to a processing facility in Grand Junction, Colorado. Both of these lines were quite successful, pointing to the viability of hydraulic transport in mining. A later pipeline, built from the Black Mesa Mine in Arizona to the Mohave power plant in Nevada, is the longest (273 mi or 439 km) in use today. Although the overland hydraulic transport of minerals is a viable technological system, the political climate for such pipelines is not favorable. A reason is the lack of an eminent domain law for pipelines, which greatly reduces their viability for long overland routes.

Perhaps the most attractive method of using hydraulic transport is to couple it with a hydraulic mining system. This has been done in a number of international mining operations. In the United States, the procedure has very seldom been used. However, one coal mine in West Virginia did employ a unique hydraulic transport system (Petry, 1982). In this mine, coal from both longwall and continuous miner faces was transported to the preparation plant via hydraulic pipeline, with the introduction of the coal into the pipeline at the working face. The mine eventually eliminated the system because of technological and economic problems. However, the system did prove that hydraulic transport could be used under these conditions.

13.4.3 Summary of Hydraulic Applications

Both hydraulic mining and hydraulic transport are proven technologies. However, the technological viability must be matched with favorable economics for the technology to be used. Clearly, this has been achieved in a limited number of cases for mining and minerals transport throughout the world. It appears that this technology has a future in mining but is restrained by the limited number of situations where an economic justification is available. The technology is thus likely to be used on a limited basis.

13.5 METHANE DRAINAGE

Methane drainage, also called coal degasification, is the practice of removing gas in a coal seam and adjoining strata through wellbores, drillholes, and pipelines. In some respects it resembles borehole mining, although operations may be conducted either from the surface or underground. Drainage is also similar to well production of natural gas, the principal constituent of which is methane. It may proceed independently of or in conjunction with mining.

Methane drainage is not a new method. Attempts to drain firedamp (methane) from coal seams date back to 1730 in Great Britain, with the first successful controlled system installed in 1943 in Germany (Buntain, 1983). Today it is heavily used in Europe and China and is rapidly becoming common in the United States. The potential supply of coalbed methane in the United States has been estimated (Potential Gas Committee, 1993) at 275 to 649 trillion ft^3 (7.8 to 18.4 trillion m^3). The world supply of methane exists largely in the hands of three countries: 40.2% in Russia, 30.8% in the United States, and 16.4% in China (Murray, 1996).

In their ventilating airstreams, U.S. underground mines emitted the following methane quantities in 1998 (Schultz, 2000):

Methane Category	Yearly Output
Ventilation system emissions	93 billion ft^3 (2.6 billion m^3)
Drainage system capture	54 billion ft^3 (1.5 billion m^3)
Total mine methane yield	147 billion ft^3 (4.2 billion m^3)
Total methane utilized	43 billion ft^3 (1.2 billion m^3)
Total emitted to atmosphere	103 billion ft^3 (2.9 billion m^3)

The methane captured in drainage systems in U.S. coal mines currently comes from about 17 mines. Most of these mines attempt to market the methane they capture, but it is not always possible to do so. Although methane drainage is a concept that has been recognized as an answer to mine safety problems for many years, it is only in the last 15 years that U.S. mines have begun to harvest significant quantities of this gas. Many mines can extract 50 to 60% of the gas using methane drainage systems, greatly reducing safety problems and the global warming effect. Economic incentives have initiated a more aggressive attitude toward methane drainage. Mining companies continue to enhance the percentage of methane that is collected, with the total volume collected increasing by a factor of 3 during the last decade.

Most of the methane generated from coal seams to date has come from the Black Warrior (Alabama), San Juan (New Mexico), and Central Appalachian (Virginia, West Virginia, Kentucky) fields. The gas from the San Juan basin has been extracted from seams not being mined; the other basins have yielded mostly gas associated with underground mines. Significant additional reserves exist in the Green River (Wyoming, Colorado), Arkoma (Arkansas, Oklahoma), Piceance (Colorado), Powder River (Wyoming), and Northern Appalachian (Pennsylvania, Ohio, West Virginia, Kentucky) basins. There are a number of coalbed methane gas plays currently in development as a result of higher natural gas prices. This may be one of our most improved fuel resources over the next few decades.

Although coalbed methane is an unconventional gas source, it is still an attractive one because many U.S. coal seams are unmined and remain as potential sources. Unlike natural gas reservoirs, coal seams are thin and relatively impermeable except for their natural fracture system (*butt* and *face cleats*). Methane occurs occluded and adsorbed as layers of gas in the cleats. The principal difference between conventional and unconventional occurrences of methane lies in the way in which the gas is transported. When a methane drainage well or mine opening is driven into the seam, a low-pressure space is created; the higher pressure of the gas causes liberation of the gas and allows it to flow to the well or opening (Ertekin, 1984).

Methane occurrence in a coal seam is expressed as the volume of gas contained per unit weight of coal (generally expressed in ft^3/ton or m^3/tonne). It indicates the potential of a seam to produce methane. Table 13.4 outlines typical methane contents and the emission rates of several high-methane U.S. coal seams. Note that the emission rates in the last column of the table are expressed in millions of ft^3/day per 100 ft of hole through the seam. Many factors control the output, including the physical properties of the coal seam (diffusivity, reservoir pressure, permeability, and gas content), mining method (if in progress), and the drainage method (Thakur and Dahl, 1982).

There are four techniques in use in the United States to drain methane from solid coal (Mutmansky, 1999):

1. Vertical wells from the surface, with hydraulic stimulation
2. Vertical gob wells from the surface
3. Inclined wells from the surface
4. Horizontal wells from existing mine openings

The traditional vertical well from the surface is drilled through a coal seam or seams and cased to predrain the methane prior to mining. The wells are normally drilled two to seven years before mining is to be initiated, and the seam is hydraulically fractured to remove as much of the methane as possible ahead of mining. Figure 13.8 shows a vertical well penetrating three different coal seams. If the mine is to extract coalbed C, the drainage of methane from coalbeds A and B will also reduce the methane in the mine because of seam A and B emissions during caving. This also enhances the methane recovery and economics. The primary advantage of these vertical wells is that the gas recovered is generally of high quality, and recovery of 50 to 80% of the methane is possible.

Vertical gob wells are usually associated with longwall mining operations. The term *gob well* refers to a coalbed methane well that extracts methane from the gob area after the mining has caved the overlying strata. Figure 13.9 shows the configuration of a gob well setup associated with a longwall. Gob wells can recover 30 to 70% of methane emissions for the gob, depending on geologic conditions and the number of gob wells within the panel. Two properties of gob wells, short production life and low gas quality, make the use of this type of well less valuable than that of a vertical well driven ahead of mining. However, gob wells may be essential to the safe and efficient operation of a longwall panel in high-methane seams. Hence, they are often used even if the methane gathered cannot be utilized.

Inclined wells from the surface require careful control of the drillhole path. They are ordinarily initiated vertically, inclined when they approach the coal

TABLE 13.4 Methane Characteristics and Emission Rates in U.S. Coal Seams

Seam	Gas Content, ft^3/ton (m^3/tonne)	Composition, %CH_4	Calorific Value Btu/ft^3 (MJ/m^3)	Depth, ft (m)	Average Emission Rates, million ft^3/day per 100 ft (million m^3/day per m)
Beckley (WV)	262	99.2	1002	900	15
	(8.2)		(37.4)	(274)	(13)
Mary Lee (AL)	430	96.1	970	1500	10
	(13.4)		(36.1)	(457)	(9)
Pittsburgh (PA, WV)	220	90.8	922	700	15
	(6.9)		(34.3)	(213)	(13)
Pocahontas (VA)	450	96.9	1003	1700	10
	(14.0)		(37.4)	(518)	(9)

Source: Thakur and Dahl, 1982. (Copyright © 1982, John Wiley & Sons, New York.)

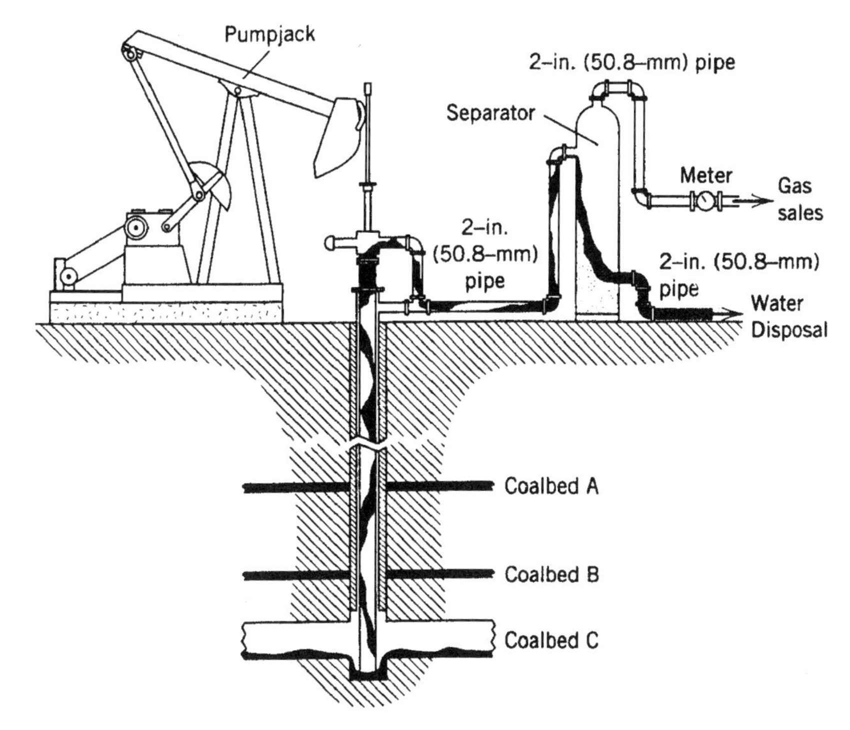

FIGURE 13.8. Typical vertical well setup for methane drainage. *Source:* Hartman et al., (1997). *Mine Ventilation and Air Conditioning.* Copyright © 1997, John Wiley & Sons, Inc., New York. Reprinted by permission of John Wiley & Sons, Inc.

seam to be drained, and kept within the coal seam for a significant length to improve the methane flow to the well. Although this technology has been around for some time, it has not been widely used. However, this procedure now seems mature enough to provide excellent drainage of coal seams before mining takes place and may thus be of greater significance in the future.

The final method of recovery of methane from underground mines uses horizontal boreholes, drilled into the coal seam from underground development openings. These holes are used to drain methane from the unmined areas shortly before mining, reducing the flow of methane into the mining section and improving the safety of mining. Because methane drainage occurs only from the mined coal seam and the period of drainage is relatively short, the recovery efficiency of this technique is low. Normally, about 10 to 20% of the methane is recovered. However, the quality of the gas is high and it can be utilized as a pipeline product in most cases. Figure 13.10 shows an example of a horizontal borehole system used in a longwall mine.

The future for coalbed methane degasification appears to be favorable. There is an increased awareness of the benefits to the mining company and a growing knowledge of efficient methods to drain the gas. In addition, our

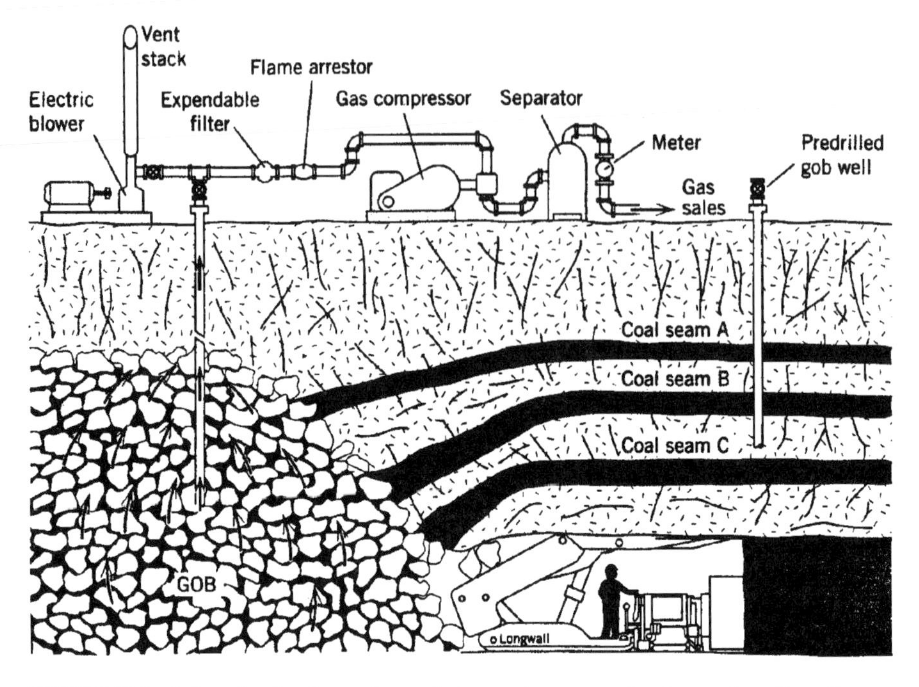

FIGURE 13.9. Gob well setup for methane drainage above an active longwall section. *Source:* Hartman et al. (1997). *Mine Ventilation and Air Conditioning.* Copyright © 1997, John Wiley & Sons, Inc., New York. Reprinted by permission of John Wiley & Sons, Inc.

seams will be deeper and higher in methane content in the future than they are at present. The approach to methane problems will likely include drainage as one of the logical solutions.

13.6 UNDERGROUND GASIFICATION

During the 1970s, the United States initiated an aggressive program aimed at reducing our energy dependence, and programs to utilize in situ energy recovery were investigated intensively. Specific research and analysis were performed to evaluate underground gasification, liquefaction, and retorting of energy sources. Chaiken and Martin (1992) present an excellent review of many of the experiments.

The idea of gasifying coal in its natural environment is not new. W. Siemens envisioned the concept in 1868, although underground experiments (by others) were not conducted until 1914, with extensive tests beginning in the 1920s (Marsden and Lucas, 1973). Many European countries as well as Canada, Japan, and the United States have experimented with underground gasification, but only Russia has exploited the technology commercially to any extent.

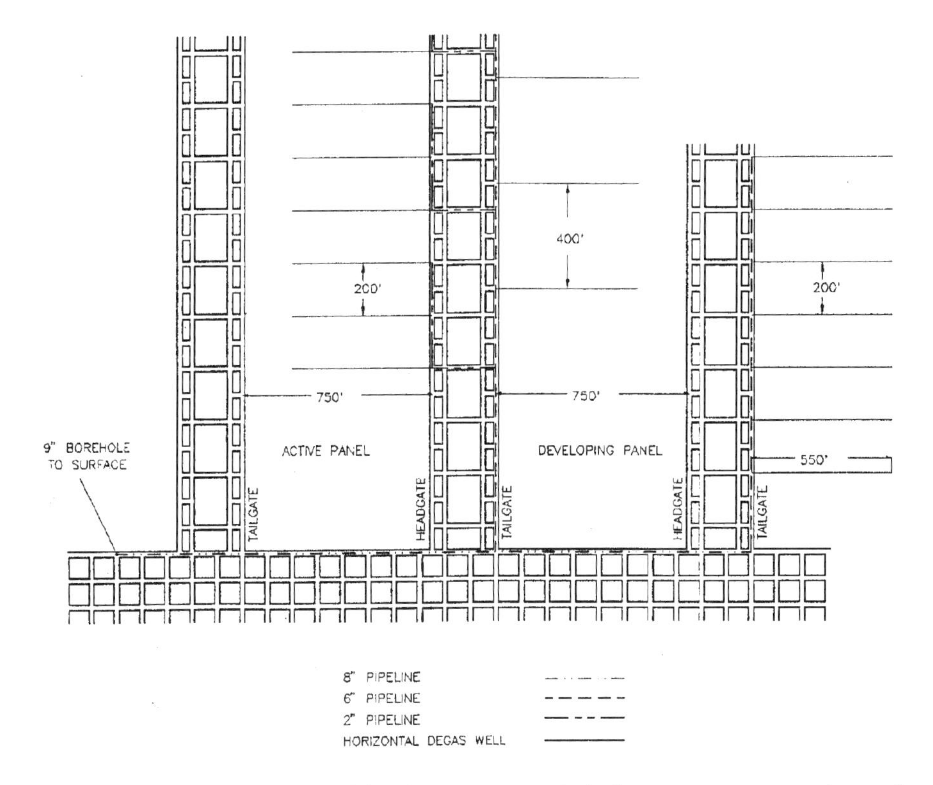

FIGURE 13.10. Typical horizontal borehole pattern and pipeline arrangement at a longwall mine. *Source:* Aul and Ray, 1991. Reprinted by permission of the Society for Mining, Metallurgy, and Exploration, Inc., Littleton, CO.

In the United States, extensive field experiments have been conducted by the U.S. Department of Energy and the private sector in Texas, Washington, West Virginia, and Wyoming (Stephens, 1980; Hill et al., 1985).

Underground gasification involves the partial combustion of coal in place, generally through boreholes, with the collection of gaseous by-products at the surface. The objective is to collect the thermal values of coal in the form of gases so as to avoid conventional mining. An alternative, shown in Figure 13.11, is to use the gases as feedstock for the production of petrochemicals or synthetic fuels, such as gasoline (Stephens, 1980).

Underground gasification of coal typically involves three stages (Zvyaghintsev, 1982):

1. Drilling of boreholes from the surface to the coal seam in pairs, one hole serving as the input of air and the other as the outlet for gaseous products.

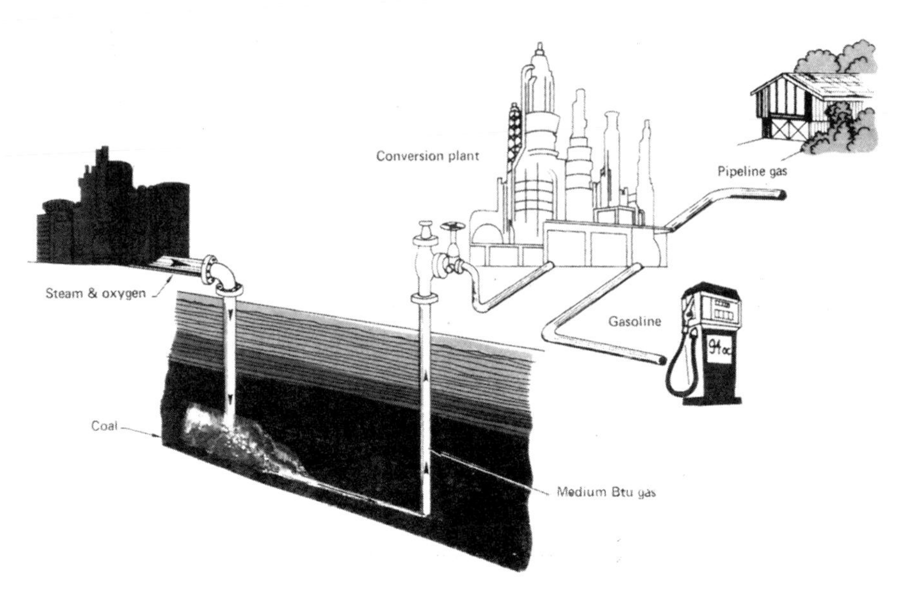

FIGURE 13.11. Conceptual diagram of underground gasification coupled with surface synthetic fuel plant. (After Stephens, 1980. By permission from Lawrence Livermore Laboratory, University of California, Livermore.)

2. Formation of reaction channels (linkages) between the injection and production holes, permitting the coal to interact with the air in a moving combustion front after ignition.
3. Gasification of the coal by supplying an air blast through the inlet hole and removing the gaseous products through the outlet hole. The two boreholes and the interconnecting channel constitute an underground gasifier.

Figure 13.12 illustrates underground gasification using percolation, one of the five methods of preparing a channel in a coal seam. Path linkage between the two boreholes is accomplished by injecting compressed air, by hydraulic fracturing or penetration, or by electrolinking using high-voltage current. After ignition, air is injected and a combustion front established. An alternative to air is a mixture of oxygen and/or steam to obtain a higher-quality product. Combustible products are carbon monoxide, hydrogen, and certain hydrocarbons, and noncombustible products are carbon dioxide and nitrogen. The system shown in Figure 13.12 is backward burning; that is, the combustion front retreats from the outlet borehole toward the inlet hole.

Additional schemes for setting up a successful underground gasification operation include the use of directional drilling technology to connect the inlet and outlet wells and to involve greater coal volumes in the extraction process (Tracy, 1988). Other gasification plans have utilized underground openings in

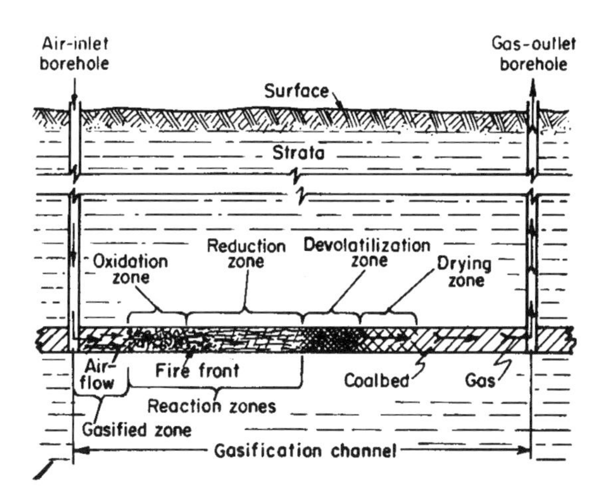

FIGURE 13.12. Borehole method of underground coal gasification, using backward-burning percolation. Shown are the seam, boreholes, linkage or gasification channel, combustion front, and reaction zones. (After Marsden and Lucas, 1973. By permission from the Society for Mining, Metallurgy, and Exploration, Inc., Littleton, CO.)

the coal seam to provide for better linkage between the inlet and outlet (Chaiken and Martin, 1992). This alternative opens up a variety of layouts to enhance the combustion process.

Much has been learned over the years about the design parameters of underground gasification projects. Summaries of these parameters can be found in Marsden and Lucas (1973), Stephens (1980), and Zvyaghintsev (1982). The parameters specified for the underground configuration will differ, depending on the specific procedures used for access and linkage. The product quality can be expected to be gases of 75 to 125 Btu/ft^3 (2.79 to 4.66 MJ/m^3) if air is injected, and about 450 Btu/ft^3 (16.8 MJ/m^3) if oxygen and steam are injected instead of air. These sources and the information by W. A. Murray (1982) are useful in defining the advantages and disadvantages of underground gasification of coal.

To summarize, the future of underground gasification technology is not bleak. However, the technology is not growing and is not likely to grow much in the future. The primary reason is the lack of economic ways of exploiting the technique. The United States is currently the home of one aboveground coal gasification plant. This government-supported plant is the Great Plains Coal Gasification Plant located northwest of Bismarck, North Dakota (U.S. Department of Energy, 2000). The plant converts North Dakota lignite into pipeline-quality gas for use in 300,000 homes and businesses in the Midwest and the eastern states. The facility also produces carbon dioxide that can have value in enhancing oil well recovery. Coal gasification is used extensively in South Africa. However, the company utilizing this technology, Sasol, employs mined coal and surface conversion plants to create its gaseous by-products.

13.7 UNDERGROUND RETORTING

During the oil crisis of the 1970s, the political atmosphere favoring the development of new energy sources required that every possibility be investigated to determine their potential in future energy production. Naturally occurring hydrocarbons were often targeted. Certain natural hydrocarbon deposits—oil shale and tar sand—are unique in that the hydrocarbons they contain occur in the solid state. Kerogen is found in oil shale, and bitumen in tar sands. Their occurrence throughout the world is widespread, but the deposits are low in grade. Because of economics, commercial exploitation is limited to deposits that can be mined at the surface and processed in surface beneficiation plants. Tar sands in Alberta and Utah are currently mined in this manner. For deeper deposits, production will depend primarily on novel methods.

The unique nature of these solid hydrocarbon deposits suggests that a novel method of applying heat would be successful in fluidizing the kerogen or bitumen. One promising method is *underground retorting*, in which the oil-bearing formation in situ is subjected to combustion or is heated by electricity or by forcing it through high-temperature, high-pressure gases or liquids. Traditional exploitation methods, either surface or underground, may be combined with surface retorting, but the mining is not novel. The discussion here is limited to underground retorting and, because of their prevalence in the United States, to oil shales.

Mineralogically and petrographically, oil shale is a fine-grained marlstone containing a solid, insoluble organic substance called kerogen (Marsden and Lucas, 1973). When heated to 900° to 950°F (480° to 510°C), kerogen decomposes to oil, gas, and a carbonaceous residue of spent shale (char). The chemical process of pyrolysis or cracking of kerogen is termed *retorting*. The main object of in situ retorting is to extract the kerogen in the form of liquid fuels and gaseous by-products. Although it is simply described, the process involves complex thermodynamic problems of heat transfer, pyrolysis kinetics, and process control. In addition, two natural characteristics of the oil shale act as deterrents: (1) its low thermal conductivity impedes efficient heat transfer, and (2) its extremely low porosity and permeability hamper fluid passage.

The inadequate permeability of oil shale is the most critical problem to overcome. It generally requires the creation of artificial fractures and channels in the formation in order to retort it in place. Alternately, the shale can be mined first and then retorted on the surface. The three most logical alternatives (Ricketts, 1992) would then be (1) mining followed by surface retorting, (2) true in situ retorting, and (3) modified in situ retorting. Because aboveground retorting requires conventional mining, the emphasis here will be on in situ methods.

True in situ operations are carried out entirely from the surface, generally using boreholes. When true in situ retorting is used, formation permeability is increased by massive fracturing (Office of Technical Assessment, 1980). This is achieved in three ways, similar to methods used in conventional oil production

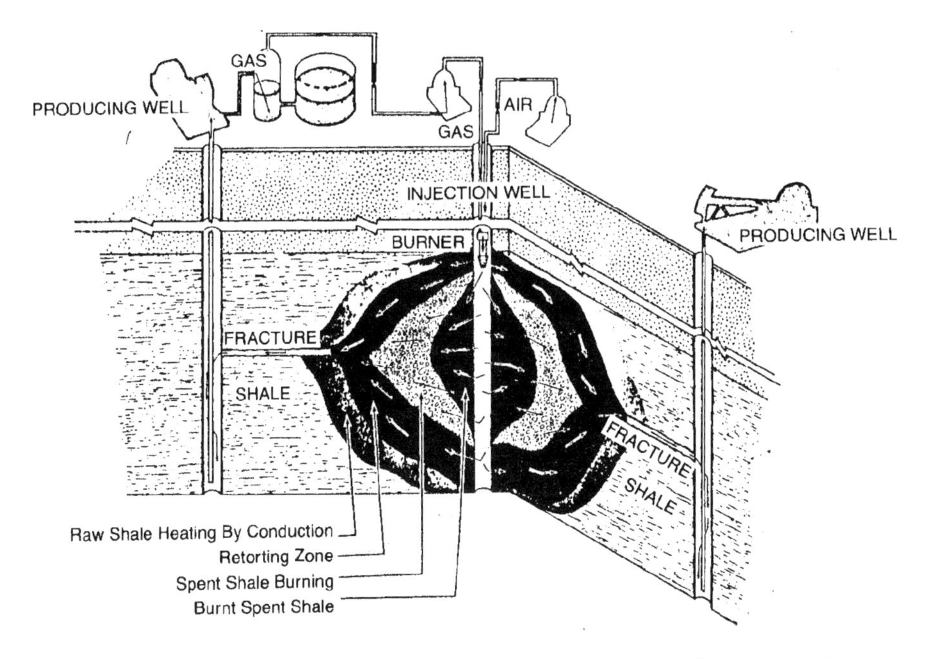

FIGURE 13.13. True in situ oil shale retorting. Conceptual diagram showing injection and production wells and underground retort. (After Grant, 1964. By permission from Colorado School of Mines Press, Copyright © 1964, Golden CO.)

or underground gasification of coal: (1) hydraulic or pneumatic fracturing by high-pressure fluid injection, (2) application of high-voltage electricity or electrolinking, and (3) explosives fracturing by large-scale detonation of liquefied blasting agents or slurries.

Various processes for true in situ retorting have been envisioned; one concept appears in Figure 13.13. Six steps are typically involved (Office of Technical Assessment, 1980; Russell, 1981; Kalia and Gresham, 1985):

1. Drilling a predetermined pattern of wells, probably three to five in number, in which an injection hole is ringed by production holes
2. Dewatering the zone to be retorted, if the deposit occurs in the water table and the method requires a dry environment
3. Fracturing the deposit to create or increase permeability
4. Igniting the bed by burner or injecting a hot fluid to provide heat for combustion and pyrolysis
5. Pumping compressed air and/or oxygen into the formation through an injection well to support combustion and forcing hot gases through the fractured rock to aid retorting
6. Recovering the retorted oil through the producing wells

Step 3 is omitted if the oil shale is sufficiently permeable for pyrolysis to be carried out by a superheated fluid (natural gas, air, or steam) in Step 4. Pyrolysis in underground shale retorting is similar to combustion in underground coal gasification.

Limited field tests of true in situ retorting have been conducted, but many uncertainties remain and commercial application is remote even if oil prices were to increase. Use appears restricted to relatively thin and permeable deposits under shallow cover (Office of Technology Assessment, 1980).

In contrast, modified in situ retorting has undergone full-scale field tests and has advanced to a moderate level of technological readiness. Although no commercial operations exist at the time of this writing, they appear feasible if the price of oil is sufficient. In *modified in situ retorting*, the permeability of the oil shale deposits is increased substantially by mining 20 to 40% of the shale to create a void or voids in the deposit and then blasting the rest of the shale into the void to rubblize it (Russell, 1981; Ricketts, 1992). The block of rubblized material then becomes the retort. Various mining methods may be used to create the voids; perhaps the most interesting is the sublevel caving method. One concept using sublevel caving is shown in Figure 13.14. Note that the sublevel caving sections are arranged in vertical blocks; each block becomes a separate retort.

The steps in modified in situ retorting are as follows (Kahlia and Gresham, 1985; Ricketts, 1992). After main and secondary development openings are driven to the block to be retorted, single or multiple voids are excavated, and the oil shale remaining in the block is blasted into the void. The shale excavated in each block is normally hoisted to the surface and retorted in a facility located aboveground. All mining access openings are sealed, and boreholes to serve as injection and production wells are drilled from the surface. The rubble is then ignited at the top of the rubblized zone, and combustion is maintained by injecting air. The burning zone progresses downward through the retort, pyrolyzing the oil, which is collected at the bottom of the retort in a sump and pumped to the surface. Shutting off the air supply extinguishes the burning when the entire block has been pyrolyzed.

Modified in situ retorts must be large to be economically as well as technologically sound. Experimental sublevel caving blocks have usually been 160 ft by 160 ft (50 m by 50 m) in cross section and a maximum of 270 ft (80 m) in height. Block heights to 500 ft (150 m) and even 700 ft (210 m) are contemplated (Dayton, 1981a, 1981b). An ultimate limit to mining height in U.S. oil shales is posed by the thickness of the Green River formation, which varies from 50 to 1870 ft (15 to 570 m).

The production scale must be similarly large to be economic. It has been projected that a modified in situ operation would have to mine and retort on surface 20,000 tons/day (18,000 tonnes/day) and blast and retort 80,000 tons/day (72,000 tonnes/day). With shale containing 30 gal/ton (125 l/tonne), the yield would be 55,000 bbl/day (8750 m^3/day) of oil. An estimated 20 retorts would be under development or exploitation at one time.

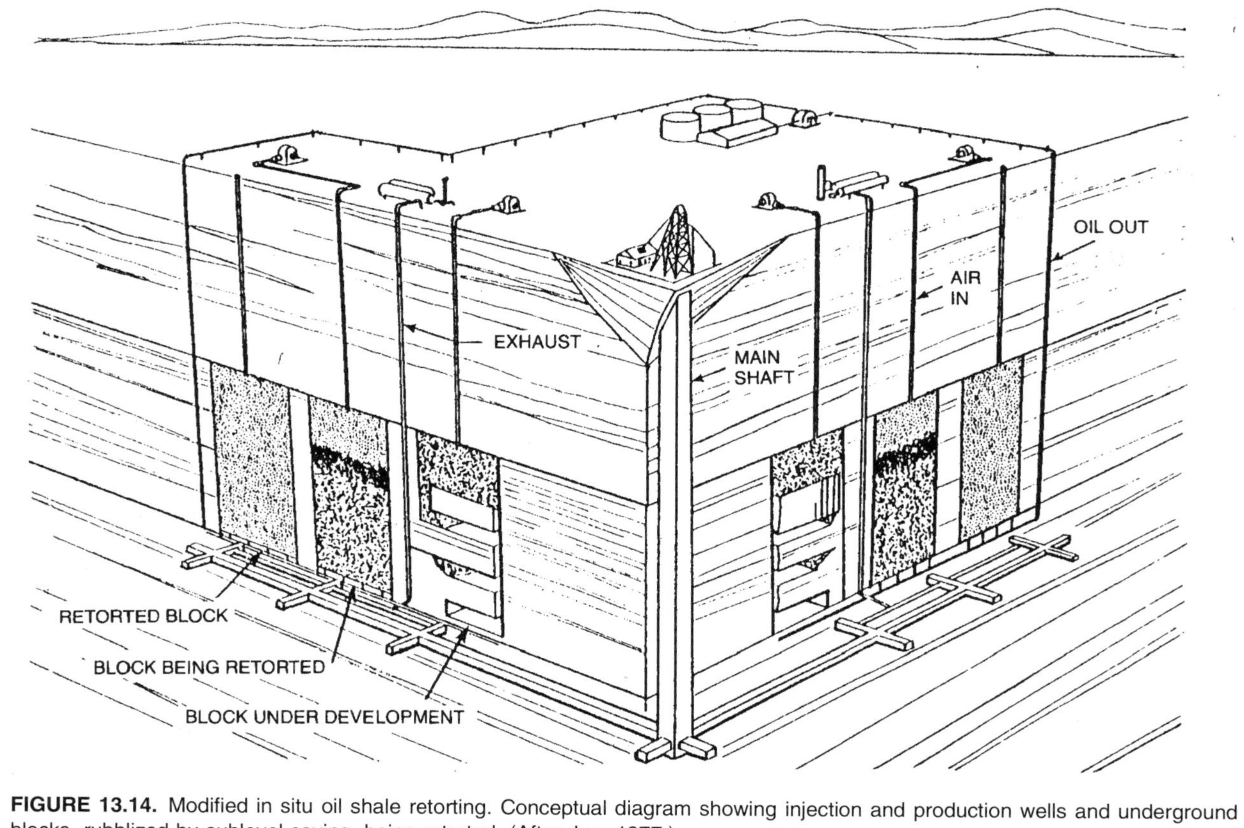

FIGURE 13.14. Modified in situ oil shale retorting. Conceptual diagram showing injection and production wells and underground blocks, rubblized by sublevel caving, being retorted. (After Jee, 1977.)

With all the experimentation and analysis done during the oil embargo period, oil shale mining has been evaluated well enough to realize that it is not going to be economic unless oil prices rise to unprecedented levels. A number of articles (Russell, 1981; DeGabriele and Aho, 1982; Rajaram, 1985) have described some of the economic problems that plague oil shale mining. In addition, the possibility of fires, gas explosions, and dust explosions in an underground oil shale facility threatens to create additional economic burdens. Suffice it to say that oil shale mining is not currently a serious consideration. Although the technology is certainly workable, the capital and operating costs are simply too high to be feasible.

13.8 MINING FOR OIL

Although oil shale mining technology may be uneconomic under current conditions, mining for oil may be a more favorable technology. The basic opportunity to utilize this mining method occurs when a shallow oil-bearing formation or tar sand can be accessed by mining openings developed below the deposit. Generally, this method will be applied to light or heavy oil reservoirs in permeable formations, to tar sands too deep for surface mining, or to other hydrocarbon sources that can be made to flow into the mine openings by gravity drainage, by thermal heating, or by chemical treatment. For a basic definition, we will say that *oil mining* is the process of extracting oil or similar petroleum products from a reservoir by using mined openings to access the hydrocarbon reservoir.

The history of oil mining is quite interesting and involves the extraction of oil in shallow excavations in Persia in 5000 B.C.E. (Kennedy et al., 1980; Dobson and Seelye, 1982). Oil mines were initiated in 1735 C.E. in France and in 1866 in California (Hutchins and Wassum, 1981). In the United States, oil mines were also attempted at one time or another in Pennsylvania, Ohio, Wyoming, Colorado, Texas, Kansas, and Utah. A variety of surface mine studies were conducted over the years to determine the feasibility of recovering oil in an economic manner (Kennedy et al., 1980). In addition, the North Tisdale oil field in Wyoming and the tar sands in Alberta have been tested for their ability to produce oil using underground mine openings and wells drilled upward into the formations containing the hydrocarbons.

The basic plan for underground oil mining is to locate mine openings under the oil-bearing formation in a rock mass that would ensure permanence of the openings over the lifetime of the project. As shown in Figure 13.15, drillholes would then be produced from the mine openings. However, unlike the holes shown in the illustration, they would typically be positioned at the base of the oil deposit so that gravity could work as the basic method of extracting the oil. Note that in tar sands, the wells would generally be supplied with steam to thin out the heavy oil and allow it to flow into the mine openings. The oil so extracted would then be gathered into sumps and pumped to the surface for processing.

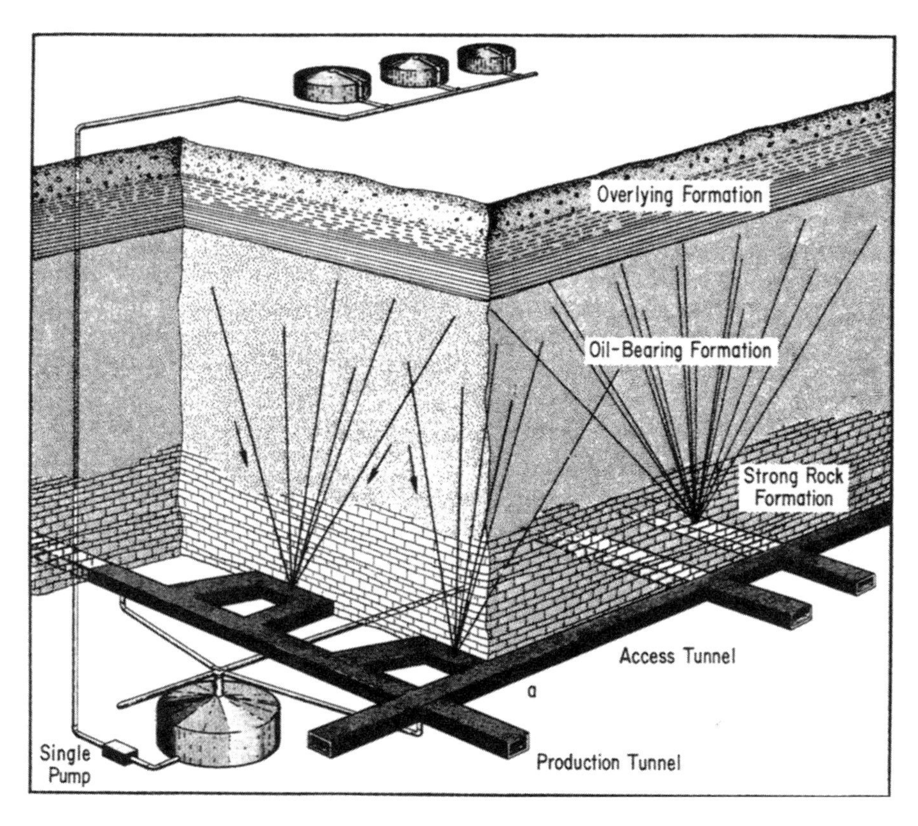

FIGURE 13.15. Basic plan for underground oil mining. *Source:* Hutchins and Wassum, 1981. By permission from the Society for Mining, Metallurgy, and Exploration, Inc., Littleton, CO.

In recent years, the number of mines involved in this process have been relatively few. One such mine to be described was developed by Conoco in the Lakota sand of the North Tisdale oil field near Casper, Wyoming (Hutchins and Wassum, 1981; Dobson and Seelye, 1982). This field was shallow, with low pressure and a poor recovery when tapped by conventional oil wells. An adit to access the Lakota sand was initiated in 1977 after confirmation of oil-saturated rock in the area. After the adit was developed, holes were carefully drilled into the bottom of the oil-saturated zone with the use of drilling tools developed for coalbed methane drainage. The holes were $2\frac{7}{8}$ in. (73 mm) in diameter, and controllable features of the drill made positioning of the holes possible. Initial drainage of the drillholes was due to residual pressure in the formation. When that dissipated, gravity was the predominant force to extract the oil. All oil was allowed to flow to a sump, where it was gathered and pumped to the surface for further processing. The mine was normally operated without personnel underground. Ventilation was more than doubled when miners were stationed underground to ensure that gases from the formation were adequately diluted. The project proved the feasibility of the oil mining procedure, but the recovery and general economics were modest.

The second project of interest was performed in the tar sands located near Fort McMurray, Alberta. This formation is a favorable target for underground oil mining because only 10% of the tar sands are mineable by surface methods. The mine was developed with two shafts through the tar sands and drifts driven through a Devonian limestone formation located below the oil-bearing sands (Stephenson and Luhning, 1988; Carter, 1991). Holes were drilled up toward the tar sands and then deflected horizontally along the bottom of the tar sands formation. The 11 in. (0.28 m) diameter holes were drilled with tricone bits and lined with special casings to allow for steam injection and bitumen collection. The project was successful, allowing the conclusion that 50% of the oil could be extracted within four years, and depths of up to 3900 ft (1200 m) appeared to be economic. Lyman and Piper (1985) describe mining projects from previous years. Literature on other oil mining properties, more recently exploited, has been published by Trent (1986) and See (1996).

To summarize the future of oil mining, it is essential to realize that the level of oil prices will determine the economics of the method. Initial economic estimates seem to indicate that the Alberta tar sands have promise for the future. Oil mining in other areas of North America will depend on the reservoir characteristics and the prevailing price of oil. The technology is currently sufficiently mature to develop properties that will produce significant quantities of oil, but the market price will determine whether the operations will be economically viable.

13.9 OCEAN MINING

Ocean mining is performed using a group of methods that have been studied for the last five decades. Most activity has been oriented toward (1) shallow unconsolidated deposits on the continental shelves, (2) deep deposits containing metallic nodules, and (3) consolidated deposits close to the shores. However, other than shallow-water projects to recover unconsolidated minerals from coastal areas, the technology is not advancing in the manner its proponents imagined when they initiated their efforts to promote the industry. Political, technological, and economic constraints have kept the industry from blossoming.

The Law of the Sea Treaty, negotiated in 1980 by the United Nations and signed in 1982 by 130 countries, was intended to resolve jurisdictional and usage disputes over the oceans, including mining rights. However, the United States (and three other nations) refused to sign or abide by the treaty. Instead, the United States in 1983 declared its jurisdiction over all mineral deposits within a so-called Exclusive Economic Zone (also acknowledged by the treaty) extending 200 miles from its coasts, mainland and possessions, and encompassing an area of 625 million mi^2 (1.6×10^9 km^2) (McGregor and Offield, 1983).

Since the treaty was signed, little mining activity has been conducted in the deeper zones of the oceans, but Russia, China, Japan, France, Germany, and certain U.S. companies have obtained undersea mining claims in the deep parts of the Pacific Ocean (Haag, 1997).

Cruickshank and Marsden (1973) and Cruickshank (1992) have identified possible ocean targets. Much of the attention of the ocean mining technologists has been centered on the shallow mineral deposits associated with the continental shelves. However, great interest has been shown in the deep deposits of manganese nodules because of their high grade, their large areal extent, and their location in unconsolidated material on the ocean bottom. There is an added advantage in that copper, nickel, cobalt, and molybdenum have also been associated with the nodules and can provide an additional economic return. Recent attention has also been paid to the metalliferous sulfides that are associated with the earth's plate movements.

Our discussion of novel mining methods will be limited to those in the aforementioned second category, unconsolidated deposits on the seabed. The technology for recovery of the first group, dissolved minerals, is rather well advanced, comprising a field of knowledge that is more akin to chemical or metallurgical engineering than to mining engineering. Not all processes for chemical recovery are economic, but many are or will be feasible in the future. The mineral deposits in the third category, consolidated deposits, will probably be recoverable by technology borrowed from offshore petroleum production or by traditional mining methods exploited from shore. The shore-based mining endeavors will likely be limited to deposits within 10 mi (16 km) of the coast.

Unconsolidated marine deposits, regardless of depth, resemble placer deposits on dry land, because they are mechanical concentrations of heavy minerals. As such, they are amenable to the principles of placer mining and, conceivably, to modifications of the practices. It is probable that standard dredging practices (Section 8.3) can be adapted to mine deeper unconsolidated deposits. However, innovative or novel adaptations may be necessary to solve the problems associated with deep-sea deposits.

An ocean-mining system conceptually consists of three components: (1) a working platform, (2) a coupling and hoist, and (3) an excavating unit. Figure 13.16 shows one possible configuration of a deep-ocean mining system, with the ship being the working platform, the pipe providing the coupling and hoisting capability, and a drag collector performing the excavation. Several possible variations of working platforms suitable for ocean mining are shown in Figure 13.17, with the platform floating, submerged, or located on the seafloor.

Excavators envisioned for ocean mining can take many forms with most of the options shown in Figure 13.18. The excavation device itself can be suspended, attached to a platform, or located on a mobile vehicle that travels the ocean bottom. Hoisting alternatives are fairly numerous, including several involving hydraulic hoisting. Some of these alternatives are shown in Figure 13.19.

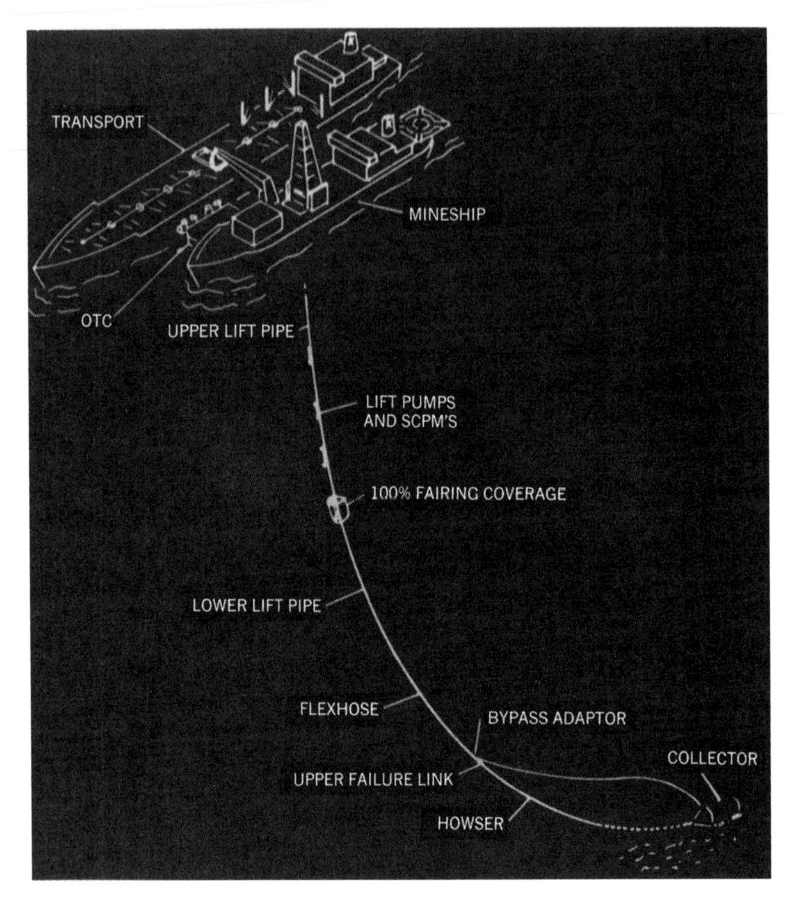

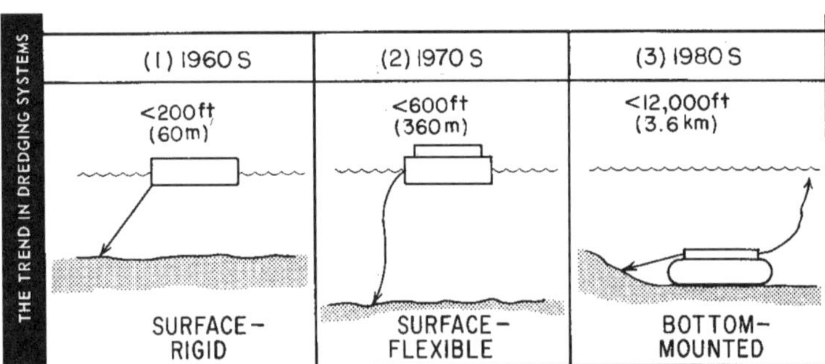

FIGURE 13.16. (*Top*) Conceptual configuration for deep-sea mining system. (After Halkyard, 1979. By permission from the National Mining Association, Washington, DC.) (*Bottom*) Transition in nature of coupling between excavating unit on ocean bottom and platform at surface: (1) surface-rigid, (2) surface-flexible, and (3) bottom-mounted. (After Cruickshank et al., 1968. Copyright © 1968, *Engineering and Mining Journal*, New York.)

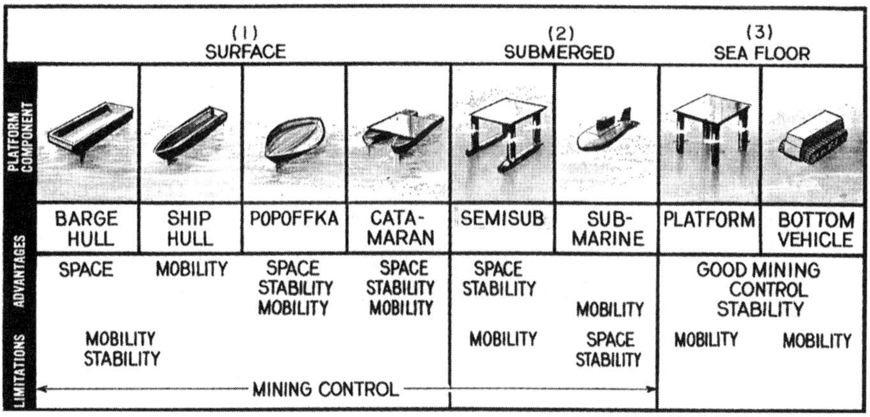

	(1) SURFACE				(2) SUBMERGED		(3) SEA FLOOR	
PLATFORM COMPONENT	BARGE HULL	SHIP HULL	POPOFFKA	CATA-MARAN	SEMISUB	SUB-MARINE	PLATFORM	BOTTOM VEHICLE
ADVANTAGES	SPACE	MOBILITY	SPACE STABILITY MOBILITY	SPACE STABILITY MOBILITY	SPACE STABILITY	MOBILITY	GOOD MINING CONTROL STABILITY	
LIMITATIONS	MOBILITY STABILITY				MOBILITY	SPACE STABILITY	MOBILITY	MOBILITY
	MINING CONTROL							

FIGURE 13.17. Types of platforms suitable for ocean mining: (1) surface, (2) submerged, and (3) seafloor. (After Cruickshank et al., 1968. Copyright © 1968, *Engineering and Mining Journal*, New York.)

Engineering design of ocean-mining systems has advanced from the study and drawing-board stages to system tests in deep water. Cruickshank (1992) lists six deep-water tests that have been conducted in the Pacific Basin with bucket-line, airlift, and slurry hoisting systems. Costs of mining were not supplied for these tests, but the general cost picture available in Flipse (1983) and Dick (1985) indicates that deep-sea mining ventures may have difficulty in keeping costs lower than those of traditional mining methods.

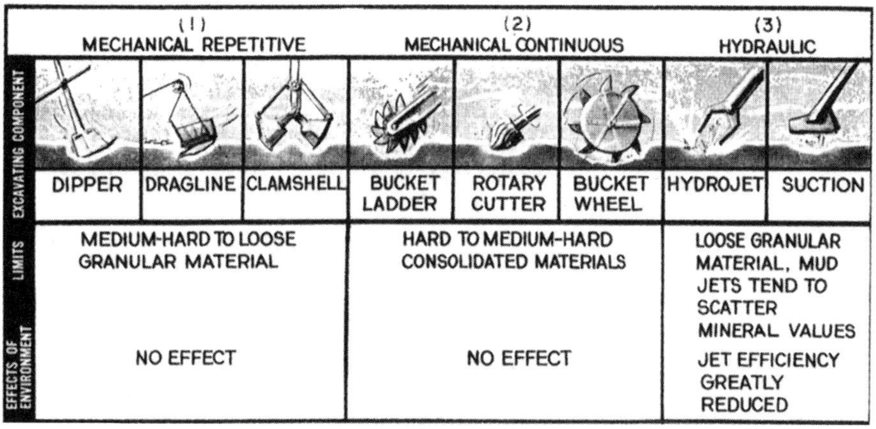

	(1) MECHANICAL REPETITIVE			(2) MECHANICAL CONTINUOUS			(3) HYDRAULIC	
EXCAVATING COMPONENT	DIPPER	DRAGLINE	CLAMSHELL	BUCKET LADDER	ROTARY CUTTER	BUCKET WHEEL	HYDROJET	SUCTION
LIMITS	MEDIUM-HARD TO LOOSE GRANULAR MATERIAL			HARD TO MEDIUM-HARD CONSOLIDATED MATERIALS			LOOSE GRANULAR MATERIAL, MUD JETS TEND TO SCATTER MINERAL VALUES	
EFFECTS OF ENVIRONMENT	NO EFFECT			NO EFFECT			JET EFFICIENCY GREATLY REDUCED	

FIGURE 13.18. Mechanisms employed by ocean-mining systems: (1) mechanical-cyclical, (2) mechanical-continuous, and (3) hydraulic. (After Cruickshank et al., 1968. Copyright © 1968, *Engineering and Mining Journal*, New York.)

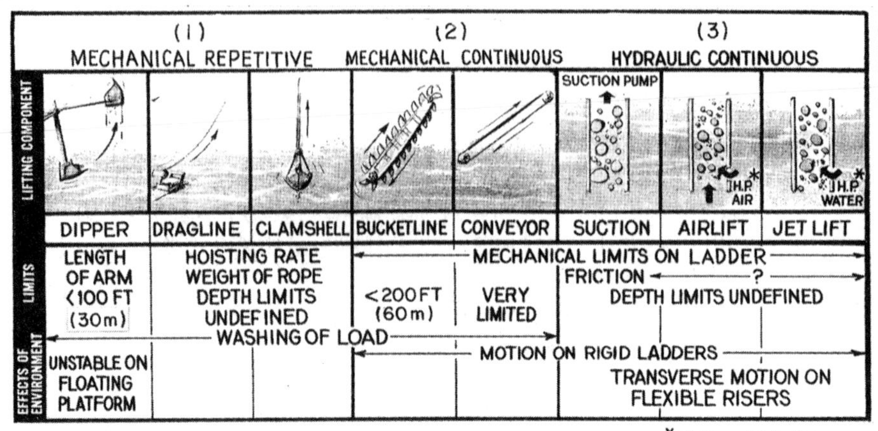

FIGURE 13.19. Hoisting mechanisms employed by ocean-mining systems: (1) mechanical-cyclical, (2) mechanical-continuous, and (3) hydraulic. (After Cruickshank et al., 1968. Copyright © 1968, *Engineering and Mining Journal*, New York.)

More recent proposals for deep-sea mining systems may have some merit to improve the cost picture. Bath (1991) has outlined a variety of nodule pickup systems, and the *Engineering and Mining Journal* (1996b) and Haag (1997) present plans for some remote-controlled, self-propelled, deep-sea miners. The development of new ideas and solutions for deep-sea mining continues. Much of the information is made available in the *Annual Offshore Technology Conference Proceedings*, published by the American Institute of Mining, Metallurgical and Petroleum Engineers (AIME). However, the industry requires a strong signal that deep-sea mining will be economic before it will blossom into a large supplier of mineral raw materials.

13.10 OTHER NOVEL METHODS

13.10.1 Nuclear Mining

The peaceful application of nuclear energy for fragmentation and excavation purposes in mining, petroleum recovery, and construction has been the subject of much scientific speculation and analysis. It formed the major thrust of Project Plowshare, the R&D program on peaceful uses of the atom sponsored by the U.S. Atomic Energy Commission (now the Nuclear Regulatory Commission) in the 1950s and 1960s. Since the ratification of the Limited Nuclear Test Ban Treaty late in that period, however, which prohibited surface detonation and underground tests venting to the surface, applications of nuclear blasting and mining are no longer seriously being advocated. In addition, society is much more prone to protest any such use of nuclear devices than it was in the 1960s.

The original thought that nuclear power could be of help to humankind is certainly understandable. Chamberlain (1992) notes that the complete fission of 1.0 lb (0.4 kg) of uranium can produce the energy equal to the detonation of about 9000 tons (8000 tonnes) of TNT. However, no other environmental risk troubles the public to the extent that radiation hazards do — and rightly so. However, the public makes little distinction between nuclear power used for weaponry and that used for blasting an ore body. Thus, nuclear power for mining is not a likely scenario in upcoming decades.

Because of the currently unfavorable image projected by nuclear energy in general, we need not discuss the possible mining uses of nuclear energy exhaustively. The proceedings of the five Plowshare Symposia, the last of which was published by the U.S. Atomic Energy Commission (1970), can provide information about the ideas considered. An excellent summary has also been presented by Russell (1973). Some of the worthwhile ideas are now ruled out under the Test Ban Treaty, including any use for surface mining applications. Underground mining applications may be more promising. Underground nuclear blasting produces a cavity, crushed rock in a rubble chamber, and a fractured zone. Some likely applications are for block caving, in situ leaching, underground coal gasification, and underground oil shale retorting. However, little in the way of experimentation has been performed to prove the usefulness of these procedures.

Based on the research done so far and a general assessment of the technological, economic, and safety aspects of nuclear energy, some general conclusions can be reached concerning the use of nuclear power for practical purposes such as mining:

1. The general engineering feasibility of nuclear blasting for mining purposes has not been demonstrated. Certain uses appear attractive, but specific tests designed to explore mining applications are needed.
2. Technological feasibility is lacking because there is insufficient knowledge about scaling relations, effects in different media, results with large yields, and effects of multiple shots; further, much technical information is classified.
3. Economic feasibility cannot be assessed conclusively from published data but appears most attractive with large yields (i.e., >10 kilotons or >9 kilotonnes of TNT equivalent).
4. Safety has not been reliably established, even for contained underground shots. Aside from serious doubts about radiation, there are environmental concerns related to groundwater contamination and seismic damage.
5. Nuclear mining is not currently a topic under consideration. Only a serious threat to world security or a major economic collapse of some magnitude will change that reality.

It is not likely that the international political scene will soon change and permit the resumption of civil underground tests of nuclear explosives. Until it does, nuclear mining will remain unproven and unexploited.

13.10.2 Extraterrestrial Mining

Easily the most exotic of all novel mining possibilities, *extraterrestrial mining* refers to the exploitation of mineral resources on the moon and elsewhere in space. Questions of a political, legal, or economic nature aside, technological feasibility remains the abiding imponderable.

Speculation about lunar mining has been rampant since the National Aeronautics and Space Administration's (NASA) Ranger and Apollo landings on the moon (Ruzic, 1964; Delinois, 1966; Penn, 1966). The nature and extent of lunar mineral resources are still not well known even after the Soviet and American exploratory landings. Analyses of surface samples collected by probes, drills, and human explorers indicated a porous, lavalike mantle on the moon. Based on similar geologic occurrences on earth, there has been speculation that iron, some of the base metals, precious metals, diamonds, and rare earths may exist on the moon.

In 1989, NASA initiated the Space Exploration Initiative to launch a sustained, self-sufficient presence in space (Chamberlain and Podnieks, 1992). The objective was to establish a lunar outpost, with possible expansion to Mars and its moons. This outpost would require that many of its resources be supplied from lunar materials. The base could be used as a refueling station for further excursions into space, with the propellant components produced on the moon.

The first lunar mining endeavors are likely to be oriented toward providing the basic necessities of life such as a supply of oxygen and hydrogen. In addition, the production of building materials may also be required (Chamberlain and Podnieks, 1992). The exploration performed so far on the moon indicates that oxygen could be derived from FeO, SiO_2, Al_2O_3, and TiO_2 close to the surface. The oxygen would probably be recovered and liquefied for a variety of uses, primarily life support. Mining activity would be concentrated in areas of the moon with high oxygen contents. Hydrogen would be needed to make water and propellants. Its source would likely be the hydrogen ions deposited in the moon's surface by solar winds. Mining operations could be limited to the upper 6 to 10 ft (2 to 3 m) and would be most productive if located in regions without boulders. Mining of building materials can probably be undertaken in many areas, with proper rock properties being the determining factor in selection of the mine site.

The first mining equipment on the moon must be delivered to the moon via rocket transport ship. Accordingly, it would not be massive and would be built from high-strength, low-density materials. It would likely be operated by remote control or automation and may utilize considerable amounts of artificial intelligence. The effects of gravity would be significant in the digging process. Excavators might be anchored into the moon's surface lest they be tossed around during digging. Dusty and cold conditions would make the equipment subject to breakdown; maintenance would be a major problem. No aspect of the mining process will be simple, and costs will be astronomical.

Extraterrestrial mining beyond the moon (planets, comets, asteroids, etc.) is even further out of the realm of realization. It appears that exploitation systems will require automation or robotic control, as the environments are too hostile for long-term human endurance. Because automation concepts have advanced significantly in the last two decades, this appears to be achievable. However, the operation of equipment in space will have many more obstacles to overcome than similar operations on earth. Significant research will be required to ensure that the procedures developed for mining are virtually indestructible and self-maintaining.

Although the ability to mine in space seems remote to us now, the research into reliable mining equipment in space may have positive effects on mining equipment used on earth. The automation and self-maintenance features should be adaptable to conventional mining equipment. In addition, the development of new materials and equipment concepts may have down-to-earth benefits. NASA's ability to spend money on appropriate research may indeed overcome the difficulties of mining in space.

13.11 SUMMARY OF INNOVATIVE AND NOVEL METHODS

Determining the future promise of the novel methods discussed in this chapter is a formidable task. However, the advancement of the methods since the first edition of this book offers some important insights. Several of the methods have evolved rapidly since the text came out. Clearly, these methods will have additional promise and may be the easiest to assess. Others, such as underground gasification and retorting, depend on the cost of energy for their possibilities. They may or may not ever become a significant part of the mineral exploitation field. The following list includes the innovative and novel in decreasing order of potential:

1. *Automation and robotics.* This field has advanced quickly in the last decade or so. Growth in this area is likely to be significant in the future as well, with many pieces of automated equipment used in the mining industry. It appears that both safety and productivity will be improved if this is the case, an incentive that will help in the advancement of this technology.
2. *Methane drainage.* Significant improvements in technology and more favorable economics of drainage should spur the field to increased usage in the foreseeable future. Both coal seams being mined and coal seams not associated with mining should see increases in methane extraction.

3. *Rapid excavation.* These processes have been undergoing improvement for four decades, with the greatest mining improvements in the areas of raising and shaft sinking. Continued though restrained growth is likely to be achieved in this area.
4. *Hydraulic mining.* Applications of hydraulic energy should see greater use in drilling, cutting, and transport in the future. However, the expansion of the field appears to be somewhat limited, indicating that growth will be modest in most cases.
5. *Oil mining.* The use of mining openings for the extraction of oil from a heavy oil reservoir appears to be economically feasible under the right conditions. Rising oil prices would obviously help this process. Otherwise, oil mining will see only occasional use.
6. *Ocean mining.* The potential for ocean mining appears to be significant. However, the political and economic climates are not particularly favorable. This leads us to predict that growth in this area may be slow.
7. *Underground gasification.* Several countries of the world utilize this method. However, in the United States the cost of coal is too low to justify any significant coal gasification, except on the surface.
8. *Underground retorting.* The most likely event to spur underground retorting of oil shale would be a significant increase in the price of oil on the world market. Barring this event, the use of underground retorting is not likely.
9. *Extraterrestrial mining.* The long-term hope for this category of mining is a research effort to overcome the many technical problems that exist. There is hope that this may be achieved, but not in the near future.
10. *Nuclear mining.* This is one category of mining that is highly unlikely in the near-term or long-term future. Society is not currently prone to accept the risks of such an endeavor.

The study of innovative methods is an interesting exploration into the possibilities for future mining practice. Some of the methods discussed here will likely enrich our society in the future, whereas others may fall by the wayside. Today's students will have an opportunity to see just how these methods advance in the future.

14

SUMMARY OF MINING METHODS AND THEIR SELECTION

14.1 INTRODUCTION

In this, the final chapter of the book, it is our objective to offer in-depth comparisons of the various methods studied and to provide an objective basis for choosing a mining method. We first summarize the characteristics of the mining methods that will allow us to choose methods that are feasible and to eliminate those that are not suited for the subject deposit. We then develop an outline of a procedure to choose a surface mining method and/or an underground method from the candidate methods. At this point in the selection process, it is important to be aware of the costs for each method we are still considering. Accordingly, the next section of the chapter discusses mining method costs. The selection procedure may then require that we choose between a surface method and an underground method. A procedure for accomplishing this is outlined. A mine investment analysis method is also introduced to ensure the economic value of the deposit and to sell the project to the financial institutions. Finally, the chapter provides a summary of the mining method selection process and a quick look at the future of mining and mining engineering.

14.2 METHOD RECAPITULATION

Surface mining methods were covered in detail in Chapters 6, 7, and 8, and underground methods were described in Chapters 9 through 12. In this section we review the important characteristics of mining methods to provide a procedure for narrowing down the number of methods that will be considered. But first we look at the production status of mining methods in the United

States to see how well utilized the various methods are at the present time. General information on production is provided in Table 14.1. Note that the usage of the methods is provided in general terms only because production statistics on mining methods are no longer readily available except for coal mining methods. Students may wish to consult the tables in Chapters 8 and 13 of Hartman (1987), which provide production statistics for the mid-1980s. More recent statistics are not readily available.

In should be noted that some of the mining methods (like leaching and longwall mining) have been increasing in usage while others (such as shrinkage stoping, stull stoping, and square-set stoping) have been decreasing. Table 14.1 provides valuable information, as the methods that are often used are normally associated with suitable technology and production characteristics that make them economic for many deposits in today's marketplace. We now turn to some of the pertinent characteristics of mining methods that will enable us to narrow the list of potential mining methods.

To accomplish the process of reducing the field of candidate methods, we refer to the general discussion in Section 4.8 of factors that determine the mining method chosen for exploitation of the deposit. These are grouped in six categories:

1. Spatial characteristics of the deposit
2. Geologic and hydrologic conditions

Table 14.1 U.S. Mineral Production by Mining Method

Method	Production
Open Pit	High
Quarrying	Low
Open cast mining	High
Auger mining	Low
Hydraulicking	Low
Dredging	Low
Borehole mining	Low
Leaching	Moderate
Room-and-pillar mining	High
Stope-and-pillar mining	Moderate
Shrinkage stoping	Low
Sublevel stoping	Low
Cut-and-fill stoping	Low
Stull stoping	Low
Square-set stoping	Low
Longwall mining	High
Sublevel caving	Low
Block caving	Low

Table 14.2 Classification of Ore and Rock Strength

Mineral or Rock	Relative Strength	Compressive Strength, lb/in.^{2} (kPa)
Coal, decomposed and badly altered rock	Very weak	<6,000 (<40,000)
Friable sandstone, mudstone, weathered rock, soft shale	Weak	6,000–14,500 (40,000–100,000)
Shale, limestone, sandstone, schistose rock	Moderate	14,500–20,000 (100,000–140,000)
Most igneous rock, strong metamorphic rock, hard limestone and dolomite	Strong	20,000–32,000 (140,000–200,000)
Quartzite, basalt, diabase	Very strong	>32,000 (>220,000)

Source: Modified after Hamrin, 1982. By permission from the Society for Mining, Metallurgy, and Exploration, Inc., Littleton, CO.

3. Geotechnical properties
4. Economic considerations
5. Technological factors
6. Environmental concerns

In this section, we attempt to outline a procedure that will eliminate methods that are not suitable for the deposit at hand, using the six factors listed. This will help to reduce our analysis in the later stages of method selection. It is also important at this point to consider whether the deposit can be mined by surface methods, by underground methods, or by a combination of both. It may be possible to limit the search for a mining method to either surface or underground methods simply by considering the depth of the deposit.

Although there are many factors that can be used in the selection of a mining method, one of the most important categories is the geotechnical properties of the ore and rock. Thus, the strength of the ore deposit and the surrounding rocks is of prime importance in the selection of the mining method. To help in this effort, we introduce Table 14.2, which defines the compressive strength of rocks for our qualitative strength descriptions from very weak to very strong. This table will help the student to determine what rock strengths are amenable for each mining method.

14.2.1 Surface Mining Methods

A variety of factors should be considered in reviewing the surface mining methods for suitability in exploiting a given ore deposit. In addition to the six categories of factors listed earlier, the selection effort should consider the sequence of development, the unit operations to be employed, the health and safety aspects of the methods, and the auxiliary operations that must be utilized in the mining operation.

In choosing or rejecting a given mining method, the first set of factors to be considered are the deposit conditions and how they relate to the various mining methods. To aid in this endeavor, information on many of the physical characteristics of the eight surface mining methods is listed in Table 14.3. The deposit characteristics can be compared with the entries in this table to choose mining methods that are compatible. Note that the initial methods singled out using Table 14.3 can be further reduced in number by referral to Table 14.4, where the advantages and disadvantages of the surface methods are outlined. A detailed study of this table may reveal additional methods that can be eliminated from the list of candidates because they have a particularly adverse set of disadvantages or a lack of advantages, as compared with other methods. It may be advisable at this point to limit the methods to the two most logical surface methods. The selection of suitable surface mining methods may be easier than choosing suitable underground methods, because the surface mining procedures are more highly dependent on what forms of geologic deposits they will economically extract.

14.2.2 Underground Mining Methods

Consideration of underground mining methods proceeds in a manner similar to that for surface methods. The first step is to compare the deposit conditions with the method characteristics in Table 14.5. The objective of this comparison is to eliminate those methods that are not suitable to the deposit conditions. The advantages and disadvantages of the underground mining methods, shown in Table 14.6, may also be used to eliminate methods that are not suited to the mineral deposit.

Note that the underground methods used for coal are relatively small in number, consisting of room-and-pillar mining, longwall mining, and occasionally shortwall mining (a subcategory of longwall mining). On the other hand, these three methods can often be used in nonmetal and metal mining along with all the other underground methods.

The choice of an underground mining method is often more difficult than that for surface mining, and mining engineers have frequently sought logical decision procedures to help in the selection. One of these selection procedures is outlined in Table 14.7. The original version of this table was produced by Peele (1941). The table was then modified by Lucas and Haycocks (1973) and by Thomas (1978). Note that the table can be used to select a couple of mining methods that may be applied to any given deposit. However, it does not always narrow the choice to a single method. Thus, the selection procedure may be continued with the idea of further narrowing the field. This will be attempted in Section 14.3.

Table 14.3 Comparison of Deposit Conditions Favorable to Surface Methods

	Mechanical Extraction				Aqueous Extraction			
Factor	Open pit	Quarrying	Open Cast	Augering	Hydraulicking	Dredging	Borehole	Leaching
1. Ore strength	Any	Any (sound structure)	Any	Any	Unconsol-idated, few boulders	Unconsoli-dated, some boulders	Consolidated	Rubblized or cavable, permeable
2. Rock strength	Any	Any	Any	Any	Unconsoli-dated	Unconsoli-dated	Competent, impervious	Competent, impervious
3. Deposit shape	Any (preferably) tabular)	Thick-bed-ded or massive	Tabular, bedded	Tabular, bedded	Tabular	Tabular	Any	Massive or thick tabular
4. Deposit dip	Any (preferably low dip)	Any, if thick	Any (preferably low dip	Low dip	Low dip	Low dip	Any (preferably low dip)	Steep
5. Deposit size	Large, thick	Large, thick	Large, moderate thicknes	Limited extent, thin	Limited extent, thin	Moderate extent, thickness	Moderate to large	Any (preferably large)
6. Ore grade	Low	High (assay not critical)	Low	Low	Very low	Very low	Intermediate	Very low
7. Ore uniformity	Uniform (or sort or blend)	Uniform	Fairly uniform	Uniform	Fairly uniform	Fairly uniform	Variable	Variable
8. Depth	Shallow to moderate	Shallow to moderate	Shallow	Shallow	Very shallow	Very shallow	Moderate to deep	Shallow to moderate

Table 14.4 Comparison of Advantages and Disadvantages of Surface Methods

	Mechanical Extraction				Aqueous Extraction			
Characteristic	Open Pit	Quarrying	Open Cast	Augering	Hydraulicking	Dredging	Borehole	Leaching
1. Mining cost	5%	100% (highest)	10%	5%	5%	<5% (lowest)	5%	10%
2. Production rate	Large-scale	Small-scale	Large-scale	Moderate	Moderate	Large-scale	Moderate	Moderate
3. Productivity	High	Very low	High	Very high	High	Highest	Very high	Very high
4. Capital investment	Large	Small	Large	Small	Small	Large	Large	Moderate
5. Development rate	Rapid	Moderate	Rapid	Rapid	Rapid	Moderate	Moderate	Moderate
6. Depth capacity	Limited	Limited	Limited	Limited	Limited	Limited	Unlimited	Limited
7. Selectivity	Low	High	Low	Low	Moderate	Low	Low	Low
8. Recovery	High	High	High	Moderate	Moderate	High	Low	Very low
9. Dilution	Moderate	Low	Low	Low	High	High	High	Very high
10. Flexibility	Moderate	Low	Moderate	Very low	Moderate	Low	Low	Low
11. Stability of openings	High	Highest	High	High	Moderate	Moderate	High	Moderate
12. Environmental risk	High	Moderate	Very high	Low	Severe	Severe	Moderate	Moderate
13. Waste disposal	Extensive	Moderate	Minor	None	Moderate	Extensive	Minor	Minor
14. Health and safety	Good	Good	Good	Good	Fair	Good	Good	Good
15. Other	Low breakage cost; rainfall and weather problems; large-scale best	Waste-intensive; labor-intensive; high breakage cost	No waste haulage; low breakage cost; large-scale best	Restrictive; used for remnant coal	Unconsolidated deposit; water required; no breakage cost; beneficiates	Unconsolidated deposit; water required; no breakage cost	Unconsolidated deposit water required; no breakage cost	Unconsolidated deposit; water required; no breakage cost

Table 14.5 Comparison of Deposit Conditions Favorable to Underground Methods

	Unsupported				Supported			Caving		
Factor	Room-and-Pillar	Stope-and-Pillar	Shrinkage	Sublevel	Cut and Fill	Stull	Square-Set	Longwall	Sublevel Caving	Block Caving
1. Ore strength	Weak to moderate	Moderate to strong	Strong (should not pack)	Moderate to strong	Moderate to strong	Fairly strong to strong	Weak to fairly weak	Any (should crush, not yield)	Moderate to fairly strong	Weak to moderate, cavable
2. Rock strength	Moderate to strong	Moderate to strong	Strong to fairly strong	Fairly strong to strong	Weak to fairly weak	Moderate	Weak to very weak	Weak to moderate, cavable	Weak to fairly strong, cavable	Weak to moderate, cavable
3. Deposit shape	Tabular	Tabular, lenticular	Tabular, lenticular	Tabular, lenticular	Tabular to irregular	Tabular to irregular	Any	Tabular	Tabular or massive	Massive or thick tabular
4. Deposit dip	Low, preferably flat	Low to moderate	Fairly steep	Fairly steep	Moderate to fairly steep	Moderate to fairly steep	Any, preferably steep	Low, preferably flat	Fairly steep	Fairly steep
5. Deposit size	Large, thin	Any, preferably large, moderately thick	Thin to moderate	Fairly thick to moderate	Thin to moderate	Thin	Any, usually small	Thin, large areal extent	Large, thick	Very large, thick
6. Ore grade	Moderate	Low to moderate	Fairly high	Moderate	Fairly high	Fairly high to high	High	Moderate	Moderate	Low
7. Ore uniformity	Fairly uniform	Variable	Uniform	Fairly uniform	Moderate, variable	Moderate, variable	Variable	Uniform	Moderate	Fairly uniform
8. Depth	Shallow to moderate	Shallow to moderate	Shallow to moderate	Moderate	Moderate to deep	Moderate	Deep	Moderate to deep	Moderate	Moderate

Table 14.6 Comparison of Advantages and Disadvantages of Underground Methods

	Unsupported				Supported			Caving		
Characteristic	Room-and-Pillar	Stope-and-Pillar	Shrinkage	Sublevel	Cut and Fill	Stull	Square-Set	Longwall	Sublevel Caving	Block Caving
1. Mining cost	20%	10%	45%	20%	55%	70%	100%	15%	15%	10%
2. Production rate	Large	Large	Moderate	Large	Moderate	Small	Small	Large	Large	Large
3. Productivity	High	High	Low	High	Moderate	Low	Low	High	Moderate	High
4. Capital investment	High	Moderate	Low	Moderate	Moderate	Low	Low	High	Moderate	High
5. Development rate	Rapid	Rapid	Rapid	Moderate	Moderate	Rapid	Slow	Moderate	Moderate	Slow
6. Depth capacity	Limited	Limited	Limited	Moderate	Moderate	Limited	Unlimited	Moderate	Moderate	Moderate
7. Selectivity	Low	High	Moderate	Low	High	High	High	Low	Low	Low
8. Recovery	Moderate	Moderate	High	Moderate	High	High	Highest	High	High	High
9. Dilution	Moderate	Low	Low	Moderate	Low	Low	Lowest	Low	Moderate	High
10. Flexibility	Moderate	High	Moderate	Low	Moderate	High	High	Low	Moderate	Low
11. Stability of openings	Moderate	High	High	High	High	Moderate	High	High	Moderate	Moderate
12. Subsidence	Moderate	Low	Low	Low	Low	Moderate	Low	High	High	High
13. Health and safety	Good	Good	Good	Good	Moderate	Moderate	Poor	Good	Good	Good
14. Other	Highly mechanized, caves with pillar recovery, good ventilation	Mechanized, fair ventilation	Gravity flow in stope, labor-intensive, simple method	Gravity flow in stope, mechanized, large blasts, good ventilation	Gravity flow in stope, mechanized, requires backfill	Gravity flow in stope labor-intensive, simple method	Gravity flow in stope, labor-intensive, high timber cost	Highly mechanized, continuous, rigid, expensive moves	Mechanized, draw control critical	Gravity flow in undercut, low breakage cost, good ventilation, draw control critical

Table 14.7 Selection Chart for All Underground Mining Methods

Deposit Shape	Deposit Orientation	Deposit Thickness	Ore Strength	Rock Strength	Applicable Method(s)
Tabular	Horizontal, flat	Thin	Strong	Strong	Room-and-pillar mining, stope-and-pillar mining
			Weak, strong	Weak	Longwall mining
		Thick	Strong	Strong	Stope-and-pillar mining
			Weak, strong	Weak	Sublevel caving
	Vertical, steep	Thin	Strong	Strong	Shrinkage stoping, sublevel stoping
			Strong	Weak	Cut-and-fill stoping, square-set stoping, stull stoping
			Weak	Strong	Square-set stoping
			Weak	Weak	Square-set stoping
		Thick	Strong	Strong	Shrinkage stoping, sublevel stoping
			Strong	Weak	Cut-and-fill stoping, sublevel caving, square-set stoping
			Weak	Strong	Sublevel caving, block caving, square-set stoping
			Weak	Weak	Sublevel caving, block caving, square-set stoping
Massive	—	—	Strong	Strong	Shrinkage stoping, sublevel stoping
			Weak	Weak, strong	Sublevel caving, block caving, square-set stoping

Sources: Modified after Peele, 1941; Lucas and Haycocks, 1973; Thomas, 1978.

14.2.3 Novel Versus Traditional Methods

Because their characteristics and applications differ significantly, it is more difficult to compare traditional methods with novel methods than surface with underground methods. By their very nature, novel methods have been devised to meet the unique requirements of a particular mineral deposit or to employ the unusual features of a new technology. In addition, they may have been utilized less than traditional methods and may not have a suitable track record for cost estimation. Accordingly, they will ordinarily be more risky than traditional mining methods.

The characteristic of the mineral deposit that is most discriminating in the selection of a novel mining method is the mineral commodity itself (see Table 13.1). Accordingly, we summarize the most likely possibilities for each of a series of deposit types:

1. *Coal.* Hydraulic mining and underground gasification are novel methods that compete with conventional underground methods like room-and-pillar and longwall mining. Both methods have been used in various places around the world, though they are not currently used in the United States. These two methods may find more acceptance in the future, but significantly increased near-term utilization is not anticipated.

2. *Coalbed methane.* As a source of natural gas, methane drainage competes more with conventional natural gas extraction than with traditional mining methods. It is currently both technologically and economically competitive and is used to make underground coal mining safer and to produce a second product to be marketed from a mining operation. It should be considered for any coal mining operation with more than 100 ft^3/ton (2.57 m^3/tonne) of methane in the seam.

3. *Hydrocarbons.* Kerogen contained in oil shale and bitumen in tar sands are natural hydrocarbons that can be converted to petroleum products following recovery. Tar sands are now being recovered in North America, and increased recovery seems possible in the future. However, both the underground retorting of oil shale and the mining and surface processing of oil shale appear to be economically infeasible. Only a large increase in oil prices will encourage oil shale utilization.

4. *Metallic and nonmetallic minerals.* Four novel methods appear to be applicable to noncoal mineral deposits: automation and robotics, rapid excavation, ocean mining, and extraterrestrial mining. The most attractive are automation and robotics, which are now coming into their own in metal mining. Also of great interest in metal mining is rapid excavation, which is limited to low- to moderate-strength rock and uniform conditions. However, it competes very well with conventional development methods, particularly for vertical openings. It should find increased use in the future. Both ocean mining and extraterrestrial mining are futuristic methods with unknown promise. Ocean mining appears more likely, particularly for the large unconsolidated deposits located in the major ocean basins. Extraterrestrial mining depends on plans to colonize the moon or other entities in space. It will be used only under those conditions because of the adverse economics of delivering materials from earth via rocket ships.

5. *All minerals.* Automation and robotics and nuclear mining are novel methods that are applicable to a wide variety of commodities. Nuclear mining, however, has no appeal at present because of its inherent hazards to human health. Automation and robotics, on the other hand, have been increasing in usage and in their level of sophistication. It is anticipated that this will result in greater use of both automation and robotics in the future, particularly for mining tasks that are hazardous or involve repetitive operations.

The use of novel methods is likely to be affected by many variables, with some methods increasing and some decreasing. The use of these methods will depend on the state of their development, the opportunities that are present, and the success of current applications. The methods that are likely to experience increased use are automation and robotics, rapid excavation, and methane drainage. Novel procedures that have promise but little usage include hydraulic mining, underground retorting, and ocean mining. These methods may find more utilization in the future. Finally, the methods that appear to be questionable in the future are nuclear mining and extraterrestrial mining.

A problem with novel methods lies in the difficulty of producing quantitative comparisons of these methods with the traditional methods. On the other hand, a new procedure may be the salvation of an operation that is not producing in an optimal manner. Accordingly, these methods should be considered as possibilities in all new mine ventures. Although they may be eliminated from consideration after careful analysis, they represent methods of improving the process of mining and reducing the costs of mineral products when they are successful.

14.3 MINING METHOD SELECTION

The final choice of a mining method should be carefully considered and the mining costs estimated before a mining company commits its monetary resources to the mine. The procedures described in the previous sections can be used to eliminate unsuitable methods and identify candidate methods for further analysis. In this section, we outline some selection methods that may be more definitive in the selection of mining methods. In doing so, we identify two procedures that are likely to isolate one method as a prime candidate or that choose one surface method and one underground method to be investigated further.

The first selection method has been designed by Hartman (1987) and is a qualitative procedure oriented toward both surface and underground mines. The procedure is outlined in Figure 14.1. For each selection, it is assumed that the analyst knows the ore strength, the rock strength, and the complete geometry of the ore body. Note that the first choice is to decide whether the mine should be located on the surface or underground. If the choice is not obvious, the logical thing to do is to choose a surface method and an underground method that best fit the ground conditions and then evaluate the two methods economically. This will be discussed in the next section. If the choice between surface and underground is obvious, then the method will result in a single choice in most cases. This choice can then be evaluated on an economic basis to allow mine management and financing agencies a review of the potential mining project so that an objective decision on the project can be made by both.

Because many engineers prefer a more quantitative selection procedure, the quest for a numerical scheme has been a logical extension to the mining method evaluation scheme. The quantitative procedure that is commonly used and quoted is that outlined in the publications by Nicholas (1981, 1992b). Not all particulars of the method are outlined here, but they can be found in the two sources cited. The essential steps in the process are as follows (Nicholas,

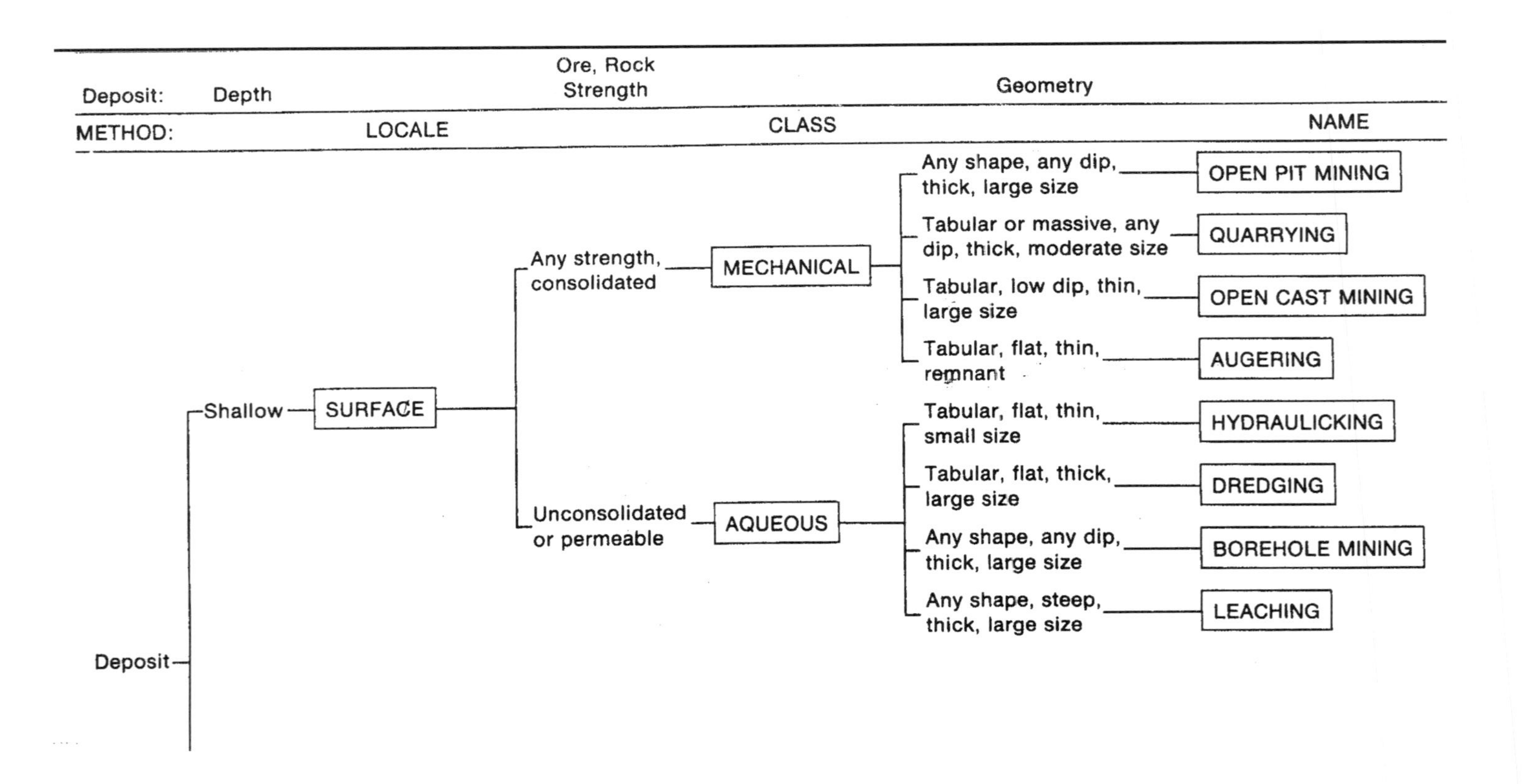
Deposit:
Depth
Ore, Rock Strength
Geometry
METHOD:
LOCALE
CLASS
NAME
Deposit
Shallow
SURFACE
Any strength, consolidated
MECHANICAL
Any shape, any dip, thick, large size
OPEN PIT MINING
Tabular or massive, any dip, thick, moderate size
QUARRYING
Tabular, low dip, thin, large size
OPEN CAST MINING
Tabular, flat, thin, remnant
AUGERING
Unconsolidated or permeable
AQUEOUS
Tabular, flat, thin, small size
HYDRAULICKING
Tabular, flat, thick, large size
DREDGING
Any shape, any dip, thick, large size
BOREHOLE MINING
Any shape, steep, thick, large size
LEACHING

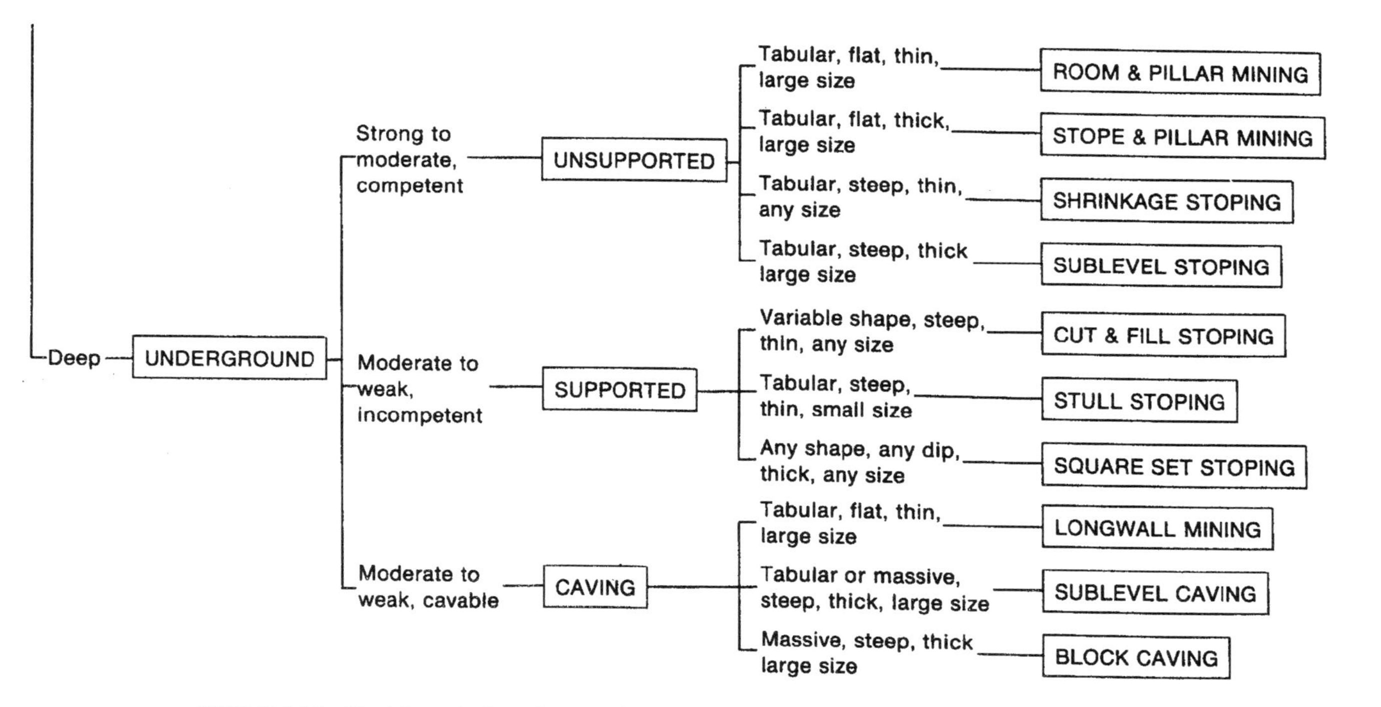

FIGURE 14.1. Chart for selection of appropriate mining method, based on deposit characteristics.

1981). In Stage 1, the mining company identifies candidate deposits in general terms, such as the rough target size, grade, and geotechnical properties. This is generally in keeping with the company's corporate philosophy and its long-term development plan. In Stage 2, mining and capital costs are developed, as well as production rate, recovery, labor supply, and environmental factors. Finally, in Stage 3, the most likely candidate methods are costed out, cutoff grades and reserves are calculated, and economic comparisons are completed to determine the most feasible mining method and overall plan of exploitation.

In this section, we outline the methodology of Stage 1. To initiate Stage 1 activity, three or four factors each are selected that are descriptive of (1) deposit geometry and grade distribution (shape, thickness, dip, depth and grade uniformity) and (2) geotechnical properties (ore strength, rock strength, fracture spacing, and fracture shear strength). Each mining method is then rated numerically as to its suitability for each category of factors. Nicholas (1981) uses four ranks: preferred, probable, unlikely, and unsuitable (eliminated). Numerical values are assigned to each rank, for example:

Preferred	3–4
Probable	1–2
Unlikely	0
Unsuitable	−49

The values used are designed to obtain answers that are definitive. For example, the value of 0 neither adds to nor subtracts from the chance of using a method, and −49 ensures elimination of a method even though other factors are rated high.

Using these values, a table of rankings is prepared for all the mining methods under consideration. Any with negative values are dropped, and the remaining are rank ordered to reveal the most attractive. The analysis is then completed in Stages 2 and 3, resulting in the quantitative selection of the most suitable method.

The following numerical example is derived from Nicholas (1981), in which ten mining methods are ranked for suitability with a given deposit. The deposit has the following characteristics:

Deposit type: tabular, flat, very thick

Depth: 425 ft (130 m)

Grade distribution: uniform

Ore characteristics:

- Moderate strength
- Close fracture spacing
- Moderate fracture shear strength

Hanging wall characteristics:

- Strong strength
- Wide fracture spacing
- Moderate fracture shear strength

Footwall characteristics:

Moderate strength
Close fracture spacing
Weak fracture shear strength

To evaluate the mining methods for suitability to various deposit types, the suitability scores outlined in Table 14.8 are used. Rankings are also assigned for a range of geotechnical conditions. For purposes of keeping the discussion brief, these are not shown here. The rankings are then gathered by category in Table 14.9 for the deposit described earlier. The methods are then listed in rank order of suitability in Table 14.10. (*Note to students:* Top slicing is a high-cost method similar to sublevel caving that is seldom used today.) Two or three of the top-ranked methods can be evaluated further using technological, economic, and environmental considerations. Assuming that open pit and block caving were to be evaluated further, the decision-making would move on to the next stage, which would involve completing detailed cost estimates for the two methods.

The beauty of a quantitative selection procedure such as this is that it lends itself to rapid, objective evaluation of a few factors or equally to computer analysis of myriad complex factors in a more precise study of the mining methods. Having narrowed the choices to two or three methods, each is now examined in Stage 2 for technological, economic, and environmental considerations. The final selection will be made only after definitive cost estimates are made for each possible method during a detailed examination in Stage 3.

14.4 MINING COSTS

14.4.1 Mining Method Costs

In estimating the costs of mining methods, it is worthwhile to establish basic definitions of the types of estimates used in industry. The American Association of Cost Engineers has established the following cost estimation terminology (Humphreys and Katell, 1981):

Type of Estimate	Accuracy
Order-of-magnitude estimate	−30% to +50%
Preliminary estimate	−15% to +30%
Definitive estimate	−5% to +15%

We will refer to these definitions throughout this section.

To this point, we have dealt only with relative costs in evaluating the economics of mining methods. Table 4.1 provides a listing of relative mining costs that can be used for comparison purposes. Those costs represent average

Table 14.8 Suitability Ratings of Various Mining Methods for Deposit Geometry and Grade Factors

Mining Method	General Shape[a]			Ore Thickness[b]				Ore Dip[c]			Grade Distribution[d]		
	M	T/P	I	N	I	T	VT	F	I	S	U	G	E
Open pit	3	2	3	2	3	4	4	3	3	4	3	3	3
Block caving	4	2	0	−49	0	2	4	3	2	4	4	2	0
Sublevel stoping	2	2	1	1	2	4	3	2	1	4	3	3	1
Sublevel caving	3	4	1	−49	0	4	4	1	1	4	4	2	0
Longwall	−49	4	−49	4	0	−49	−49	4	0	−49	4	2	0
Stope-and-pillar	0	4	2	4	2	−49	−49	4	1	0	3	3	3
Shrinkage stoping	2	2	1	1	2	4	3	2	1	4	3	2	1
Cut-and-fill	0	4	2	4	4	0	0	0	3	4	3	3	3
Top slicing	3	3	0	−49	0	3	4	4	1	2	4	2	0
Square-set	0	2	4	4	4	1	1	2	3	3	3	3	3

[a] M = massive, T/P = tabular or platy, I = irregular.
[b] N = narrow, I = intermediate, T = thick, VT = very thick.
[c] F = flat, I = intermediate, S = steep.
[d] U = uniform, G = gradational, E = erratic.

Source: Nicholas, 1981. (By permission from Society for Mining, Metallurgy, and Exploration, Inc., Littleton, CO).

Table 14.9 Numerical Ratings of Various Mining Methods for a Mineral Deposit with Specified Characteristics

Mining Method	Geometry/Grade Distribution	Geomechanics Characteristics				Grand Total
		Ore	Hanging Wall	Footwall	Subtotal	
Open pit	12	9	11	8	28	40
Block caving	13	8	6	7	21	34
Sublevel stoping	10	5	7	2	14	24
Sublevel caving	13	7	6	3	16	29
Longwall	−37	8	5	6	19	−18
Stope-and-pillar	−38	7	8	3	18	−20
Shrinkage stoping	10	6	6	8	20	30
Cut-and-fill	7	8	7	10	25	32
Top slicing	15	6	6	7	19	34
Square-set	8	8	7	10	25	33

Source: Nicholas, 1981. (By permission from the Society for Mining, Metallurgy, and Exploration, Inc., Littleton, CO).

relative costs, with the 100% value being somewhere around \$100/ton (\$91/tonne). However, the costs will vary considerably within each method depending on the ore deposit conditions, the equipment used, and the environmental measures utilized (Mutmansky et al., 1992). The costs listed can be considered only order-of-magnitude estimates at best because of the way in which they were estimated. It will therefore be necessary to provide better estimates of

Table 14.10 Rank Order and Rating of Mining Methods Compiled in Table 14.8 for a Specified Mineral Deposit

Rank	Method	Rating
1	Open pit mining	40
2	Block caving	34
3	Top slicing	34
4	Square-set stoping	33
5	Cut-and-fill stoping	32
6	Shrinkage stoping	30
7	Sublevel caving	29
8	Sublevel stoping	24
9	Longwall mining	−18
10	Stope-and-pillar mining	−20

Source: Modified after Nicholas, 1981. (By Permission from the Society for Mining, Metallurgy, and Exploration, Inc., Littleton, CO.)

TABLE 14.11 Open Pits Costs as a Function of Production and Stripping Ratio

Daily Ore Production tonnes/day (tons/day)	Operating Costs, $/tonne ($/ton) Stripping Ratio, tonnes/tonne (tons/ton)			
	1:1	2:1	4:1	8:1
1,000 (1,103)	8.88 (8.06)	11.54 (10.48)	13.54 (12.29)	19.82 (17.99)
5,000 (5,513)	3.64 (3.30)	4.81 (3.70)	6.73 (6.11)	10.70 (9.71)
20,000 (22,050)	2.38 (2.16)	3.43 (3.11)	5.10 (4.64)	8.81 (8.00)
80,000 (88,200)	1.71 (1.55)	2.60 (2.36)	4.44 (4.03)	7.80 (7.08)
	Capital Costs ($)			
1,000 (1,103)	4,395,200	5,420,800	8,164,400	13,648,400
5,000 (5,513)	12,783,100	15,551,100	23,365,100	40,755,600
20,000 (22,050)	35,222,400	52,910,300	83,441,300	141,713,300
80,000 (88,200)	114,321,100	179,239,000	324,196,300	616,154,600

Source: Western Mine Engineering, 1998. By permission from Western Mine Engineering, Inc., Spokane, WA.

capital and operating costs for the potential mine. Additional information on mining costs can be found in publications by Hoskins (1982, 1986), U.S. Bureau of Mines Staff (1987a, 1987b), O'Hara and Suboleski (1992), Mutmansky et al. (1992), Western Mine Engineering (1998, 1999), and Stebbins and Schumacher (2001). The information found in Western Mine Engineering (1998) is of particular importance. The cost data found therein are updated yearly and include mining methods, equipment, power, and supply costs. Also included are estimated costs for various mining methods as a function of the production level and type of access. For example, Table 14.11 shows some of the costs for open pit mining as a function of the production level and stripping ratio. Note that the data illustrate that the mining method has a wide range of costs.

The underground mining costs outlined by Western Mine Engineering (1998) are limited to a few typical metal mining methods. A cost summary for these methods is provided in Table 14.12. Note that the cost data in the table indicate that the costs are again a function of the production level. Primarily using the information presented by Mutmansky et al. (1992) and Western Mine Engineering (1998), the average and range of costs for the mining methods studied are compiled in Table 14.13.

The range of costs presented in Tables 14.11 and 14.12 indicates the obvious need for a detailed cost estimate for any mining project that is being seriously considered for development. The costs outlined in Table 4.1 and in the tables in this chapter represent only order-of-magnitude costs for the mining methods, and definitive cost estimates must be developed for each of the

TABLE 14.12 Underground Mine Cost Comparison as a Function of Entry Type and Production Level

Mining System	Production Rate in tonnes per day (tons/day)		
Stoping Method and Primary Method of Access	Operating and Capital Costs Upper Numbers are Total Operating Costs in $/tonne Ore Lower Numbers are Total Capital Costs		
Cut-and Fill	*200 (220)*	*1000 (1103)*	*2000 (2205)*
Adit entry	$70.58	$41.83	$33.76
	$7,643,210	$13,675,850	$20,912,890
Shaft entry	$74.56	$43.79	$35.15
	$11,257,030	$19,396,850	$28,980,190
Shrinkage	*200 (220)*	*1000 (1103)*	*2000 (2205)*
Adit entry	$60.11	$30.34	$29.12
	$7,609,530	$13,388,520	$22,075,700
Shaft entry	$63.32	$32.62	$31.00
	$11,436,330	$18,923,720	$30,945,600
End Slice	*800 (882)*	*2000 (2205)*	*4000 (4410)*
Adit entry	$20.94	$15.82	$13.31
	$12,236,660	$19,956,240	$32,163,400
Shaft entry	$22.95	$17.29	$14.82
	$16,674,150	$27,218,540	$50,953,400
Vertical Crater Retreat	*800 (882)*	*2000 (2205)*	*4000 (4410)*
Adit entry	$30.97	$23.55	$21.63
	$16,393,200	$29,854,400	$50,780,800
Shaft entry	$33.33	$27.32	$22.76
	$22,063,600	$39,146,800	$77,869,800
Sublevel Longhole	*800 (882)*	*4000 (4410)*	*8000 (8820)*
Adit entry	$19.94	$11.61	$10.82
	$11,342,730	$27,454,290	$43,986,280
Shaft entry	$21.69	$12.81	$11.50
	$14,809,560	$38,411,630	$65,450,980
Room-and-Pillar	*1200 (1323)*	*8000 (8820)*	*14,000 (15,435)*
Adit entry	$21.01	$11.02	$8.64
	$19,870,370	$61,777,300	$78,351,700
Shaft entry	$23.33	$12.97	$10.21
	$24,979,400	$76,399,100	$91,624,400

Note: Costs in $/ton can be obtained by multiplying the cost in $/tonne by 0.9078.

Source: Western Mine Engineering, 1998. By permission from Western Mine Engineering, Inc., Spokane, WA.

TABLE 14.13 Estimated Overall Mining Costs for Traditional Mining Methods, on Relative and Absolute Bases[a]

Mining Method	Average Relative Cost,[b] Percent	Range of Absolute Mining Cost,[c]	
		$/ton	($/tonne)
Surface			
Open pit mining	5	2–20	(2–22)
Quarrying	100	25–150	(28–165)
Open cast mining	10	4–20	(4–22)
Hydraulicking	5	2–10	(2–11)
Dredging	<5	1–5	(1–6)
Borehole mining	5	2–10	(2–11)
Leaching	10%	4–20	(4–22)
Underground			
Room-and-pillar mining	20	10–25	(11–28)
Stope-and-pillar mining	10	5–15	(6–17)
Shrinkage stoping	45	30–70	(33–77)
Sublevel stoping	20	12–35	(13–39)
Cut-and-fill stoping	55	30–70	(33–77)
Stull stoping	70	20–65	(22–72)
Square-set stoping	100	50–150	(55–165)
Longwall mining	15	10–20	(11–22)
Sublevel caving	15	10–30	(11–33)
Block caving	10	5–15	(6–17)

[a]Mining costs include prospecting, exploration, development, and exploitation, but exclude processing, transportation, taxes, royalties, etc.
[b]Values taken from Table 4.1.
[c]Values based on run-of-mine (raw) material.

methods to be used. The cost estimates should take into account the production rate, geotechnical conditions, stripping ratio (if applicable), and the amount of development necessary to open and sustain the mine. To this end, the mining system must be accurately defined so that the labor, equipment, stope dimensions, production rate, and supply requirements can all be used in the cost estimation process.

The most reliable cost estimates will normally entail a detailed study of the total labor, equipment, and supplies required by the proposed mining system. To accomplish this task, the planner must outline the entire labor force required to operate the mine, the total capital investment in the mine, and the cost of supplies. These estimates are then used to determine the cost per ton (tonne) of ore extracted from the mine. Because development costs are a function of the depth of an underground mine, these costs must be specific to the actual layout of the mine. Accordingly, the mine layout must be completed before the mine costs can be accurately estimated. This will enable the analyst to gather the costs of shafts, slopes, drifts, crosscuts, and any other required openings that must be developed. The analyst must also have reliable data on production tonnage per worker and supply cost per ton to accurately complete the cost estimation for the particular method. This will often require that the analyst have data from similar mines to adequately estimate the costs associated with the proposed operation.

The capital and production costs are illustrated here by summarizing a cost determination for a shaft-access vertical crater retreat (VCR) stoping operation (Western Mine Engineering, 1998) that is producing 2200 tons (2000 tonnes) of ore per day from stopes located in a vein that averages 33 ft (10 m) in width and extends 2200 ft (675 m) along the strike. The production is to be conducted using down-the-hole drills in the stope with sand filling for support after the stope is completed. The mine openings are assumed to be 2700 ft (823 m) deep and are served by two shafts that are about 244 ft^2 (22.6 m^2) in cross section. Preproduction development consists of 3230 ft (984 m) of drifts and 1550 ft (472 m) of crosscuts, each about 150 ft^2 (13.9 m^2) in cross section; 1615 ft (492 m) of ore pass openings, each 27 ft^2 (2.5 m^2) in area; and 2699 ft (792 m) of ventilation raises, each 57 ft^2 (5.3 m^2) in cross section. Total costs of development openings are listed on page 511.

To provide personnel to the mine, the following numbers of hourly workers are assumed:

Stope miners	24
Development miners	14
Equipment operators	2
Hoist operators	4
Support miners	2
Diamond drillers	2
Backfill plant workers	4
Electricians	6
Mechanics	15
Maintenance workers	6
Helpers	7
Underground laborers	8
Surface laborers	6
Total hourly personnel	100

Note that the number of support personnel for the stope miners is relatively large. The cost estimation process will not be accurate unless this number is realistic, making the estimate of support workers an important element in the cost estimation.

The salaried personnel have been similarly estimated, as follows:

Mine manager	1
Superintendents	2
Forepersons	4
Engineers	3
Geologists	3
Shift bosses	8
Technicians	6
Accountants	2
Purchasing	4
Personnel	5
Secretaries	6
Clerks	8
Total salaried personnel	52

The salaried personnel are also significant in number, being approximately half of the total number of hourly personnel. Estimation of this personnel pool is therefore very important as well.

The cost of supplies will not be detailed here. However, the supply requirements include blasting materials, drilling supplies, pipes for air and water, ventilation tubing, rock bolts, cement, and timber. Western Mine Engineering (1998) estimated these costs to total \$6.05/ton (\$6.67/tonne). The equipment requirements are significant, with down-the-hole drills, diesel load-haul-dump (LHD) devices, and development drills requiring most of the investment. The equipment list is as follows:

Down-the-hole drills	8
Stope LHDs	7
Horizontal development drills	4
Development LHDs	3
Raise borers	1
Production hoists	1
Rock bolt drills	1
Drain pumps	9
Freshwater pumps	2
Backfill mixers	1
Backfill pumps	2
Service vehicles	10
Compressors	1
Ventilation fans	1
Exploration drills	1

Without going into all the calculations, the operating costs in $/ton are given by Western Mine Engineering (1998) as follows:

Equipment operation	$2.44
Supplies	6.05
Hourly labor	9.47
Administration	4.57
Sundries	2.24
Total operating costs	$24.77

It is also necessary to determine the capital cost of putting the mine into production. Without performing the calculations, the resulting figures are as follows:

Equipment purchase	$15,290,000
Development openings	9,862,800
Surface facilities	3,832,000
Working capital	2,914,000
Engineering and management	4,349,000
Contingency funds	2,899,000
Total capital costs	$39,146,800

This total results in the capital cost requirement of approximately $17,800 per ton of daily capacity. Most of the capital costs are incurred only once during the life of the mine, but equipment costs will be incurred periodically during its lifetime if the mine lasts longer than the lifetimes of the equipment.

The estimation of the costs for this example mine provides only an outline of the estimation process. The entire process is revealed in Western Mine Engineering (1998). This exercise in cost estimation is dependent on the accurate portrayal of labor, production, development, and supply costs. Useful indices for these costs can often be obtained from a number of sources in the literature, but the best data are obtained from similar operations. The pitfall here is that the information is likely to be somewhat inaccurate as applied to a proposed project because of dissimilar geologic conditions. Insightful interpretation of the match between deposit conditions and the mining method is therefore necessary in the use of any data for cost estimation purposes.

Note that the preceding cost estimate can be considered a definitive estimate if the conditions assumed are based on measurements made in an actual deposit and similar operations have been studied to accurately determine the number of support personnel, the production and productivity values, and the supply requirements of the proposed mining operation. Otherwise, the values determined can be considered only preliminary cost estimates.

14.4.2 Budgeting and Cost Control

The cost estimation for the hypothetical mine leads us to the topic of budgeting and cost control. Cost categories that must be controlled can be determined from knowledge of the mining system. Blasting, drilling, haulage, hoisting, and backfilling costs are just some of the categories that should be monitored and controlled. The management of the mine will normally have accounting personnel develop a reporting system for all the important activities to periodically ensure that the mining process is conducted with proper concern for costs and cost escalations.

Table 14.14 compiles the elements of a typical budget for an underground coal mine using the longwall system. Customary cost categories, broken down into direct and indirect mining and other production costs, are tabulated for as many cost centers as desired if the costs can be isolated. In this case there are four: mine development, mine exploitation, underground auxiliary, and surface auxiliary activities. The equipment for each center is normally identified on a capital and operating basis. Then other items of operating expense, categorized in the first column, are costed for each center, as applicable. An entire mine budget would itemize all these costs for each center in a computerized tabulation, usually on both a total and unit cost basis, for the period under consideration (month, year, long range, etc.). Thus, actual costs can be compared with budgeted amounts to provide for effective control of costs. For a more comprehensive treatment of mine budgeting, see Halls (1982) and Gentry and O'Neil (1984).

14.5 SURFACE VERSUS UNDERGROUND

The choice between a surface or an underground mining method is another important decision in many mine development scenarios. In some cases, the deposit is so shallow that only a surface method need be considered. In others, the deposit is so deep that only an underground method would be chosen. It is, however, important to consider the choice that can or should be made for a deposit that is of modest depth, amenable to either surface or underground mining methods. In this case, the mine may be developed as a surface mine, an underground mine, or a mine that is initiated as a surface mine and continued as an underground mine. This is an important consideration in many operations.

Normally, we would assume that underground mining costs exceed surface mining costs for any deposit close to the surface. Hedberg (1981) cites reasons for the cost-effectiveness of surface mining, based on hypothetical deposits in the hard-rock industry: larger equipment, lower capital intensity, simpler development, higher energy efficiency, less expensive auxiliary operations, and better health and safety factors. His analysis of operating costs, capital costs, and overall costs is summarized in Figure 14.2. The analysis of operating costs

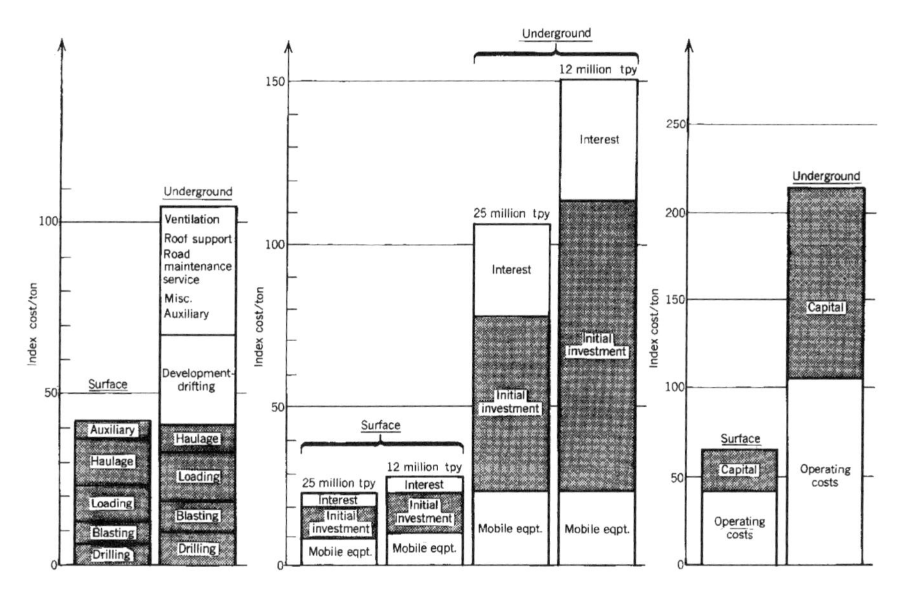

FIGURE 14.2. Unit cost comparisons on relative basis for surface and underground mining; hypothetical but similar conditions. (*Left*) Operating costs. (*Center*) Capital costs. (*Right*) Overall costs. Note: scales differ. (Modified after Hedberg, 1981. By permission from *Mining Magazine*, London.)

was performed on a unit operations basis. The comparison shows that material and supply costs for underground mines are 50% more than those of surface mining, and that labor costs are five times higher.

Hedberg's (1981) analysis of capital costs also shows higher costs for underground mines. In Figure 14.2, two production rates were chosen and the capital costs estimated for both. In each case, the unit capital cost was about five times higher for underground than for surface mining. The overall cost of underground mining was found to be about four times higher than that of surface mining, thus showing the advantage of surface operations in many ore deposits.

Although it can be shown that surface mining has many advantages over underground mining for relatively shallow deposits, those advantages disappear quickly at depth. As the overburden thickness increases, the cost of removing the overburden can go up exponentially with the depth. A depth is soon reached at which the cost of underground mining is lower than that of surface mining. However, Nilsson (1992) indicates with his examples that the optimum strategy of surface versus underground mining is different for various forms of the deposit. A study of his examples leads to the following conclusions:

1. In a buried horizontal deposit, the deposit is normally optimally mined with either a surface method or an underground method, but not both.

TABLE 14.14 Mine Budget Format for Hypothetical Underground Coal Mine

	Cost-Profit Centers: Cost Items			
Cost Category	Mining Unit A: Development (Entry Driving)	Mining Unit B: Exploitation (Longwall)	Underground Auxiliary Operations	Surface Auxiliary Operations
Direct Ownership Cost	*Equipment*	*Equipment*	*Equipment*	*Equipment*
Purchase depreciation	Continuous miner	Shearer	Locomotives	Shaft hoist
Interest charge	Shuttle cars	Shields	Mine cars	Slope belt
Insurance, tax, etc.	Roof drill	Jacks	Jeeps, cars	Bins, hoppers
Equipment lease	Scoop tram	Panline	Main conveyor	Feeders
Equipment rental	Feeder	Scoop tram	Roof drill	Trucks, cars
Total ownership	Section conveyer	Feeder	Scoop trams	Main fan
		Section conveyer	Cages	Monitoring system
			Face fans	Buildings
			Pumps	Other structures
				Waste disposal
Direct/Operating Cost	*Consumable Supplies*	*Consumable Supplies*	*Consumable Supplies*	*Consumable Supplies*
Repairs, maintenance	Roof bolts	Arches	Track	Plant supplies
Consumable supplies	Timber	Timber	Conveyor belt	Office supplies
Fuel, power	Rock dust	Rock dust	Timber, trusses	Safety supplies, clothing
Lubrication, hydraulics	Brattice cloth, tubing	Brattice cloth, tubing	Roof bolts	Training materials
Labor, benefits	Miner bits	Shearer bits	Concrete blocks	Miscellaneous
Special	Drill bits	Water sprays	Pipe	
Total operating	Hand tools	Hand tools	Rock dust	
Total exploitation	Miscellaneous	Miscellaneous	Hand tools	
			Miscellaneous	

Other Mining Costs	*Fuel, Power*	*Fuel, Power*	*Fuel, Power*	*Fuel, Power*
Prospecting				
Exploration	Diesel fuel	Electricity	Electricity	Gasoline, diesel fuel
Development	Electricity	Water	Water	Electricity
(premining)	Water		Compressed Air	Natural gas
Total other				Water
				Telephone
Indirect Cost	*Special*	*Special*	*Special*	*Special*
Administration	Freight	Freight	Freight	Freight
Clerical support	Travel	Travel	Travel	Travel
Engineering	Entertainment	Entertainment	Entertainment	Entertainment
Computing				
Consultants				
Total indirect				
Total mining				
Other Production Costs				
Reclamation (postmining)				
Processing				
Transportation				
Union welfare				
Royalties				
Taxes				
Total other				
Grand total expense				

2. For a steeply dipping vein or massive deposit that outcrops on the surface and extends to depth, the optimum strategy is often to mine first using surface methods, then switching to an underground method.
3. The point at which surface mining should be switched to underground mining is normally achieved when the surface mining cost reaches the underground mining cost, if ore production rates do not change at that point.
4. If ore production changes upon the switch from surface to underground mining, the switch point must be obtained by maximizing the net present value (NPV) of the profit on the deposit.

(*Note to students:* The net present value, often abbreviated NPV, is a term representing the value of all income amounts, each discounted back to the present, minus the value of all annual costs, each of which is also discounted back to the present. The NPV thus tells the current value of all the potential costs and benefits accruing from an investment.)

It is logical that the decision to choose a surface or underground mining method for a buried horizontal deposit will normally be made by a simple cost comparison between the surface method and the underground method. However, for a steeply dipping deposit that extends to depth, the analysis of the strategy involves both surface and underground mining costs as well as a thorough analysis of the optimal point of switching from surface to underground mining.

Social and environmental considerations may also affect the decision to choose an underground mine in preference to a surface operation. With environmental problems causing higher costs for reclamation and public relations problems as well, many mining companies are now choosing underground operations in coal, aggregates, and certain other minerals to reduce the public outcry and gain acceptance for their mineral operations. In this case, the reduced costs of surface mining may not be sufficient to overcome the social costs of conducting the operation on the surface. This appears to be a trend that will increase in the future.

14.6 MINE INVESTMENT ANALYSIS

When the mining method and production plan have finally been determined by the mine planning personnel, the next step will be to provide an investment analysis that accurately portrays the costs and revenues expected. Gentry and O'Neil (1984) employ the term *mine investment analysis* for the myriad financial tasks to be addressed in project evaluation. In this case, we will use the term to mean the final financial report prepared to sell the project to the mining company officers, stockholders, and lending agencies. It is crucial at this point that the planning staff present neither an optimistic nor pessimistic picture of the project, but one that can be considered to be a definitive estimate of the costs and profits expected from the mine. The report will be carefully evaluated by the company officers with due regard for analysis of changing markets, escalation in labor and supply costs, possibility of competitor entry into the market, taxation and regulatory problems, and any other imponderables that may affect the success of the operation. The report should thus address the risk factors and how the mine will be able to cope with each change in the current conditions. The following example will suffice to indicate how differences in the price of gold can affect the NPV of an underground gold operation.

Consider an underground gold mine, characterized as a high-grade, low-tonnage deposit (Glanville, 1984). Production rates of 100 to 1000 tons/day (90 to 900 tonnes/day), comparable to mine lives of 15 to 1.5 years, respectively, are under consideration. This problem will be subjected to mine investment analysis to determine the mine's net present value as a function of the production rate and market price for gold. After an estimate of costs and profits is performed, a "sensitivity graph" of net present value versus production rate is prepared for several gold prices. This graph is shown as Figure 14.3. For this graph, the optimal production rate is about 600 tons/day, equivalent to a mine life of 2.5 years, regardless of the market price of the gold. Thus, the outcome of the mine is not sensitive to the production rate. However, the NPV of the mine is clearly sensitive to the gold price, and a positive NPV will not

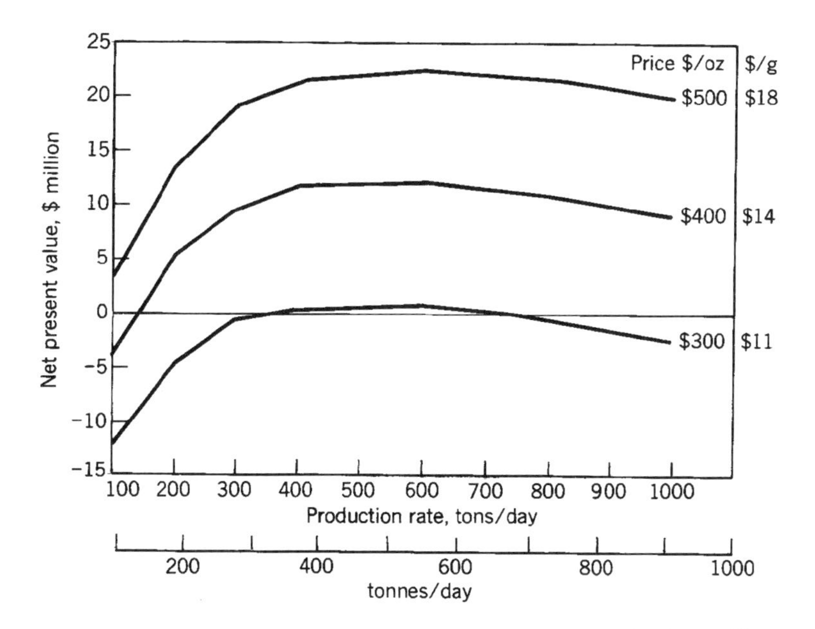

FIGURE 14.3. Relationship of net present value to production rate for three different gold prices. (After Glanville, 1984. By permission from the Society for Mining, Metallurgy, and Exploration, Inc., Littleton, CO.)

be obtained if the market price is not above $300/oz. Thus, the outcome of the mine is sensitive to the price of gold, and this variable is one that can make or break the mine.

When a mine is started, the first profit may be many years away, and the final payback of all the invested funds may take a much longer time. The White Pine Copper Mine serves as an actual case history of mine investment analysis (Boyd, 1967). Table 14.15 summarizes the annual cash flows (cash investment, cash returns, and net returns) over the 36-year history of the mine. It was not until 1966—14 years after development commenced and 11 years after exploitation began—that the mine's net return was positive. Although revenue was obtained in 1955 and an operating profit was achieved in 1956, the original investments were not paid back until 1966, a long wait for a positive net return.

When the mining company officers have been sold on the viability of the mining plan, the next step may be to seek loans to enable the mine to begin production. The mine investment analysis may therefore undergo another review by financial institutions that are interested in providing funds to initiate mining. The lending company analysts will look over the mining plan to determine how well prepared the company is to ensure that the mine will make money and to be capable of paying back the loan. The lending agency may also suggest improvements or backup plans to help the company in the successful operation of the mine.

Mine investment analysis has surfaced here in only a few paragraphs, but it is an extremely important topic for every mining analyst who is involved in planning for the operation of a mine. The selection of a mining method and the associated mining plan cannot be adequately addressed without a thorough knowledge of this field. Accordingly, the student should become aware of the concepts of mine investment analysis by reviewing the material in Gentry and O'Neil (1984).

14.7 SUMMARY

14.7.1 Review

The 14 chapters of this book are arranged in eight broad parts:

	Topic	Chapter(s)
I.	Introduction	1
II.	Mining Consequences	2
III.	Stages of Mining	3, 4
IV.	Unit Operations	5
V.	Surface Mining	6, 7, 8
VI.	Underground Mining	9, 10, 11, 12
VII.	Novel Methods	13
VIII.	Mining Methods Selection	14

We now review and summarize the logic of each section.

TABLE 14.15 Cash Investments and Returns for White Pine Copper Mine, 1929–1965

	Cash Flow, $ million		
	Cash Investment ($)	Cash Return from Operations ($)	Net Annual Return ($)
Property purchased			
1929	$0.1		($0.1)
Exploration period			
1930/43	0.1		(0.1)
1944/45	0.1		(0.1)
1946	0.4		(0.4)
1947	0.4		(0.4)
1948	0.1		(0.1)
1949	0.1		(0.1)
1950	0.2		(0.2)
1951	0.1		(0.1)
Development period			
1952	9.6		(9.6)
1953	31.6		(31.6)
1954	18.9		(18.9)
Exploitation started			
1955	12.5	9.7	(2.8)
1956	1.8	11.9	10.1
1957	1.7	4.2	2.5
1958	0.7	5.2	4.5
1959	1.6	5.2	3.6
1960	2.2	3.6	1.4
Operation reached maturity			
1961	1.5	6.0	4.5
1962	2.8	7.5	4.7
1963	2.1	8.3	6.2
1964	2.0	9.4	7.4
1965	2.6	13.5	10.9
Total:	$93.2	$84.5	($8.7)

Source: Boyd, 1967. (By permission from the Society for Mining, Metallurgy, and Exploration, Inc., Littleton, CO.)

14.7.1.1 Introduction. Chapter 1 began by introducing some basic concepts, beginning with mining's contribution to our modern civilization. Mining terminology was introduced; a short history of mining technology followed. The five major stages in the life of a mine — prospecting, exploration, development, exploitation, and reclamation — were then outlined. Unit operations of mining and economics of the minerals industry were the topics discussed next. Finally, computer technology and applications were mentioned to emphasize the importance of this field in modern engineering practice.

14.7.1.2 Consequences of Mining. Chapter 2 is a new chapter in the book; it is meant to gather and formalize all the social and environmental concerns related to the mining process. Regulation of mining is studied first. The important mining laws are summarized, and their importance in mining practice is assessed. Health and safety issues are addressed in the next section, with emphasis on the improvement that has been achieved in these areas over the last century. Finally, environmental concerns are addressed, particularly with regard to the reclamation of mines.

14.7.1.3 Stages of Mining. The stages of mining were introduced, with two — prospecting and exploration — being precursors to the actual mining process. Prospecting is the search for potentially valuable mineral deposits, and exploration is the evaluation of any deposits that are located. Their objectives are to narrow the search, provide valuable information on the deposits, and furnish the mining engineer with suitable deposits for mining. Both direct and indirect methods are employed; the principal tools are geology, geophysics, and geochemistry. Drilling or excavation is normally the final step in this process, as it is the only way of providing proof of existence of a viable ore deposit.

Mining proper commences with the development stage of mining. The cardinal rule during development is to conduct both development and exploitation so that the total profit is maximized for each new mine project. In opening up a mine, three major groups of factors govern development. As many as 11 sequential steps, sometimes scheduled by a CPM or PERT program, may be required in developing the mine. Exploitation is the production stage of mining and commences near the completion of development. In selecting a mining method, we choose one that best matches the unique characteristics of the deposit, within limits imposed by safety, technology, and economics, to return the greatest profit. Mining methods have been outlined here with some ideas on the costs for the five stages of mining.

14.7.1.4 Unit Operations. The rest of the book is devoted to engineering aspects of mine development and exploitation. We deal first with certain fundamental operations that are performed to free and transport the mineral from the deposit. They are called the unit operations of mining and include both production and auxiliary processes. In the production cycle, the common unit operations are drilling, blasting, loading, and hauling. These mining steps are supported by auxiliary operations such as health and safety, environmental control, ground control, and various other support activities. Because mining is heavily mechanized today (essentially 100% in the United States), unit operations are closely associated with equipment. The basic equipment types are discussed, as well as the formation of cycles and systems, for both production and auxiliary unit operations.

14.7.1.5 Surface Mining. Employing our classification of mining methods, we discuss those used at the surface. Development of surface mining is directed especially at three targets: advanced stripping, waste disposal, and land reclamation. Pit planning and design — partly because of the immensity and scale of some of the operations — is crucial to the success of a surface mine. It is predicated on several objectives and broken down into short-range and long-range planning. In both phases, the calculation of stripping ratios and the location of pit limits are required.

Two classes of surface methods are employed in mining: mechanical extraction and aqueous extraction. The former is by far the more prevalent (constituting about 90% of U.S. surface production), the latter being limited to applications where water can be used for exploitation. Open pit mining and open cast mining, both mechanical, are the most commonly used of all mining methods. Quarrying (dimension stone variety) is employed to extract intact blocks of decorative or architectural stone. Of the aqueous methods, placer mining (dredging and hydraulicking) is applied to near-surface deposits of unconsolidated minerals, and solution mining (well methods and leaching) is applied to either surface or deeper deposits that can be recovered by dissolving, melting, or slurrying minerals.

In importance, surface mining clearly ranks ahead of underground mining, if we compare tonnage of current annual production (about 85% vs. 15%) or value. In spite of its many attractions, however, there are some serious limitations to surface mining, not the least of which are depth, selectivity and flexibility, and environmental concerns.

14.7.1.6 Underground Mining. Mine development in the underground locale is more specialized, extensive, and expensive than on the surface. Development openings are classified (by rank order of importance) as primary or main, level or zone, and lateral or panel. Primary access is provided by a shaft, slope (decline or incline), or adit. Design factors to be taken into account in mine development are the type of mining method, production rate, mine life, and interval between levels. The physical plant required to conduct subsurface mining has three components: surface, shaft, and underground. A unique plant is the hoist plant, a major task of engineering design.

We have outlined three classes of underground mining methods, differentiated by the nature and extent of ground support provided. Unsupported methods are essentially self-supporting, relying on pillars and localized means of support (roof bolts, timber, light sets, etc.) to maintain the openings.

Room-and-pillar and stope-and-pillar mining, two of the unsupported methods, are responsible for much of the production from underground mining. They are widely used in coal, metal, and nonmetallic mining. The supported class of methods consists of cut-and-fill stoping, stull stoping, and square-set mining. These methods require substantial amounts of support (such as waste rock or hydraulic backfill material) to supplement the localized means of support. Only cut-and-fill finds any substantial use today; the other methods are often too expensive for ordinary utilization. The third underground mining class is caving, consisting of longwall mining, sublevel caving, and block caving. In these methods, controlled caving of the ore body is conducted in a systematic extraction process. These methods are finding increased usage in coal and metal mining today because they are more cost-effective than many of the other underground metal mining methods.

Underground mining is gaining in usage in some of the mineral commodities, such as coal in the eastern states and certain nonmetallic commodities. In part, this is because the easy-to-exploit surface deposits have largely been mined out. However, the decision to go underground may also be the result of the social and environmental problems associated with surface mining operations.

14.7.1.7 Novel Methods. The novel methods of mining are new procedures that have not been widely practiced or methods that are unique in their approach to mineral extraction. We classify them in three categories, based on the likelihood of early commercial application. Those in use today include automation and robotics, rapid excavation (in hard-rock mining), hydraulic mining, methane drainage, and oil mining. Methods showing promise but not now in use include underground gasification, underground retorting, and ocean mining. Ocean mining appears to be quite attractive for future metal mining, but its economic and legal problems need to be solved. The third group has a questionable future; it consists of nuclear mining and extraterrestrial mining. During the last decade, automation and robotics, rapid excavation, and methane drainage have grown substantially and will in all likelihood continue to grow. The second group may find more application, but energy prices may have to increase substantially to encourage their adoption. Government funding may make extraterrestrial mining a reality, but only in the distant future. Nuclear mining appears to be politically a dead issue.

14.7.1.8 Mining Method Selection. The last chapter of this book addresses one of the most fundamental decisions facing a mining engineer: choosing a suitable mining method for any deposit. The chapter begins with a comparison of the surface and underground mining methods' characteristics, advantages, and disadvantages. It then discusses strategies for choosing a method, emphasizing one qualitative and one quantitative method. The discussion then turns to mining method costs plus budgeting and cost control. The choice between surface mining and underground mining (or both) is discussed. The final choice is based on a thorough mine investment analysis. The chapter and the book conclude with a review of the mining industry and a look to the future.

14.7.2 The Role of the Mining Engineer

In Chapter 1 we said that if lumbering is considered part of agriculture and oil production is considered part of mining, then all our consumer products are derived from either agriculture or mining or both. Mining engineers thus have important status in the modern world. To accomplish the task of mining, mining engineers plan, design, develop, exploit, evaluate, operate, and reclaim mines. With additional experience or education, they may also prospect and explore for mines, get involved in mineral processing, or manage mines. And should they enter manufacturing or sales—or go into education, research, consulting, or government agencies that relate to mining—they are still involved in the practice of mining engineering. It is important to note that while holding a key position in our society, mining engineers have a responsibility to operate their mines with a minimum of negative impacts. Indeed, they must be environmentalists to be good mining engineers.

The mining engineer was once referred to with pride as a jack-of-all-trades. This designation came from the broad education that a mining engineer often receives. In practice, the mining engineer frequently becomes a specialist through on-the-job training in designing underground openings, calculating pit limits, choosing equipment, selecting mining methods, ventilating mines, making cost estimates, writing environmental impact statements, and coping with the social, economic, and political consequences of mining. They thus become specialists in a given area.

Jobs have been plentiful for mining engineers during the last decade. This will likely continue into the future, because young people often shun the mining field and the number of graduating mining engineers is not increasing. The only requirement to practice in the field is an undergraduate degree. Registration as a professional engineer is encouraged but not required. Advanced degrees are often required to perform research functions or teach at a university. Second degrees in business administration or law also open up additional opportunities for the mining engineer.

14.7.3 Future Outlook

To close this textbook, it is perhaps fitting that we project the practice of mining engineering into the future. Here are some of the important areas and the trends we see that will shape the mining industry in upcoming decades:

1. *U.S. self-sufficiency in mineral production.* Continuing present trends, the United States will increasingly become a net importer of minerals, particularly metals, as it enters further into Period 5 of Lovering's chart (Figure 1.3). This does not mean the demise of the domestic mining industry, but that U.S.

mining will grow more selective and competitive. It is likely that coal and other energy sources (tar sands, oil mining, and methane from coal seams) will continue their growth. Aggregates seem to have a similar fate. However, more of our copper, lead, zinc, and uranium are likely to come from abroad.

2. *Government regulation.* The role and influence of government will likely increase, particularly in the areas of environmental protection and protection of workers' health. All of the industry will be under constant pressure to clean up its atmospheric emissions and the environment in which its personnel must work.

3. *Sustainable development.* In Section 1.4, we addressed the worldwide trend toward promoting the concepts of sustainable development. Using these as guidelines, citizens and governments alike will encourage mining and other industries to consider the effects of their operations and design their businesses to minimize future impacts.

4. *Prospecting and exploration.* High-technology methods of locating and proving mineral deposits, especially airborne geophysics, will improve in resolution and precision and largely replace direct methods. Ore estimation will increasingly be conducted by computer-based geostatistical techniques.

5. *Development and exploitation.* Construction tasks for development and exploitation will be performed increasingly by contractors who are specialists in the field, for either underground or surface mining. Automation and robotics will become routine in many mining operations, with hazardous jobs being targeted as ideal applications for these technologies.

6. *Unit operations and equipment.* Surface equipment may continue to grow in size and in computer monitoring of all its important components. Automation will increase, and driverless trucks may become a reality in surface mining. Underground equipment will also become more automated, with many pieces of equipment controlled by semiautomated or automated systems. The number of miners underground will be decreased because of the advance of automation and robotics. Computers will be everywhere in mining systems to aid personnel and equipment in performing their jobs better.

7. *Surface mining methods.* Open pit mines will be reduced in the metals industry, being replaced by foreign operations. Open cast mining will increase in the western states and decrease in eastern states. Solution mining, both heap and in situ varieties, will increase. Environmental regulations will likely be more stringent in surface mines. However, new procedures and technology are likely to emerge to allow mining companies to restore mined lands to useful purposes.

8. *Underground mining methods.* The number of metal mining methods in common use will likely decrease, with the emphasis shifting to the bulk mining procedures (stope-and-pillar mining, sublevel stoping, sublevel caving, and block caving). In coal, room-and-pillar mining will continue to be replaced by longwall mining. Automation and robotics will continue a steady increase in underground mining.

9. *Novel mining methods.* Some of the novel methods, like automation and robotics, rapid excavation, and methane drainage, will continue to increase. Ocean mining, oil mining, and hydraulic mining may enjoy modest increases as well. Of these, ocean mining has the greatest potential for the future. Underground retorting or gasification will find use only if energy prices rise significantly. Nuclear mining and extraterrestrial mining will most likely remain as concepts rather than adopted technology.

10. *Computer applications.* Computers will be employed in all aspects of the mining business, and the only restraint on their utility will be the inventiveness of computer engineers and users. Computers will be used in controlling automated equipment, monitoring maintenance variables in equipment, controlling production costs, performing hazardous jobs, processing minerals, and controlling environmental problems. Computers will displace some people, primarily in the hazardous job categories.

11. *Demand for mining engineers.* Well-educated, capable mining engineers will remain in demand, particularly those who are proficient with the computer. The aggregates and coal industries will likely increase their need for engineers, but metal mining jobs may exist primarily in operations abroad. The engineer today should be prepared to move as the need arises; working all through a career for a single company is becoming increasingly rare. Versatility and seeking additional areas of expertise will help every engineer to be more useful to his or her employer.

14.8 SPECIAL TOPIC: MINE DESIGN PROBLEMS

As the culminating exercise for an introductory text in mining engineering, it is appropriate that we consider an open-ended design problem, one that spans all five stages in the life of a mine. The problem must be open-ended because, in real life, mine design is seldom a well-defined task. Information provided is often incomplete, requiring assumptions on the part of the engineer during analysis. The design procedure is generally left up to the engineer, with only a rough outline available of the requirements of the design. Similar circumstances must be considered by students in the following problems. Satisfactory solutions to the problems can benefit from the procedures outlined in this book, but the student may be better able to outline a solution to a specific problem by designing a different approach. However, a satisfactory solution to each problem requires the following as a minimum:

1. Complete description of the steps involved in bringing a given mineral deposit through the four stages into a producing mine
2. Specification of the unit operations and equipment required for development and exploitation

3. Selection of the most suitable mining method, with justification and explanation of operating details
4. Preparation of an appropriate order-of-magnitude cost estimate
5. A premining plan for reclaiming the mine after its useful life is over

The instructor can determine the amount of detail that is required for each of these requirements in setting up the selection exercise. Students should note that detailed mining plans—complete with maps, development schedules, and other important details of the mining system—are often required for senior projects. The problems that follow will thus be appropriate practice for such a project.

PROBLEMS

14.1 For the mineral deposit described in the following list, prepare a specific, detailed outline of the steps involved in bringing a prospect through the four operational stages into a producing mine and what would be required to reclaim the mine site after mining (Stage 5). Select the mining method that would be best suited, most economical, and safest for the given conditions. Defend your choice briefly, and cite the main advantages and disadvantages of the method. Sketch the overall layout of the mine and the details of the mining method. Identify the cycle of operations and major pieces of equipment for mining, and state the relative cost. Design conditions are as follows:

Deposit	Coal seam, steam quality, calorific value 13,500 Btu/lb (31,400 kJ/kg), thin rock parting Dip 2°, thickness 5.5 ft (1.7 m), average depth of cover 300 ft (91 m) Deposit area 20 mi^2 (5180 ha), no outcrop Coal compressive strength 4000 lb/in.^{2} (27.6 MPa) Roof (sandstone and shale) average compressive strength 25,000 lb/in.^{2} (172 MPa) Floor (shale) compressive strength 9000 lb/in.^{2} (62.1 MPa)
Topography	Hilly terrain, elevation 800 ft (245 m)
Climate	Warm temperate, mean annual rainfall 60 in. (1.5 m)
Production rate	15,000 tons/day (13,600 tonnes/day) over a 20-year lifetime

14.2 For the mineral deposit described in the following list, prepare a specific, detailed outline of the steps involved in bringing a prospect through the four operational stages into a producing mine, plus a plan for reclamation of the property after mining is complete. Select the mining method that would be best suited, most economical, and safest for the given conditions, defending your choice briefly and citing the main advantages and disadvantages of the method. Sketch the overall mine layout and the mining method. Identify the cycle of operations and major equipment selected for mining, and state the relative cost. Outline a rational plan for mine reclamation. Design conditions are as follows:

Deposit	Limestone bed, average grade 98.5% $CaCO_3$, uniform, some mud seams
	Dip 5°, thickness 40 ft (12.2 m), average depth of cover 300 ft (91 m)
	Deposit area 4 mi^2 (1040 ha), no outcrop
	Ore compressive strength 20,000 lb/$in.^2$ (138 MPa)
	Hanging wall (sandstone) average compressive strength 26,000 lb/$in.^2$ (179 MPa)
	Footwall competent limestone
Topography	Low mountain ranges, rough terrain, elevation 1800 ft (550 m)
Climate	Temperate, mean annual rainfall 40 in. (1.0 m)
Production rate	10,000 tons/day (9100 tonnes/day) over a 25-year life

14.3 For the mineral deposit described in the following list, prepare a specific, detailed outline of the steps involved in bringing a prospect through the four operational stages into a producing mine. Select the mining method that would be best suited, most economical, and safest for the given conditions, defending your choice briefly and citing the main advantages and disadvantages of the method. Sketch the overall mine layout and the mining method. Identify the cycle of operations and major equipment selected for mining, and state the relative cost. Then give a summary of the requirements necessary to reclaim the mine site after mining is complete. Design conditions are as follows:

Deposit	Chalcocite/chalcopyrite ore minerals disseminated in quartz porphyry intrusive, average grade 2.0% Cu, fairly uniform
	Massive irregular deposit, reserves 100 million tons (90 million tonnes), average depth of cover 3000 ft (915 m)
	Ore compressive strength 12,000 lb/$in.^2$ (82.7 MPa), frequent joints
	Metamorphosed capping and adjoining rock (schist), compressive strength 8000 lb/$in.^2$ (55.2 MPa), fractured, distinguishable boundary with ore

Topography	Desert and low mountain ridges, rugged terrain, elevation 4500 ft (1370 m)
Climate	Arid, mean annual rainfall 10 in. (254 mm)
Production rate	35,000 tons/day (31,700 tonnes/day) over a 25-year life

14.4 Solve problem 14.1, given the same deposit, except for the following changes in conditions:

Deposit	Average depth of cover 2000 ft (610 m) Coal compressive strength 2000 lb/in.^{2} (13.8 MPa) Roof average compressive strength 8000 lb/in.^{2} (55.2 MPa) Floor compressive strength 8000 lb/in.^{2} (55.2 MPa)
Topography	Controlled subsidence permitted
Production rate	25,000 tons/day (22,700 tonnes/day)

14.5 Prepare an approximate cost estimate for the development and exploitation operations for the mine of the following problems:

a. Problem 14.1
b. Problem 14.2
c. Problem 14.3
d. Problem 14.4

Calculate on a unit basis, using the estimating values given in Table 1.3 for prospecting and exploration. Then determine the range of overall mining costs for each mine.

APPENDIX

CONVERSION TABLE, ENGLISH TO METRIC UNITS (SI, OR INTERNATIONAL SYSTEM) USEFUL IN MINING ENGINEERING

Acceleration of gravity	32.174 ft/sec^2 = 9.8066 m/sec^2
Angular measure	1 deg = 0.01745 rad
Area	1 in.^{2} = 645.16 mm^2
	1 ft^2 = 0.09290 m^2
	1 yd^2 = 0.83613 m^2
	1 acre = 0.40469 hectare
	1 mi^2 = 259.0 hectares
Barometric pressure	1 mm Hg = 133.32 Pa
	1 atm = 101.325 kPa
Blasting factor	1 ft^2/drillhole = 0.09290 m^2/drillhole
Drilling factor	1 ft/ton = 0.33598 m/tonne
Enery or work	1 Btu = 1.0551 kJ
	1 kW-hr = 3600 kJ
	1 cal* = 4.1868 J
	1 erg = 0.10000 μJ
Force	1 lb (force) = 4.4482 N
	1 dyn = 1 × 10^{-5} N
Friction factor, airflow	1 lb-min^2/ft^4 = 1.8550 × 10^6 kg/m^3
Gas constant, air	53.35 ft-lb/lb (mass)-°R = 287.045 J/kg-K
Grade of ore	1 oz/ton = 31.25 g/tonne

*Former metric unit not approved in SI usage.

Head or pressure difference, airflow	1 in. water = 25.4 mm water*
	1 in. water = 248.84 Pa @ 60°F
	1 in. Hg = 3.3768 kPa @ 60°F
Heat content or enthalpy	1 Btu/lb = 2.3260 kJ/kg
Heat flow	1 Btu/hr = 0.29307 W
Length	1 in. = 25.40 mm
	1 ft = 0.30480 m
	1 mi = 1.6093 km
Loading factor, explosives	1 lb/ft = 1.4882 kg/m
Mass	1 lb (mass) = 0.45359 kg
	1 grain = 64.799 mg
Mass density	1 lb-sec^2/ft^4 = 1 slug/ft^3 = 515.38 kg/m^3
Mass per unit area	1 lb/ft^2 = 4.8824 kg/m^2
	1 ton/ft^2 = 9.7652 tonnes/m^2
Moment	1 ft-lb = 0.13825 m-kg
	1 ft-ton = 0.27651 m-tonne
Powder factor	1 lb/ton = 0.50000 kg/tonne
Power	1 hp = 0.74570 kW
Pressure or stress	1 lb/in.^{2} = 6.8948 kPa
	1 lb/ft^2 = 47.880 Pa
Quantity, airflow	1 ft^3/min = 0.47195×10^{-3} m^3/sec
Resistance, airflow	1 in.-min^2/ft^6 = 1.1170×10^9 N-sec^2/m^8
Specific volume	1 ft^3/lb = 0.06243 m^3/kg
Specific weight	1 lb/ft^3 = 16.018 kg/m^3
	1 ton/yd^3 = 1.1866 tonnes/m^3
Stripping ratio	1 yd^3/ton = 0.84278 m^3/tonne
Temperature	$°C = \frac{5}{9}(°F - 32)$
	K = °C + 273.15
Tonnage factor	1 ft^3/ton = 0.03121 m^3/tonne
Velocity	1 in./min = 0.42333 mm/sec
	1 ft/min = 0.00508 m/sec
	1 mi/hr = 1.6093 km/hr
Volume	1 in.^{3} = 16,387 mm^3
	1 ft^3 = 0.02832 m^3
	1 yd^3 = 0.76455 m^3
	1 gal = 3.7854 L
	1 acre-ft = 1233.5 m^3
Volume flow rate	1 ft^3/min = 0.47195 L/sec
	1 ft^3/min = 471.95 cm^3/sec*
	1 gal/min = 63.09×10^{-6} m^3/sec
	1 gal/min = 0.06309 L/sec
Weight	1 oz (avoir.) = 28.35 g
	1 lb = 0.45359 kg
	1 ton = 0.90718 tonne

*Former metric unit not approved in SI usage.

	1 ton (long) = 1.0160 tonnes
Weight flow rate	1 lb/hr = 0.12600×10^{-3} kg/sec

ANSWERS TO SELECTED PROBLEMS

Chapter 3

3.1 468,000 long tons (475,500 tonnes)
56.7% Fe
1,244,400 yd^3 (951,400 m^3)
2.65 yd^3/long ton (2.03 m^3/tonne)

3.3 **a.** 3.0 ft (0.92 m)
b. 253 ft (77 m)
c. 1.28% S
d. 2,546,000 tons (2,309,700 tonnes)
e. 84.1 ft/ft (m/m)

3.4 31.3% Fe

Chapter 5

5.2 **a.** −10.6414 g-atoms/kg
b. No, it is oxygen-deficient
c. Increase O_0 or decrease C

Chapter 6

6.1 6.00 yd^3/ton (5.06 m^3/tonne)
99 ft (30 m)
300 ft (91 m)
2.25 yd^3/ton (1.90 m^3/tonne)

6.3 357 ft (109 m)
20.0 yd^3/ton (16.9 m^3/tonne)
6.06 yd^3/ton (5.11 m^3/tonne)

Chapter 7

7.2 Fails
FS = 0.96

7.4 **a.** 1.44 (Safe)
b. 0.18 (Fails)

7.6 **a.** 29 ft; 7 holes
b. Reduce parameters

Chapter 8

8.1 **a.** Loader capacity = 1395 tons/hr; truck capacity = 1500 tons/hr
b. The loader governs.
c. Improve t_1 by better blasting practice.

8.3 $n = 6$ trucks
a. The trucks control the capacity.
b. Total truck capacity = 23,940 tons/shift

Chapter 9

9.1 **a.** 10,944 tons (9928 tonnes)
b. 16.64 kW-hr/skip

9.3 145 sec/cycle
6.0 tons (5.4 tonnes)
4 ropes @ 0.75 in. (19 mm)
5.0-ft sheave (1.5 m)
Peak 2043 hp (1523 kW)
Average 909 hp (678 kW)
RMS 1038 hp (774 kW)
2.14 kW-hr/ton (2.36 kW-hr/tonne)
$0.107/ton ($0.118/tonne)

Chapter 10

10.2 **a.** 25.5%
b. 79.5%
c. 31.0%

10.4 25.0 ft (7.6 m)

10.6 **a.** 3600 lb/in.^{2} (24.8 MPa)
1200 lb/in.^{2} (8.3 MPa)
c. Top: 1080, 360, 1440, 1800 lb/in.^{2} (7.5, 2.5, 9.9, 12.4 MPa)
Corner: −11,160, 14,760, 20,160, 31,320 lb/in.^{2}
(−77.0, 101.8, 139.0, 216.0 MPa)
d. $R/h \geqslant 4$, fails
e. $R/h = 1$, FS = 1.11
$R/h = 2$, FS = 1.52
f. $R/h = 1$ and 2, metal mine drift
$R/h = 4$, coal mine opening
$R/h = 8$, block caving undercut

10.8 **a.** Fails
b. No
c. None

10.10 14 ft (4.3 m)

Chapter 11

11.1 8.5 in./min (216 mm/min)

11.3 **a.** 4
b. $6.72/hr
c. $38.72/hr
d. $0.15/ft ($0.49/m)

11.5

Pneumatic (4 drills)	Hydraulic (3 drills)
$179.91/hr	$133.77/hr
$0.57/ft ($1.87/m)	$0.42/ft ($1.38/m)
$1.07/ton ($1.18/tonne)	$0.80/ton ($0.88/tonne)

Hydraulic more economical

11.7 **a.** 5 ft (1.5 m)
b. Per hole 7 ft (2.1 m)
5 ft (1.5 m)
7 lb (3.2 kg)

11.9 **b.** Per hole, cut, 4 @ 4.1 lb (1.9 kg)
Relievers, 4 @ 0.6 lb (0.3 kg)
No. 1 enlargers, 4 @ 1.6 lb (0.7 kg)
No. 2 enlargers, 4 @ 3.3 lb (1.5 kg)
Trimmers, 4 @ 3.3 lb (1.5 kg)
c. 20 holes
d. 140.4 ft (42.8 m)
e. 9.4 ft/ton (3.2 m/tonne)
f. 50 lb (23.6 kg)
g. 3.3 lb/ton (1.7 kg/tonne)

Chapter 12

12.2 89 ft^3/min (0.042 m^3/sec)

12.4 **a.** 0.19 in. water (47 Pa)
b. 0.60 in. water (149 Pa)

12.6 **a.** 9.02 in. water (2.24 Pa)
c. 110,000 ft^3/min (51.9 m^3/sec)
7.0 in. water (1.74 kPa)
d. 1140 rpm

REFERENCES

Ackerman, J. M. 1992. Main pass — Frasch sulfur mine development. *Mining Engineering* 44(3): 222–226.

Ahlness, J. K., D. R. Tweeton, W. C. Larson, D. J. Millenacker, and R. D. Schmidt. 1992. In situ mining of hard-rock ores, Sec. 15.3.3 in *SME Mining Engineering Handbook*, 2d ed., edited by H. L. Hartman, pp. 1515–1528. Littleton, CO: Society for Mining, Metallurgy, and Exploration.

Ahrens, E. H. 1983. *Practical Mining Geology*, Preprint 83–371. Fall Meeting. SME-AIME, Salt Lake City, UT. 11 pp.

Alexander, C. 1999. Tunnel boring at Stillwater's East Boulder project. *Mining Engineering* 51(9): 15–24.

American Conference of Governmental Industrial Hygienists. 2000. *2000 TLVs and BEIs*. Cincinnati, OH: American Conference of Governmental Industrial Hygienists. 184 pp. (updated annually).

American Geological Institute. 1997. *A Dictionary of Mining, Mineral, and Related Terms*, 2d ed. Alexandria, VA: American Geological Institute. 646 pp.

American Mining Congress. 1981. Solution trona mining. *Mining Congress Journal* 67(4): 12.

American National Standards Institute. 1992. *American National Standard for Metric Practice*. ANSI/IEEE, Standard 268–82. New York: American National Standards Institute.

Anderson, G. P., and S. J. Kirk. 1992. Monolithic overburden, thick horizontal coal seams: Jacobs Ranch Mine, Sec. 14.3.2 in *SME Mining Engineering Handbook*, 2d ed., edited by H. L. Hartman, pp. 1411–1414. Littleton, CO: Society for Mining. Metallurgy, and Exploration.

Anderson, O. E. 1981. White Pine separates safety and MSHA compliance functions. *Mining Congress Journal* 67(5): 29–33.

Antonides, L. E. 1999. Stone (dimension). *Mineral Commodity Surveys*. U.S. Geological Surveys, pp. 164–165.

Aplan, F. F. 1973. Evaluation to indicate processing approach, Sec. 27.3 in *SME Mining Engineering Handbook*, edited by A. B. Cummings and I. A. Givens, pp. 27-15–27-28. New York: AIME.

Aplan, F. F. 1999. *Mineral Preparation* 301 *Notes*. University Park, PA: Pennsylvania State University.

Arnold, E. 1999. The new wilderness land grab. *Outside* 24(9): 66–68.

Arnold, T. D. 1996. Underground mining: A challenge to established open-pit operations. *Mining Engineering* 48(4): 25–29.

Arrouet, D. 1992. Investment strategy for mining projects, Sec. 13.3 in *SME Mining Engineering Handbook*, 2d ed., edited by H. L. Hartman, pp. 96–116. Littleton, CO: Society for Mining, Metallurgy, and Exploration.

Atkinson, T. 1983. Surface mining and quarrying. *Proceedings of the 2nd International Surface Mining and Quarrying Symposium*. Bristol, U.K.: Institute of Mining and Metallurgy. 449 pp.

Atkinson, T. 1992a. Selection and sizing of mining equipment, Sec. 13.3 in *SME Mining Engineering Handbook*, 2d ed., edited by H. L. Hartman, pp. 1311–1333. Littleton, CO: Society for Mining, Metallurgy and Exploration.

Atkinson, T. 1992b, Design and layout of haul roads, Sec. 13.4 in *SME Mining Engineering Handbook*, 2d ed., edited by H. L. Hartman, pp. 1334–1342. Littleton, CO: Society for Mining Metallurgy and Exploration.

Atlas Powder Company. 1987. *Explosives and Rock Blasting*. Dallas: Atlas Powder Company. 662 pp.

Aul, G., and R. Ray Jr. 1991. Optimizing methane drainage systems to reduce mine ventilation requirements in *Proceedings of the 5th U.S. Mine Ventilation Symposium*, edited by Y. J. Wang, pp. 638–646. Littleton, CO: Society for Mining Metallurgy, and Exploration.

Baase, R. A., W. D. Diment, and A. J. Petrina. 1982. Sublevel caving at Craigmont Mines Ltd. In *Underground Mining Methods Handbook*, edited by W. A. Hustrulid, pp. 898–915. New York: SME-AIME.

Baiden, G. R. 1994. Combining teleoperation with vehicle guidance for improving LHD productivity at Inco Limited. *CIM Bulletin* 87(981): 36–39.

Baiden, G. R. 1996. Future robotic mining at INCO Limited—The next 25 years. *CIM Bulletin* 89(996): 36–40.

Baiden, G. R., and E. Henderson. 1994. LHD operation and guidance proven productivity improvement tools. *CIM Bulletin* 87(984): 47–51.

Bailly, P. A. 1966. Mineral exploration and mine development problems. *Proceedings of the Public Land Law Conference*. Moscow: University of Idaho, pp. 51–99.

Bailly, P. A. 1968. Exploration methods and requirements, Sec. 2.1 in *Surface Mining*, edited by E. P. Pfleider, pp. 19–42. New York: AIME.

Barczak, T. M., and D. F. Gearhart. 1994. Engineering method of the design and placement of wood cribs. In *New Technology for Longwall Ground Control*, edited by C. Mark, R. J. Tuchman, R. C. Repsher, and C. L. Simon, pp. 103–116. Special Publication 01-94. U.S. Bureau of Mines.

Barker, R. M., and T. R. Harter. 1997. Underground development at Getchell and Turquoise Ridge. *Mining Engineering* 49(8): 33–41.

Bartlett, R. W. 1998. *Solution Mining*. Amsterdam: Gordon and Breach Science Publishers. 443 pp.

Barton, W. R. 1968. *Marble*. IC 8391, pp. 83–103. U.S. Bureau of Mines.

Bath, A. R. 1991. Deep sea mining technology: Recent developments and future projects. *Mining Engineering* 43(1): 125–128.

Bauer, A., J. L. Workman, and W. A. Crosby. 1983. *Principles and Applications of Displacing Overburden in Strip Mines by Explosives Casting*, Preprint 83-426. Fall Meeting. SME-AIME, Salt Lake City, UT. 22 pp.

Bernard, G. M. 1995. Rio Algom's Cerro Colorado begins copper heap-leach operation in Chile. *Mining Engineering* 47(4): 323–326.

Bhappu, R. B. 1982. Past, present, and future of solution mining. In *Interfacing Technologies in Solution Mining*, edited by W. J. Schlitt, pp. 3–12. New York: SME-SPE-AIME.

Bibb, T. C., and K. H. Hargrove. 1992. Coal mining: Method selection, Sec. 21.4 in *SME Mining Engineering Handbook*, 2d ed., edited by H. L. Hartman, pp. 1854–1866. Littleton, CO: Society for Mining, Metallurgy, and Exploration.

Bieniawski, Z. T. 1984. *Rock Mechanics Design in Mining and Tunneling*. Boston: Balkema. 272 pp.

Biles, G., R. Anand, and K. Loughran. 2000. Eskay Creek Mine and Mill—A continuing success. *Mining Engineering* 52(1): 19–23.

Bingham, N. 1994. Mining's image—What does the public really think? *Mining Engineering* 46(3): 200–203.

Blakely, J. W. 1975. New triple-head auger proves its worth in low-seam coals. *Coal Mining and Processing* 12(10): 66–67.

Bluekamp, P. R. 1981. Block cave mining at the Mather Mine, Chap. 23 in *Design and Operation of Caving and Sublevel Stoping Mines*, edited by D. R. Stewart, pp. 321–327. New York: SME-AIME.

Boshkov, S. V., and F. D. Wright. 1973. Underground mining systems and equipment, Sec. 12.1 in *SME Mining Engineering Handbook*, edited by A. B. Cummings and I. A. Givens, pp. 12-2–12-13. New York: AIME.

Bossard, F. C. (ed.). 1983. *Manual of Mine Ventilation Design Practices*, 2d ed. Butte, MT: Floyd C. Bossard & Associates. 350 pp.

Bourne, H. L. 1996. What it's worth: A review of mineral royalty information. *Mining Engineering* 48(7): 35–38.

Bowles, O. 1958. *Dimension stone*. IC 7829, pp. 8, 21–28. U.S. Bureau of Mines.

Boyd. J. 1967. The influence of the minerals industry on general economics. *Mining Engineering* 19(3): 54–59.

Boyd, J. 1973. Administration and management, Sec. 29 in *SME Mining Engineering Handbook*, edited by A. B. Cummings and I. A. Givens, pp. 29-1–29-11. New York: AIME.

Brackebusch, F. W. 1992a. Underground mining: Unsupported methods, Sec. 19.0 in *SME Mining Engineering Handbook*, 2d ed., edited by H. L. Hartman, pp. 1741–1742. Littleton, CO: Society for Mining, Metallurgy, and Exploration.

Brackebusch, F. W. 1992b. Cut-and-fill stoping, Sec. 19.1 in *SME Mining Engineering Handbook*, 2d ed., edited by H. L. Hartman, pp. 1743–1748. Littleton, CO: Society for Mining, Metallurgy, and Exploration.

Brady, B. H. G., and E. T. Brown. 1993. *Rock Mechanics for Underground Mining*, 2d ed. London: Chapman and Hall. 571 pp.

Bray, D. N., A. C. Alexander, W. Strickland, and D. Eirnarson. 2001. Mining of PGMs at the Stillwater Mine. In *Underground Mining Methods: Engineering Fundamentals and International Case Studies*, edited by W. A. Hustrulid and R. L. Bullock, pp. 299–307. Littleton, CO: Society for Mining, Metallurgy, and Exploration.

Bray, R. N., A. D. Bates, and J. M. Land. 1997. *Dredging*, 2d ed. New York: John Wiley & Sons, Inc. 434 pp.

Brechtel, C. E., G. R. Struble, and B. Guenther. 2001. Underhand cut-and-fill mining at the Murray Mine, Jerritt Canyon Joint Venture. In *Underground Mining Methods: Engineering Fundamentals and International Case Studies*, edited by W. A. Hustrulid and R. L. Bullock, pp. 333–337. Littleton, CO: Society for Mining, Metallurgy, and Exploration.

Breeds, C. D., and J. J. Conway. 1992. Rapid excavation. In *SME Mining Engineering Handbook*, 2d ed., edited by H. L. Hartman, pp. 1871–1907. Littleton, CO: Society for Mining, Metallurgy, and Exploration.

Bricker, M. L. 1992. Monolithic overburden, horizontal coal seams: Bridger Mine, Sec. 14.3.1 in *SME Mining Engineering Handbook*, 2d ed., edited by H. L. Hartman, pp. 1407–1410. Littleton, CO: Society for Mining, Metallurgy, and Exploration.

Bristol, W. W., and B. W. Guenther. 1997. Jerritt Canyon's transition to underground mining. *Mining Engineering* 49(11): 47–49.

Britton, S. G. 1981. *Practical Coal Mine Management*. New York: John Wiley & Sons, Inc.

Britton, S. G., and B. A. Brasfield. 1992. Mine development, Sec. 7 in *SME Mining Engineering Handbook*, 2d ed., edited by H. L. Hartman, pp. 483–528. Littleton, CO: Society for Mining, Metallurgy, and Exploration.

Brockway, J. E. 1983. Incline/decline boring with tunnel boring machines, Chap. 43 in *Proceedings of Rapid Excavation and Tunnelling Conference*, edited by H. Sutcliffe and J. W. Wilson, pp. 743–760. New York: SME-AIME.

Brooks, A. B., and D. S. Hursh. 1982. Acquiring capital funds for coal mining. *Mining Engineering* 34(9): 1332–1335.

Brooks, R. R. 1995. Geobotanical prospecting. Part 1. *Biological Systems in Mineral Exploration and Processing*, edited by R. R. Brooks, C. E. Dunn, and G. E. M. Hall, pp. 7–116. New York: Ellis Horwood Ltd.

Brooks, R. R., C. E. Dunn, and G. E. M. Hall (eds.). 1995. *Biological Systems in Mineral Exploration and Processing*. New York: Ellis Horwood Ltd. 538 pp.

Brooks, W. E. 1991. Accident to, and repair of, Dredge 21 at Yuba, California. In *Alluvial Mining*, pp. 19–35. London: The Institute of Mining and Metallurgy.

Brophey, D. G., and D. W. Euler. 1994. The Opti-Trak system, a system for automating today's LHDs and trucks. *CIM Bulletin* 87(984): 52–57.

Brown, N. A. 1995. Union Pacific instrumental in developing Wyoming trona. *Mining Engineering* 47(2): 135–141.

Bruce, C. 1982. Ore body evaluation. In *Underground Mining Methods Handbook*, edited by W. A. Hustrulid, pp. 3–16. New York: SME-AIME.

Brucker, D. S. 1975. *Faster and Deeper—The Sign of the Times in Hoisting*. Report, ASEA. Los Angeles: Swedish Trade Commission. 29 pp.

Buchan, G. 1998. Long panels for longwall mining at Cyprus Twentymile Coal. *Mining Engineering* 50(12): 21–27.

Bucyrus-Erie Company. 1976. *Mining Supervisory Training Program*, pp. 1.1–5.13. South Milwaukee, WI: Bucyrus-Erie, Co.

Bullock, R. L. 1982. General mine planning. In *Underground Mining Methods Handbook*, edited by W. A. Hustrulid, pp. 113–137. New York: SME-AIME.

Bullock, R. L. 1994. Underground hard rock mechanical mining. *Mining Engineering* 46(11): 1254–1258.

Bullock, R. L., and W. A. Hustrulid. 2001. Planning the underground mine on the basis of mining method. In *Underground Mining Methods: Engineering Fundamentals and International Case Studies*, edited by W. A. Hustrulid and R. L. Bullock, pp. 29–48. Littleton, CO: Society for Mining, Metallurgy, and Exploration.

Bullock, R. L. and C. D. Mann (eds.). 1982. Stopes requiring minimum support, Sec. 2 in *Underground Mining Methods Handbook*, edited by W. A. Hustrulid, pp. 227–482. New York: SME-AIME.

Buntain, D. 1983. Methane extraction and utilization in the United Kingdom. *Coal Journal* (Sydney) (4): 63–71.

Burrows, J. (ed.). 1982. *Environmental Engineering in South African Mines*. Marshalltown, South Africa: Mine Ventilation Society of South Africa. 987 pp.

Butler, D. W., and A. G. Schneyderberg. 1982. Headframe selection: Steel vs. concrete. *Mining Congress Journal* 68(1): 15–19.

Call, R. D. 1986. Cost-benefit design of open pit slopes, pp. 1–18. *Proceedings of the 1st Open Pit Mining Symposium*, Antofogosta, Chile.

Callahan, W. H. 1982. Prospecting in the USA: Promise and problems. *Mining Engineering* 34(6): 673–676.

Capp, F. M. 1962. Factors in rotary drill evaluation. *Mining Congress Journal* 48(12): 20–23.

Cardwell, G. J. 1984. Analytical methods for applied geology. In *Applied Mining Geology*, edited by A. J. Erickson Jr., p. 203. New York: SME-AIME.

Carson, R. 1964. *Silent Spring*. Greenwich, CT: Fawcett Publications. 304 pp.

Carter, R. A. 1990. Kennecott Utah Copper modernization pays off. *Engineering and Mining Journal* 191(1): 22–29.

Carter, R. A. 1991. Mining methods could unlock Canada's deep tar sands. *Engineering and Mining Journal* 192(4): 39–45.

Carter, R. A. 1992. Snip's first year is a success. *Engineering and Mining Journal* 193(6): WW29–WW31.

Carter, R. A. 1997. Florida Canyon plays a new hand. *Engineering Mining Journal* 198(6): 38WW–43WW.

Caterpillar, Inc. 1997. *Caterpillar Performance Handbook*. Peoria, IL: Caterpillar Inc.

Caudle, R., and G. B. Clark. 1955. *Stresses Around Mine Openings in Some Simple Geologic Structures*. Engineering Experiment Station. Bulletin No. 430. Urbana: University of Illinois. 42 pp.

Chadwick, J. 1993. Highwall mining. *Mining Magazine* (London) 169 (Dec.): 347–353.

Chaiken, R. F., and J. W. Martin. 1992. In situ gasification and combustion of coal, Chap. 22.6 in *SME Mining Engineering Handbook*, 2d ed., edited by H. L. Hartman. pp. 1954–1970. Littleton, CO: Society for Mining, Metallurgy, and Exploration.

Chamberlain, P. G. 1992. Nuclear-assisted mining, Sec. 22.9.7 in *SME Mining Engineering Handbook*, 2d ed., edited by H. L. Hartman, pp. 2045–2047. Littleton, CO: Society for Mining, Metallurgy, and Exploration.

Chamberlain, P. G., and E. R. Podnieks. 1992. Lunar and planetary mining, Sec. 22.9.6 in *SME Mining Engineering Handbook*, 2d ed., edited by H. L. Hartman, pp. 2042–2045. Littleton, CO: Society for Mining, Metallurgy, and Exploration.

Chironis, N. P. 1980. Casting overburden by blasting. *Coal Age* 85(5): 172–180.

Chironis, N. P. 1983. Shooting coal pays off. *Coal Age* 88(9): 86–91.

Church, H. K. 1981. *Excavation Handbook*. New York: McGraw-Hill. 913 pp.

Clark, G. R. 1968. Explosives, Sec. 7.1 in *Surface Mining*, edited by E. P. Pfleider, pp. 341–354. New York: AIME.

Clark, G. R. 1987. *Principles of Rock Fragmentation*. New York: John Wiley & Sons, Inc. 610 pp.

Clark, I. 1979. *Practical Geostatistics*. London: Applied Science Publishers. 129 pp.

Coal Age. 1962. Hydraulic mining with a rotary drill unit. *Coal Age* 67(7): 96–99.

Code of Federal Regulations. 2000. Title 30, *Mineral Resources*, Parts 1–699. Washington, DC: U.S. Government Printing Office (Updated annually).

Cokayne, E. W. 1982. Sublevel caving: Introduction. In *Underground Mining Methods Handbook*, edited by W. A. Hustrulid, pp. 872–879. New York: SME-AIME.

Cook, D. J. 1983. *Placer Mining in Alaska*. Report No. 65. Fairbanks, AK: Mineral Industry Research Laboratory, University of Alaska. 157 pp.

Cook, D. R. 1986. Analysis of significant mineral discoveries in the last 40 years and future trends. *Mining Engineering* 38(2): 87–94.

Cook, M. A. 1974. *The Science of Industrial Explosives*. Salt Lake City: IRECO Chemicals. 449 pp.

Coopersmith, H. G. 1997. Kelsey Lake: First diamond mine in North America. *Mining Engineering* 49(4): 30–33.

Copier, G. D., and E. C. Meza. 1998. Robotization at El Teniente Mine. In *Latin American Perspectives: Exploration, Mining, and Processing*, edited by O. A. Bascur, pp. 113–122. Littleton, CO: Society for Mining, Metallurgy, and Exploration.

Copur, H., L. Ozdemir, and J. Rostami. 1998. Roadheader applications in mining and tunneling. *Mining Engineering* 50(3): 38–42.

Cording, E. J., and A. F. Cepeda-Diaz. 1992. Soil mechanics, Sec. 10.1, in *SME Mining Engineering Handbook*, 2d ed., edited by H. L. Hartman, pp. 809–828. Littleton, CO: Society for Mining, Metallurgy, and Exploration.

Cox, K. C. 1973. Opening and development, Sec. 10.4 in *SME Mining Engineering Handbook*, edited by A. B. Cummings and I. A. Givens, pp. 10-61–10-81. New York: AIME.

Crandall, W. E. 1992. Backfilling methods, Sec. 19.3 in *SME Mining Engineering Handbook*, 2d ed., edited by H. L. Hartman, pp. 1756–1778. Littleton, CO: Society for Mining, Metallurgy, and Exploration.

Crawford, J. T., and W. A. Hustrulid (eds.). 1979. *Open Pit Mine Planning and Design.* New York: SME-AIME. 367 pp.

Crooks, T. 2000. Telephone conversation with author. Washington, PA: R. G. Johnson Co.

Crowson, P. 1998. *Inside Mining: The Economics of the Supply and Demand of Minerals and Metals.* London: Mining Journal Books. 230 pp.

Cruickshank, M. J. 1992. Marine mining, Chap. 22.8 in *SME Mining Engineering Handbook*, 2d ed., edited by H. L. Hartman, pp. 1985–2027. Littleton, CO: Society for Mining, Metallurgy, and Exploration.

Cruickshank, M. J., and R. W. Marsden. 1973. Marine mining, Sec. 20 in *SME Mining Engineering Handbook*, edited by A. B. Cummins and I. A. Givens. New York: SME-AIME. 200 pp.

Cruickshank, M. J., C. M. Romanowitz, and M. P. Overall. 1968. Offshore mining: Present and future. *Engineering Mining Journal* 169(1): 84–91.

Culp, S. L. 2002. E-mail messages to author. Montreal, Canada: ABB Mining Systems.

Daily, A. F. 1968a. Placer mining, Sec. 13.5 in *Surface Mining*, edited by E. P. Pfleider, pp. 928–954. New York: AIME.

Daily, A. F. 1968b. Dredges and hydraulicking, Sec. 8.5 in *Surface Mining*, edited by E. P. Pfleider, pp. 503–527. New York: AIME.

Danielson, V., and J. Whyte. 1997. *Bre-X: Gold Today, Gone Tomorrow.* Toronto: The Northern Miner. 304 pp.

Das, B. M. 1983. *Advanced Soil Mechanics.* New York: McGraw-Hill. 511 pp.

Das, B. M. 1990. *Principles of Geotechnical Engineering*, 2d ed. Boston: PWS-Kent Publishing. 665 pp.

Dayton, S. H. 1981a. Cathedral Bluffs: Pushing to the outer limits for 94,000 bbl/day of shale oil. *Engineering and Mining Journal* 182(6): 78–84.

Dayton, S. H. 1981b. Rio Blanco pursues MIS tests as well as open-pit studies. *Engineering and Mining Journal* 182(6): 85–89.

DeGabriele, R. M., and G. D. Aho. 1982. Potential for reducing oil shale mining costs. *Mining Congress Journal* 68(12): 72–75.

Delinois, S. L. 1966. The challenge of the 70s: Mining on the moon. *Mining Engineering* 18(1): 63–69.

DeWolfe, V. 1981. Draw control in principle and practice at Henderson Mine, Chap. 56 in *Design and Operation of Caving and Sublevel Stoping Mines*, edited by D. R. Stewart, pp. 729–735. New York: SME-AIME.

Dick, R. A. 1973. Fragmentation, Sec. 11 in *SME Mining Engineering Handbook*, edited by A. B. Cummings and I. A. Givens, pp. 78–99. New York: AIME.

Dick, R. A., L. R. Fletcher, and D. V. D'Andrea. 1983. *Explosives and Blasting Procedures Manual.* IC 8925. U.S. Bureau of Mines. 95 pp.

Dick, R. 1985. Deep-sea mining versus land-based mining: A cost comparison, Chap. 1 in *The Economics of Deep-Sea Mining*, Berlin: Springer-Verlag. pp. 2–60.

Dobson, W. F., and D. R. Seelye. 1982. Mining technology assists oil recovery from Wyoming field. *Journal of Petroleum Technology* 34(2): 259–265.

Donner, W. S., and R. O. Wornat. 1973. Mining through boreholes—Frasch sulfur mining system, Sec. 21.6, in *SME Mining Engineering Handbook*, edited by A. B. Cummings and I. A. Givens, pp. 21-60–21-70. New York: AIME.

Dravo Corporation. 1974. *Analysis of Large-Scale Noncoal Underground Mining Methods*. U.S. Bureau of Mines, OFR 36-74, Dravo Corporation. Washington, DC: U.S. Government Printing Office. 605 pp.

Driscoll, J. M. 1997. From surface to underground: Newmont Gold's Carlin East Mine. *Mining Engineering* 49(8): 23–28.

Dunker, R. E., R. I. Barnhisel, and R. G. Darmody (eds.). 1992. *Proceedings of the* 1992 *National Symposium on Prime Farmland Reclamation*. Urbana: Department of Agronomy, University of Illinois. 284 pp.

Dwyer, R. T. 1997. Mining wastes and materials, Sec. 19.3, in *Mining Environmental Handbook*, edited by J. J. Marcus, pp. 726–728. London: Imperial College Press.

Dyas, K., and J. Marcus. 1998. Stillwater plans to triple PGM production by 2003. *Engineering and Mining Journal* 199(12): WW20–WW25.

E. I. duPont de Nemours and Co. 1977. *Blaster's Handbook*. 175th anniversary edition, Wilmington, DE: E. I. duPont de Nemours and Co. 494 pp.

Elkington, P. A., M. J. Scoble, J. R. Browne, and Y. V. Muftuoghe. 1983. Geophysical techniques applied to surface mine design in stratified deposits. *Mining Science and Technology* (Amsterdam) 1(1): 3–19.

Emerick, J. 2000. Telephone conversation with author. Ebensburg, PA: Central Cambria Drilling.

Emerson, D. W. 1982. Australian exploration and development—Comments and costs. In *Mineral Industry Costs*, edited by J. R. Hoskins, pp. 231–240. Spokane, WA: Northwest Mining Association.

Energy Information Administration. 1983. *Annual Energy Review*. Washington, DC: U.S. Department of Energy. 259 pp.

Energy Information Administration. 2000. *Coal Industry Annual* 1998. Washington, DC: U.S. Department of Energy. 256 pp.

Engineering and Mining Journal. 1969. Sublevel caving is now firmly established at Inco's Stobie Mine. *Engineering and Mining Journal* 170(5): 110–112.

Engineering and Mining Journal. 1991. A closer look at mining risk. *Engineering and Mining Journal* 192(1): 20–25.

Engineering and Mining Journal. 1992a. The Space Age comes to earth. *Engeering and Mining Journal* 193(7): 26–29.

Engineering and Mining Journal. 1992b. Non-electric blast initiation: How to select the right system. *Engineering and Mining Journal* 193(5): 16FF–16GG.

Engineering and Mining Journal. 1992c. Swedish Mining Group: Total technology for mining and processing. *Engineering and Mining Journal* 193(4) :S4-WW–S20-WW.

Engineering and Mining Journal. 1993. Refractory gold. *Engineering and Mining Journal* 194(6): WW20–WW24.

Engineering and Mining Journal. 1995a, Annual project survey *Engineering and Mining Journal* 196(1): 18–22.

Engineering and Mining Journal. 1995b. Mechanical mining. *Engineering and Mining Journal* 197(7): WW44–WW47.

Engineering and Mining Journal. 1996a. Annual project survey. *Engineering and Mining Journal* 197(1): 20–25.

Engineering and Mining Journal. 1996b. Seabed solutions to mining offshore. *Engineering and Mining Journal* 197(1): 16B–16D.

Engineering and Mining Journal. 1997. Annual project survey. *Engineering and Mining Journal* 198(1): 25–28.

Engineering and Mining Journal. 1998. Annual project survey. *Engineering and Mining Journal* 199(1): 15–20.

Engineering and Mining Journal. 1999a. Annual project survey. *Engineering and Mining Journal* 200(1): 22–25.

Engineering and Mining Journal. 1999b. Chile displaced by Nevada. *Engineering and Mining Journal* 200(1): 11-FF–16-II.

Erickson, A. J. 1992. Geologic interpretation, modeling, and representation, Sec. 5.5 in *SME Mining Engineering Handbook*, 2d ed., edited by H. L. Hartman, pp. 333–343. Littleton, CO: Society for Mining, Metallurgy, and Exploration.

Ertekin, T. 1984. Flow dynamics of coalbed methane in the vicinity of degasification wells. *Earth and Mineral Sciences* (Pennsylvania State University) 53(2): 17–19.

Espley, S., and G. Tan. 1994. Final recovery at Inco's Little Stobie Mine: A case study. *CIM Bulletin* 87(976): 66–73.

Evers, J. L. 1983. *Hydraulic Mining*, Preprint 83-384. SME Fall Meeting. Society for Mining, Metallurgy, and Exploration, Littleton, CO. 7 pp.

Farmer, I. 1992. Room-and-pillar mining, Sec. 18.1 in *SME Mining Engineering Handbook*, 2d ed., edited by H. L. Hartman, pp. 1681–1701. Littleton, CO: Society for Mining, Metallurgy, and Exploration.

Farquharson, G., and J. Marshall. 1996. The mining town—An endangered species. *CIM Bulletin* 89(1005): 74–76.

Filas, B. 1997. Coal, Chap. 12 in *Mining Environmental Handbook*, edited by J. J. Marcus, pp. 569–598. London: Imperial College Press.

Filas, B., and J. T. Gormley. 1997. The Summitville Mine: Buildup to disaster, Sec. 18.3 in *Mining Environmental Handbook*, edited by J. J. Marcus, pp. 687–697. London: Imperial College Press.

Fiscor, S. 2000. U.S. longwall census 2000. *Coal Age* 105(2): 32–35.

Flipse, J. E. 1983. Deep ocean mining economics. *Proceedings of the Offshore Technology Conference*. pp. 415–418. Houston, TX: Society of Petroleum Engineers.

Folinsbee, J. C., and R. W. Clarke. 1981. Selecting a mining method, Chap. 5 in *Design and Operation of Caving and Sublevel Stoping Mines*, edited by D. R. Stewart, pp. 55–65. New York: SME-AIME.

Ford, Bacon, and Davis, Inc. 1975. *Technology of Auger Mining*. Report PB259316. Springfield, VA: National Technical Information Center. 78 + pp.

Fortney, S. J. 1996. Advanced minewide automation in potash. *CIM Bulletin* 89(996)): 41–46.

Fortney, S. J. 2001. Advanced minewide automation in potash. In *Underground Mining Methods: Engineering Fundamentals and International Case Studies*, edited by W. A. Hustrulid and R. L. Bullock, pp. 143–148. Littleton, CO: Society for Mining, Metallurgy, and Exploration.

Foster, F. J. 1994. Australian mineral sands — Problems but confidence. *Engineering and Mining Journal* 195(1): WW56–WW62.

Fourie, G. A., and G. C. Dohm. 1992. Open pit planning and design, Sec. 13.1 in *SME Mining Engineering Handbook*, 2d ed., edited by H. L. Hartman, pp. 1274–1278, Littleton, CO: Society for Mining, Metallurgy, and Exploration.

Fowkes, R. S., and J. J. Wallace. 1968. *Hydraulic Coal Mining Research*. RI 7090. U.S. Bureau of Mines. 23 pp.

Gallagher, J., and J. S. Teuscher. 1998. Henderson Mine: Preparing for the future. *Mining Engineering* 50(8): 47–52.

Gardner, E. D., and W. O. Vandenburg. 1982. Square-set system of mining. In *Underground Mining Methods Handbook*, edited by W. A. Hustrulid, pp. 667–729. Littleton, CO: Society for Mining, Metallurgy, and Exploration.

Garnett, R. H. T. 1997. Problems with dredging in offshore Alaska. *Mining Engineering* 49(3): 27–33.

Gentry, D. W., and T. J. O'Neil. 1984. *Mine Investment Analysis*. New York: SME-AIME. 502 pp.

Gertsch, R. E. 1994. Mechanical mining: Challenges and directions. *Mining Engineering* 46(11): 1250–1253.

Glanville, R. 1984. *Optimum Production Rate for High-Grade/Low-Tonnage Mines*, Preprint 84-355. Fall Meeting, SME-AIME, Denver. 15 pp.

Goodman, R. E. 1980. *Introduction to Rock Mechanics*. New York: John Wiley & Sons, Inc. 478 pp.

Graham, M. 1998. Longwall mining systems evolve. *Coal Age* 103(3): 42–46.

Grant, B. F. 1964. Retorting oil shale underground: Problems and possibilities. *Colorado School of Mines Quarterly* 59(9): 86–89.

Green, P. 1976. Big hole blasting at Inco. *Mining Congress Journal* 62(12): 21–28.

Green, P. 1985. Sublevel caving method extracts coal from 55° pitching seam. *Coal Age* 90(9): 86–89.

Greene, D. 1999. Computer-aided earthmoving systems. *Mining Engineering* 51(2): 49–52.

Gregory, C. E. 1979. *Explosives for North American Engineers*, 2d ed. Rockport, MA: Trans Tech Publication. 303 pp.

Gregory, C. E. 1980. *A Concise History of Mining*. Oxford: Pergamon. 259 pp.

Guilbert, J. M., and C. F. Park Jr. 1986. *The Geology of Ore Deposits*. New York: W. H. Freeman and Company. 985 pp.

Gupta, R. P. 1991. *Remote Sensing Geology*. Berlin: Springer-Verlag. 356 pp.

Gustafsson, H. R. 1981. Field test of sublevel shrinkage caving (MSTM) at LKAB Kiruna, Chap. 30 in *Design and Operation of Caving and Sublevel Stoping Mines*, edited by D. R. Stewart, pp. 419–424. New York: SME-AIME.

Gustafsson, R. 1981. *Blasting Technique*. Vienna: Dynamite Nobel Wien Gesellschaft. 327 pp.

Haag, T. 1997. The sky's no limit. *Mining Voice* 3(1): 32–35.

Halkyard, J. E. 1979. Ocean mining technology and environmental studies. *Mining Congress Journal* 65(8): 25–28.

Hallof, P. G. 1992. Electrical addendum, Chap. 2 in *Practical Geophysics II*, edited by R. Van Blaricom, pp. 139–176. Spokane, WA: Northwest Mining Association.

Halls, J. L. 1982. Financial considerations, Sec. 6 in *Underground Mining Methods Handbook*, edited by W. A. Hustrulid, pp. 1422–1449. New York: SME-AIME.

Hamilton, L. H., and B. S. Trasker. 1984. Practical aspects of drilling for coal and stratiform deposits on triangular grids. *Coal Journal* (Sydney) (March): 67–73.

Hamrin, H. 1982. Choosing an underground mining method, Sec. 1.6 in *Underground Mining Methods Handbook*, edited by W. A. Hustrulid, pp. 88–112. New York: SME-AIME.

Hamrin, H. 1998. Choosing an underground mining method, Chap. 2 in *Techniques in Underground Mining*, edited by R. E. Gertsch and R. L. Bullock, pp. 45–85. Littleton, CO: Society for Mining, Metallurgy, and Exploration.

Hamrin, H. 2001. Underground mining methods and applications. In *Underground Mining Methods: Engineering Fundamentals and International Case Studies*, edited by W. A. Hustrulid and R. L. Bullock, pp. 3–14. Littleton, CO: Society for Mining, Metallurgy, and Exploration.

Haptonstall, J. 1992. Shrinkage stoping, Sec. 18.3 in *SME Mining Engineering Handbook*, 2d ed., edited by H. L. Hartman, pp. 1712–1716. Littleton, CO: Society for Mining, Metallurgy, and Exploration.

Harmon, J. H. 1973. Hoists and hoisting systems, Sec. 15 in *SME Mining Engineering Handbook*, edited by A. B. Cummins and I. A. Given. New York: SME-AIME, pp. 15-1 to 15-69.

Hartman, H. L. 1970. Shall we look toward the socio-engineer. *Professional Engineer* 40(5): 36–39.

Hartman, H. L. 1974. *A Social Direction for Engineering Education*, Preprint 74-WA/TS-5. New York Meeting, American Society of Mechanical Engineers. 13 pp.

Hartman, H. L. 1987. *Introductory Mining Engineering*. New York: John Wiley & Sons, Inc. 633 pp.

Hartman, H. L. 1990. Drilling principles, Sec. 6.1.1 in *Surface Mining*, 2d ed., edited by B. A. Kennedy, pp. 513–523. Littleton, CO: Society for Mining, Metallurgy, and Exploration.

Hartman, H. L. (ed.). 1992. *SME Mining Engineering Handbook*, 2d ed. 2 vols. Littleton, CO: Society for Mining, Metallurgy, and Exploration. 2260 pp.

Hartman, H. L., J. M. Mutmansky, R. V., Ramani, and Y. J. Wang. 1997. *Mine Ventilation and Air Conditioning*, 3d ed. New York: John Wiley & Sons, Inc. 730 pp.

Haycocks, C. 1992. Stope-and-pillar mining, Sec. 18.2 in *SME Mining Engineering Handbook*, 2d ed., edited by H. L. Hartman, pp. 1702–1711. Littleton, CO: Society for Mining, Metallurgy, and Exploration.

Haycocks, C., and R. C. Aelick. 1992. Sublevel stoping, Sec. 18.4 in *SME Mining Engineering Handbook*, 2d ed., edited by H. L. Hartman, pp. 1717–1731. Littleton, CO: Society for Mining, Metallurgy, and Exploration.

Hedberg, B. 1981. Large-scale underground mining: An alternative to open cast mining. *Mining Magazine* (London) 148(9): 177–183.

Heden, H., K. Lidin, and R. Malenstrom. 1982. Sublevel caving at LKAB's Kiirunavaara Mine. In *Underground Mining Methods Handbook*, edited by W. A. Hustrulid, pp. 928–944. New York: SME-AIME.

Heinz, W. F. 1989. *Diamond Drilling Handbook.* Halfway House, Republic of South Africa: W. F. Heinz. 525 pp.

Hemphill, G. B. 1981. *Blasting Operations.* New York: McGraw-Hill. 258 pp.

Henderson, K. J. 1982. Shrinkage stoping at the Crean Hill Mine. In *Underground Mining Methods Handbook* edited by W. A. Hustrulid, pp. 490–494. New York: SME-AIME.

Henderson, M. E. 1997. Heap and dump leach design, Sec. 8.7 in *Mining Environmental Handbook*, edited by J. J. Marcus, pp. 463–476. London: Imperial College Press.

Herbich, J. B. 1992. *Handbook of Dredging Engineering.* New York: McGraw-Hill.

Hewett, D. F. 1929. Cycles in metal production. *American Institute of Mining, Metallurgical, and Petroleum Engineers, Transactions* 85: 65–93.

Hill, R. W., C. B. Thorness, R. J. Cena, and D. R. Stephens. 1985. Results of the Centralia underground coal gasification field test. *In Situ* 9(3): 233–259.

Hoek, E., and E. T. Brown. 1980. *Underground Excavations in Rock.* Brookfield, VT: Institute of Mining and Metallurgy, North American Publication Center. 532 pp.

Hood, M. C., and F. F. Roxborough. 1992. Rock breakage: Mechanical, Sec. 9.1 in *SME Mining Engineering Handbook*, 2d ed., edited by H. L. Hartman, pp. 680–721. Littleton, CO: Society for Mining, Metallurgy, and Exploration.

Hopler, R. B. (ed.). 1998. *Blasters' Handbook*, Cleveland: International Society of Explosive Engineers. 742 pp.

Hoppe, R. W. 1976. Phosphates are vital to agriculture—And Florida mines for one-third the world. *Engineering and Mining Journal* 177(9): 79–89.

Hoppe, R. W. 1978. Fine-tuned room-and-pillar stoping yields iron ore at $4 per ton. *Engineering and Mining Journal* 170(6): 157–165.

Hornsby, B., and Staff. 1982. Cut-and-fill mining at Mt. Isa Mines Ltd. In *Underground Mining Methods Handbook*, edited by W. A. Hustrulid, pp. 531–538. New York: SME-AIME.

Hoskins, J. R. (ed.). 1982. *Mineral Industry Costs.* Spokane, WA: Northwest Mining Association. 248 pp.

Hoskins, J. R. (ed.). 1986. *Mineral Industry Costs.* Spokane, WA: Northwest Mining Association.

Hower, J. C., and B. K. Parekh. 1991. Chemical/physical properties and marketing, Chap. 1 in *Coal Preparation*, 5th ed. edited by J. W. Leonard and B. C. Hardinge, pp. 1–94. Littleton, CO: Society for Mining, Metallurgy, and Exploration.

Hrabik, J. A. 1986. Economic and environmental comparison: Borehole mining versus conventional mining of phosphate. *Mining Engineering* 38(1): 33–39.

Humphreys, K. K., and S. Katell. 1981. *Basic Cost Engineering.* New York: Marcel Dekker. 218 pp.

Hunt, D. K. 1992. Environmental protection and permitting, Sec. 7.3 in *SME Mining Engineering Handbook*, 2d ed., edited by H. L. Hartman, pp. 502–519. Littleton, CO: Society for Mining, Metallurgy, and Exploration.

Huston, J. 1970. *Hydraulic Dredging.* Cambridge, MD: Cornell Maritime Press. 304 pp.

Hustrulid, W. A. (ed.). 1982. *Underground Mining Methods Handbook.* Littleton, CO: Society for Mining, Metallurgy, and Exploration.

Hustrulid, W. A. 1999. *Blasting Principles for Open Pit Mining*. 2 vols. Rotterdam: Balkema.

Hustrulid, W. A., and R. L. Bullock (eds.). 2001. *Underground Mining Methods: Engineering Fundamentals and International Case Studies*. Littleton, CO: Society for Mining, Metallurgy, and Exploration. 718 pp.

Hustrulid, W. A., and M. Kuchta. 1995. *Open Pit Mine Planning and Design*, Vol. 1. Rotterdam: Balkema. 636 pp.

Hutchins, J. S., and D. L. Wassum. 1981. Oil mining: An emerging technology. *Mining Engineering* 53(12): 1695–1698.

Hutchinson, I. P., and R. D. Ellison (eds.). 1992. *Mine Waste Management*. Chelsea, MI: Lewis Publishers. 654 pp.

Illinois Department of Mines and Minerals. 1985. *Citizen's Guide to Coal Mining and Reclamation in Illinois*. Springfield: Illinois Department of Mines and Minerals. 43 pp.

Jackson, C. F., and J. H. Hedges. 1939. *Metal Mining Practice*. U.S. Bureau of Mines, Bulletin No. 419. Washington, DC: U.S. Government Printing Office. 512 pp.

Jackson, D. 1980. Hydromining comes of age. *Coal Age* 85(11): 50–56, 59–60, 63.

Jaeger, J. C., and N. G. Cook. 1976. *Fundamentals of Rock Mechanics*, 2d ed. New York: Halsted/John Wiley & Sons, Inc.; London: Chapman and Hall, 1985.

Jee, C. K. 1977. *Review and Analysis of Oil Shale Technology*. Vol. 3, *Modified In-Situ Technology*, p. 6. U.S. Department of Energy, Report No. EX-76-C01-2343. Boston, MA: Booz-Allen and Hamilton.

Jenkins, J. G. 1994. Copper heap leaching at San Manuel. *Mining Engineering* 46(9): 1094–1098.

Jenkins, R. 1997. Job satisfaction—Its effect on safety and turnover on a FIFO Operation. *AusIMM Bulletin* (4, June): 17–23.

Jensen, M. L., and A. M. Bateman. 1981. *Economic Mineral Deposits*. New York: John Wiley & Sons, Inc. 519 pp.

Jeremic, M. L. 1979. Hydraulic mining: Possible method for Rocky Mountain coal. *Mining Magazine* (London) 141(4): 330–339.

Jeremic, M. L. 1987. *Ground Mechanics in Hard Rock Mining*, pp. 218–256. Rotterdam: Balkema.

Johns, J. H. 1981. Rubber-tired mining equipment at Climax, Chap. 51 in *Design and Operation of Caving and Sublevel Stoping Mines*, edited by D. R. Stewart, pp. 675–681. New York: SME-AIME.

Johnson, D. D., and T. G. Penny. 1997. Case study—NWT diamonds project. *Proceedings of Mindev '97*, pp. 255–273. Sydney: Australasian Institute of Mining and Metallurgy.

Johnson, R., and G. Buchan. 1999. *Optimization of Longwall Mining with Gateroad Planning*, Preprint. SME Annual Meeting, Denver, CO.

Jones, E. A., and W. T. Pettijohn. 1973. Examinations, valuation, and reports, Sec. 32 in *SME Mining Engineering Handbook*, edited by A. B. Cummings and I. A. Givens, pp. 32-2–32-56. New York: AIME.

Jones, P. C. 1993. Can the mining industry survive Summitville? *Mining Engineering* 45(11): 1377–1381.

Journel, A. G., and C. J. Huijbregts. 1978. *Mining Geostatistics*. New York: Academic Press. 600 pp.

Julin, D. E. 1992. Block caving, Chap. 20.3 in *SME Mining Engineering Handbook*, 2d ed., edited by H. L. Hartman, pp. 1815–1836. Littleton, CO: Society for Mining, Metallurgy, and Exploration.

Kaas, L. M. 1992. Major federal environmental laws and regulation, Sec. 3.4.2, in *SME Mining Engineering Handbook*, 2d ed. edited by H. L. Hartman, pp. 175–182. Littleton, CO: Society for Mining, Metallurgy, and Exploration.

Kahle, M. B., and C. A. Moseley. 1983. Development of mining methods in Gulf Coast lignites. *Mining Engineering* 35(8): 1163–1166.

Kalia, H. N., and J. B. Gresham. 1985. Commercialization of oil shales via in-situ retorting: Some considerations. *Mining Engineering* 37(9): 1141–1148.

Karanam, U. M. R., and B. Misra. 1998. *Principles of Rock Drilling*. Rotterdam: Balkema. 265 pp.

Katen, K. P. 1992. Health and safety standards, Sec. 3.3 in *SME Mining Engineering Handbook*, 2d ed., edited by H. L. Hartman, pp. 162–173. Littleton, CO: Society for Mining, Metallurgy, and Exploration.

Kaufman, W. W., and J. C. Ault. 1977. *Design of Surface Mine Haulage Roads*. IC 8758. U.S. Bureau of Mines. 68 pp.

Keen, A. J. 1992. Polaris update. *CIM Bulletin* 85(961): 51–57.

Kennedy, B. A. (ed.). 1990. *Surface Mining*, 2d ed. Littleton, CO: Society for Mining, Metallurgy, and Exploration. 1194 pp.

Kennedy, B. A., O. Nair, and L. A. Readdy. 1980. The mining of oil. *Mining Magazine* (London) 143(1): 26–37.

Kim, Y. C. 1979. Production scheduling, technical overview, Sec. 4.3.1 in *Computer Methods for the 80s in the Mineral Industry*, edited by A. Weiss, pp. 610–614. New York: SME-AIME.

Kim, Y. C., F. Martino, and I. K. Chopra. 1981. Application of geostatistics in a coal deposit. *Mining Engineering* 33(10): 1476–1481.

King, R. H. 1992. Automation and robotics, Sec. 22.2 in *SME Mining Engineering Handbook*, 2d ed., edited by H. L. Hartman, pp. 1908–1917. Littleton, CO: Society for Mining, Metallurgy, and Exploration.

Klein, J., and J. J. Lajoie. 1992. Electromagnetics, Chap. 6 in *Practical Geophysics II*, edited by R. Van Blaricom, pp. 383–437. Spokane, WA: Northwest Mining Association.

Knudsen, H. P., Y. C. Kim, and E. Mueller. 1978. Comparative study of the geostatistical ore reserve estimation method over the conventional methods. *Mining Engineering* 30(1): 54–58.

Kostick, D. S. 1982. The influence of solution mining on the world soda ash market. In *Interfacing Technologies in Solution Mining*, edited by W. J. Schlitt, pp. 21–30. New York: SME-SPE-AIME.

Kostner, F. 1976. Surface milling of granite. *Industrial Diamond Review* 36 (July): 242–246.

Kral, S. 1997. Mining industry beginning to rediscover Alaska. *Mining Engineering* 49(1): 45–50.

Kramer, D. A. 1999. *Explosives*. Commodity Report. Washington, DC: U.S. Geological Survey. 6 pp. (available at http://minerals. usgs,gov/minerals/pubs/commodity/explosives/600498. pdf).

Kump, D., and T. Arnold. 2001. Underhand cut-and-fill at the Barrick Bullfrog Mine. In *Underground Mining Methods: Engineering Fundamentals and International Case Studies*, edited by W. A. Hustrulid and R. L. Bullock, pp. 345–350. Littleton, CO: Society for Mining, Metallurgy, and Exploration.

Kvapil, R. 1982. The mechanics and design of sublevel caving systems. In *Underground Mining Methods Handbook*, edited by W. A. Hustrulid, pp. 880–897. New York: SME-AIME.

Kvapil, R. 1992. Sublevel caving, Chap. 20.2 in *SME Mining Engineering Handbook*, 2d ed., edited by H. L. Hartman, pp. 1789–1814. Littleton, CO: Society for Mining, Metallurgy, and Exploration.

Lacy, W. C., and J. C. Lacy. 1992. History of mining, Sec. 1.1 in *SME Mining Engineering Handbook*, 2d ed., edited by H. L. Hartman, pp. 5–23. Littleton, CO: Society for Mining, Metallurgy, and Exploration.

Laflamme, M., S. Planeta, and C. Bourgoin. 1994. Technological aspects of narrow vein mining: Suggested modifications and new developments. *CIM Bulletin* 87(978): 145–149.

Langefors, U., and B. Kihlstrom. 1978. *The Modern Technique of Rock Blasting*, 3d ed. New York: Halsted/Wiley. 438 pp.

Le Roy, R. 2001. Evolution of undercut-and-fill at SMJ's Jouac Mine, France. In *Underground Mining Methods: Engineering Fundamentals and International Case Studies*, edited by W. A. Hustrulid and R. L. Bullock, pp. 355–357. Littleton, CO: Society for Mining, Metallurgy, and Exploration.

Lewis, A. 1984. The comeback of California placers. *Engineering and Mining Journal* 185(2): 36–41.

Lewis, R. S., and G. B. Clark. 1964. *Elements of Mining*, 3d ed. New York: John Wiley & Sons, Inc. 768 pp.

Li, T. M. 1976. Caland opens new iron ore reserve with hydraulic slurry system. *Mining Engineering* 28(5): 37–45.

Lineberry, G. T., and A. P. Paolini. 1992. Equipment selection and sizing, Sec. 17.2 in *SME Mining Engineering Handbook*, 2d ed., edited by H. L. Hartman, pp. 1550–1571. Littleton, CO: Society for Mining, Metallurgy, and Exploration.

Link, J. M. 1982. Pipelining bulk materials. *Mining Engineering* 34(10): 1444–1447, 1456.

Livingston, C. W. 1956. Fundamental concepts of rock failure: Discussion. *Proceedings of 1st Symposium on Rock Mechanics*, publ. in *Colorado School of Mines Quarterly* 51(3): 1–11, 226–228.

Lopez Jimeno, C. L., E. L. Lopez Jimeno, and F. J. Ayala Carcedo. 1995. *Drilling and Blasting of Rocks*. Rotterdam: Balkema. 391 pp.

Lovering, T. S. 1943. *Minerals in World Affairs*., Englewood Cliffs, NJ: Prentice-Hall. 394 pp.

Lucas, J. R., and L. Adler (eds.). 1973. Roof and ground control, Sec. 13 in *SME Mining Engineering Handbook*, edited by A. B. Cummins and I. A. Given. New York: SME-AIME. pp. 13-1 to 13-196.

Lucas, J. R., and C. Haycocks (eds.). 1973. Underground mining systems and equipment, Sec. 12 in *SME Mining Engineering Handbook*, edited by A. B. Cummins and I. A. Given. New York: SME-AIME. 262 pp.

Ludlow, M. 2000. Twentymile Coal Company—A case study. Third Annual Longwall Summit. *Australian Journal of Mining* (Yeppon, Queensland, Australia).

Lyman, T. J., and E. M. Piper. 1985. Heavy oil mining—An overview. *Journal of Technical Topics in Civil Engineering* 111(1): 20–32.

Lyman, W. 1982. Shrinkage stoping. In *Underground Mining Methods Handbook*, edited by W. A. Hustrulid, pp. 485–489. Littleton, CO: Society for Mining, Metallurgy, and Exploration.

Macdonald, E. H. 1983. *Alluvial Mining*. London: Chapman & Hall. 508 pp.

Mace, B. C. 2000. Telephone conversation with author. Houston, TX: Hitachi Construction Machinery Corporation.

MacKenzie, B. W., and M. L. Bilodeau. 1984. *Economics of Mineral Exploration in Australia—Guidelines for Corporate Planning and Government Policy*. Glenside, S. Australia: Australian Mineral Foundation. 171 pp.

Mackenzie, S., and G. Hare. 1994. Equipment selection for narrow vein operations. *Engineering Mining Journal* 195(1): WW65–WW67.

Madigan, R. T. 1981. *Of Minerals and Man*. Parkville, Australia: Australasian Institute of Mining and Metallurgy. 138 pp.

Madsen, D., A. Moss, B. Salamondra, and D. Etienne. 1991. Stope development for raise mining at the Namew Lake Mine. *CIM Bulletin* 84(949): 33–39.

Malenka, W. T. 1968. *Hydraulic Mining of Anthracite*. RI 7120. U.S. Bureau of Mines. 19 pp.

Maley, T. S. 1996. *Mineral Law*, 6th ed. Boise, ID: Mineral Land Publications. 936 pp.

Maloney, W. J. 1993. Blind drilling becoming an accepted technology for ventilation shafts. *Mining Engineering* 45(11): 1374–1376.

Maloney, W. J. 2000. Telephone conversation with author. Morgantown, WV: North American Drillers, Inc.

Mann, C. D. 1982. Introduction to sublevel stoping. In *Underground Mining Methods Handbook*, edited by W. A. Hustrulid, pp. 362–363. New York: SME-AIME.

Marchand, R., P. Godin, and C. Doucet. 2001. Shrinkage stoping at the Mouska Mine. In *Underground Mining Methods: Engineering Fundamentals and International Case Studies*, edited by W. A. Hustrulid and R. L. Bullock, pp. 189–194. Littleton, CO: Society for Mining, Metallurgy, and Exploration.

Marcus, J. J. (ed.). 1997a. *Mining Environmental Handbook*. London: Imperial College Press. 785 pp.

Marcus, J. J. 1997b. The Briggs Mine. *Engineering and Mining Journal* 198(9):16GG–1600.

Marsden, R. W., and J. R. Lucas (eds.). 1973. Specialized underground extraction systems, Sec. 21 in *SME Mining Engineering Handbook*, edited by A. B. Cummins and I. A. Givens. New York: SME-AIME. pp. 21-1 to 21-118.

Martens, C. D. 1982. Mining and quarrying trends in the metallic and nonmetallic industries. In *Minerals Yearbook*, vol. 1, pp. 1–25. U.S. Bureau of Mines. Washington, DC: U.S. Government Printing Office.

Martin, D. C., M. E. Smith, and F. A. Chesworth. 1995. Cerro Colorado on-off leach pad built in six months. *Mining Engineering* 47(4): 327–332.

Martin, J. W., T. J. Martin, T. P. Bennett, and K. M. Martin. 1982. *Surface Mining Equipment*. Golden, CO: Martin Consultants, Inc. 455 pp.

Mathieson, G. A. 1982. *Open Pit Sequencing and Scheduling*, Preprint 82-368. Fall meeting, SME-AIME, Honolulu. 15 pp.

Mathtech, Inc. 1976. *Evaluation of Current Surface Coal Mining Overburden Handling Techniques and Reclamation Practices: Phase III, Eastern U.S.* Report. USBM Contract S0144081. Princeton, NJ: Mathtech, Inc.

Maurer, W. C. 1967. The state of rock mechanics knowledge in drilling. In *Proceedings of 8th Symposium on Rock Mechanics*, edited by C. Fairhurst, pp. 355–395. New York AIME.

Maurer, W. C. 1968. *Novel Drilling Techniques*. New York: Pergamon Press. 114 pp.

Maurer, W. C. 1980. *Advanced Drilling Techniques*. Tulsa: Petroleum Publishing. 698 pp.

McCarter, M. K., and H. M. Smolnikar. 1992. Auger mining, Chap. 14.4 in *SME Mining Engineering Handbook*, 2d ed., edited by H. L. Hartman, pp. 1447–1452. Littleton, CO: Society for Mining, Metallurgy, and Exploration.

McClanahan, E. A. 1995. Coalbed methane: Myths, facts, and legends of its history and the legislative and regulatory climate into the 21st century. *Oklahoma Law Review* 48(3): 471–561.

McDonald, D. B. 1992. Variable interburden, horizontal multiple coal seams: Colowyo Coal Company, Sec. 14.3.3 in *SME Mining Engineering Handbook*, 2d ed., edited by H. L. Hartman, pp. 1415–1421. Littleton, CO: Society for Mining, Metallurgy, and Exploration.

McGee, E. S. 2000. *Colorado Yule Marble—Building Stone of the Lincoln Memorial*. Bulletin 2162, Washington, DC: U.S. Geological Survey (available at http://publ. usgs. gov/pdf/bulletin/b2162/).

McGinn, R. E. 1991. *Science, Technology, and Society*. Englewood Cliffs, NJ: Prentice-Hall. 302 pp.

McGregor, B. A., and T. W. Offield. 1983. *The Exclusive Economic Zone: An Exciting New Frontier*. U.S. Geological Survey. Washington, DC: U.S. Government Printing Office. 20 pp.

McKee, W. N. 1992. Athabasca oilsands: Syncrude mine, Sec. 14.3.7 in *SME Mining Engineering Handbook*, 2d ed., edited by H. L. Hartman, pp. 1435–1441. Littleton, CO: Society for Mining, Metallurgy, and Exploration.

McLaughlin, R. 1998. Diamonds from the Far North. *Engineering and Mining Journal* 199(8): 34–40.

McLean, C. A. 1992. Placer mining costs, Sec. 15.1.7 in *SME Mining Engineering Handbook*, 2d ed., edited by H. L. Hartman, pp. 1471–1473. Littleton, CO: Society for Mining, Metallurgy, and Exploration.

McLean, C. A., A. W. K. McDowell, and T. D. McWaters. 1992. Dredging and placer mining, Chap. 15.1 in *SME Mining Engineering Handbook*, 2d ed., edited by H. L. Hartman, pp. 1454–1473. Littleton, CO: Society for Mining, Metallurgy, and Exploration.

McLean, M. E. 2001. Mechanizing sunshine mining operations. In *Underground Mining Methods: Engineering Fundamentals and International Case Studies*, edited by W. A.

Hustrulid and R. L. Bullock, pp. 309–311. Littleton, CO: Society for Mining, Metallurgy, and Exploration.

McMahon, B. K., and R. F. Kendrick. 1969. *Predicting the Block Caving Behavior of Ore Bodies*, Preprint 69-AU-51. New York: SME-AIME. 15 pp.

McPherson, M. J. 1993. *Subsurface Ventilation and Environmental Engineering*. London: Chapman and Hall. 905 pp.

McQuiston, F. W. Jr., and L. J. Bechaud, Jr. 1968. Metallurgical sampling and testing, Sec. 3.2 in *Surface Mining*, edited by E. P. Pfleider, pp. 101–121. New York: AIME.

Meade, L. P. 1992. Dimension stone: Friendsville Quarry, Sec. 14.2.1 in *SME Mining Engineering Handbook*, 2d ed., edited by H. L. Hartman, pp. 1400–1402. Littleton, CO: Society for Mining, Metallurgy, and Exploration.

Meiklejohn, T. C., and E. A. Meiklejohn. 1997. Visiting a Scottish gold mine. *AUSIMM Bulletin* 7: 53–55.

Metz, R. A. 1992. Sample collection, Sec. 5.3 in *SME Mining Engineering Handbook*, 2d ed., edited by H. L. Hartman, pp. 314–326. Littleton, CO: Society for Mining, Metallurgy, and Exploration.

Mine Safety and Health Administration. 2000. Home page, http://www. msha. gov/.

Mining Survey. 1983. Shaft sinking: The years ahead. *Mining Survey* (Johannesburg) (3/4): 39–45.

Mitchell, S. T. 1981. Vertical crater retreat stoping as applied at the Homestake Mine, Chap. 44 in *Design and Operation of Caving and Sublevel Stoping Mines*, edited by D. R. Stewart, pp. 609–626. New York: SME-AIME.

Molloy, P. M. 1986. *The History of Mining and Metallurgy*. New York: Garland. 319 pp.

Moore, J. K. 1997. *Mining and Quarrying Trends*. U.S. Geological Survey, Minerals Information. 13 pp. (available as http://minerals. usgs. gov/minerals/pubs/commodity/m&9/873497. pdf).

Morrison, R. G., and P. L. Russell. 1973. Selecting a mining method: Rock mechanics, other factors, Sec. 9 in *SME Mining Engineering Handbook*, edited by A. B. Cummings and I. A. Givens, pp. 9-1–9-22. New York: AIME.

Mudder, T., and S. Miller. 1998. Closure alternatives for heap-leach facilities. *Mining Environmental Management* 6(1): 17–20.

Murdy, W. M., and R. R. Bhappu. 1997. Risk mitigation in global mining development. *Mining Engineering* 49(9): 31–43.

Murphy, J. N. 1985. Remote control promises safety and productivity gains. *Coal Mining* 22(10): 38–42.

Murray, D. K. 1996. Coalbed methane in the USA: Analogues for worldwide development. In *Coalbed Methane and Coal Geology*, edited by R. Gayer and I. Harris, pp. 1–12. Special Publication No. 109. London: The Geological Society.

Murray, J. W. 1982. Introduction to undercut-and-fill mining. In *Underground Mining Methods Handbook*, edited by W. A. Hustrulid, pp. 631–638. Littleton, CO: Society for Mining, Metallurgy, and Exploration.

Murray, W. A. 1982. *Environmental Impacts of Underground Coal Gasification*, Preprint 82-259. Fall Meeting, SME-AIME. 13 pp.

Mutmansky, J. M. 1999. *Guidebook on Coalbed Methane Drainage for Underground Coal Mines.* White paper. Washington, DC: U.S. Environmental Protection Agency. 46 pp.

Mutmansky, J. M., and C. J. Bise (eds.). 1992. Auxiliary operations, Sec. 12 in *SME Mining Engineering Handbook*, 2d ed., edited by H. L. Hartman, pp. 1155–1269. Littleton, CO: Society for Mining, Metallurgy, and Exploration.

Mutmansky, J. M., S. C. Suboleski, T. A. O'Hara, and K. V. K. Prasad. 1992. Cost comparisons, Sec. 23.3 in *SME Mining Engineering Handbook*, 2d ed., edited by H. L. Hartman, pp. 2070–2089. Littleton, CO: Society for Mining, Metallurgy, and Exploration.

National Mining Association. 1998 September. *The Future Begins with Mining: A Vision of the Mining Industry of the Future.* Washington, DC: National Mining Association. 16 pp.

National Society of Professional Engineers. 2000. Home page, http://www.nspe.org/.

Naylor, D. J., and G. N. Pande. 1981. *Finite Elements in Geotechnical Engineering.* Swansea, UK.: Pineridge Press. 245 pp.

Nelson, P. P., Y. A. Al-Jalil, and C. Laughton. 1994. *Tunnel Boring Machine Project Data Bases and Construction Simulation.* Geotechnical Engineering Report GR94-4. Austin: Geotechnical Engineering Center, Department of Civil Engineering, University of Texas.

Nicholas, D. S. 1981. Method selection: A numerical approach. In *Design and Operation of Caving and Sublevel Stoping Mines*, edited by D. R. Stewart, pp. 39–53. New York: SME-AIME.

Nicholas, D. S. 1992a. Selection variables, Sec. 23.1 in *SME Mining Engineering Handbook*, 2d ed., edited by H. L. Hartman, pp. 2051–2057. Littleton, CO: Society for Mining Metallurgy, and Exploration.

Nicholas, D. S. 1992b. Selection procedure, Sec. 23.4 in *SME Mining Engineering Handbook*, 2d ed., edited by H. L. Hartman, pp. 2090–2106. Littleton, CO: Society for Mining, Metallurgy, and Exploration.

Nilsson, D. 1982a. Choosing an underground mining method. In *Underground Mining Methods Handbook*, edited by W.. A. Hustrulid, pp. 88–112. Littleton, CO: Society for Mining, Metallurgy, and Exploration.

Nilsson, D. 198b. Planning economics of sublevel caving. In *Underground Mining Methods Handbook*, edited by, W. A. Hustrulid, pp. 953–972. Littleton, CO: Society for Mining, Metallurgy, and Exploration.

Nilsson, D. 1992. Surface vs. underground methods, Sec. 23.2 in *SME Mining Engineering Handbook*, 2d ed., edited by H. L. Hartman, pp. 2058–2069. Littleton, CO: Society for Mining, Metallurgy, and Exploration.

Noble, A. C. 1992. Ore reserve/resource estimation, Sec. 5.6 in *SME Mining Engineering Handbook* 2d ed., edited by H. L. Hartman, pp. 344–359. Littleton, CO: Society for Mining, Metallurgy, and Exploration.

Norquist, B. 2001. Shrinkage stoping practices at the Schwartzwalder Mine. In *Underground Mining Methods: Engineering Fundamentals and International Case Studies*, edited by W. A. Hustrulid and R. L. Bullock, pp. 195–204. Littleton, CO: Society for Mining, Metallurgy, and Exploration.

Ober, J. A. 2000a. Lithium. *Mineral Commodities Survey*, pp. 100–101. Washington, DC: U.S. Geological Survey.

Ober, J. A. 2000b. Sulfur. *Mineral Commodities Survey*, pp. 164–165. Washington, DC: U.S. Geological Survey.

Obert, L. (ed.). 1973a. Elements of soil and rock mechanics, Sec. 6 in *SME Mining Engineering Handbook*, edited by A. B. Cummings and I. A. Givens, pp. 6-1–6-52. New York AIME.

Obert, L. 1973b. Design and stability of excavations, Sec. 7 in *SME Mining Engineering Handbook*, edited by A. B. Cummings and I. A. Givens, pp. 7-1–7-49. New York: AIME.

Obert, L. 1973c. Opening and development, Sec. 10 in *SME Mining Engineering Handbook*, edited by A. B. Cummings and I. A. Givens, pp. 10-1–10-4. New York: AIME.

Obert, L., W. I. Duvall, and R. H. Merrill. 1960. *Design of Underground Openings in Competent Rock*. Bulletin 587. U.S. Bureau of Mines. 36 pp.

Office of Surface Mining. 2000. Home page, http://www.osmre.gov/.

Office of Technology Assessment. 1980. *An Assessment of Oil Shale Technologies*. U.S. Congress, Office of Technology Assessment. New York: McGraw-Hill. 517 pp.

O'Hara, T. A., and S. C. Suboleski. 1992. Costs and cost estimation, Sec. 6.3 in *SME Mining Engineering Handbook*, 2d ed., edited by H. L. Hartman, pp. 405–424. Littleton, CO: Society for Mining, Metallurgy, and Exploration.

Olson, S. I. 1992. Florida phosphate: Ft. Green Mine, Sec. 14.3.8 in *SME Mining Engineering Handbook*, 2d ed., edited by H. L. Hartman, pp. 1442–1446. Littleton, CO: Society for Mining, Metallurgy, and Exploration.

O'Neil, T. 1996. Barrick goes underground with Meikle. *Mining Engineering* 48(11): 20–26.

Pana, M. T., and R. K. Davey. 1973. Pit planning and design, Sec. 17.2.2 in *SME Mining Engineering Handbook*, edited by A. B. Cummings and I. A. Givens, pp. 17-10–17-19. New York: AIME.

Panek, L. A. 1951. *Stresses About Mine Openings in Homogeneous Rock*. Ann Arbor, MI: Edwards Bros. 50 pp.

Paraszczak, J. 1992. Mechanized mining of narrow veins — Problems and equipment options. *Mining Engineering* 44(2): 147–150.

Paroni, W. A. 1992. Excavation techniques, Sec. 19.2 in *SME Mining Engineering Handbook*, 2d ed., edited by H. L. Hartman, pp. 1749–1755. Littleton, CO: Society for Mining, Metallurgy, and Exploration.

Parr, C. J. 1992. Mining law, Sec. 3.2 in *SME Mining Engineering Handbook*, 2d ed., edited by H. L. Hartman, pp. 140–161. Littleton, CO: Society for Mining, Metallurgy, and Exploration.

Parr, C. J., and N. Ely. 1973. Mining law, Sec. 2 in *SME Mining Engineering Handbook*, edited by A. B. Cummings and I. A. Givens, pp. 2-2–2-52. New York: AIME.

Payne, A. L. (ed.). 1973. Exploration for mineral deposits, Sec. 5 in *SME Mining Engineering Handbook*, edited by A. B. Cummings and I. A. Givens, pp. 5-1–5-105. New York: AIME.

Pearson, N. B. 1981. The development and control of block caving at the Chingola

Division of Nchanga Consolidated Copper Mines Ltd. Zambia, Chap. 18 in *Design and Operation of Caving and Sublevel Stoping Mines*, edited by D. R. Stewart, pp. 211–222. New York: SME-AIME.

Pease, R. C., and W. S. Watters, 1996. San Juan Ridge Gold Mine begins production. *Mining Engineering* 48(12): 26–31.

Peck, J., and J. Gray. 1995. The total mining system (TMS): The basis for open pit automation. *CIM Bulletin* 88(993): 38–44.

Peele, R. (ed.). 1941. *Mining Engineer's Handbook*, 3d ed., 2 vols. New York: John Wiley & Sons, Inc. 45 sections.

Peng, S. S., and H. S. Chiang. 1984. *Longwall Mining*. New York: John Wiley & Sons, Inc. 708 pp.

Peng, S. S., and H. S. Chiang. 1992. Longwall mining, Sec. 20.1 in *SME Mining Engineering Handbook*, 2d ed., edited by H. L. Hartman, pp. 1780–1788. Littleton, CO: Society for Mining, Metallurgy, and Exploration.

Penn, S. H. 1966. A preliminary look at lunar natural resources. *Mining Engineering* 18(3): 66–70.

Peppin, C., T. Fudge, K. Hartman, D. Bauer, and T. DeVoe. 2001. Underhand cut-and-fill mining at the Lucky Friday Mine. In *Underground Mining Methods: Engineering Fundamentals and International Case Studies*, edited by W. A. Hustrulid and R. L. Bullock, pp. 313–318. Littleton, CO: Society for Mining, Metallurgy, and Exploration.

Perkins, P. E. 1997. *World Metal Markets: The United States Strategic Stockpile and Global Market Influence*., Westport, CT: Praeger Publishers. 165 pp.

Peters, W. C. 1980. The environment of ore genesis. Paper 3. *Proceedings of the Institute on Mining Exploration Technology for Lawyers and Landmen*. Denver CO: Rocky Mountain Mineral Law Foundation. pp. 3–7.

Peters, W. C. 1987. *Exploration and Mining Geology*. New York: John Wiley & Sons, Inc. 685 pp.

Petry, E. F. 1982. Coarse coal transport. *Mining Congress Journal* 68(10): 35–37.

Petty, P. C. (ed.). 1981. *Metals, Minerals, Mining*, 2d ed., Special Issue. Westminster, CO: Colorado Section. American Institute of Professional Geologists. 36 pp.

Pfarr, J. D. 1991. Mechanized cut-and-fill mining as applied at the Homestake Mine. *Mining Engineering* 43(11): 1437–1439.

Pfleider, E. P. (ed.). 1968. *Surface Mining*. New York: AIME. 1061 pp.

Pfleider, E. P. (ed.). 1973. Open-pit and surface mining systems and equipment, Sec. 17 in *SME Mining Engineering Handbook*, edited by A. B. Cummings and I. A. Givens, pp. 17-2–17-180. New York: AIME.

Phelps, R. W. 1991. Hecla Mining Centennial, 1981–1991. *Engineering and Mining Journal* 192(9): 19–25.

Phelps, R. W. 1994. Centromin. *Engineering and Mining Journal* 195(5): WW62–WW68.

Phelps, R. W. 1999. A million barrels a day, plus. *Engineering and Mining Journal* 200(1): 26–30.

Piche, P., and P. Gaultier. 1996. Mining automation technology—The first frontier. *CIM Bulletin* 89(996): 51–54.

Pillar, C. L. 1981. A comparison of block caving methods, Chap. 8 in *Design and Operation of Caving and Sublevel Stoping Mines*, edited by D. R. Stewart, pp. 87–97. New York: SME-AIME.

Pillar, C. L., A. J. Petrina, and E. W. Cokayne. 1971. How Craigmont chose sublevel caving and why it proved successful. *World Mining* 24(6): 26–33.

Piotte, M., C. Coache, A. Drouin, and P. Gaultier. 1997. Automatic control of underground vehicles at high speed. *CIM Bulletin* 90(1006): 78–81.

Placer Development Ltd. 1980. *The Mine Development Process*. Annual report. Vancouver, BC: Placer Dome Inc. 49 pp.

Pond, R. 2000. Telephone conversation with author. Evansville, IN: Frontier-Kemper Constructors.

Poole, R. A., P. V. Golde, and G. R. Baiden. 1996. Remote operation from surface of Tamrock DataSolo drills at INCO's Stobie Mine. *CIM Bulletin* 89(996): 47–50.

Poole, R. A., P. V. Golde, G. R. Baiden, and M. Scoble. 1998. A review of INCO's mining automation efforts in the Sudbury Basin. *CIM Bulletin* 91(1016): 68–74.

Potential Gas Committee. 1993. *Potential Supply of Natural Gas in the USA*. Report. Golden, CO: Potential Gas Agency, Colorado School of Mines.

Power, W. R. 1975. Construction materials: Dimension and cut stone. In *Industrial Minerals and Rocks*, 4th ed., edited by S. J. Lefond, pp. 157–174. New York: SME-AIME.

Power Crane and Shovel Association. 1976. *Operating Cost Guide for Estimating Costs of Owning and Operating Cranes and Excavators*. Milwaukee, WI: Power Crane and Shovel Association. 23 pp.

Prager, S. 1997. Changing North America's mind-set about mining. *Engineering and Mining Journal* 198(2): 36–44.

Pugh, G. M., and D. G. Rasmussen. 1982. Cost calculations for highly mechanized cut-and-fill mining. In *Underground Mining Methods Handbook*, edited by W. A. Hustrulid, pp. 610–630. Littleton, CO: Society for Mining, Metallurgy, and Exploration.

Quinteiro, C., M. Quinteiro, and O. Hedstrom. 2001. Underground ore mining at LKAB, Sweden. In *Underground Mining Methods: Engineering Fundamentals and International Case Studies*, edited by W. A. Hustrulid and R. L. Bullock, pp. 361–368. Littleton, CO: Society for Mining, Metallurgy, and Exploration.

Rajaram, V. 1985. Commercialization of eastern U.S. oil shales: A review. *Mining Engineering* 37(12): 1381–1385.

Rand, G. 1992. Lignite: Big Brown mine, Sec. 14.3.6 in *SME Mining Engineering Handbook*, 2d ed., edited by H. L. Hartman, pp. 1431–1435. Littleton, CO: Society for Mining, Metallurgy, and Exploration.

Rausch, D., O. Strandberg, and W. E. Hawes. 1982. Ground support systems for heavy, squeezing ground at the Burgin Mine. In *Underground Mining Methods Handbook*, edited by W. A. Hustrulid, pp. 738–743. Littleton, CO: Society for Mining, Metallurgy, and Exploration.

Ravasz, K. 1984. New caving method boosts output. *Coal Age* 89(1): 60–61.

Raymond, R. 1984. *Out of the Fiery Furnace*. Melbourne, Australia: Macmillan. 274 pp.

Readdy, L. A., D. S. Bolin, and G. A. Mathieson. 1982. Ore reserve estimation, Sec. 1.3 in *Underground Mining Methods Handbook*, edited by W. A. Hustrulid, pp. 17–38. New York: SME-AIME.

Rech, W. D. 2001. Henderson Mine. In *Underground Mining Methods: Engineering Fundamentals and International Case Studies*, edited by W. A. Hustrulid and R. L. Bullock, pp. 385–403. Littleton, CO: Society for Mining, Metallurgy, and Exploration.

Reid, B. 1997a. 1997 International longwall census. Part 1. *Coal Age* 102(9): 28–34.

Reid, B. 1997b. 1997 International longwall census. Part 2. *Coal Age* 102(10): 34–39.

Rendu, J. M. 1981. *An Introduction to Geostatistical Methods of Mineral Evaluation*, 2d ed. Monograph Series. Johannesburg: South African Institute of Mining and Metallurgy. 84 pp.

Rhealt, J., and D. Bronkhurst. 1994. Backfill practices at the Williams Mine. *CIM Bulletin* 87(979): 44–48.

Rhoades, B. 1998. Typical underground drilling costs. *Aggregates Manager* 3(7): 21–23.

Richardson, M. P. 1981. Area of draw influence and drawpoint spacing for block caving mines, Chap. 13 in *Design and Operation of Caving and Sublevel Stoping Mines*, edited by D. R. Stewart, pp. 149–156. New York: SME-AIME.

Richner, D. R. 1992. In-situ mining of soluble salts, Sec. 15.3.1 in *SME Mining Engineering Handbook*, 2d ed., edited by H. L. Hartman, pp. 1493–1512. Littleton, CO: Society for Mining, Metallurgy, and Exploration.

Rickard, T. A. 1932. *Man and Metals*, vols. 1 and 2. New York: McGraw-Hill 1068 pp.

Ricketts, T. E. 1992. Underground retorting, Chap. 22.7 in *SME Mining Engineering Handbook*, 2d ed., edited by H. L. Hartman, pp. 1971–1984. Littleton, CO: Society for Mining, Metallurgy, and Exploration.

Ripley, E. A., R. E. Redmann, and A. A. Crowder. 1996. *Environmental Effects of Mining*. Delray Beach, FL: St. Lucie Press. 356 pp.

Robbins, R. J. 1984. Future of mechanical excavation in underground mining. *Mining Engineering* 36(6): 617–627.

Rojas, E., R. Molina, and P. Cavieres. 2001. Preundercut caving in El Teniente Mine, Chile. In *Underground Mining Methods: Engineering Fundamentals and International Case Studies*, edited by W. A. Hustrulid and R. L. Bullock, pp. 417–423. Littleton, CO: Society for Mining, Metallurgy, and Exploration.

Roman, R. J. 1973. *Simulation of an In-Situ Leaching Operation as an Aid to the Feasibility Study*. Socorro, NM: New Mexico Bureau of Mines and Mineral Resources.

Ropchan, D., F. -D. Wang, and J. Wolgamott. 1980. *Application of Water Jet Assisted Drag Bit and Pick Cutter for the Cutting of Coal Measure Rocks*. Final Report, U.S. Bureau of Mines. Golden, CO: Colorado School of Mines. Contract No. ET-77-G-01-9082.

Rose, A. W., H. E. Hawkes, and J. S. Webb. 1979. *Geochemistry in Mineral Exploration*, 2d ed. London: Academic Press. 657 pp.

Russell, P. L. 1973. Nuclear-device mining systems, Sec. 21-10 in *SME Mining Engineering Handbook*, edited by A. B. Cummins and I. A. Given, pp. 21-96–21-106. New York: SME-AIME.

Russell, P. L. 1981. An oil shale perspective. *Mining Engineering* 53(1): 29–47.

Russell, P. L. (ed.). 1982. Underground equipment. In *Underground Mining Methods Handbook*, edited by W. A. Hustrulid, pp. 998–1421. Littleton, CO: Society for Mining, Metallurgy, and Exploration.

Ruzic, N. P. 1964. The case for mining the moon. *Industrial Research* 6(11): 86–110.

Rylatt, M. G., and G. M. Popplewell. 1999. *Mining Engineering* 51(1): 37–43.

Sabins, L. F. 1997. *Remote Sensing*, 3d ed. New York: W. H. Freeman and Company. 494 pp.

Sager, J. W., J. B. Griffiths, and N. J. Fargo. 1984. Spirit Lake outlet tunnel. *Tunneling Technology Newsletter* (48):1–5. Washington, DC: U.S. National Committee on Tunneling Technology NAS/NAE.

Savanick, G. A. 1992. Hydraulic mining: Borehole slurrying, Chap. 22.4 in *SME Mining Engineering Handbook*, 2d ed., edited by H. L. Hartman, pp. 1930–1938. Littleton, CO: Society for Mining, Metallurgy, and Exploration.

Schanz, J. J. 1992. Social-legal-political-economic impacts, Sec. 3.1 in *SME Mining Engineering Handbook*, 2d ed., edited by H. L. Hartman, pp. 125–139. Littleton, CO: Society for Mining, Metallurgy, and Exploration.

Schlitt, W. J. 1982. *Interfacing Technologies in Solution Mining*. New York: SME-SPE-AIME. 370 pp.

Schlitt, W. J. 1992. Solution mining: Surface techniques, Chap. 15.2 in *SME Mining Engineering Handbook*, 2d ed., edited by H. L. Hartman, pp. 1474–1492. Littleton, CO: Society for Mining, Metallurgy, and Exploration.

Schroder, J. L, Jr. 1973. Modern mining methods — Underground, Chap. 14 in *Elements of Practical Coal Mining*, edited by S. M. Cassidy, pp. 346–476. New York: SME-AIME.

Schultz, K. H. 2000. E-mail message to author. Washington, DC: U.S. Environmental Protection Agency.

Scott, F. E. 1985. New technology improves roof control safety. *Coal Mining* 22(8): 24–26.

Seaton, A. 1975. Silicosis. In *Occupational Lung Diseases*, edited by W. K. C. Morgan and A. Seaton, pp. 80–111. Philadelphia: Saunders.

See, M. 1996. Oil mining field test to start in east Texas. *World Oil* 217(11): 71–73.

Sharp, W. E., E. R. Kennedy, and W. E Little. 1983. Estimating tunneling costs using an interactive computer model, Chap. 67 in *Proceedings of the Rapid Excavation and Tunneling Conference*, edited by H. Sutliffe and J. W. Wilson, pp. 1079–1094. New York: SME-AIME.

Sherman, W. C. 1973. Soil mechanics, Sec. 6.1 in *SME Mining Engineering Handbook*, edited by A. B. Cummings and I. A. Givens, pp. 6-2–6-13. New York: AIME.

Shock, D. 1992 Frasch sulfur mining, Sec. 15.3.2 in *SME Mining Engineering Handbook*, edited by A. B. Cummings and I. A. Givens, pp. 1512–1515. New York: AIME.

Shuey, S. 1998. Sao Bento: El Dorado's 1M-oz Brazilian crown jewel. *Engineering and Mining Journal* 199(10): WW28–WW32.

Shuey, S. 1999. Mining technology for the 21st century. *Engineering and Mining Journal* 200(4): WW-18–WW-24.

Silver, D. B. 1997. Gold Road Mine: Anatomy of a turnaround. *Mining Engineering* 49(8): 28–32.

Singh, S. P. 1988. Blast design in monumental stone quarries. *International Journal of Surface Mining* 2(1): 33–36.

Singleton, R. H. 1980. Stone. In *Mineral Facts and Problems.* Bulletin 671, pp. 853–868 U.S. Bureau of Mines.

Skelly and Loy, Inc. 1979. *Illustrated Surface Mining Methods.* New York: McGraw-Hill. 87 pp.

Skodack, T. 1982. Computerized drilling analysis speeds quarry startup. *Pit & Quarry* 74(6): 80–82.

Sloan, D. A. 1983. *Mine Management.* London: Chapman and Hall. 495 pp.

Snyder, M. T. 1994. Boring for the lower K. *Engineering and Mining Journal* 195(4): WW20–WW24.

Sobering, J. G. 2001. The Carlin underground mine. In *Underground Mining Methods: Engineering Fundamentals and International Case Studies,* edited by W. A. Hustrulid and R. L. Bullock, pp. 339–343. Littleton, CO: Society for Mining, Metallurgy, and Exploration.

Society for Mining, Metallurgy, and Exploration. 1999a. *Guide for Reporting Exploration Information, Mineral Resources, and Mineral Reserves.* Littleton, CO: Society for Mining, Metallurgy, and Exploration. 17 pp.

Society for Mining, Metallurgy, and Exploration. 1999b. Guide for reporting exploration information, mineral resources, and mineral reserves—An abstract. *Mining Engineering* 51(6): 82–84.

Soderberg, A., and D. O. Rausch. 1968. Pit planning and layout, Sec. 4.1 in *Surface Mining,* edited by E. P. Pfleider, pp. 141–165. New York: AIME.

Songstad, J. 1982. Square-set timber in load-haul-dump stopes at the Bunker Hill Mine, Kellogg, Idaho. In *Underground Mining Methods Handbook,* edited by W. A. Hustrulid, pp. 744–748. Littleton, CO: Society for Mining, Metallurgy, and Exploration.

Souder, W. E., and R. J. Evans. 1983. Water jet coal cutting: The resurgence of an old technology, Chap. 24 in *Proceedings of the Rapid Excavation and Tunneling Conference,* edited by H. Sutcliffe and J. W. Wilson. pp. 719–739. New York: SME-AIME.

Souza, P., R. Mira, and S. Rosa. 1998. Development and mining planning of Fazenda Brasileiro Gold Mine. In *Latin American Perspectives: Exploration, Mining, and Processing,* edited by O. A. Bascur, pp. 79–85. Littleton, CO: Society for Mining, Metallurgy, and Exploration.

Spickelmier, K. 1993. Round Mountain halves its cutoff grade. *Mining Engineering* 45(1): 41–48.

Spielvogel, E. 1978. Mechanical systems improve stone sawing productivity. *Industrial Diamond Review* 38(July): 250–252.

Sprouls, M. W. 1997. Opposites attract. *Mining Voice* 3(1): 38–40.

Stack, B. 1982. *Handbook of Mining and Tunneling Machinery.* New York: John Wiley & Sons, Inc. 742 pp.

Stebbins, S. A., and O. L Schumacher. 2001. Cost estimating for underground mines. In *Underground Mining Methods: Engineering Fundamentals and International Case*

Studies, edited by W. A. Hustrulid and R. L. Bullock, pp. 49–72. Littleton, CO: Society for Mining, Metallurgy, and Exploration.

Stefanko, R., and C. Bise. 1983. *Coal Mining Technology Theory and Practice*. New York: SME-AIME. 410 pp.

Stephens, D. R. 1980. *An Introduction to Underground Coal Gasification*. Livermore, CA: Lawrence Livermore National Laboratory. Department of Energy, Report UCID-18801. 27 pp.

Stephenson, H. G., and R. H. Luhning. 1988. Underground mining and tunneling techniques for in-situ oil recovery. *Tunnels & Tunneling* 20(9): 41–46.

Stevenson, G. W. 1999. Empirical estimates of TBM performance in hard rock, Chap. 56 in *Proc. Rapid Excavation and Tunneling Conference*, edited by D. E. Hilton and K. Samuelson, pp. 994–1009. Littleton, CO: Society for Mining, Metallurgy, and Exploration.

Stewart, D. R. (ed.). 1981. *Design and Operation of Caving and Sublevel Stoping Mines*. New York: SME-AIME. 843 pp.

Stoces, B. 1954. *Introduction to Mining*. London: Lange, Maxwell, and Springer. 1068 pp.

Stout, K. S. 1980. *Mining Methods and Equipment*. New York: McGraw-Hill. 218 pp.

Strauss, S. D. 1986. *Trouble in the Third Kingdom: The Mineral Industry in Transition*. London: Mining Journal Books. 227 pp.

Stubblefield, G. M., and R. W. Fish. 1992. Monolithic overburden, inclined coal seams: Trapper Mine, Sec. 14.3.4 in *SME Mining Engineering Handbook*, 2d ed., edited by H. L. Hartman, pp. 1421–1426. Littleton, CO: Society for Mining, Metallurgy, and Exploration.

Summers, D. A. 1992. Hydraulic mining: Jet-assisted cutting, Chap. 22.3 in *SME Mining Engineering Handbook*, 2d ed., edited by H. L. Hartman, pp. 1918–1929. Littleton, CO: Society for Mining, Metallurgy, and Exploration.

Sumner, J. S. 1992. Geophysical prospecting, Sec. 4.4 in *SME Mining Engineering Handbook*, 2d ed., edited by H. L. Hartman, pp. 233–242. Littleton, CO: Society for Mining, Metallurgy, and Exploration.

Suttill, K. R. 1991a. Toqui stimulates interest in southern zinc. *Engineering and Mining Journal* 172(10): 25–27.

Suttill, K. R. 1991b. Eiffel Tower stopes. *Engineering and Mining Journal* 192(7): 37–39.

Suttill, K. R. 1993. Modernizing Porco. *Engineering and Mining Journal* 174(1): 32–35.

Suttill, K. R. 1994. Olympic Dam strides ahead. *Engineering and Mining Journal* 195(1): WW45–WW54.

Sweigard, R. J. 1992a. Reclamation, Sec. 12.3 in *SME Mining Engineering Handbook*, 2d ed., edited by H. L. Hartman, pp. 1181–1197. Littleton, CO: Society for Mining, Metallurgy, and Exploration.

Sweigard, R. J. 1992b. Materials handling: Loading and hauling, Sec. 9.3 in *SME Mining Engineering Handbook*, 2d ed., edited by H. L. Hartman, pp. 761–782. Littleton, CO: Society for Mining, Metallurgy, and Exploration.

Taylor, W. J. 1982. Stull support in flat-bedded ore bodies. In *Underground Mining Methods Handbook*, edited by W. A. Hustrulid, p. 749, Littleton, CO: Society for Mining, Metallurgy, and Exploration.

Terex Corporation. 1981. *Production and Cost Estimation of Material Movement with Earthmoving Equipment*, Hudson, OH: Terex Corporation. 82 pp.

Terzaghi, K., R. B. Peck, and G. Mesri. 1996. *Soil Mechanics in Engineering Practice.* New York: John Wiley & Sons, Inc. 549 pp.

Thakur, P. C., and J. G. Dahl. 1982. Methane drainage, Chap. 4 in *Mine Ventilation and Air Conditioning*, 2d ed., New York: John Wiley & Sons, Inc. pp. 69–83.

Thomas, L. J. 1973. *An Introduction to Mining.* Sydney: Hicks Smith & Sons. 436 pp.

Thomas, L. J. 1978. *An Introduction to Mining*, rev. ed. Sydney: Methuen of Australia. 471 pp.

Thrush, P. W. (ed.). 1968. *A Dictionary of Mining, Mineral, and Related Terms.* U.S. Bureau of Mines. Washington, DC: Government Printing Office. 1269 pp.

Tien, J. C. 1999. *Practical Mine Ventilation Engineering.* Chicago: Intertec Publishing. 460 pp.

Tilley, C. M. 1991. Mechanical mine development systems at the Stillwater Mine. *Mining Engineering* 43(7): 721–723.

Tinsley, R. L. 1992. Mine financing, Sec. 6.6 in *SME Mining Engineering Handbook*, 2d ed., edited by H. L. Hartman, pp. 470–481. Littleton, CO: Society for Mining, Metallurgy, and Exploration.

Tinsley, R. L., M. Emerson, and R. Eppler. 1985. *Finance for the Minerals Industry.* New York: SME-AIME. 883 pp.

Tobie, R. L., and D. E. Julin. 1982. Block caving. In *Underground Mining Methods Handbook*, edited by W. A. Hustrulid, pp. 967–972. Littleton, CO: Society for Mining, Metallurgy, and Exploration.

Torres, R., V. Encina, and C. Segura. 1981. Damp mineral and its effect on block caving with gravity transfer on the Andina Mine, Chap. 20 in *Design and Operation of Caving and Sublevel Stoping Mines*, edited by D. R. Stewart, pp. 251–282. New York: SME-AIME.

Tracy, P. B. 1988. *Lateral Drilling Technology Tested on UCG Projects*, Paper IADC/SPE 17237, IADC/SPE Drilling Conference, Dallas, TX.

Trent, R. H. 1986. Quartenary oil mining at North Tisdale. *AAPG Bulletin* 70(8): 1059.

Trent, R. H., and W. Harrison. 1982. Longwall mining: Introduction. In *Underground Mining Methods Handbook*, edited by W. A. Hustrulid, pp. 790–815. Littleton, CO: Society for Mining, Metallurgy, and Exploration.

Trepanier, M. L., and A. H. Underwood. 1981. Block caving at King Beaver Mine, Chap. 22 in *Design and Operation of Caving and Sublevel Stoping Mines*, edited by D. R. Stewart, pp. 299–318. New York: SME-AIME.

Turner, J., and D. Carey. 1993. Automation of a mobile miner, Chap. 10 in *Proceedings of the Rapid Excavation and Tunneling Conference*, edited by L. D. Bowerman and J. E. Monsees, pp. 147–167. Littleton, CO: Society for Mining, Metallurgy, and Exploration.

Turner, T. M. 1996. *Fundamentals of Hydraulic Dredging*, 2d ed. New York:ASCE Press. 258 pp.

Tussey, I. J. 1992. Appalachian contour mining, horizontal coal seams: Martin County Coal Corporation, Sec. 14.3.5 in *SME Mining Engineering Handbook*, 2d ed., edited by H. L. Hartman, pp. 1426–1431. Littleton, CO: Society for Mining, Metallurgy, and Exploration.

Tygesen, J. D. 1992. Porphyry copper: Bingham Canyon Mine, Sec. 14.1.1 in *SME Mining Engineering Handbook*, 2d ed., edited by H. L. Hartman, pp. 1371–1376. Littleton, CO: Society for Mining, Metallurgy, and Exploration.

U.S. Army Corps of Engineers. 1953. *The Unified Soil Classification System*. Technical Memorandum 3-357. Vicksburg, MS: U.S. Corps of Engineers (also see ASTM Designation D2487-83, *Annual Book of ASTM Standards*).

U.S. Atomic Energy Commission. 1970. *Proc. on Eng. with Nuclear Explosives*. Las Vegas: U.S. Atomic Energy Commission and American Nuclear Society. 1785 pp.

U.S. Bureau of Mines. 1980. *Principles of a Resource/Reserve Classification for Minerals*. Geological Circular 831. U.S Bureau of Mines/U.S. Geological Survey. 5 pp.

U.S. Bureau of Mines. 1985. A water jet mining system for underground sandstone mines. U.S. Bureau of Mines. *Technical News* (222). 2 pp.

U.S. Bureau of Mines Staff, 1987a. *Bureau of Mines Cost Estimating Handbook: Surface and Underground Mining*. IC 9142. U.S. Bureau of Mines. 631 pp.

U.S. Bureau of Mines Staff. 1987b. *Bureau of Mines Cost Estimating Handbook: Mineral Processing*. IC 9143. U.S. Bureau of Mines. 566 pp.

U.S. Department of Energy. 1982. *Coal Data: A Reference*. U.S. Department of Energy, Energy Information Administration. Washington, DC: U.S. Government Printing Office. 118 pp.

U.S. Department of Energy. 2000. Richardson announces agreement to keep North Dakota's Great Plains plant operating. News release, August 30. 2 pp.

U.S. Geological Survey. 1995. *Minerals Yearbook*, vol. 1. Washington, DC: U.S. Government Printing Office. 939 pp.

U.S. Geological Survey. 1999. *Mineral Commodity Summaries* 1998. Washington, DC: U.S. Government Printing Office. 197 pp.

Van Blaricom, R. (ed.). 1992. *Practical Geophysics II*. Spokane, WA: Northwest Mining Association. 570 pp.

VanDerPas, E., and R. Allum. 1995. TBM technology in a deep underground copper mine, Chap. 8 in *Proceedings of the Rapid Excavation and Tunneling Conference*, edited by G. E. Williamson and I. M. Gowring, pp. 129–143. Littleton, CO: Society for Mining, Metallurgy, and Exploration.

Vogely, W. A. (ed.). 1985. *Economics of the Mineral Industries*, 4th ed. New York: SME-AIME. 660 pp.

Volkwein, J. C., and J. P. Ulery. 1993. *A Method to Eliminate Explosion Hazards in Auger Highwall Mining*. RI 9462. U.S. Bureau of Mines. 14 pp.

Voynick, S. 1998. The secret to staying young. *Mining Voice* 4(1): 44–47.

Walker, S. 1991a. Las Cuevas: Mechanizing Mexico's major fluorite mine has significant cost benefits. *Engineering and Mining Journal* 192(7): 24–25.

Walker, S. 1991b. Greens Creek keeps it clean. *Engineering and Mining Journal* 192(11): 20–23.

Walker, S. 1992. La Aurora means dawn of new mining era at Charcas. *Engineering and Mining Journal* 193(5): 16HH–16KK.

Walker, S. 1997. Raising the stakes. *World Mining Equipment* 21(6): 33–38.

Wanless, R. M. 1984. *Finance for Mine Management* London: Chapman and Hall/ Methuen. 209 pp.

Ward, M. H. 1981. Technical and economical considerations of the block caving mine, Chap. 11 in *Design and Operation of Caving and Sublevel Stoping Mines*, edited by D. R. Stewart, pp. 119–142. New York: SME-AIME.

Ward, M. H., and S. G. Britton. 1992. Management and administration, Sec. 8.6 in *SME Mining Engineering Handbook*, 2d ed., edited by H. L. Hartman, pp. 641–658. Littleton, CO: Society for Mining, Metallurgy, and Exploration.

Warner, R. C. 1992. Design and management of water and sediment control systems, Sec. 12.1 in *SME Mining Engineering Handbook*, 2d ed., edited by H. L. Hartman, pp. 1158–1169. Littleton, CO: Society for Mining, Metallurgy, and Exploration.

Waterland, J. 1982. Introduction to cut-and-fill stoping. In *Underground Mining Methods Handbook*, edited by W. A. Hustrulid, pp. 523–525. Littleton, CO: Society for Mining, Metallurgy, and Exploration.

Watson, D., and K. Keskimaki. 1992. Shotcreting, concreting at Climax Molybdenum's Henderson Mine. *Mining Engineering* 44(11): 1330–1332.

Weiss, P. F., G. B. Fettweis, I. M. Moschitz, A. Olsacher, and H. Riedler. 1981. Relevant factors for development and draw control of block caving, Chap. 54 in *Design and Operation of Caving and Sublevel Stoping Mines*, edited by D. R. Stewart, pp. 705–714. New York: SME-AIME.

Werniuk, G. 1996. Meikle Mine opens. *Engineering and Mining Journal* 197(10): WW37–WW34.

Western Mine Engineering. 1998. *Mining Cost Service*, Cost Models Section. Spokane, WA: Western Mine Engineering, Inc. (updated annually).

Western Mine Engineering. 1999. *Mine and Mill Equipment Costs: An Estimator's Guide*. Spokane, WA: Western Mine Engineering, Inc.

White, J. W., and L. T. Zoschke. 1994. Automating surface mines. *Mining Engineering* 46(6): 510–511.

White, L. 1975. Texasgulf readies for '75 Startup at Comanche Creek Frasch Mine in Texas. *Engineering and Mining Journal* 176(8): 83–88.

White, L. 1979. Middle Tennessee zinc. *Engineering and Mining Journal* 180(8): 66–76.

White, T. G. 1992. Hard-rock mining: Method advantages and disadvantages, Sec. 21.2 in *SME Mining Engineering Handbook*, 2d ed., edited by H. L. Hartman, pp. 1843–1849. Littleton, CO: Society for Mining, Metallurgy, and Exploration.

Wilhelm, G. L. 1982. Timber supported system—introduction. In *Underground Mining Methods Handbook*, edited by W. A. Hustrulid, p. 666. Littleton, CO: Society for Mining, Metallurgy, and Exploration.

Wolf, T. 1999. Western aggregates, a real gold mine. *Aggregates Manager* 3(10): 27–30.

Wolfe, L. M. 1946. *Son of the Wilderness: The Life of John Muir*. New York: Knopf. 254 pp.

Wood, P. A. 1980. *Less-Conventional Underground Coal Mining*. Report No. ICTIS/TR 12. London: IEA Coal Research. 58 pp.

World Commission on Environment and Development. 1987. *Our Common Future*. Oxford: Oxford University Press. 400 pp.

World Mining Equipment. 1997. Stepping stone. *World Mining Equipment* 21(5): 32–34.

Wyllie, R. J. M. 1994. 21st Century arrives early at Kiruna. *Engineering and Mining Journal* 195(10): WW20–WW45.

Wyllie, R. J. M. 1996. LKAB invests in the future. *Engineering and Mining Journal* 197(11): 35–50.

Young, G. J. 1946. *Elements of Mining*, 4th ed. New York: McGraw-Hill. 755 pp.

Zahl, E. G., F. Biggs, C. M. K. Boldt, R. E. Connolly, L. Gertsch, R. H. Lambeth, B. M. Stewart, and J. D. Vickery. 1992. Waste disposal and contaminant control. In *SME Mining Engineering Handbook*. 2d ed., edited by H. L. Hartman, pp. 1170–1180. Littleton, CO: Society for Mining, Metallurgy, and Exploration.

Zappia, M. A. 1981. Gravity caving and production hoisting at the San Manuel Mine, Chap. 16 in *Design and Operation of Caving and Sublevel Stoping Mines*, edited by D. R. Stewart, pp. 189–193. New York: SME-AIME.

Zeni, D. R. 1995. Large-diameter shaft drilling: A versatile tool for advanced mining. *Mining Engineering* 47(9): 847–849.

Zvyaghintsev, K. N. 1982. Underground coal gasification. *Coal International* (Redhill, Surrey, U.K.) 1(2): 15–23.

INDEX